Essential Oils and Nanotechnology for Treatment of Microbial Diseases

Essential Oils and Nanotechnology for Treatment of Microbial Diseases

Editors

Mahendra Rai
Department of Biotechnology
Sant Gadge Baba Amravati University
Amravati
Maharashtra
India

Susana Zacchino
Area Farmacognosia
Facultad de Ciencias Bioquímicas y Farmacéuticas
Universidad Nacional de Rosario
Rosario (Santa Fe)
Argentina

Marcos G. Derita
Area Farmacognosia
Facultad de Ciencias Bioquímicas y Farmacéuticas
Universidad Nacional de Rosario
Rosario (Santa Fe)
Argentina

and

CONICET
Cátedra de Cultivos Intensivos
Facultad de Ciencias Agrarias
Universidad Nacional del Litoral
Esperanza (Santa Fe)
Argentina

CRC Press is an imprint of the
Taylor & Francis Group, an **informa** business
A SCIENCE PUBLISHERS BOOK

Cover illustrations:

Top right image: Antibacterial mechanism of nanoparticles. Figure 9.1 from Chapter 9. Reproduced by kind courtesy of Dr. Mahendra Rai (book editor/chapter author).

Bottom left image: Antimicrobial strategies in anti-infectious coatings. Figure 10.1 from Chapter 10. Reproduced by kind courtesy of Dr. Juan Bueno (chapter author).

CRC Press
Taylor & Francis Group
6000 Broken Sound Parkway NW, Suite 300
Boca Raton, FL 33487-2742

First issued in paperback 2020

ISBN-13: 978-1-138-63072-7 (hbk)
ISBN-13: 978-0-367-78181-1 (pbk)

Library of Congress Cataloging-in-Publication Data

Names: Rai, Mahendra, editor. | Zacchino, Susana, editor. | Derita, Marcos G., editor.
Title: Essential oils and nanotechnology for treatment of microbial diseases / editors, Mahendra Rai, Department of Biotechnology, S.G.B. Amravati University, Amravati, Maharashtra, India, Susana Zacchino, Pharmacognosy Area School of Pharmaceutical and Biochemical, National University of Rosario, Rosario, Argentina, Marcos G. Derita, CONICET, Universidad Nacional de Rosario, Facultad de Ciencias Bioquâimicas y Farmacâeuticas, Câatedra de Farmacognosia, Rosario, Santa Fe, Argentina, and Universidad Nacional del Litoral/Facultad de Ciencias Agrarias, Câatedra de Cultivos Intensivos, Esperanza, Santa Fe, Argentina.
Description: Boca Raton, FL : CRC Press/Taylor & Francis Group, 2017. | Includes bibliographical references and index.
Identifiers: LCCN 2017026614| ISBN 9781138630727 (hardback : alk. paper) | ISBN 9781315209241 (e-book)
Subjects: LCSH: Bacterial diseases--Alternative treatment. | Essences and essential oils--Therapeutic use. | Nanotechnology. | Drug resistance in microorganisms.
Classification: LCC RC112 .E88 2017 | DDC 616.9/041--dc23
LC record available at https://lccn.loc.gov/2017026614

Visit the Taylor & Francis Web site at
http://www.taylorandfrancis.com

CRC Press Web site at
http://www.crcpress.com

Science Publishers Web site at
http://www.scipub.net

Preface

There has been emergence of multidrug resistance all over the world due to overuse or underuse of antibiotics. Most microbes including bacteria, fungi, protozoans and others have developed resistance to antibiotics, and therefore, this problem is now recognized as a global problem. Ubiquitous occurrence of multidrug-resistant bacteria decreases effectiveness of current treatment, which results in thousands of deaths all over the world. Hence, investigations for new alternatives and novel strategies are urgently needed to address the problem of multidrug resistance. The antimicrobial potential of essential oils and metallic nanoparticles represent an effective solution for microbial resistance. Moreover, the use of essential oils in combination with metallic nanoparticles may exert synergistic antimicrobial effects and would be the novel approach.

Essential Oils (EOs) are volatile, natural, aromatic oily liquids that can be obtained from several parts of plants especially the aerial ones as leaves and flowers. They are derived from complex metabolic pathways in order to protect plants from diverse pathogenic microorganisms. In fact, the bioactivity of EOs have been confirmed by several studies, including antibacterial, antiviral, anti-inflammatory, antifungal, antimutagenic, anticarcinogenic and antioxidant. These bioactivities can be enhanced by emerging technologies like nanotechnology.

Nanotechnology is one of the most important and emerging technologies, which has generated a technological revolution in the world. It has enormous applications in the field of medicine. Nanoparticles are very important tools in curing different diseases in general and microbial diseases in particular due to their significantly novel and improved chemical, physical and biological properties and high surface area-to-volume ratio. Among these, metal nanoparticles are known to play pivotal role in various biomedical applications. In this context, nanoparticles like silver have demonstrated its potential and hoped to be the new generation of antimicrobials. Silver nanoparticles have broad-spectrum biological activities and hence used in many biomedical applications. Many biomedical applications of silver nanoparticles such as for the treatment of wounds, burns, in water-disinfecting systems, in nanobased bone implantations, in dentistry for the development of dental materials and as antibacterial, antivirals, anti-protozoals, anti-arthropods and anticancerous agents. Apart from silver, other metal nanoparticles like gold and platinum and copper, oxides of different metals, etc. have been also the materials of choice for many scientists for their biological applications.

The book would be very useful for a diverse group of readers including chemists, microbiologists, biotechnologists, food technologists, nanotechnologists, pharmacologists, clinicians and those who are interested in a natural cure. The students should find this book useful and reader friendly.

Mahendra Rai
Susana Zacchino
Marcos G. Derita

Contents

Preface v

List of Contributors ix

Section I: Essential Oils for Treatment of Different Microbial Diseases

1. **Essential Oils with Antimicrobial Properties Formulated in Lipid Nanoparticles—Review of the State of the Art** 3
   Patrícia Severino, Flavia Resende Diniz Acioli, Juliana Cardoso Cordeiro, Maria do Céu Teixeira, Antonello Santini, Andjelka B. Kovačević and *Eliana B. Souto*
2. **Antimicrobial Activities of Plant Essential Oils and Their Components against Antibiotic-Susceptible and Antibiotic-Resistant Foodborne Pathogens** 14
   Mendel Friedman
3. **Essential Oils Constituted by Prop-1(2)-enylbenzene Derivatives Used for Treatment of Microbial Infections** 39
   Amner Muñoz-Acevedo, María del Carmen González, Erika Amparo Torres, Martha Cervantes-Díaz and *Elena E. Stashenko*
4. **Antibacterial and Anti-biofilm Activities of Essential Oils and Their Components Including Modes of Action** 99
   Julio A. Zygadlo, María P. Zunino, Romina P. Pizzolitto, Carolina Merlo, Alejandra Omarini and *José S. Dambolena*
5. **Role of Essential Oils for the Cure of Human Pathogenic Fungal Infections** 127
   Melina G. Di Liberto, Laura A. Svetaz, María V. Castelli, Mahendra Rai and *Marcos G. Derita*
6. **Essential Oils against Microbial Resistance Mechanisms, Challenges and Applications in Drug Discovery** 143
   Juan Bueno, Fatih Demirci and *K. Husnu Can Baser*
7. **Essential Oils and Nanoemulsions: Alternative Tool to Biofilm Eradication** 159
   Z. Aumeeruddy-Elalfi and *F. Mahomoodally*
8. **Nano-Ag Particles and Pathogenic Microorganisms: Antimicrobial Mechanism and its Application** 175
   JiEun Yun and *Dong Gun Lee*

Section II: Nanotechnology for Treatment of Different Microbial Diseases

9. **Nanoparticles as Therapeutic Agent for Treatment of Bacterial Infections** 191
   Mahendra Rai, Raksha Pandit, Priti Paralikar, Sudhir Shende, Swapnil Gaikwad, Avinash P. Ingle and *Indarchand Gupta*

10. Anti-Adhesion Coating with Natural Products: When the Nanotechnology Meet the Antimicrobial Prevention **209**
Juan Bueno

11. Nanotechnologies for the Delivery of Water-Insoluble Drugs **221**
Omar M. Najjar and *Rabih Talhouk*

12. Potential of Oils in Development of Nanostructured Lipid Carriers **242**
Anisha A. D'Souza and *Ranjita Shegokar*

13. Essential Oil-Based Nanomedicines against Trypanosomatides **258**
Maria Jose Morilla and *Eder Lilia Romero*

14. Combining Inorganic Antibacterial Nanophases and Essential Oils: Recent Findings and Prospects **279**
Mauro Pollini, Alessandro Sannino, Federica Paladini, Maria Chiara Sportelli, Rosaria Anna Picca, Nicola Cioffi, Giuseppe Fracchiolla and *Antonio Valentini*

Section III: Antimicrobial Activity Testing

15. Antimicrobial Activity Testing Techniques **297**
Estefanía Butassi, Marcela Raimondi, Agustina Postigo, Estefanía Cordisco and *Maximiliano Sortino*

Index **311**

List of Contributors

Flavia Resende Diniz Acioli, University of Tiradentes (Unit) and Institute of Technology and Research (ITP), Av. Murilo Dantas, 300, 49010-390 Aracaju, Brazil.

Zaahira Aumeeruddy-Elalfi, Department of Health Sciences, Faculty of Science, University of Mauritius, Réduit, Mauritius.

K. Husnu Can Baser, Department of Pharmacognosy, Faculty of Pharmacy, Near East University, Department of Pharmacognosy, Nicosia, N. Cyprus.

Juan Bueno, Research Center of Bioprospecting and Biotechnology for Biodiversity Foundation (BIOLABB), Colombia.

Estefanía Butassi, Farmacognosia, Facultad de Ciencias Bioquímicas y Farmacéuticas, Universidad Nacional de Rosario. Suipacha 531, Rosario, Argentina.

María V. Castelli, CONICET, Universidad Nacional de Rosario/Facultad de Ciencias Bioquímicas y Farmacéuticas/Cátedra de Farmacognosia, Suipacha 531, Rosario, Santa Fe, Argentina.

Martha Cervantes-Díaz, Environmental Research Group for Sustainable Development – Environmental Chemistry Faculty – Universidad Santo Tomás, Bucaramanga, Colombia.

Nicola Cioffi, University of Bari "Aldo Moro", Department of Chemistry, Via Orabona 4, 70126 Bari, Italy.

Juliana Cardoso Cordeiro, University of Tiradentes (Unit) and Institute of Technology and Research (ITP), Av. Murilo Dantas, 300, 49010-390 Aracaju, Brazil.

Estefanía Cordisco, Farmacognosia, Facultad de Ciencias Bioquímicas y Farmacéuticas, Universidad Nacional de Rosario. Suipacha 531, Rosario, Argentina.

José S. Dambolena, Universidad Nacional de Córdoba, Instituto Multidisciplinario de Biologia Vegetal, CONICET, Córdoba, Argentina. Avenida Velez Sarsfiled 1611.

Fatih Demirci, Faculty of Health Sciences, Anadolu University, Eskişehir, Turkey and; Department of Pharmacognosy, Faculty of Pharmacy, Anadolu University, Eskişehir, Turkey.

Marcos G. Derita, CONICET, Universidad Nacional de Rosario/Facultad de Ciencias Bioquímicas y Farmacéuticas/Cátedra de Farmacognosia, Suipacha 531, Rosario, Santa Fe, Argentina and CONICET, Universidad Nacional del Litoral/Facultad de Ciencias Agrarias/Cátedra de Cultivos Intensivos, Kreder 2805, Esperanza, Santa Fe, Argentina.

Melina G. Di Liberto, CONICET, Universidad Nacional de Rosario/Facultad de Ciencias Bioquímicas y Farmacéuticas/Cátedra de Farmacognosia, Suipacha 531, Rosario, Santa Fe, Argentina.

Anisha A. D'Souza, Indian Institute of Technology-Bombay (IIT-B), Department of Bioscience and Biomedical Engineering, Powai, Mumbai, India.

Giuseppe Fracchiolla, University of Bari "Aldo Moro", Dipartimento di Farmacia - Scienze del Farmaco, Via Orabona 4, 70126 Bari, Italy.

Mendel Friedman, Healthy Processed Foods Research, United States Department of Agriculture, Agricultural Research Service, Western Regional Research Center, Albany, California 94710 USA.

Swapnil Gaikwad, Dr. D.Y. Patil Biotechnology and Bioinformatics Institute, Tathawade, Pune 411 033, Maharashtra, India.

María del Carmen González, Chemistry and Biology Research Group – Department of Chemistry and Biology – Universidad del Norte, Barranquilla, Colombia.

Indarchand Gupta, Nanobiotechnology Lab., Department of Biotechnology, SGB Amravati University, Amravati-444 602, Maharashtra.

Avinash Ingle, Nanobiotechnology Lab., Department of Biotechnology, SGB Amravati University, Amravati-444 602, Maharashtra.

Andjelka B. Kovačević, Department of Pharmaceutical Technology, Faculty of Pharmacy, University of Coimbra (FFUC), Pólo das Ciências da Saúde, Azinhaga de Santa Comba, 3000-548 Coimbra, Portugal.

Dong Gun Lee, School of Life Sciences, College of Natural Sciences, Kyungpook National University, 80 Daehakro, Bukgu, Daegu 41566, Republic of Korea.

Fawzi Mahomoodally, Department of Health Sciences, Faculty of Science, University of Mauritius, Réduit, Mauritius.

Carolina Merlo, Universidad Nacional de Córdoba, Instituto Multidisciplinario de Biologia Vegetal, CONICET, Córdoba, Argentina, Avenida VelezSarsfiled 1611.

Maria Jose Morilla, Programa de Nanomedicinas, Departamento de Ciencia y Tecnología, Universidad Nacional de Quilmes. Roque Saenz Peña 352, Bernal, B1876BXD, Buenos Aires, Argentina.

AmnerMuñoz-Acevedo, Chemistry and Biology Research Group – Department of Chemistry and Biology – Universidad del Norte, Barranquilla, Colombia.

Omar M. Najjar, Department of Biology, Faculty of Arts and Sciences, American University of Beirut, P.O. Box 11-0236, Beirut, Lebanon.

Alejandra Omarini, Universidad Nacional de Córdoba, Instituto Multidisciplinario de Biologia Vegetal, CONICET, Córdoba, Argentina, Avenida VelezSarsfiled 1611.

Federica Paladini, University of Salento, Department of Engineering for Innovation; Via per Monteroni, 73100 Lecce, Italy.

Raksha Pandit, Nanobiotechnology Lab., Department of Biotechnology, SGB Amravati University, Amravati-444 602, Maharashtra.

Priti Paralikar, Nanobiotechnology Lab., Department of Biotechnology, SGB Amravati University, Amravati-444 602, Maharashtra.

Rosaria Anna Picca, University of Bari "Aldo Moro", Department of Chemistry, Via Orabona 4, 70126 Bari, Italy.

Romina P. Pizzolitto, Universidad Nacional de Córdoba, Instituto Multidisciplinario de Biologia Vegetal, CONICET, Córdoba, Argentina, Avenida VelezSarsfiled 1611.

Mauro Pollini, University of Salento, Department of Engineering for Innovation, Via per Monteroni, 73100 Lecce, Italy.

Agustina Postigo, Farmacognosia, Facultad de Ciencias Bioquímicas y Farmacéuticas, Universidad Nacional de Rosario. Suipacha 531, Rosario, Argentina.

Mahendra Rai, Department of Biotechnology, Sant Gadge Baba Amravati University, Amravati 444602, Maharashtra, India.

Marcela Raimondi, Farmacognosia, Facultad de Ciencias Bioquímicas y Farmacéuticas, Universidad Nacional de Rosario. Suipacha 531, Rosario, Argentina.

Eder Lilia Romero, Programa de Nanomedicinas, Departamento de Ciencia y Tecnología, Universidad Nacional de Quilmes. Roque Saenz Peña 352, Bernal, B1876BXD, Buenos Aires, Argentina.

Alessandro Sannino, University of Salento, Department of Engineering for Innovation, Via per Monteroni, 73100 Lecce, Italy.

Antonello Santini, Department of Pharmacy, University of Napoli "Federico II", Via D. Montesano 49, 80131 Napoli, Italy.

Patricia Severino, University of Tiradentes (Unit) and Institute of Technology and Research (ITP), Av. Murilo Dantas, 300, 49010-390 Aracaju, Brazil.

Ranjita Shegokar, Freie Universität Berlin, Institute of Pharmacy, Department Pharmaceutics, Biopharmaceutics, and Nutricosmetics, Berlin, Germany.

Sudhir Shende, Nanobiotechnology Lab., Department of Biotechnology, SGB Amravati University, Amravati-444 602, Maharashtra.

Maximiliano Sortino, CONICET/Farmacognosia, Facultad de Ciencias Bioquímicas y Farmacéuticas, Universidad Nacional de Rosario. Suipacha 531, Rosario, Argentina.

Eliana B. Souto, Department of Pharmaceutical Technology, Faculty of Pharmacy, University of Coimbra (FFUC), Pólo das Ciências da Saúde, Azinhaga de Santa Comba, 3000-548 Coimbra, Portugal.

Maria Chiara Sportelli, University of Bari "Aldo Moro", Department of Chemistry, Via Orabona 4, 70126 Bari, Italy.

Elena E. Stashenko, Mass Spectrometry and Chromatography Centre - CIBIMOL - CENIVAM, Universidad Industrial de Santander, Bucaramanga, Colombia.

Laura A. Svetaz, CONICET, Universidad Nacional de Rosario/Facultad de Ciencias Bioquímicas y Farmacéuticas/Cátedra de Farmacognosia, Suipacha 531, Rosario, Santa Fe, Argentina.

Rabih Talhouk, Department of Biology, Faculty of Arts and Sciences, American University of Beirut, P.O. Box 11-0236, Beirut, Lebanon.

Maria do Céu Teixeira, Department of Pharmaceutical Technology, Faculty of Pharmacy, University of Coimbra (FFUC), Pólo das Ciências da Saúde, Azinhaga de Santa Comba, 3000-548 Coimbra, Portugal.

Erika Amparo Torres, Chemistry and Biology Research Group – Department of Chemistry and Biology – Universidad del Norte, Barranquilla, Colombia.

Antonio Valentini, University of Bari "Aldo Moro", Department of Physics "M. Merlin", Via Amendola 173, 70126 Bari, Italy.

JiEun Yun, School of Life Sciences, College of Natural Sciences, Kyungpook National University, 80 Daehakro, Bukgu, Daegu 41566, Republic of Korea.

María P. Zunino, Universidad Nacional de Córdoba, Instituto Multidisciplinario de Biologia Vegetal, CONICET, Córdoba, Argentina, Avenida VelezSarsfiled 1611.

Julio A. Zygadlo, Universidad Nacional de Córdoba, Instituto Multidisciplinario de Biologia Vegetal, CONICET, Córdoba, Argentina, Avenida VelezSarsfiled 1611.

Section I

Essential Oils for Treatment of Different Microbial Diseases

1

Essential Oils with Antimicrobial Properties Formulated in Lipid Nanoparticles

Review of the State of the Art

Patrícia Severino,[1] *Flavia Resende Diniz Acioli,*[1] *Juliana Cardoso Cordeiro,*[1] *Maria do Céu Teixeira,*[2] *Antonello Santini,*[3] *Andjelka B. Kovačević,*[2] and *Eliana B. Souto*[2,4,*]

Introduction

Plant-based therapies have been used since the dawn of humanity, where the treatment of diseases was based on the primitive instinct of the need to relieve body injuries. The first historical event observed to relieve pain was to put cold water with fresh leaves onto the injury or protecting it with mud (Pan et al. 2014). Through these experiences, humans understood that certain plants were more effective than others, and these contributed to the development of drug therapy. Previously, the tribes recognized who knew the healing qualities of plants, acquired skills through experience or inherited them from ancestors, and were called to treat the sick and injured preparing medicinal products (Erichsen-Brown 2013).

The use of plants is still being explored nowadays, which are considered very promising for the treatment of various diseases. With the evolution of technology, the isolation of bioactives from plants became a reality (Gurib-Fakim 2006). To elucidate the production of promising molecules in plants it is important to understand the physiology of the same.

The primary metabolism occurring in the cells of plants meet their nutritional requirements of energy (i.e., ATP), NADPH and biosynthetic substances (e.g., carbohydrates, lipids, proteins, chlorophyll and nucleic acids). The secondary metabolism, which produces important molecules for the development of bioactives offer advantages for maintenance, adjustment and development of plants (Alinian et al. 2016). The secondary metabolism is also responsible for defence against pathogens and herbivores, attraction of pollinators, allows tolerance of extreme temperatures and capacity to adapt

[1] University of Tiradentes (Unit) and Institute of Technology and Research (ITP), Av. Murilo Dantas, 300, 49010-390 Aracaju, Brazil.
[2] Department of Pharmaceutical Technology, Faculty of Pharmacy, University of Coimbra (FFUC), Pólo das Ciências da Saúde, Azinhaga de Santa Comba, 3000-548 Coimbra, Portugal.
[3] Department of Pharmacy, University of Napoli "Federico II", Via D. Montesano 49, 80131 Napoli, Italy.
[4] REQUIMTE/LAQV, Group of Pharmaceutical Technology, Faculty of Pharmacy, University of Coimbra, Coimbra, Portugal.
* Corresponding author: ebsouto@ff.uc.pt

to water stress or deficiency of nutrients and minerals from the soil. The main secondary metabolites are tannins (Engström et al. 2016), flavonoids (Tsou et al. 2016), phenols (Kuppusamy et al. 2016), alkaloids (Schläger and Dräger 2016), terpenoids (Mukherjee et al. 2016), among other classes.

Although, synthetic chemistry has largely advanced in recent years, the development of molecules is still facings several challenges, including the need for patenting complex molecules. Since the 90s, the pharmaceutical industry started to invest in this area by applying bioassays in the development of drug molecules (Beutler 2009).

From the World Health Organization (WHO) data, about 25% of sold medicines are derived from medicinal plants. Considering the class of antimicrobial and antitumour drugs these numbers rise up to 60% (Calixto 2000). Some examples of medicines derived from plants are acetylsalicylic acid, the antimalarial quinine, the antineoplastic drugs vincristine and vinblastine and the cardiotonic digoxin.

The aim of this chapter is to discuss the main essential oils obtained from plants, which are recognized to have antimicrobial activity, highlighting the recent nanotechnological advances in lipid nanoparticles to improve the loading of antimicrobials and their therapeutic effects.

Definitions and Properties of Essential Oils

The International Standard Organization (ISO) defines 'essential oils' as the products obtained by parts of plants by extraction, effleurage, drag steam vapour, extraction with organic solvents, pressing or by supercritical fluid extraction (Baser and Buchbauer 2016). The cold process is preferred to minimize the risk of degradation of sensitive bioactives (Richter and Schellenberg 2007, Tomaino et al. 2005). Essential oils differ from fixed oils due to the physical and chemical characteristics, i.e., essential oils are complex mixtures of volatile liquid and lipophilic compounds. Volatility is what differs essential oils from fixed oils. The important features of the essential oils are their pleasant aroma being also called essences (Hüsnü et al. 2007).

The plants rich in essential oils are the angiosperms dicotyledons. Essential oils may be produced by specialized structures of the plant as the gland, the parenchymal cells and differentiated channels (Araújo and Meira 2016). They may also be stored in plant organs such as flowers, leaves, stems, wood, berries, rhizomes and seeds (De Filippis 2016). Besides that, the location, time of collection, extraction process, weather conditions and sun, are factors influencing the chemical and physical characteristics of different odours, even from the same plant species. Moreover, essential oil extracted from different parts of the same plant, although visually similar, may exhibit significant chemical differences (Organization 2003).

Most of the essential oils are derived from a mixture of phenylpropanoids and terpenes (monoterpenes, sesquiterpenes and their derivatives), which originate from the secondary metabolism of plants. Such substances are related to biological functions of plants as a defence against herbivores and microorganisms, UV protection, attraction of pollinators and animal dispersers of seeds, protection and inhibition of germination. Thus, essential oils are related to adaptation, and ecological requirements (Astani et al. 2010).

The biosynthesis of essential oils begin with glucose metabolism, which may be metabolized to shikimic acid and acetate. The shikimic acid gives the phenylpropanoid precursor of aromatic amino acids. The main known phenylpropanoids include eugenol, methyl eugenol, myristicin, elemicina, chavicol, methyl chavicol, dillapiol, anethole, estragole, apiol (Ferrer et al. 2008). Terpenoids (monoterpenes and sesquiterpenes) are originated from acetate. Terpenoids belong to the largest family of natural products and are found in all kinds of organisms. Monoterpenes, terpenes or terpenoids can be classified according to the number of isoprene, which include: hemiterpenoids, monoterpenoids, sesquiterpenoids, diterpenoids, triterpenoids and tetraterpenoids polyterpenoids (Champagne and Boutry 2016).

Essential oils are volatile substances extracted from plants and are an important raw material for pharmaceutical, cosmetic, food, cosmetic and agricultural industries (Al-Haj et al. 2010). There is a great interest from pharmaceutical companies in the secondary metabolites because they have important pharmacological activities. The main components of the essential oils offer antimicrobial characteristics (Knezevic et al. 2016, Miranda et al. 2016), anti-inflammatory (Giovannini et al. 2016) and antioxidant (Vilas et al. 2016), antispasmodic (Silva et al. 2016), are stimulant of the digestive tract (Aumeeruddy-Elalfi et al. 2016), have cardiovascular and anaesthetic actions (Cherkaoui-Tangi et al. 2016). Essential oils are extremely promising in pharmacotherapy; however, their toxicological profile cannot be underestimated (Fabiane et al. 2008).

The literature reports the antimicrobial activity of several essential oils, showing activity against Gram-positive and Gram-negative bacteria and fungi. The antimicrobial activity of essential oils is especially focused on the treatment of strains resistant to conventional antibiotics. Several studies have reported the antimicrobial activity of essential oils *Eucalyptus camaldulensis* (Knezevic et al. 2016), *Artemisia santonica* (Hassanshahian and Khosravi 2016), *Zataria multiflora* (Fereshteh et al. 2016). There are, however, numerous examples of plants that have not been identified yet, which can be useful for the treatment of resistant bacteria and fungi (São Pedro et al. 2013).

Essential oils can act as enhancers of antimicrobial activity, therefore, essential oils may improve the antibiotic profile, for which the action is limited by mechanisms of multidrug resistance (González-Lamothe et al. 2009). Essential oils have activity against a variety of microorganisms, including viruses, fungi (Flores et al. 2016), protozoa and bacteria (Miranda et al. 2016).

The antimicrobial activity of essential oils is mainly dependent on their lipophilicity. Due to the small chain carbonic molecules, these interact with membranes of the microorganism carrying out changes in the membrane potential, with inhibition of cell wall synthesis and consequent outer membrane fluidity, leading to rupture and loss of cytoplasmic material. Inhibition of the respiratory chain occurs by disrupting mitochondrial. The essential oil molecules can bind or inhibit specific proteins and cause damage to DNA and RNA inhibiting chromosomal replication and transcription (Hammer et al. 1999). Furthermore, the essential oils can act as a virulence attenuating exposure to terpenes, and can interfere with the expression of genes encoding virulence factors, such as strains of *Staphylococcus aureus* enterotoxin producers (Qiu et al. 2011).

The advantages of using the antimicrobial activity of essential oils, compared to the synthetic commercial antibiotics, rely on fact that the pharmacological activity cannot be attributed only to one component, because several components may interact synergistically in the promotion of this activity, and by the multi-target mechanisms of phytochemicals from essential oils (Bassolé and Juliani 2012).

Encapsulation of Essential Oils

While essential oils have several uses in different fields (e.g., pharmaceutical, chemical and food industry), they are physicochemical labile, justifying the need of their loading in nanoparticles with the aim to improve their properties (Table 1.1). Nanoparticles play an essential role in drug delivery and targeting, since they can release their payload in specific organs, tissues and cells. Natural polymers and lipid compounds are commonly used as raw materials in the production of nanoparticles due to their availability, biocompatibility, absence of toxicity and biodegradability.

The primary purpose of the loading of drugs in nanoparticles is to improve their bioavailability and physicochemical protection. The therapeutic response happens when the drug concentration at the site of action is sufficient to promote its pharmacological effect. Nanotechnology allows obtaining these results through the control of the physicochemical characteristics of the formulation that allows the release of drugs in a controlled manner in the target tissue (Bonifacio et al. 2014, Severino et al. 2015b).

Table 1.1. Therapeutic applications of essential oils loaded in lipid nanoparticles.

Lipid nanoparticles	Essential oil	Therapeutic applications	References
Liposomes	*Salvia officinalis* (Salvia oil)	*Listeria monocytogenes*	(Cui et al. 2016c)
	Cinnamomum zeylanicum oil	*Staphylococcus aureus*	(Cui et al. 2016b)
	Myristica fragrans (Nutmeg) oil	*Listeria monocytogenes*	(Lin et al. 2016)
	Zanthoxylum tingoassuiba oil	Multidrug-resistant *Staphylococcus aureus;* dermatophytes	(Detoni et al. 2009)
Solid Lipid Nanoparticles	*Copaifera langsdorffii* oil	*Trichophyton rubrum* *Microsporum canis*	(Svetlichny et al. 2015)
	Artemisia arborescens oil	Herpes Simplex Virus-1	(Lai et al. 2007)
	Boswellia rivae and *Commiphora myrrha* (Frankincense and myrrh) essential oils	Anti-tumour activity (H22-bearing Kunming mice)	(Shi et al. 2012)
Nanoemulsions	*Thymus vulgaris* oil	*Zygosaccharomyces bailii* *Escherichia coli*	(Chang et al. 2015, Moghimi et al. 2016)
	Helianthus annuus oil	*Escherichia coli* *Lactobacillus delbrueckii* *Saccharomyces cerevisiae*	(Donsì et al. 2012)
	Eucalyptus globulus oil	*Staphylococcus aureus*	(Sugumar et al. 2015)

In order to achieve site-specific antimicrobial action, the control of the particle size and morphology is mandatory. In addition, since the aim is to avoid the evaporation of the essential oil, the production method needs to be accurately chosen and the long-term stability of the formulation secured (Severino et al. 2011). The protection of the essential oil from oxidation, volatilization and exposure to high temperatures and UV light, is achieved by loading in lipid nanoparticles, which contribute for the modification of the release profile of the essential oils (El Asbahani et al. 2015, Severino et al. 2015a). Figure 1.1 illustrates three different types of lipid nanoparticulate carriers, namely, solid lipid nanoparticles, liposomes and nanoemulsions.

Solid lipid nanoparticles

Solid lipid nanoparticles (SLN) were first described in 1990s (Schwarz et al. 1994, Gasco 1993, Bunjes et al. 1996), as spherical particles, composed of physiological lipids, with a diameter ranging between 1 nm and 1000 nm. The SLN consist of a hydrophobic core, solid at room and body temperatures, and stabilized by a surfactant layer embedded in their surface. The active compound is dissolved or dispersed in the solid lipid matrix, which offers the opportunity to modify the release profile of loaded compounds (Souto and Muller 2010). The methods used for the production and physicochemical characterization are relatively simple and can be scaled-up (Marengo et al. 2000, Marengo et al. 2003). The most popular methods for the production of SLN are the high pressure homogenization and the high shear homogenization (by ultrasound probe or by mechanical stirring) (Souto and Muller 2006). SLN are of low production costs, which can also stimulate the pharmaceutical market (Attama and Umeyor 2015).

Being composed of physiological lipids, the risk of toxicity is also reduced (Doktorovova et al. 2014, Doktorovova et al. 2016, Severino et al. 2011). The raw materials are physiologically compatible, attributed to the structural similarity with the cell membranes, providing *in vivo* tolerability and biodegradation of the SLN. Various types of lipids can be used, including triglycerides, partial glycerides, fatty acids, steroids and waxes. The surfactants are used to stabilize the SLN, giving surface electrical charge, and determining the first interaction of the SLN with tissues and cells. The surface charge may be used to increase absorption and determine the different interactions with intestinal epithelium (Severino et al. 2012b).

The literature reports several examples of essential oils loaded in SLN, such as *Artemisia arborescens* (Lai et al. 2006a, Lai et al. 2006b, Lai et al. 2007), *Zataria multiflora* (Moghimipour et al. 2013), and *Copaifera* sp. (Svetlichny et al. 2015). The *in vitro* antiviral activity and skin permeation of *Artemisia arborescens* essential oil loaded SLN, composed of Compritol 888 ATO as solid lipid, were studied by Lai et al. (Lai et al. 2007). Two different formulations were produced by hot-pressure homogenization technique, using Poloxamer 188 or Miranol Ultra C32 as surfactants, while the antiviral activity of oil-free and oil loaded-SLN incorporated was tested *in vitro* against Herpes Simplex Virus-1. *In vitro* antiviral assays showed that SLN incorporation did not influence the anti-herpetic activity of *Artemisia arborescens,* whereas *in vitro* skin permeation experiments demonstrated the capability of SLN to increase the oil accumulation into the skin.

SLN containing frankincense and myrrh (*Boswellia rivae* and *Commiphora myrrha*) essential oils (FMO) for oral delivery have been produced by high-pressure homogenization method using Compritol 888 ATO as solid lipid and soybean lecithin and Tween 80 as surfactants (Shi et al. 2012). The authors reported the increased antitumour efficacy of FMO in H22-bearing Kunming mice when loaded in SLN.

Garg and Singh (2011) produced SLN by high pressure homogenization loading eugenol. Eugenol is a bioactive phenylpropene of clove oil, nutmeg, cinnamon, basil and bay leaf with reported antifungal activity (Choudhury et al. 2014, El-Soud et al. 2015, Perez-Roses et al. 2016, Radwan et al. 2014). The *in vitro* antifungal activity was found to be similar between the loaded and non-loaded eugenol (Garg and Singh 2011). However, when testing the formulations for the treatment of oral candidiasis in immunosuppressed rats, increased efficacy was demonstrated when using eugenol-loaded SLN.

Tea Tree Oil (TTO) or Melaleuca alternifolia essential oil (7.5% m/V) was formulated in acetyl palmitate SLN and its biological activity was studied against biofilm of *Pseudomonas aeruginosa* in buccal epithelial cells (Comin et al. 2016). Results demonstrated that both oil and nanoparticles interfered with the mobility of the microorganisms, decreasing their adhesion to buccal cells, compromising the formation of the biofilm. *Melaleuca alternifolia* essential oil is composed of terpene hydrocarbons, mainly monoterpenes, sesquiterpenes, and their associated alcohols. The mechanism of action of terpenes is based on the disruption of the membrane causing permeability, which may contribute to act against resistant fungal strains.

Moghimipour et al. (2013) produced SLN loaded with *Zataria multiflora* essential oil, reporting 38.66% of encapsulation efficiency. The authors observed 93.2% of the essential oil released after 24 hours of assay performed in dialysis bag. Apart from the capacity to modify the release profile, advantages of using SLN have also been attributed to the reduction of the fast evaporation of essential oil in comparison to reference emulsions (Lai et al. 2006a).

Liposomes

Liposomes are spherical vesicles comprising one or more concentric phospholipid bilayers, and are recognized as biocompatible, biodegradable and non-immunogenic drug delivery systems, being the first to be approved for clinical applications (Zylberberg and Matosevic 2016). Liposomes are widely versatile, with capacity to encapsulate hydrophilic and/or lipophilic compounds. Hydrophilic substances are placed within their aqueous core, whereas lipophilic have affinity to the lipid bilayer. The formulation is primarily composed of natural phospholipids derived from egg, soybean and synthetic. Layers can be multiple or single, and it can be sorted. Vesicles may vary in size being Small Unilamellar Vesicle (SUV) or Large Unilamellar Vesicle (LUV), Large Unilamellar Vesicle (LUV) and Large Multilamellar Vesicle (LMV) (Kostarelos 2007). The most popular methods for the production of liposomes are the Bangham method (Severino et al. 2012a), reverse-phase (Tang et al. 2016), evaporation by organic solvent (Liu et al. 2016), injection method (Sebaaly et al. 2015) and homogenization (Li et al. 2016b). The main features of liposomes rely on the possibility to modulate their properties by selection of raw material, volume and type of production process (Sherry et al. 2013).

Liposomes may be used to load essential oils to protect them against degradation and improve solubilization (Musthaba et al. 2009). Clove essential oil-loaded oligolamellar liposomes were produced by Sebaaly et al. (2015b) who tested their capacity to protect eugenol from degradation induced by UV exposure. The liposomes also demonstrated capacity to keep the DPPH-scavenging activity of free eugenol.

Eucalyptus camaldulensis essential oil is rich in phenol, cineole, limonene, alcohol, pinene and terpinene molecules and has been loaded in liposomes-containing hydroxyethyl cellulose hydrogels (Moghimipour et al. 2012). The antifungal activity of the obtained liposomal gels was tested against *Microsporum canis*, *M. gypseum*, *Trichophyton rubrum* and *T. verrucosum*. The loading of *E. camaldulensis* oil in liposomes enhanced its stability and may also lead to improved antifungal activity.

The anti-biofilm activity of cinnamon oil-loaded liposomes was tested against methicillin-resistant *Staphylococcus aureus* (MRSA) biofilms on stainless steel, gauze, nylon membrane and non-woven fabrics, by monitoring the colony forming (Cui et al. 2016a). Liposome improved the stability of antimicrobial agents and prolonged the time of action, while exhibited a satisfactory antibacterial performance on MRSA and its biofilms. Besides its use in the pharmaceutical and medical fields, cinnamon oil is also employed in food (Wu et al. 2015) and cosmetic (McCarthy 2004) industries for preservation.

Celia et al. (2013) described anticancer activity of bergamot essential oil loaded-liposomes. The authors reported improved water solubility of the phytocomponents of the oil and increased anticancer activity *in vitro* against human SH-SY5Y neuroblastoma cells. While the antifungal activity of *Ligusticum chuanxiong* is attributed to ligustilide and butylidene phthalide, which are its major oil components (Sim and Shin 2008), the essential oil was formulated in liposomes for the treatment of hypertrophic scarring (Zhang et al. 2012). Liposomal formulations of different oil concentration were tested (2.5, 5, and 10%) once daily to the scars of rabbit ear model for 28 days. The scar tissue was excised after postoperative 56 days, for further determination of the Scar Elevation Index (SEI), and detection of fibroblast apoptosis, collagen level, and analysis of mRNA expression of matrix metalloproteinase-1 (MMP-1), caspase-3 and -9, and transforming growth factor beta 1 (TGF-beta(1)). The loading of *Ligusticum chuanxiong* essential oil in liposomes significantly alleviated hypertrophic scars *in vivo*, whereas SEI was reduced. The levels of TGF-beta(1), MMP-1, collagen I, and collagen III decreased, while caspase-3 and -9 levels and apoptosis cells increased. The authors demonstrate the therapeutic effects *Ligusticum chuanxiong* essential oil on formed hypertrophic scars.

As reported above, *Melaleuca alternifolia* essential oil (TTO) is known for its antibacterial, antifungal, antiviral and antiprotozoal broad-spectrum activities. However, it is sensitive to degradation by oxygen, light and temperature. Liposomes were also used to load the oil and its antimicrobial activity of was evaluated against three common microbial species, *Staphylococcus aureus*, *Escherichia coli* and *Candida albicans* (Ge and Ge 2016). The results demonstrated that the systems caused perforation of the cell membrane, followed by cell autolysis. While the structure of bacteria (Gram positive versus Gram negative) influence their interaction with liposomes, the loading of TTO significant increased its antimicrobial activity.

Nanoemulsions

Nanoemulsions are composed of two immiscible liquids, usually oil and water, stabilized by a surfactant film placed in the oil/water interface. A suitable mixture of surfactants with hydrophilic-lipophilic balance condition provides the maximum 'solubilization' of the oil and water. If small and spherical oil droplets are surrounded by surfactant molecules in a continuous aqueous medium, an o/w nanoemulsion type is obtained. When aqueous phase are located inside small spherical droplets surrounded by surfactant molecules in a continuous oil medium, nanoemulsions of opposite signal w/o are obtained, i.e., where the aqueous phase is dispersed in the outer oil phase. However, bicontinuous nanoemulsions can also be obtained, when no structured spherical droplets are formed as inner phase, which can occur: (i) when increasing gradually, by titration, the volume of the internal phase of the systems; (ii) during the migration of o/w to w/o or w/o to o/w; and (iii) when the volume of the two phases is similar (de Oliveira et al. 2004, Formariz et al. 2005).

Nanoemulsions have been widely used in food (Moghimi et al. 2016, Salvia-Trujillo et al. 2015, Sugumar et al. 2015) and pharmaceutical (Li et al. 2016a, Sugumar et al. 2015) industries to load essential oils. They may be used to improve the solubilization of lipophilic bioactives present in essential oils, to protect them from enzymatic hydrolysis, and to increase the absorption due to the presence of the surfactant (Rossi et al. 2007).

An o/w nanoemulsion, of 143 nm mean droplet size, containing *Thymus daenensis* essential oil was developed and tested against *Escherichia coli* (Moghimi et al. 2016). The antibacterial activity against relevant food-borne pathogen bacterium was determined by measuring the Minimum Inhibitory Concentration (MIC) and Minimum Bactericidal Concentration (MBC). The nanoemulsion significantly improved the antibacterial activity of the essential oil, attributed to the increased ability to disrupt cell membrane integrity. Salvia-Trujillo et al. (2015) tested several essential oils loaded nanoemulsions (e.g., lemongrass, clove, tea tree, thyme, geranium, marjoram, palmarosa, rosewood, sage and mint) and their antibacterial activity against *Escherichia coli*. The authors demonstrated the potential applications of nanoemulsions for food preservation and improve taste.

Donsi et al. (2012) described the development and characterization of nanoemulsions composed of sunflower oil, loaded with carvacrol, limonene and cinnamaldehyde and stabilized with different surfactants, namely, lecithin, pea proteins, sugar ester and a combination of Tween 20 and glycerol monooleate. *Escherichia coli, Lactobacillus delbrueckii* and *Saccharomyces cerevisiae* were used to evaluate the antimicrobial activity of the developed formulations, which was found to be affected by the composition of the nanoemulsions. Indeed, the antimicrobial activity was correlated to the concentration of the essential oil components in the water phase in equilibrium with the inner oil droplets, which suggests that the capacity of interacting with bacterial cells might be dependent on the dissolution of the essential oil components in the outer aqueous phase of the nanoemulsion.

Eucalyptus globulus essential oil shows wide application against bacterial pathogens; however, it is instable, volatile and is irritating to the skin. Eucalyptus oil (*Eucalyptus globulus*) loaded-nanoemulsions were formulated in chitosan biopolymer (1%) to develop a membrane for wound healing, commonly infected by opportunistic microorganisms, such as *Staphylococcus aureus* (Sugumar et al. 2015). The authors reported that nanoemulsion-chitosan film demonstrated higher antibacterial activity than chitosan film (without nanoemulsion).

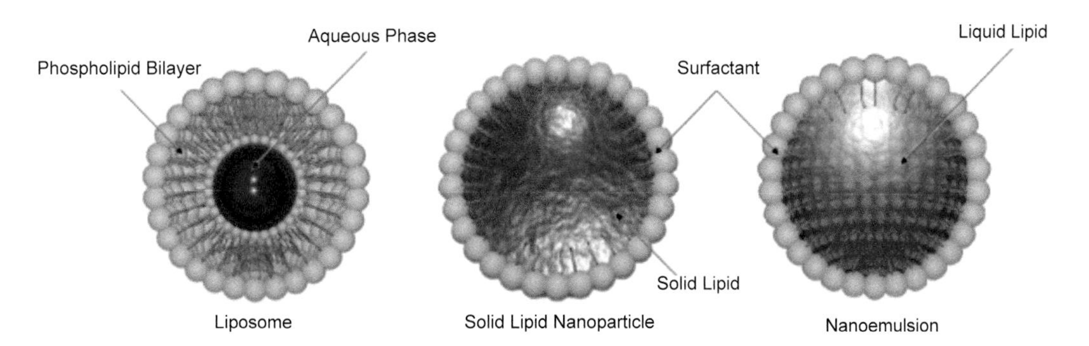

Figure 1.1. Schematic representation of the three types of lipid nanoparticulate carriers.

Conclusions

Essential oils have demonstrated antimicrobial activity, with promising applications in pharmaceutical, biomedical and food industries. Synergistic effects may be observed when loading these natural products in lipid nanoparticles, with particular emphasis for their use against multi-drug resistant bacteria. Lipid nanoparticles are biocompatible, biodegradable and biotolerable drug delivery systems. They are used to improve water solubility, stability, light protection, organoleptic characteristics and antimicrobial action. Among Solid Lipid Nanoparticles (SLN), liposomes and nanoemulsions, those involving lower energy input are the nanoemulsions, while scale up facilities are recognized for SLN. Liposomes are versatile systems with the capacity to load a variety of compounds. The selection of the most appropriate system depends on the physicochemical properties of the essential oil compounds and administration route.

Acknowledgements

The authors would like to acknowledge the financial support received through the projects M-ERA-NET/0004/2015 and UID/QUI/50006/2013, from the Portuguese Science and Technology Foundation, Ministry of Science and Education (FCT/MEC) through national funds, and co-financed by FEDER, under the Partnership Agreement PT2020.

References

Al-Haj, N.A., Shamsudin, M.N., Alipiah, N., Zamri, H.F., Bustamam, A., Ibrahim, S. and Abdullah, R. 2010. Characterization of *Nigella sativa* L. essential oil-loaded solid lipid nanoparticles. Am. J. Pharmacol. Toxicol. 5: 52–57.

Alinian, S., Razmjoo, J. and Zeinali, H. 2016. Flavonoids, anthocynins, phenolics and essential oil produced in cumin (*Cuminum cyminum* L.) accessions under different irrigation regimes. Ind. Crops Prod. 81: 49–55.

Araújo, J.S. and Meira, R.M.S.A. 2016. Comparative anatomy of calyx and foliar glands of Banisteriopsis CB Rob. (*Malpighiaceae*). Acta Bot. Bras. 30: 112–123.

Astani, A., Reichling, J. and Schnitzler, P. 2010. Comparative study on the antiviral activity of selected monoterpenes derived from essential oils. Phytother. Res. 24: 673–679.

Attama, A.A. and Umeyor, C.E. 2015. The use of solid lipid nanoparticles for sustained drug release. Ther. Deliv. 6: 669–684.

Aumeeruddy-Elalfi, Z., Gurib-Fakim, A. and Mahomoodally, M. 2016. Kinetic studies of tyrosinase inhibitory activity of 19 essential oils extracted from endemic and exotic medicinal plants. South Africa. J. Bot. 103: 89–94.

Baser, K.H.C. and Buchbauer, G. 2016. Handbook of Essential Oils: Science, Technology, and Applications. CRC Press, Taylor and Francis Group. Boca Raton, FL.

Bassolé, I.H.N. and Juliani, H.R. 2012. Essential oils in combination and their antimicrobial properties. Molecules 17: 3989–4006.

Beutler, J.A. 2009. Natural products as a foundation for drug discovery. Curr. Protoc. Pharmacol. 46: 1–21.

Bonifacio, B.V., Silva, P.B.D., Ramos, S., Silveira Negri, K.M., Bauab, T.M. and Chorilli, M. 2014. Nanotechnology-based drug delivery systems and herbal medicines: a review. Int. J. Nanomed. 9: 1–15.

Bunjes, H., Westesen, K. and Koch, M.H. 1996. Crystallization tendency and polymorphic transitions in triglyceride nanoparticles. Int. J. Pharma. 129: 159–173.

Calixto, J. 2000. Efficacy, safety, quality control, marketing and regulatory guidelines for herbal medicines (phytotherapeutic agents). Braz. J. Med. Biol. Res. 33: 179–189.

Celia, C., Trapasso, E., Locatelli, M., Navarra, M., Ventura, C.A., Wolfram, J., Carafa, M., Morittu, V.M., Britti, D., Di Marzio, L. and Paolino, D. 2013. Anticancer activity of liposomal bergamot essential oil (BEO) on human neuroblastoma cells. Colloids Sur. B: Biointerfaces 112: 548–553.

Champagne, A. and Boutry, M. 2016. Proteomics of terpenoid biosynthesis and secretion in trichomes of higher plant species. Biochim. Biophys. Acta (BBA)-Proteins and Proteomics 1864: 1039–1049.

Chang, Y., McLandsborough, L. and McClements, D.J. 2015. Fabrication, stability and efficacy of dual-component antimicrobial nanoemulsions: essential oil (thyme oil) and cationic surfactant (lauric arginate). Food Chem. 172: 298–304.

Cherkaoui-Tangi, K., Israili, Z.H. and Lyoussi, B. 2016. Vasorelaxant effect of essential oil isolated from *Nigella sativa* L. seeds in rat aorta: Proposed mechanism. Pak. J. Pharma. Sci. 29: 1–8.

Choudhury, S.S., Bashyam, L., Manthapuram, N., Bitla, P., Kollipara, P. and Tetali, S.D. 2014. Ocimum sanctum leaf extracts attenuate human monocytic (THP-1) cell activation. J. Ethnopharmacol. 154: 148–155.

Comin, V.M., Lopes, L.Q., Quatrin, P.M., de Souza, M.E., Bonez, P.C., Pintos, F.G., Raffin, R.P., Vaucher, R.d.A., Martinez, D.S. and Santos, R.C. 2016. Influence of *Melaleuca alternifolia* oil nanoparticles on aspects of *Pseudomonas aeruginosa* biofilm. Microb. Pathog. 93: 120–125.

Cui, H., Zhou, H. and Lin, L. 2016a. The specific antibacterial effect of the Salvia oil nanoliposomes against *Staphylococcus aureus* biofilms on milk container. Food Control 61: 92–98.

Cui, H., Li, W., Li, C., Vittayapadung, S. and Lin, L. 2016b. Liposome containing cinnamon oil with antibacterial activity against methicillin-resistant *Staphylococcus aureus* biofilm. Biofouling 32: 215–225.

De Filippis, L. 2016. Plant secondary metabolites: from molecular biology to health products. pp. 263–299. *In*: Mahgoub Azooz, Mohamed and Ahmad, Parvaiz (eds.). Plant-Environment Interaction: Responses and Approaches to Mitigate Stress.Wiley-Blackwell.

de Oliveira, A.G., Scarpa, M.V., Correa, M.A., Cera, L.F.R. and Formariz, T.P. 2004. Microemulsões: estrutura e aplicações como sistema de liberação de fármacos. Química Nova 27: 131–138.

Detoni, C., Cabral-Albuquerque, E., Hohlemweger, S., Sampaio, C., Barros, T. and Velozo, E. 2009. Essential oil from *Zanthoxylum tingoassuiba* loaded into multilamellar liposomes useful as antimicrobial agents. J. Microencapsul. 26: 684–691.

Doktorovova, S., Souto, E.B. and Silva, A.M. 2014. Nanotoxicology applied to solid lipid nanoparticles and nanostructured lipid carriers—a systematic review of *in vitro* data. Eur. J. Pharma. Biopharma. 87: 1–18.

Doktorovova, S., Kovačević, A.B., Garcia, M.L. and Souto, E.B. 2016. Pre-clinical safety of solid lipid nanoparticles and nanostructured lipid carriers: current evidence from *in vitro* and *in vivo* evaluation. Eur. J. Pharma. Biopharma. 108: 235–252.

Donsi, F., Annunziata, M., Vincensi, M. and Ferrari, G. 2012. Design of nanoemulsion-based delivery systems of natural antimicrobials: effect of the emulsifier. J. Biotechnol. 159: 342–350.

El-Soud, N.H., Deabes, M., El-Kassem, L.A. and Khalil, M. 2015. Chemical composition and antifungal activity of *Ocimum basilicum* L. essential oil. Maced. J. Med. Sci. 3: 374–379.

El Asbahani, A., Miladi, K., Badri, W., Sala, M., Addi, E.A., Casabianca, H., El Mousadik, A., Hartmann, D., Jilale, A. and Renaud, F. 2015. Essential oils: From extraction to encapsulation. Int. J. Pharma. 483: 220–243.

Engström, M.T., Karonen, M., Ahern, J.R., Baert, N., Payré, B., Hoste, H. and Salminen, J.-P. 2016. Chemical structures of plant hydrolyzable tannins reveal their *in vitro* activity against egg hatching and motility of *Haemonchus contortus* nematodes. J. Agri. Food Chem. 64: 840–851.

Erichsen-Brown, C. 2013. Medicinal and other uses of North American plants: a historical survey with special reference to the eastern Indian tribes. Dover Publications Inc.

Fabiane, K.C., Ferronatto, R., Santos, A.C.d. and Onofre, S.B. 2008. Physicochemical characteristics of the essential oils of *Baccharis dracunculifolia* and *Baccharis uncinella* DC (*Asteraceae*). Rev. Bras. Farmacogn. 18: 197–203.

Fereshteh, E., Samin, Z., Morteza, Y., Javad, H. and Samad, N.E. 2011. Antibacterial activity of *Zataria multiflora* Boiss essential oil against extended spectrum β lactamase produced by urinary isolates of Klebsiella pneumonia. Jundishapur J. Microbiol. 4(Supplement 1): S43–S49.

Ferrer, J.-L., Austin, M., Stewart, C. and Noel, J. 2008. Structure and function of enzymes involved in the biosynthesis of phenylpropanoids. Plant Physiol. Biochem. 46: 356–370.

Flores, F.C., Beck, R.C. and da Silva, C.d.B. 2016. Essential oils for treatment for onychomycosis: a mini-review. Mycopathologia 181: 9–15.

Formariz, T.P., Urban, M.C.C., Silva-Júnior, A., Gremião, M.P.D. and Oliveira, A. 2005. Microemulsões e fases líquidas cristalinas como sistemas de liberação de fármacos. Rev. Bras. Ciências Farma. 41: 301–313.

Garg, A. and Singh, S. 2011. Enhancement in antifungal activity of eugenol in immunosuppressed rats through lipid nanocarriers. Colloids Surf. B: Biointerfaces 87: 280–288.

Gasco, M.R. 1993. Method for producing solid lipid microspheres having a narrow size distribution. (US5250236 A).

Ge, Y. and Ge, M. 2016. Distribution of *Melaleuca alternifolia* essential oil in liposomes with Tween 80 addition and enhancement of *in vitro* antimicrobial effect. J. Exp. Nanosci. 11: 345–358.

Giovannini, D., Gismondi Basso, A., Canuti, L., Braglia, R., Canini, A., Mariani, F. and Cappelli, G. 2016. *Lavandula angustifolia* Mill. essential oil exerts antibacterial and anti-inflammatory effect in macrophage mediated immune response to *Staphylococcus aureus*. Immunol. Invest. 45: 11–28.

González-Lamothe, R., Mitchell, G., Gattuso, M., Diarra, M.S., Malouin, F. and Bouarab, K. 2009. Plant antimicrobial agents and their effects on plant and human pathogens. International J. Mol. Sci. 10: 3400–3419.

Gurib-Fakim, A. 2006. Medicinal plants: traditions of yesterday and drugs of tomorrow. Mol. Aspect. Med. 27: 1–93.

Hammer, K.A., Carson, C. and Riley, T. 1999. Antimicrobial activity of essential oils and other plant extracts. J. Appl. Microbiol. 86: 985–990.

Hassanshahian, M. and Khosravi, F. 2016. Study the antimicrobial effects of artemisia santonica extract on some pathogenic bacteria. Adv. Herbal Med. 1: 43–46.

Hüsnü, K., Başer, C. and Demirci, F. 2007. Chemistry of essential oils, Flavours and Fragrances. Springer, pp. 43–86.

Knezevic, P., Aleksic, V., Simin, N., Svircev, E., Petrovic, A. and Mimica-Dukic, N. 2016. Antimicrobial activity of *Eucalyptus camaldulensis* essential oils and their interactions with conventional antimicrobial agents against multi-drug resistant *Acinetobacter baumannii*. J. Ethnopharmacol. 178: 125–136.

Kostarelos, K. 2007. Construction of nanoscale multicompartment liposomes for combinatory drug delivery. Int. J. Pharm. 331: 182–185.

Kuppusamy, S., Thavamani, P., Megharaj, M., Nirola, R., Lee, Y.B. and Naidu, R. 2016. Assessment of antioxidant activity, minerals, phenols and flavonoid contents of common plant/tree waste extracts. Ind. Crops Prod. 83: 630–634.

Lai, F., Wissing, S.A., Muller, R.H. and Fadda, A.M. 2006a. *Artemisia arborescens* L. essential oil-loaded solid lipid nanoparticles for potential agricultural application: preparation and characterization. AAPS Pharm. Sci. Tech. 7: E2.

Lai, F., Wissing, S.A., Müller, R.H. and Fadda, A.M. 2006b. *Artemisia arborescens* L. essential oil-loaded solid lipid nanoparticles for potential agricultural application: preparation and characterization. AAPS Pharm. Sci. Tech. 7: E10–E18.

Lai, F., Sinico, C., De Logu, A., Zaru, M., Muller, R.H. and Fadda, A.M. 2007. SLN as a topical delivery system for *Artemisia arborescens* essential oil: *in vitro* antiviral activity and skin permeation study. Int. J. Nanomed. 2: 419–425.
Li, M., Zhu, L., Liu, B., Du, L., Jia, X., Han, L. and Jin, Y. 2016a. Tea tree oil nanoemulsions for inhalation therapies of bacterial and fungal pneumonia. Colloids Surf. B: Biointerfaces 141: 408–416.
Li, W.-Z., Hao, X.-L., Zhao, N., Han, W.-X., Zhai, X.-F., Zhao, Q., Wang, Y.-E., Zhou, Y.-Q., Cheng, Y.-C. and Yue, Y.-H. 2016b. Propylene glycol-embodying deformable liposomes as a novel drug delivery carrier for vaginal fibrauretine delivery applications. J. Controlled Release 226: 107–114.
Lin, L., Zhang, X., Zhao, C. and Cui, H. 2016. Liposome containing nutmeg oil as the targeted preservative against *Listeria monocytogenes* in dumplings. RSC Advances 6: 978–986.
Liu, Z.-W., Zeng, X.-A. and Han, Z. 2016. Effects of the Content of Cholesterol on the Permeability of Vesicles Membranes Induced by Pulsed Electric Fields, 1st World Congress on Electroporation and Pulsed Electric Fields in Biology, Medicine and Food & Environmental Technologies. Springer, pp. 179–182.
Marengo, E., Cavalli, R., Caputo, O., Rodriguez, L. and Gasco, M.R. 2000. Scale-up of the preparation process of solid lipid nanospheres. Part I. Int. J. Pharma. 205: 3–13.
Marengo, E., Cavalli, R., Rovero, G. and Gasco, M.R. 2003. Scale-up and optimization of an evaporative drying process applied to aqueous dispersions of solid lipid nanoparticles. Pharma. Develop. Technol. 8: 299–309.
McCarthy, K. 2004. Cosmetic compositions and methods. (US20060002871 A1).
Miranda, C.A.S.F., Cardoso, M.d.G., Batista, L.R., Rodrigues, L.M.A. and Figueiredo, A.C.d.S. 2016. Essential oils from leaves of various species: antioxidant and antibacterial properties on growth in pathogenic species. Rev. Ciência Agronôm. 47: 213–220.
Moghimi, R., Ghaderi, L., Rafati, H., Aliahmadi, A. and McClements, D.J. 2016. Superior antibacterial activity of nanoemulsion of *Thymus daenensis* essential oil against *E. coli*. Food Chem. 194: 410–415.
Moghimipour, E., Aghel, N., Zarei Mahmoudabadi, A., Ramezani, Z. and Handali, S. 2012. Preparation and characterization of liposomes containing essential oil of *Eucalyptus camaldulensis* Leaf. Jundishapur J. Nat. Pharma. Prod. 7: 117–122.
Moghimipour, E., Ramezani, Z. and Handali, S. 2013. Solid lipid nanoparticles as a delivery System for *Zataria multiflora* essential oil: formulation and characterization. Curr. Drug Deliv. 10: 151–157.
Mukherjee, C., Samanta, T. and Mitra, A. 2016. Redirection of metabolite biosynthesis from hydroxybenzoates to volatile terpenoids in green hairy roots of *Daucus carota*. Planta 243: 305–320.
Musthaba, S.M., Baboota, S., Ahmed, S., Ahuja, A. and Ali, J. 2009. Status of novel drug delivery technology for phytotherapeutics. Expert Opin. Drug Deliv. 6: 625–637.
Organization, W.H. 2003. WHO guidelines on good agricultural and collection practices (GACP) for medicinal plants. Geneva: World Health Organization.
Pan, S.-Y., Litscher, G., Gao, S.-H., Zhou, S.-F., Yu, Z.-L., Chen, H.-Q., Zhang, S.-F., Tang, M.-K., Sun, J.-N. and Ko, K.-M. 2014. Historical perspective of traditional indigenous medical practices: the current renaissance and conservation of herbal resources. Evid. -Based Complementary Alt. Med., Article ID 525340, p. 20.
Perez-Roses, R., Risco, E., Vila, R., Penalver, P. and Canigueral, S. 2016. Biological and Nonbiological antioxidant activity of some essential oils. J. Agri. Food Chem. 64: 4716–4724.
Qiu, J., Zhang, X., Luo, M., Li, H., Dong, J., Wang, J., Leng, B., Wang, X., Feng, H. and Ren, W. 2011. Subinhibitory concentrations of perilla oil affect the expression of secreted virulence factor genes in *Staphylococcus aureus*. PloS One 6: e16160.
Radwan, M.M., Tabanca, N., Wedge, D.E., Tarawneh, A.H. and Cutler, S.J. 2014. Antifungal compounds from turmeric and nutmeg with activity against plant pathogens. Fitoterapia 99: 341–346.
Richter, J. and Schellenberg, I. 2007. Comparison of different extraction methods for the determination of essential oils and related compounds from aromatic plants and optimization of solid-phase microextraction/gas chromatography. Anal. Bioanal. Chem. 387: 2207–2217.
Rossi, C., Dantas, T., Neto, A.A.D. and Maciel, M.A.M. 2007. Microemulsões: uma abordagem básica e perspectivas para aplicabilidade industrial. Revista Universal Rural Série Ciências Exatas Terra 26: 45–66.
Salvia-Trujillo, L., Rojas-Graü, A., Soliva-Fortuny, R. and Martín-Belloso, O. 2015. Physicochemical characterization and antimicrobial activity of food-grade emulsions and nanoemulsions incorporating essential oils. Food Hydrocoll. 43: 547–556.
São Pedro, A., Santo, I., Silva, C., Detoni, C. and Albuquerque, E. 2013. The use of nanotechnology as an approach for essential oil-based formulations with antimicrobial activity. *In*: Méndez-Vilas, A. (ed.). Microbial Pathogens and Strategies for Combating Them Formatex Research Center Publisher 2: 1364–1374.
Schläger, S. and Dräger, B. 2016. Exploiting plant alkaloids. Curr. Opin. Biotechnol. 37: 155–164.
Schwarz, C., Mehnert, W., Lucks, J. and Müller, R. 1994. Solid lipid nanoparticles (SLN) for controlled drug delivery. I. Production, characterization and sterilization. J. Controlled Release 30: 83–96.
Sebaaly, C., Jraij, A., Fessi, H., Charcosset, C. and Greige-Gerges, H. 2015a. Preparation and characterization of clove essential oil-loaded liposomes. Food Chem. 178: 52–62.
Severino, P., Andreani, T., Macedo, A.S., Fangueiro, J.F., Santana, M.H.A., Silva, A.M. and Souto, E.B. 2011. Current state-of-art and new trends on lipid nanoparticles (SLN and NLC) for oral drug delivery. J. Drug Deliv. Article ID 750891, p. 10.
Severino, P., Moraes, L.F., Zanchetta, B., Souto, E.B. and Santana, M.H. 2012a. Elastic liposomes containing benzophenone-3 for sun protection factor enhancement. Pharma.Develop.Technol. 17: 661–665.

Severino, P., Santana, M.H.A. and Souto, E.B. 2012b. Optimizing SLN and NLC by 2 2 full factorial design: Effect of homogenization technique. Mater. Sci. Eng. C 32: 1375–1379.
Severino, P., Andreani, T., Chaud, M.V., Benites, C.I., Pinho, S.C. and Souto, E.B. 2015a. Essential oils as active ingredients of lipid nanocarriers for chemotherapeutic use. Curr. Pharma. Biotechnol. 16: 365–370.
Severino, P., Andreani, T., Chaud, M.V., Benites, C.I., Pinho, S.C. and Souto, E.B. 2015b. Essential oils as active ingredients of lipid nanocarriers for chemotherapeutic use. Curr. Pharma. Biotechnol. 16: 365–370.
Sherry, M., Charcosset, C., Fessi, H. and Greige-Gerges, H. 2013. Essential oils encapsulated in liposomes: a review. J. Liposome Res. 23: 268–275.
Shi, F., Zhao, J.H., Liu, Y., Wang, Z., Zhang, Y.T. and Feng, N.P. 2012. Preparation and characterization of solid lipid nanoparticles loaded with frankincense and myrrh oil. Inter. J. Nanomed. 7: 2033–2043.
Silva, F.S., Menezes, P.M.N., Sá, P.G.S.d., Oliveira, A.L.d.S., Souza, E.A.A., Almeida, J.R.G.d.S., Lima, J.T.d., Uetanabaro, A.P.T., Silva, T.R.d.S. and Peralta, E.D. 2016. Chemical composition and pharmacological properties of the essential oils obtained seasonally from *Lippia thymoides*. Pharma. Biol. 54: 25–34.
Sim, Y. and Shin, S. 2008. Combinatorial anti-Trichophyton effects of *Ligusticum chuanxiong* essential oil components with antibiotics. Arch. Pharm. Res. 31: 497–502.
Souto, E.B. and Muller, R.H. 2006. Investigation of the factors influencing the incorporation of clotrimazole in SLN and NLC prepared by hot high-pressure homogenization. J. Microencapsul. 23: 377–388.
Souto, E.B. and Muller, R.H. 2010. Lipid nanoparticles: effect on bioavailability and pharmacokinetic changes. Handbook Exp. Pharmacol. 197: 115–141.
Sugumar, S., Mukherjee, A. and Chandrasekaran, N. 2015. Eucalyptus oil nanoemulsion-impregnated chitosan film: antibacterial effects against a clinical pathogen, *Staphylococcus aureus*, *in vitro*. Int. J. Nanomed. 10(Suppl 1): 67–75.
Svetlichny, G., Külkamp-Guerreiro, I., Cunha, S., Silva, F., Bueno, K., Pohlmann, A., Fuentefria, A. and Guterres, S. 2015. Solid lipid nanoparticles containing copaiba oil and allantoin: development and role of nanoencapsulation on the antifungal activity. Die Pharmazie 70: 155–164.
Tang, J., Huang, Y., Liu, H., Zhang, C. and Tang, D. 2016. Novel glucometer-based immunosensing strategy suitable for complex systems with signal amplification using surfactant-responsive cargo release from glucose-encapsulated liposome nanocarriers. Biosens. Bioelectron. 79: 508–514.
Tomaino, A., Cimino, F., Zimbalatti, V., Venuti, V., Sulfaro, V., De Pasquale, A. and Saija, A. 2005. Influence of heating on antioxidant activity and the chemical composition of some spice essential oils. Food Chem. 89: 549–554.
Tsou, L.K., Lara-Tejero, M., Rose Figura, J., Zhang, Z.J., Wang, Y.-C., Yount, J.S., Lefebre, M., Dossa, P.D., Kato, J. and Guan, F. 2016. Anti-bacterial flavonoids from medicinal plants covalently inactivate type III protein secretion substrates. J. Am. Chem. Soc. 138: 2209–2218.
Vilas, V., Philip, D. and Mathew, J. 2016. Essential oil mediated synthesis of silver nanocrystals for environmental, anti-microbial and antioxidant applications. Mater. Sci. Eng. C 61: 429–436.
Wu, J., Liu, H., Ge, S., Wang, S., Qin, Z., Chen, L., Zheng, Q., Liu, Q. and Zhang, Q. 2015. The preparation, characterization, antimicrobial stability and *in vitro* release evaluation of fish gelatin films incorporated with cinnamon essential oil nanoliposomes. Food Hydrocoll. 43: 427–435.
Zhang, H., Ran, X., Hu, C.L., Qin, L.P., Lu, Y. and Peng, C. 2012. Therapeutic effects of liposome-enveloped *Ligusticum chuanxiong* essential oil on hypertrophic scars in the rabbit ear model. PloS One 7: e31157.
Zylberberg, C. and Matosevic, S. 2016. Pharmaceutical liposomal drug delivery: a review of new delivery systems and a look at the regulatory landscape. Drug Deliv. 1–11.

2

Antimicrobial Activities of Plant Essential Oils and Their Components against Antibiotic-Susceptible and Antibiotic-Resistant Foodborne Pathogens

Mendel Friedman

Introduction

Food processors, regulatory agencies, microbiologists, physicians, and the general public have been increasingly concerned with the growing number of foodborne illness outbreaks caused by pathogens. Numerous foodborne diseases are syndromes that result from ingesting foods that are contaminated with either infectious microorganisms or toxins produced by microorganisms. Foodborne pathogenic bacteria include strains of *Bacillus cereus*, *Campylobacter jejuni*, *Clostridium botulinum*, *C. perfringens*, *Escherichia coli*, *Listeria monocytogenes*, *Mycobacterium aviumparatuberculosis, Salmonella enterica* and *Staphylococcus aureus*, among others. The antibiotic resistance of some pathogens associated with foodborne illness is of particular concern. Bacteria can exert adverse effects in tissues of animals and humans in at least one of two ways: adhesion to and penetration into cells; and release of cellular toxins. Cellular and molecular targets for the action of natural antimicrobial compounds include cell membranes, cell wall peptidoglycans, enzymes, genetic material and essential nutrients for pathogens such as zinc and iron. Understanding the molecular basis of the action of bacteria and of bacterial toxins *in vitro* and *in vivo* will facilitate devising appropriate food-compatible strategies to inactivate the pathogens.

There has been considerable interest in developing new types of effective and non-toxic antimicrobial compounds that have the potential to inhibit the growth of both susceptible and antibiotic-resistant pathogens (Friedman 2015). Some promising natural antimicrobials include polyphenols from food plants such as tea, apples or grapes, chitosan from crustaceans and essential oils from aromatic plants.

Healthy Processed Foods Research, United States Department of Agriculture, Agricultural Research Service, Western Regional Research Center, Albany, California 94710 USA.
E-mail: Mendel.Friedman@ars.usda.gov

Essential Oils (EO) are oily mixtures of lipophilic and volatile constituents produced by aromatic plants which give herbs, spices and flowers their distinctive olfactory characteristics (i.e., the essence of the plant). Many EO-containing plants and their extracts are commonly used in cooking, perfumery and in traditional medicines. EOs are typically extracted by steam distillation, but other methods include expression, supercritical fluid extraction, solvent extraction, and various combinations of these techniques. Distillates are commonly available at health food stores, marketed as natural perfumes and medicines. EOs are necessarily strongly flavored and scented and are indispensable for creating distinctive culinary foods worldwide.

EOs major components consist of terpenes (5-carbon isoprene derivatives), terpenoids (modified terpenes), and other aromatic compounds known as phenylpropanoids (cinnamic acid derivatives) (Fig. 2.1). Numerous EOs are composed mostly of one to three of these components with the

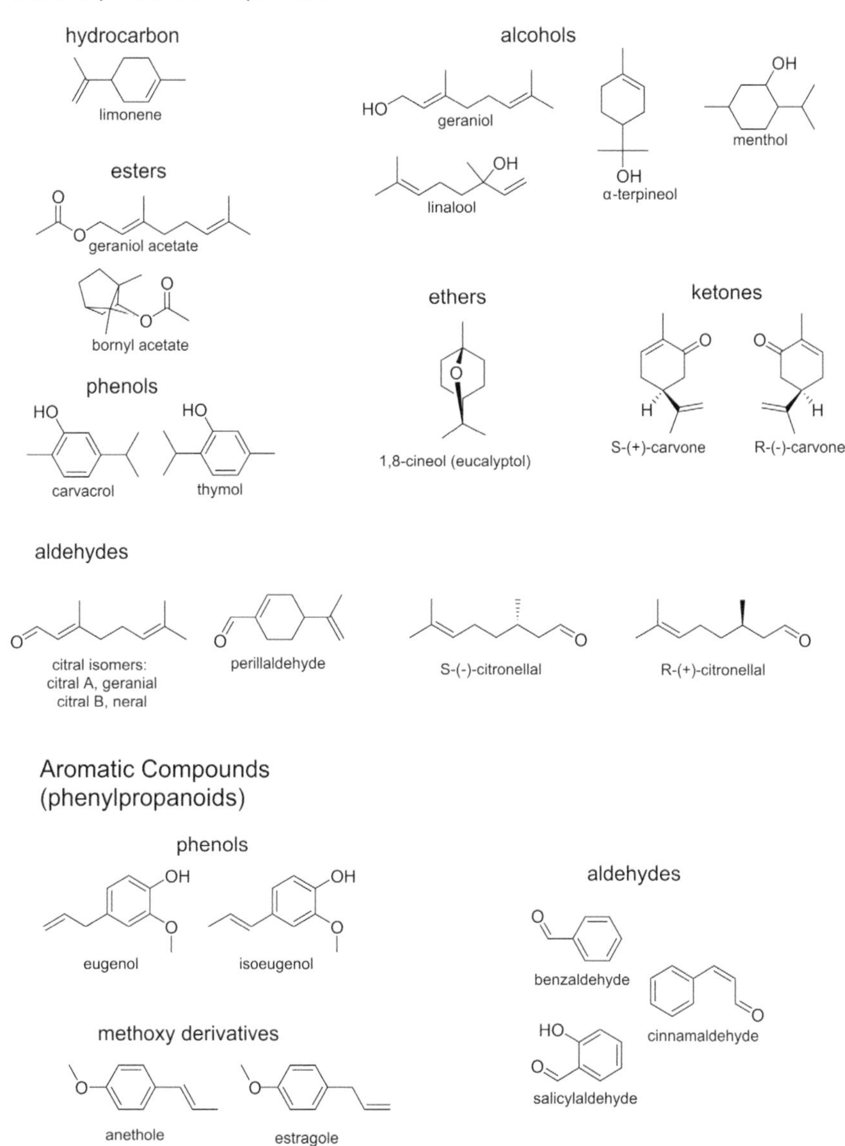

Figure 2.1. Structures of some of antimicrobial EO compounds.

remainder being large numbers of minor components. Sometimes different plants contain nearly the same components but in different proportions, giving them distinctive characteristics. Even within a species, variety, growing conditions and maturity can significantly influence EO composition in a plant (Dudai et al. 2001). For example, we analyzed for the mono-terpenes, carvacrol and thymol, in the EOs of the two plants, oregano and thyme. We had two EOs obtained from different sources for each plant. Sample 1 of oregano oil contained 68.7% carvacrol and 8.6% thymol, while sample 2 contained 63.1% carvacrol and no thymol. Sample 1 of thyme oil contained 6.8% carvacrol and 50.0% thymol and sample 2 contained 3.2% carvacrol and 55.1% of thymol (Friedman et al. 2004b). Thyme and oregano have distinctly different sensory properties, so both proportion and content of EOs is integral to their characteristic differences.

There is much variation in even the major components of EOs. Minor components can add additional variations. Gardeners and cooks may be familiar with the fact that cultivated thyme is available in a multitude of scents: common, lemon, caraway, lime, etc. These interesting varieties undoubtedly contain EO components plentiful in the other plants such as limonene (lemon) or S-(+)-carvone (caraway). This commonality is found in many other herbs as well. Cinnamon and cloves both contain cinnamaldehyde and eugenol as major components (Friedman et al. 2000).

Chirality is important to our perception of the odors. For instance, the two optical isomers of carvone smell very differently. S-(+)-carvone smells like caraway and dill, while R-(–)-carvone smells of spearmint. An odorous molecule has a stereostructure complementary to the sites of the odor receptors (Ravid 2008). This suggests the possibility that the isomers may have different biological potencies at other sites as well.

Many of the EOs from herbs, having been used in cooking for many years, are classified by the U.S. FDA as Generally Recognized As Safe (GRAS) and thus may be exempted from the burden of regulatory approval. This means they may be incorporated into foods without extensive testing. The European Union has somewhat stricter standards, excluding estragole and methyl eugenol from their list of approved flavorings due to safety concerns (Burt 2004). However, these substances are still classified as GRAS in the USA.

Evidence that humans used herbs for health purposes dates back to as early as 5000 BCE (Tapsell et al. 2006). Today, there is considerable interest EOs for food preservation uses, making use of their antioxidant and antimicrobial properties.

The primary objective of our research effort is to develop new ways to reduce the human pathogen burden in foods with the aid of naturally occurring, plant-derived antimicrobial compounds. In the course of our research we screened 120 naturally occurring and food-compatible plant-derived EOs and EO components for their antimicrobial activities against four foodborne pathogens. Some of the more promising EOs were also tested against antibiotic-resistant bacteria and in bacterially contaminated foods. The ultimate goal of these studies is to develop a better understanding of plant-derived compounds that possess antibacterial activities as well as devising practical food formulations that use the active compounds to reduce pathogens in foods, feeds, and possibly also in animals and humans after consumption. Studies with foods may require assessment of safety, sensory properties and cost of added test compounds as well as the influence of the food matrix and of storage temperatures and times on antibacterial activities. This chapter is largely limited to summarizing our own studies designed to define relative activities (potencies) of essential oils and oil constituents against foodborne pathogens in buffers and in food.

Bactericidal Assessment

The ability of a substance to inhibit the growth of bacteria is usually measured by the Minimum Inhibitory Concentration (MIC) test. This measures the lowest concentration of the test substance which inhibits the bacteria. The MIC is not an ideal assay for a variety of reasons. In particular, natural substances, which are multi-component and may act by a variety of mechanisms, do not always behave predictably *in vitro*. For instance, they may need to reach a concentration threshold before inhibition is observed, after which inhibition may rapidly increase. Conversely, inhibition may be concentration dependent at low levels, but fall off with increasing concentrations. For some substances, no amount will inactivate the bacteria completely. Yet these substances may be very effective when combined with other substances. Synergism and additive effects are often observed with mixtures of natural antimicrobials.

In practice, the MIC is imprecise because it is simply the first concentration where no growth is observed. The true MIC value resides somewhere between this and the last concentration that supported any growth. Because of the inexact nature of the MIC, Lambert and Pearson (2000) suggested that all points in the MIC test be used to create a curve to more exactly pinpoint the MIC. While this approach certainly improves the quality of the calculated MIC, we suggest that an approach, using the midpoint rather than the endpoint, as it is the most sensitive point in the data set. With these concerns in mind, we developed the Bactericidal Activity value (BA_{50}), analogous to the Inhibitory Concentration (IC_{50}) values in enzyme and cell inhibition assays or the Lethal Dose (LD_{50}) in toxicology studies (Friedman et al. 2002). BA_{50} is defined as the concentration of antimicrobial in a dilution series that kills 50% of the initial amounts of bacteria. BA_{50} values offer maximum precision with a dose-response analysis. This is because with regression analysis, maximum precision is achieved at the mean response, since the corresponding dose represents the center of the information upon which the design matrix is based (dose levels).

We determined the BA_{50} values of numerous natural substances (Friedman et al. 2002, 2003, 2006b,c, 2007a). Each compound was tested in a dilution series, from 0.00065 to 0.67% in the reaction mixture. A 3-parameter regression dose-response curve was drawn using SigmaPlot 11 (Systat Software, Inc., Chicago, IL). Bactericidal activities (BA_{50}) were defined as the percentage of test compound that kills 50% of the bacteria under the test conditions. The BA_{50} values were interpolated from the halfway point between no inhibition and maximal inhibition (Fig. 2.2). The lower the BA_{50} (and conversely, the higher the $1/BA_{50}$ value) the higher the activity. The bactericidal assay was simple

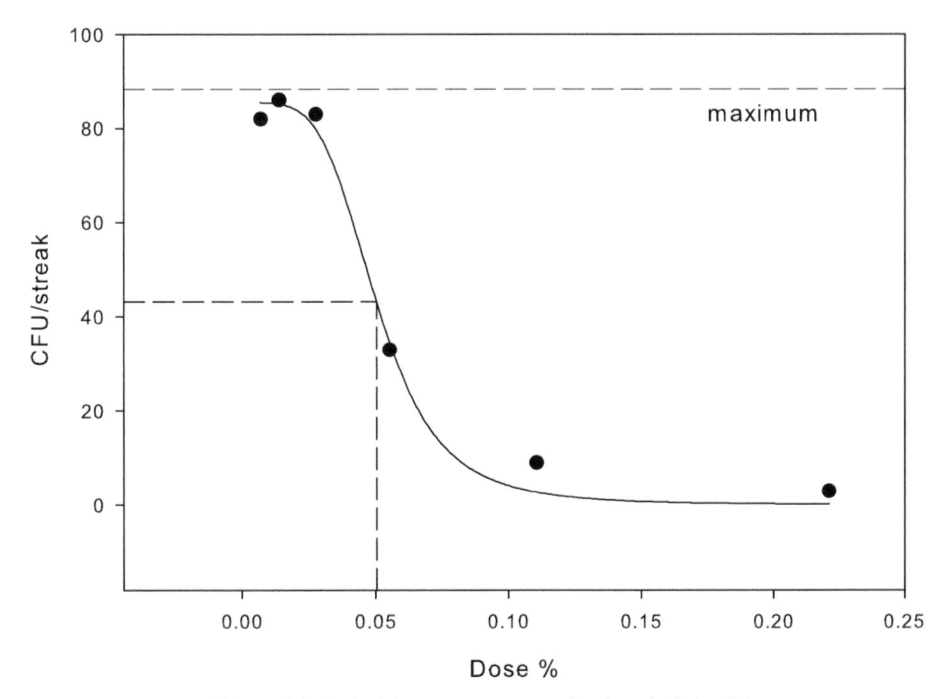

Figure 2.2. Typical dose-response regression for calculating BA_{50}.

to perform and the results were found to be reproducible in assays with the same samples tested at different times. The units of BA_{50} are usually in per cent of antimicrobial. However, with compounds of known molecular weights, the units can be in nanomolar, micromolar or millimolar concentrations.

In Vitro Antimicrobial Effects of Essential Oils and Oil Components

Effects against the foodborne pathogens Escherichia coli, Salmonella enterica, Campylobacter jejuni, and L. monocytogenes

Table 2.1 lists the EOs we tested against four bacteria, Gram-negative *E. coli*, *S. enterica*, and *C. jejuni*, and Gram-positive *L. monocytogenes*, including two strains of *L. monocytogenes*. The highest concentration tested was 0.67% (v/v). The BA_{50} of those test substances which had no effect are listed as > 0.67. Those boxes are grayed for easier viewing. It is apparent from the table that *C. jejuni* was more susceptible to inactivation by the oils than the other bacteria.

Table 2.2 lists the effectiveness of the individual EO components which we tested against the same bacteria. Again, *C. jejuni* was the most susceptible to inactivation.

Table 2.3 lists the tested EOs or EO components from Tables 2.1 and 2.2 which proved active against all four strains of bacteria. The top 10 (or more if equal to the top 10) most effective test components for each bacterial strain are shown in bold font. The following test substances were universally highly antimicrobial: carvacrol, cinnamaldehyde, eugenol, Spanish oregano oil and thymol.

The antimicrobial activities under food processing conditions such as baking, cooking, frying and microwaving are mostly unknown. The taste of oils can vary widely (sweet cinnamon oil, savory oregano oil), but all are strongly flavored. Compatibility with foods would have to be considered. The most active compounds provide candidates for studies of flavor, taste and antibacterial activity in foods and against infectious diseases of animals and humans.

Antimicrobial activity against mycobacterium avium paratuberculosis

Mycobacterium avium subspecies *paratuberculosis* (MAP) is responsible for serious infections in cattle, causing Johne's disease. Its role in humans is not well elucidated, but it may be associated with Crohn's disease and diabetes (Scanu et al. 2007, 2008). The bacteria can pass into the milk of infected cows either systemically or through fecal contamination, and can survive pasteurization (Grant et al. 2005). Currently there is no approved treatment for infection with this organism other than culling the infected animals.

Several EOs were tested against three strains of MAP (Wong et al. 2008). Cinnamaldehyde and cinnamon oil were found to be highly effective against the bacteria, followed by oregano oil and carvacrol.

Figure 2.3 shows that the inhibitory effect of added cinnamaldehyde continued over a 42-day incubation period when concentrations of 25.9 μg/ml or greater were used.

Antimicrobial activities against antibiotic-resistant bacteria

Antibiotics are widely used as animal feed supplements to fight infections, to promote growth of livestock and poultry, and to reduce production costs. Antibiotics are also used in horticulture as aerosols to fruit trees for controlling bacterial infection. They are present in animal waste, often contaminating groundwater, surface water, irrigation water, fruits and vegetables. They can then disseminate through the food chain and enter the human intestinal tract after the produce or undercooked meat is eaten. Resistant microorganisms often arise from administration of sub-therapeutic levels of antibiotics in animal feeds. There is a need to develop alternatives for standard antibiotics that can be effective against antibiotic-resistant bacteria (Davis and Lederberg 2001, Walsh 2003).

Table 2.1. BA_{50} values of EOs against *E. coli* O157:H7, *S. enterica*, *C. jejuni*, and *L. monocytogenes*. Adapted from Friedman et al. (2002).

	E. coli RM1484	*S. enterica* RM1309	*C. jejuni* RM1221	*L. monocytogenes* RM2199	*L. monocytogenes* RM2388
Allspice	0.14	0.13	0.023	0.089	0.076
Almond Bitter	>0.67	>0.67	0.042	>0.67	0.25
Almond Sweet	>0.67	>0.67	>0.67	>0.67	>0.67
Aloe Vera	>0.67	>0.67	>0.67	>0.67	>0.67
Anise Seed	>0.67	>0.67	0.10	>0.67	0.31
Anise Star	>0.67	>0.67	0.22	>0.67	>0.67
Apricot	>0.67	>0.67	>0.67	>0.67	>0.67
Balsam Peru	>0.67	>0.67	0.15	>0.67	>0.67
Basil	0.41	0.42	0.023	0.089	0.12
Bay Leaf	0.13	0.13	0.034	0.070	0.073
Benzoin Gum	>0.67	>0.67	0.031	0.35	>0.67
Bergamot	>0.67	0.51	0.081	>0.67	>0.67
Birch	>0.67	>0.67	0.29	>0.67	>0.67
Cajeput	0.19	0.36	0.065	>0.67	>0.67
Caraway	0.46	0.47	0.029	0.33	0.24
Cardamom	>0.67	>0.67	0.022	0.58	0.40
Carrot Seed	>0.67	>0.67	0.0078	0.15	0.052
Cedarwood	>0.67	>0.67	0.0075	0.067	0.028
Celery Seed	>0.67	>0.67	0.0085	0.29	0.13
Chamomile Roman	>0.67	>0.67	0.022	0.32	0.29
Cinnamon Bark	0.18	0.14	0.021	0.085	0.079
Cinnamon Cassia	0.11	0.066	0.014	0.19	0.15
Cinnamon Leaf	0.13	0.084	0.027	0.087	0.090
Citronella	0.41	0.49	0.086	0.40	0.18
Clove Bud	0.13	0.13	0.016	0.074	0.092
Coriander	0.40	0.48	0.081	0.665	0.50
Cumin Seed	0.30	0.36	0.099	0.37	0.25
Cypress	>0.67	>0.67	0.084	0.11	0.27
Dill Weed	0.40	0.48	0.087	>0.67	0.66
Elemi	0.40	0.44	0.011	0.26	0.22
Eucalyptus	>0.67	>0.67	0.028	>0.67	0.34
Evening Primrose	>0.67	>0.67	0.32	>0.67	>0.67
Fennel Seed	0.56	0.38	0.10	>0.67	>0.67
Fir Needle Balsam	>0.67	>0.67	0.047	0.37	0.17
Fir Needle Siberian	0.48	0.61	0.014	0.13	0.082
Frankincense	>0.67	>0.67	0.025	>0.67	0.27
Gardenia	>0.67	>0.67	0.007	0.057	0.038
Ginger Root	>0.67	>0.67	0.005	0.50	0.23
Grape Seed	>0.67	>0.67	>0.67	>0.67	>0.67
Hazelnut	>0.67	>0.67	>0.67	>0.67	>0.67

Table 2.1 contd. ...

... *Table 2.1 contd.*

	***E. coli* RM1484**	***S. enterica* RM1309**	***C. jejuni* RM1221**	***L. monocytogenes* RM2199**	***L. monocytogenes* RM2388**
Helichrysum	>0.67	>0.67	0.10	0.50	0.09
Hyssop	0.57	0.41	0.096	0.33	0.18
Jasmine	>0.67	>0.67	0.006	0.30	0.36
Jojoba	>0.67	>0.67	>0.67	>0.67	>0.67
Juniper Berry	>0.67	0.43	0.034	0.33	0.19
Lavender	0.41	0.41	0.061	0.48	0.34
Lavender Spike	0.43	0.28	0.083	>0.67	>0.67
Lemon	>0.67	>0.67	0.045	0.35	0.22
Lemongrass	0.14	0.16	0.018	0.12	0.12
Lemon Verbena	0.49	>0.67	0.012	0.24	0.086
Lime	0.50	>0.67	0.044	0.22	0.056
Marigold Calendula	>0.67	>0.67	0.02	0.37	0.35
Marigold Tagetes	>0.67	0.40	0.003	>0.67	0.18
Marjoram	0.43	0.14	0.026	>0.67	>0.67
Mugwort	0.57	0.40	0.009	>0.67	0.56
Myrrh Gum	>0.67	>0.67	0.026	>0.67	0.069
Myrtle	>0.67	>0.67	0.23	0.11	0.10
Nutmeg	0.55	0.44	0.18	0.27	0.20
Oakmoss	>0.67	>0.67	0.22	0.37	0.40
Orange Bitter	0.47	>0.67	0.009	0.095	0.075
Orange Mandarin	0.41	0.64	0.010	0.18	0.10
Orange Neroli Blossom	0.45	>0.67	0.016	0.12	0.21
Orange Sweet	>0.67	>0.67	0.077	0.097	0.040
Oregano Origanum	0.048	0.050	0.019	0.078	0.098
Oregano Spanish	0.046	0.049	0.011	0.074	0.077
Palmarosa	0.12	0.14	0.067	0.17	0.27
Patchouli	>0.67	>0.67	0.007	0.092	0.029
Pennyroyal	0.25	0.41	0.12	>0.67	0.53
Pepper Black	>0.67	>0.67	0.034	>0.67	0.13
Peppermint	0.47	0.53	0.07	>0.67	0.33
Petitgrain	0.49	>0.67	0.036	>0.67	0.21
Pine Needle	>0.67	>0.67	0.059	>0.67	0.35
Ravensara	0.40	0.40	0.087	>0.67	0.57
Rose Damask	0.55	0.44	0.11	0.55	0.36
Rose French	0.43	0.50	0.05	0.44	0.29
Rose Geranium	0.41	0.40	0.088	0.60	0.32
Rosemary	0.38	0.45	0.060	>0.67	0.61
Rosewood	0.50	0.43	0.38	>0.67	0.26
Sage Clary	>0.67	>0.67	0.084	>0.67	0.33
Sage White Dalmatian	>0.67	>0.67	0.079	>0.67	>0.67

Table 2.1 contd. ...

...Table 2.1 contd.

	E. coli RM1484	*S. enterica* RM1309	*C. jejuni* RM1221	*L. monocytogenes* RM2199	*L. monocytogenes* RM2388
Sage White Desert	>0.67	>0.67	0.070	>0.67	0.31
Sandalwood Indian	>0.67	>0.67	0.028	0.22	0.083
Sassafras	>0.67	0.38	0.059	0.38	0.19
Sesame	>0.67	>0.67	>0.67	>0.67	>0.67
Spearmint	0.28	0.29	0.030	0.31	0.57
Spikenard	>0.67	>0.67	0.009	0.21	0.02
Spruce	>0.67	>0.67	0.10	0.58	0.44
Tangerine	>0.67	>0.67	0.10	0.665	0.20
Tarragon	>0.67	>0.67	0.017	>0.67	>0.67
Tea Tree	0.42	0.18	0.10	0.41	>0.67
Thuja	>0.67	0.40	0.10	>0.67	>0.67
Thyme	0.047	0.045	0.022	0.091	0.22
Tuberose	>0.67	>0.67	0.015	>0.67	>0.67
Vanilla Oleo Resin	>0.67	>0.67	0.10	>0.67	>0.67
Wintergreen	>0.67	0.54	0.24	>0.67	0.11
Wormwood	0.55	0.52	0.38	0.50	0.097
Ylang-ylang	>0.67	>0.67	0.21	0.65	>0.67

Antibiotic resistant Bacillus cereus, E. coli, and Staphylococcus aureus

Our expectation that naturally occurring compounds would kill antibiotic-resistant bacteria was demonstrated by collaborative studies with the Department of Agriculture and Rural Development for Northern Ireland using the antibiotic-resistant pathogens *B. cereus* (NCTC10989) vegetative cells and spores (tetracycline-resistant), *E. coli* (NCTC1186) (multi-resistant including ampicillin and chloramphenicol), and *S. aureus* (ATCC12715) (tetracycline- and streptomycin-resistant), as well as the non-pathogenic antibiotic-resistant bacteria *Micrococcus luteus*, reported to contaminate foods and feeds (Friedman et al. 2004a, 2006a). We evaluated seven natural products: three EOs (cinnamon, oregano and thyme); two EO components (carvacrol and perillaldehyde); and two phenolic compounds (β-resorcylic acid and dopamine); for their ability to inhibit growth of these organisms. Comparison of activities of the test compounds against the three pathogens revealed the following approximate order of effectiveness: oregano oil > thyme oil > carvacrol > cinnamon oil > perillaldehyde > dopamine > β-resorcylic acid (Fig. 2.4). The order of susceptibilities of the pathogens to inactivation was: *B. cereus* (vegetative) >> *S. aureus* >> *E. coli* >> *B. cereus* (spores). The oils and pure compounds exhibited exceptional activity against *B. cereus* vegetative cells, with oregano oil being active at nanograms/ml levels. In contrast, activities against *B. cereus* spores were low.

Oregano oil was the most effective antimicrobial of the compounds tested against all of the bacteria. Since carvacrol was less active than oregano oil, either minor constituents of oregano oil contribute to antibacterial activity, or possibly other components acted synergistically to produce a greater antimicrobial effect.

Since the compounds evaluated were active against both 'non-pathogenic' antibiotic-resistant *Mycobacterium luteus* bacteria as well as against pathogenic strains of *B. cereus*, *E. coli*, and *S. aureus*, future studies of the effectiveness of new antibiotic agents with the resistant *M. luteus* strain may predict their effectiveness against resistant pathogenic bacteria. The availability of plant-derived safe antibiotics provides expanded options for treatment of livestock, poultry and produce and reduces the exposure of humans to resistant bacteria.

Table 2.2. BA50 values of EO components against *E. coli* O157:H7, *S. enterica*, *C. jejuni* and *L. monocytogenes*. Adapted from Friedman et al. (2002, 2003).

Oil compound	*E. coli* RM1484	*S. enterica* RM1309	*C. jejuni* RM1221	*L. monocytogenes* RM2199	*L. monocytogenes* RM2388	Common Sources
Anethole, trans	>0.67	>0.67	0.12	>0.67	>0.67	anise, fennel, licorice
Benzaldehyde	0.48	0.36	0.019	0.46	0.36	bitter almond, almond, licorice
Bornyl Acetate	>0.67	>0.67	0.10	>0.67	>0.67	lavender, rosemary
Carvacrol	0.057	0.054	0.011	0.083	0.086	thyme, oregano, bergamot
Carvone R	0.45	0.41	0.031	>0.67	>0.67	spearmint
Carvone S	0.48	0.39	0.044	0.35	0.17	caraway and dill seeds
Cineol	>0.67	>0.67	0.10	>0.67	>0.67	eucalyptus, bay
Cinnamaldehyde	0.057	0.033	0.0028	0.019	0.008	cinnamon
Citral	0.22	0.23	0.021	0.099	0.20	lemongrass, ginger, basil
Citronellal, R	>0.67	>0.67	0.22	>0.67	0.45	citronella, lemongrass
Citronellal, S	>0.67	>0.67	0.050	>0.67	0.44	lemon, mandarin orange
Estragole	0.28	0.21	0.004	0.36	0.35	tarragon, basil
Eugenol	0.11	0.087	0.022	0.061	0.081	clove, allspice
Geraniol	0.15	0.15	0.10	0.28	0.51	grape leaves, palmarosa, rose
Geranyl acetate	>0.67	>0.67	0.034	>0.67	>0.67	citronella, palmarosa, thyme
Isoeugenol	0.15	0.16	0.35	>0.67	>0.67	ylang-ylang, mace, basil
Limonene	>0.67	>0.67	0.35	>0.67	0.25	citrus, celery
Linalool	0.40	0.37	0.35	>0.67	>0.67	(S)-coriander, palmarosa, orange, (R)-lavender
Menthol	0.53	0.50	0.40	0.57	0.48	peppermint, sunflower
Perillaldehyde	0.27	0.20	0.03	0.35	0.30	citronella, perilla, cumin
Salicylaldehyde	0.13	0.12	0.040	0.43	0.45	buckwheat, almond
Terpineol	0.39	0.18	0.10	0.56	>0.67	pine, tea tree, turmeric
Thymol	0.060	0.034	0.024	0.077	0.077	lemon, thyme, oregano

Antibiotic resistant Campylobacter jejuni

In another collaborative study with scientists at the Department of Veterinary Science and Microbiology at the University of Arizona in Tucson, 63 *C. jejuni* isolates were screened for their resistance against widely used commercial antibiotics (Ravishankar et al. 2008). Based on this screen, two resistant strains and one nonresistant strain were evaluated for their susceptibility to inactivation by cinnamaldehyde and carvacrol. The inhibition of microbial growth was related to both concentration of antimicrobials and incubation time. The data suggest that plant-derived compounds can inactivate at about the same

Table 2.3. BA_{50} values of 39 EO/EO-components active against *E. coli* O157:H7, *S. enterica*, *C. jejuni*, and *L. monocytogenes*. Adapted from Friedman et al. (2002).

Oil/oil compound	***E. coli***	***S. enterica***	***C. jejuni***	***L. monocytogenes***
Allspice	0.14	0.13	**0.02**	**0.08**
Basil	0.41	0.42	**0.02**	0.12
Bay Leaf	**0.13**	0.13	0.03	**0.07**
Benzaldehyde	0.48	0.36	**0.02**	0.36
Caraway	0.46	0.47	0.03	0.24
Carvacrol	**0.06**	**0.05**	**0.01**	**0.09**
Carvone S	0.48	0.39	0.04	0.17
Cinnamaldehyde	**0.06**	**0.04**	**0.003**	**0.01**
Cinnamon Bark	0.18	0.14	**0.02**	**0.08**
Cinnamon Cassia	**0.11**	**0.07**	**0.01**	0.15
Cinnamon Leaf	**0.13**	**0.08**	0.03	**0.09**
Citral	0.22	0.23	**0.02**	0.2
Citronella	0.41	0.49	0.09	0.18
Clove Bud	**0.13**	0.13	**0.02**	**0.09**
Coriander	0.4	0.48	0.08	0.5
Cumin Seed	0.3	0.36	0.1	0.25
Elemi	0.4	0.44	**0.01**	0.22
Estragole	0.28	0.21	**0.01**	0.35
Eugenol	**0.11**	**0.09**	**0.02**	**0.08**
Fir Needle	0.48	0.61	**0.01**	**0.08**
Geraniol	0.15	0.15	0.1	0.51
Hyssop	0.57	0.41	0.1	0.18
Lavender	0.41	0.41	0.06	0.34
Lemongrass	0.14	0.16	**0.02**	0.12
Menthol	0.53	0.5	0.4	0.48
Nutmeg	0.55	0.44	0.18	0.2
Orange Mandarin	0.41	0.64	**0.01**	0.1
Oregano Origanum	**0.05**	**0.05**	**0.02**	0.1
Oregano Spanish	**0.05**	**0.05**	**0.01**	**0.08**
Palmarosa	**0.12**	0.14	0.07	0.27
Perillaldehyde	0.27	0.2	0.03	0.3
Rose Damask	0.55	0.44	0.11	0.36
Rose French	0.43	0.5	0.05	0.29
Rose Geranium	0.41	0.4	0.09	0.32
Salicylaldehyde	0.13	**0.12**	0.04	0.45
Spearmint	0.28	0.29	0.03	0.57
Thyme	**0.05**	**0.05**	**0.02**	0.22
Thymol	**0.06**	**0.03**	**0.02**	**0.08**
Wormwood	0.55	0.52	0.38	0.1

The top 10 (or equal to top 10) most effective test substances for each bacteria strain are shown in bold font.

rate both antibiotic-resistant and nonresistant strains of foodborne *Campylobacter* bacteria. These studies provide candidates for incorporation into formulations that can protect food and consumers against antibiotic resistant *C. jejuni*.

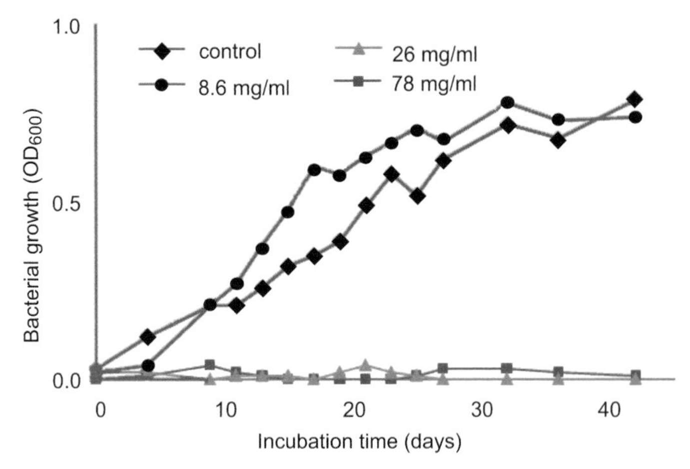

Figure 2.3. Effects of different concentrations (μg/ml) of *trans*-cinnamaldehyde and non-inhibitory concentrations of ethanol (0.4%, negative control) on the growth of *M. avium* subsp. *Paratuberculosis* NCTC 8578 in Middlebrook 7H9 broth. Modified from Wong et al. (2008).

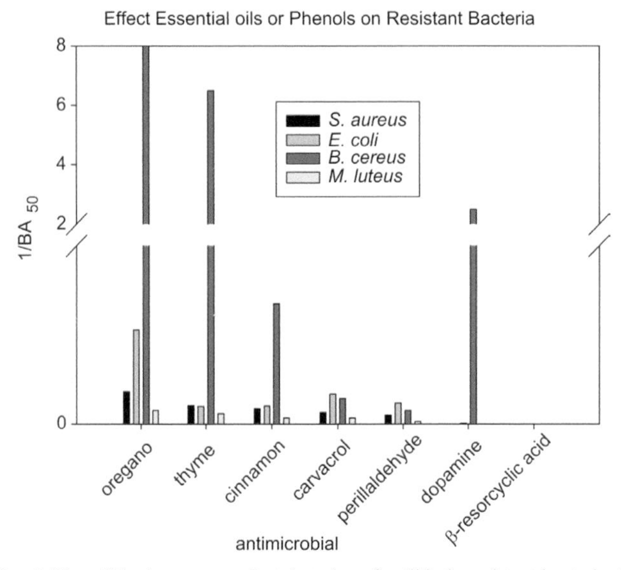

Figure 2.4. Bactericidal activities of 7 substances against 4 strains of antibiotic resistant bacteria. Modified from Friedman et al. (2004a, 2006a).

Antibiotic resistant Salmonella enterica

Antimicrobial activities of cinnamaldehyde and carvacrol against antibiotic resistant *S. enterica* were also tested (Ravishankar et al. 2010). Out of 23 isolates screened for resistance to seven antibiotics, two antibiotic resistant and two antibiotic susceptible strains were chosen for the study. The results indicate that the extent of inhibition was related to both the nature and concentration of antimicrobials and the incubation time. Both cinnamaldehyde and carvacrol showed complete inactivation of *S. enterica* at 0.3 and 0.4% concentrations at all time points tested. Cinnamaldehyde at 0.1% showed no survivors of *Salmonella* at 5 hours. Both cinnamaldehyde and carvacrol have potential antimicrobial activity against *Salmonella* at concentrations of > 0.1%. The section below on leafy greens covers related studies.

Mechanisms of Antimicrobial Effects

Phenolics

Phenolic EOs apparently exert their lethal effect by permeabilizing the membrane of cells (Lambert et al. 2001, Xu et al. 2008). Carvacrol and thymol are the most studied phenolic terpenes (Friedman 2014b). Ultee et al. (2002) showed that the consequences of exposing *B. cereus* to carvacrol included depletion of the intracellular ATP pool, changes in membrane potential, and increases in permeability of the cytoplasmic membrane to proteins, and potassium ions, presumably through incorporation of the EO into the membrane. Membrane changes apparently occur at much lower concentrations than required for lethality.

In a collaborative study with colleagues at Oklahoma State University, we utilized several techniques to study the membrane integrity of *E. coli* C600 in the presence of non-lethal concentrations of carvacrol: autofluorescence, membrane potential and ATP flux (Friedman 2006). The data showed that autofluorescence spectra were able to measure membrane leakage at much lower concentrations than were membrane potential or ATP flux measurements. A related study with model membranes describes the creation of quantitative parameters that indicate that the antimicrobial compounds such as carvacrol could modify the lipid monolayer structure by forming aggregates of antimicrobials and lipids. Figure 2.5 schematically illustrates the disruption of a cell membrane by natural antimicrobials (Nowotarska et al. 2014).

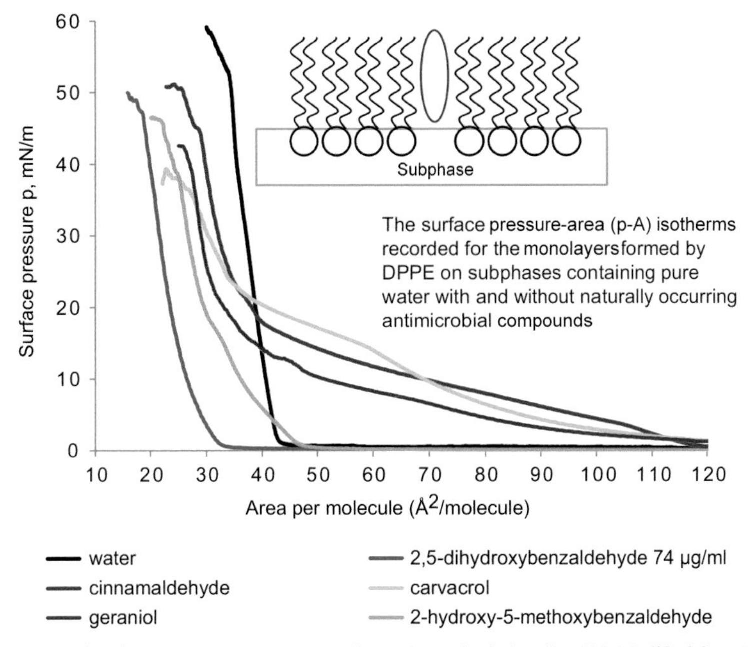

Figure 2.5. Schematics of surface pressure events on a cell membrane by induced an EO Modified from Nowotarska et al. (2014).

Aldehydes

Cinnamaldehyde may exert its effect by a different mechanism. For example, Helander et al. (1998) found that while both carvacrol and thymol disintegrated the cell membrane in the Gram-negative bacteria *S. typhimurium*, cinnamaldehyde did not. Di Pasqua et al. (2007) also found that cinnamaldehyde did not cause collapse of the cell membrane, although it did cause a profound change in the composition of the fatty acids in the membrane resulting in alteration of its external structure. This change could facilitate cellular incorporation of cinnamaldehyde or even other interstitial compounds. This mechanism, of facilitating entrance into the cell, could explain the synergism sometimes seen when using multiple antimicrobials.

Relative susceptibilities of the bacteria to inactivation

There are notable differences in susceptibilities to inactivation of different foodborne pathogens. For example, Cox et al. (2000) noted that although the MIC values for tea tree oil against *E. coli* and *S. aureus* were similar, *S. aureus* was initially slower in responding, suggesting that the oil diffused through the membrane more slowly. Cristani et al. (2007) found that the Gram-positive *S. aureus* was more susceptible to membrane damage by the more hydrophilic thymol, while the Gram-negative *E. coli* was more susceptible to the lipophilic carvacrol. They speculated that the compound-specific antimicrobial effect is influenced by both the net surface charge of the membrane and the hydrophobicity of the monoterpene. In our studies, we found that *S. enterica*, but not *E. coli*, had very different responses to EOs added to clarified vs. unfiltered apple juice (Friedman et al. 2004b). We do not know whether this is a difference in the mechanism of EO activity or a difference in the sensitivity to the media.

Susceptibility is also influenced by the growth stage of the cells. Stationary-phase cells were found to be less susceptible to terpenes than were exponential-phase cells, possibly reflecting a decrease in the incorporation of these molecules (Gustafson et al. 1998, Kwiecinski et al. 2009).

Different strains of a bacterium may also not be equally susceptible to EOs. For example, Ravishankar et al. (2010) compared the susceptibility of antibiotic resistant vs. sensitive isolates of *Salmonella enterica* to carvacrol and cinnamaldehyde. The highly resistant *Salmonella* Newport showed different susceptibilities to carvacrol and cinnamaldehyde than did the other strains.

Relative activities of isomeric antimicrobials

The fact that isomers of the same compound can exhibit different antimicrobial potencies implies that specific rather than purely physical mechanisms are involved in EO activity. If the relative effectiveness of terpenes were dependent solely on their ability to be absorbed into the cell membrane as previously suggested (Sikkema et al. 1994), then isomerization should not be a factor. Lis-Balchin et al. (1999) found that 19 out of 20 *L. monocytogenes* strains were more affected by (+)-α-pinene than by (–)-α-pinene. We found that the activities of isomeric carvones and citronellal were significantly different (Table 2.2) in some bacteria but not others. Thus, whereas S-(+)-carvone was effective against all bacteria tested, R-(–)-carvone was ineffective against *L. monocytogenes*. The isomers exhibited similar activities against *E. coli*, *S. enterica* and *C. jejuni*. For citronellal, BA_{50} were significantly different only for the Gram-positive bacteria *C. jejuni*. Additional studies with isomers could lead to a better understanding of antimicrobial mechanisms.

Application to Foods

Thus far we have shown that EOs are effective in laboratory media. The real need is to find practical applications to protect our food supply. Here, we briefly mention studies designed to ascertain antimicrobial effects of EOs in selected foods.

Liquid Foods

Apple juice

Contaminated raw apple juice was responsible for a deadly *E. coli* O157:H7 outbreak of 1996 (Centers for Disease Control and Prevention 1996). Although pasteurization or irradiation can kill the bacteria, it can also cause undesired compositional and organoleptic changes. We evaluated 17 plant EOs and 9 EO components for antibacterial activity (BA_{50}) against *E. coli* O157:H7 and *S. enterica* in apple juice (Friedman et al. 2004b). The data show that many of the test compounds were highly active in juice for both *E. coli* and *S. enterica* (Fig. 2.6). The activity against *S. enterica* was greater than against *E. coli*. We found the purified EO components were more active than the EOs from which they are derived. Carvacrol, cinnamaldehyde, geraniol, linalool and terpineol inactivated the bacteria almost on contact. The compounds were found to be stable in apple juice when stored both at room temperature and under refrigeration. Additionally, we found that their potencies in the acid environment of the juice are significantly higher than in pH 7 buffer.

The above tests were executed with the commonly available clarified apple juice. Tests were repeated with a commercially available unfiltered variety, containing suspended solids. Test results were similar in the two juices for *E. coli*, but for *S. enterica*, some of the BA_{50} values were significantly reduced in the unfiltered juice (Fig. 2.7). This suggests that the two bacteria are inhibited by a different mechanism or they have different sensitivities to the medium.

Cinnamon oil may be a better EO to use in apple juice than oregano oil. To make products acceptable to consumers, sometimes compromises between flavor and activity need to be made. Cinnamaldehyde may have potential for commercial acceptability in apple juice, as cinnamon is a

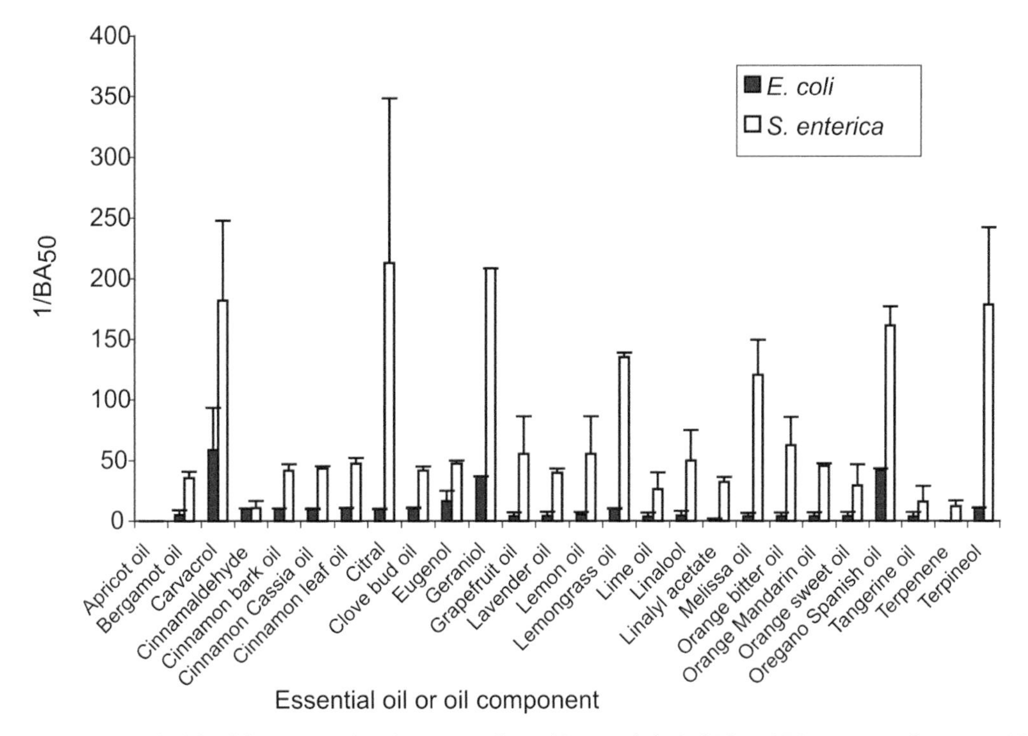

Figure 2.6. Bactericidal activity (expressed as $1/BA_{50}$ to graph a positive correlation) of EOs and EO components in commercial apple juice, incubated for 30 min at 37°C.

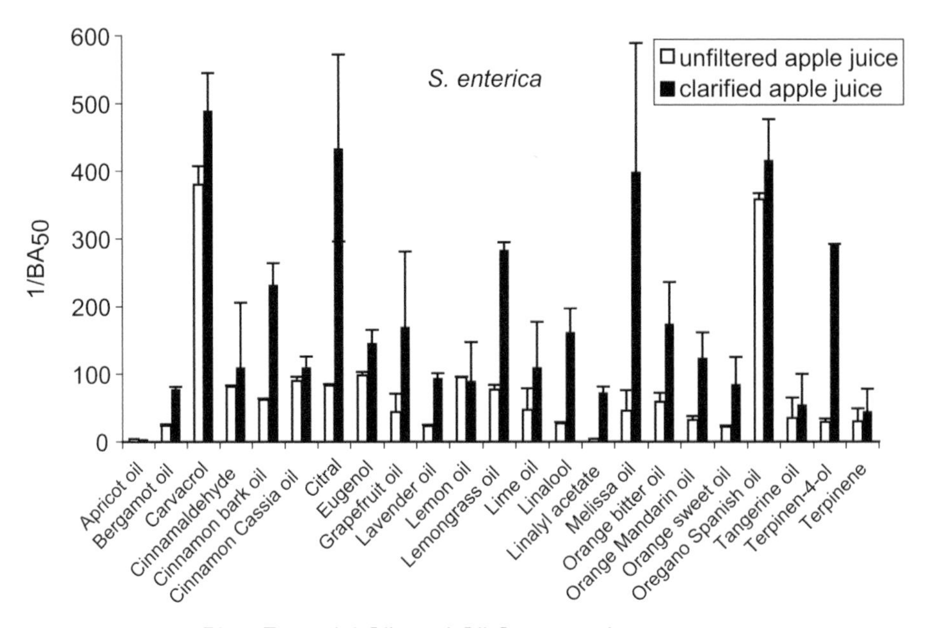

Figure 2.7. Bactericidal activity (expressed as $1/BA_{50}$ to graph a positive correlation) of selected EOs and EO components in commercial apple juice, incubated for 30 min at 37°C against *S. enterica*. Modified from Friedman et al. (2004b).

spice often pared with apples (Friedman et al. 2000). In terms of sensory properties more cinnamon oil could be added to apple juice than oregano oil. We may be able to further decrease the sensory impact on foods by taking advantage of synergism of mixtures of antimicrobials, thus lowering the required dose. For example, eugenol and cinnamaldehyde were found to act synergistically in apple juice against the extremophile spoilage bacteria, *Alicycloba cillusacidoterrestris* (Bevilacqua et al. 2010). EO-induced inhibition of this bacterium that survives pasteurization of apple juice could be an additional benefit. These studies provide information about new ways to protect apple juice and possibly other liquid and solid foods against contamination by human pathogens.

Tomato juice

The BA_{50} values in tomato juice for all the compounds evaluated were similar (Fig. 2.8). We expect the flavor of carvacrol to be more compatible with tomato than with apple juice.

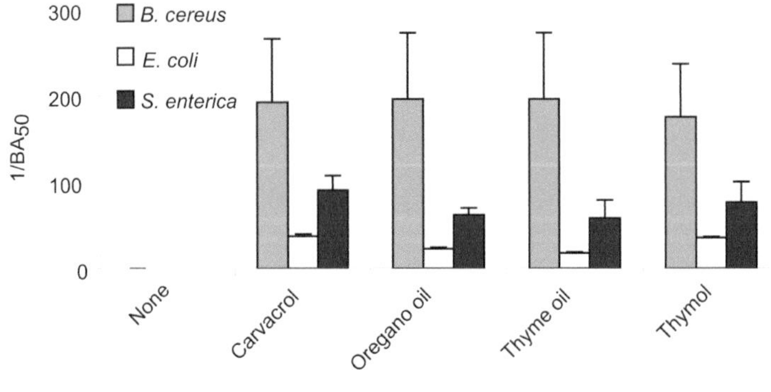

Figure 2.8. Select EOs added to commercial tomato juice and tested for antimicrobial activity against 3 bacteria. Unpublished results.

Wine marinades

Wines are frequently used as a marinade for meats and other grilled foods. In a simulation of conditions possibly found in the kitchen, we evaluated bactericidal activities against *B. cereus*, *E. coli* O157:H7, *L. monocytogenes* and *S. enterica* of several antimicrobial wine recipes, each consisting of red, white or sherry wine with added herbs and garlic or EOs (Friedman et al. 2007a). Several combinations of the naturally occurring plant-derived ingredients rapidly inactivated the four pathogens. We showed that (a) incubation temperature affected activities in the following order: 37°C > 21°C > 4°C; (b) varying the initial bacterial concentrations from 10^3 to 10^4 to 10^5 CFU/well did not significantly affect BA_{50} values; (c) storage of three marinades up to two months did not change their effectiveness against *S. enterica* (Fig. 2.9); and (d) polyphenolic compounds isolated from red wine by chromatography exhibited exceptional activity at nanogram levels against two strains of *B. cereus* (Table 2.4). A later study confirmed the value of wines as a solvent for natural antimicrobials (Friedman et al. 2015). These observations suggest that antimicrobial wine formulations have the potential to improve the microbiological safety of foods.

Antibacterial activities of plant EOs, EO components, and phenolic benzaldehydes against *E. coli* O157:H7, *S. enterica*, and *L. monocytogenes* were higher when dissolved in Chardonnay, Pinot Noir or Sherry wines than in a buffer of the same pH and ethanol level. Therefore, compounds other than the alcohol content of wines may be responsible for the enhanced antibiotic effects. Enhanced activity was not apparent in matched solutions of aqueous ethanol. Additionally, measurement of the EO components in the wines and the controls showed that the amount of EO dissolved increased with increasing alcohol content. A crude extract of polyphenolic compounds from red wine exhibited exceptional activity at nanogram levels against two strains of *B. cereus*, similar to that of tea catechins (Friedman et al. 2006c, Friedman 2007). It is likely that flavonoids and the EOs act synergistically.

Fats added to the marinades had a negative effect on antimicrobial activity. Wine marinades containing carvacrol or oregano oil and garlic powder prepared with canola oil had very low activity

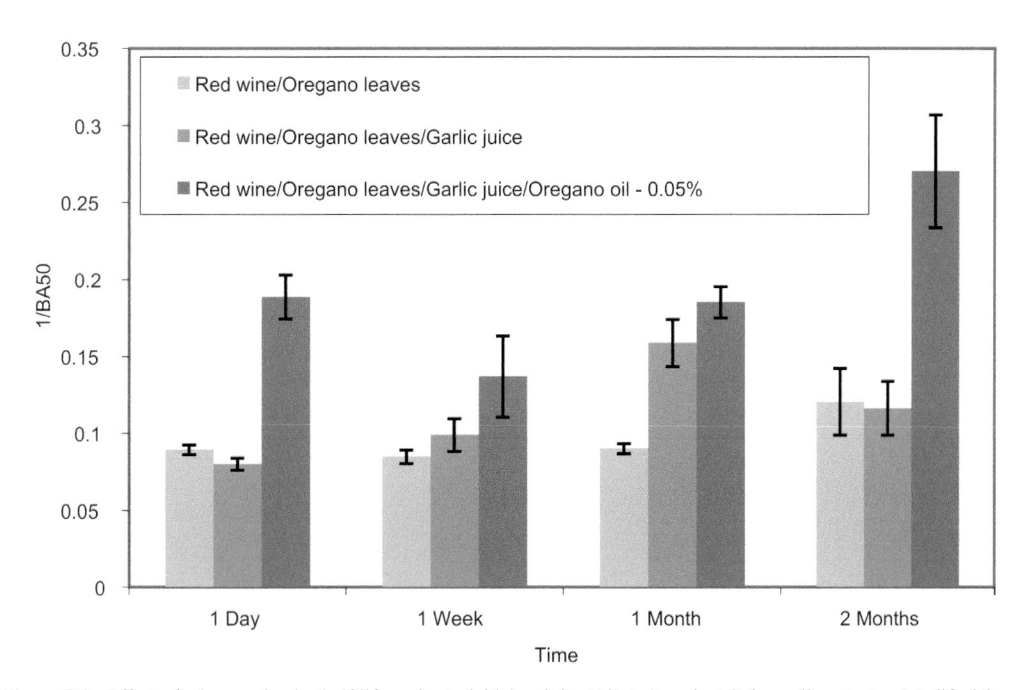

Figure 2.9. Effect of wine-marinade shelf life on bactericidal activity ($1/BA_{50}$) against *Salmonella enterica*. Modified from Friedman et al. (2007a).

Table 2.4. BA_{50} for *B. cereus* in wines and isolated red wine phenolics, 21°C, 60 min. Adapted from Friedman et al. (2007a).

	Diluent	BA_{50}, %
Beringer Pinot noir	Saline pH 3.7	0.013
Beringer Pinot noir	PBS pH 7.0	0.033
Kenwood Pinot noir	Saline pH 3.7	0.017
Kenwood Pinot noir	PBS pH 7.0	0.048
Red Wine polyphenolics	Saline pH 3.7	0.000059
Red Wine polyphenolics	PBS pH 7.0	0.00012

against both *E. coli* and *S. enterica*, while the marinades without the canola oil exhibited strong antimicrobial activity. Wine extracts of store-bought oregano and thyme leaves widely used in culinary practice provide a ubiquitous source of natural antimicrobials for home use.

Solid foods

Meats

Campylobacter jejuni

Edible antimicrobial apple films inactivated susceptible and resistant *C. jejuni* strains on chicken breast (Mild et al. 2011).

Clostridium perfringens spores

The presence of *Cl. perfringens* spores in foods is particularly hazardous to human health as they are resistant to heat treatment. In collaborative studies, we evaluated the inhibition of germination and outgrowth from spores of *Cl. perfringens* by natural antimicrobials in ground beef during chilling of cooked meat (Juneja et al. 2006, Juneja and Friedman 2007). At concentrations of 0.1–0.5%, carvacrol, cinnamaldehyde, thymol and oregano oil completely prevented the growth of this pathogen during cooling over a period of 15 hours. Concentrations of 0.5–2.0% were required for longer cooling times (15, 18, and 21 hours). Cinnamaldehyde was the most effective of the four test substances. It may be possible to prevent the germination and outgrowth of the spores in meat and poultry products by incorporating natural antimicrobials, thus minimizing risks to the consumer.

Escherichia coli O157:H7

Carvacrol and cinnamaldehyde added to beef reduced D values (time in minutes to kill 90% of the bacteria) 40–100% in sous-vide cooked beef (Juneja and Friedman 2008). Thermal death times from these studies will assist the retail food industry to design cooking regimes that ensure safety of beef contaminated with *E. coli* O157:H7. Additional benefits include improved nutrition from antioxidative antimicrobials and reduced energy used needed to bring the meat to a temperature that kills the bacteria.

Salmonella enterica

In a collaborative study (Juneja et al. 2010), we assessed the heat resistance of a cocktail of eight *Salmonella* serotypes in raw ground beef in the presence of sodium lactate, oregano oil, and in combinations of these two GRAS-listed ingredients. The outgrowth of the bacteria in these samples were then determined during post-thermal treatment storage at 15°C for up to 30 days. Related studies describe the inhibition of multiple *Salmonella* serotypes in ground chicken (Juneja et al. 2012, 2013).

Cinnamaldehyde added to the feed reduced chicken-egg-borne transmission of *Salmonella enterica*, suggesting that the EO compound has the potential to reduce egg-born transmission of the pathogen to humans (Upadhyaya et al. 2015).

The results indicate EOs may be used in combination at low concentrations to inhibit growth of multiple *Salmonella* strains that survive heat treatment. This information is expected to aid in designing cooking regimes to ensure food safety.

Concurrent inactivation E. coli and carcinogenic heterocyclic amines

Insufficient thermal treatment to inactivate pathogens such as *E. coli* O157:H7 has been considered as one of the major factors contributing to foodborne illness outbreaks following consumption of meats. Hence, meat products need to be sufficiently heated to inactivate foodborne pathogens. However, high-temperature heat treatment of meat is a cause for concern due to the formation of potentially carcinogenic heterocyclic amines. There is, therefore, a need to develop methods to cook the meats in such a way to inactivate the foodborne bacteria, while concurrently reducing or eliminating the formation of potentially carcinogenic amines.

In collaborative studies, we found that carvacrol added to ground beef patties contaminated with *E. coli* O157:H7 facilitated heat inactivation of the bacteria during grilling of the patties (hamburgers) and concurrently induced reduction in the formation of heat-induced heterocyclic amines (Friedman et al. 2009) (Fig. 2.10).

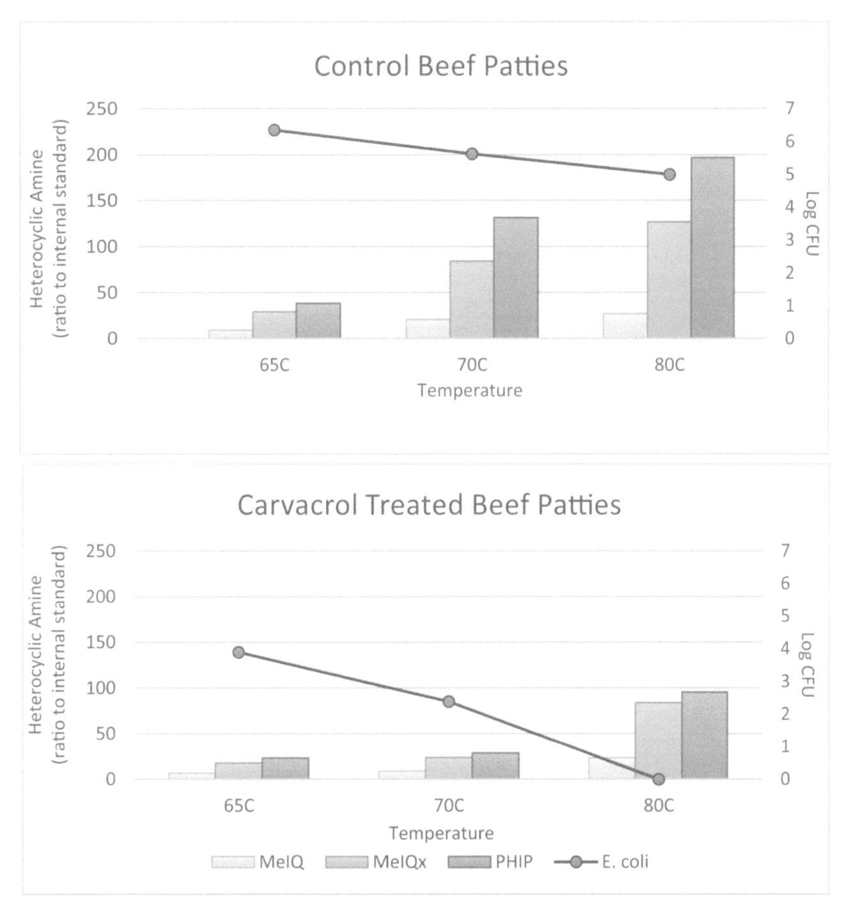

Figure 2.10. Carvacrol inhibits the formation of carcinogenic heterocyclic amines in grilled beef patties (hamburgers). Modified from Friedman et al. (2009).

Formation of amines was reduced in two ways: directly by the carvacrol and indirectly by cooking at the lower temperatures. Additional studies further demonstrated the potential of inhibiting pathogenic organisms and potential toxic heterocyclic amines in cooked meats (Rounds et al. 2012, 2013). These results imply that it may be possible to reduce the need to consume well-done meat. The recommended change in culinary practice will benefit microbial food safety and human health.

Oysters

In collaborative studies, we tested the antimicrobial activities cinnamaldehyde and carvacrol against the antibiotic resistant *Salmonella* Newport on contaminated oysters (Ravishankar et al. 2010). Cinnamaldehyde and carvacrol showed similar results; the 60 minutes dipping combined with holding for 3 days almost completely inactivated *Salmonella* (Fig. 2.11). Carvacrol and cinnamaldehyde have the potential to inactivate *S.* Newporton surfaces of contaminated seafood.

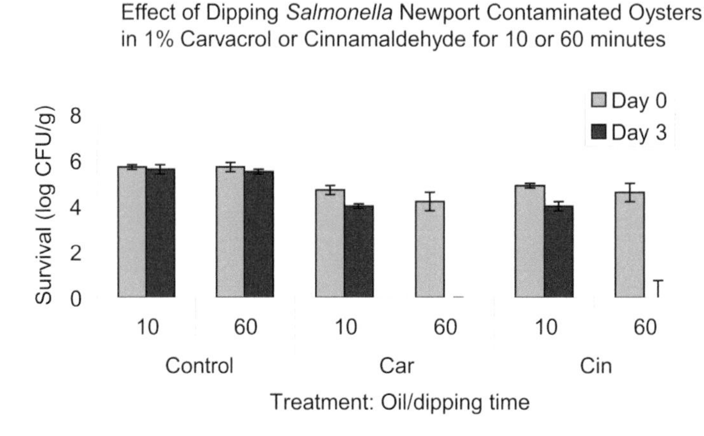

Figure 2.11. Inactivation of antibiotic-resistant *Salmonella* Newport on contaminated oysters by carvacrol and cinnamaldehyde. Modified from Ravishankar et al. (2010).

Leafy Greens—Salads

We tested the effectiveness of dipping *S.* Newport contaminated celery in carvacrol and cinnamaldehyde suspensions (Ravishankar et al. 2010). The data show that samples dipped in 1% carvacrol for 10 minutes induced complete inactivation of the bacteria at day 0 and day 3 (Fig. 2.12). By contrast, celery samples dipped in 1% cinnamaldehyde showed only 1 and 2.3 log reductions at day 0 and day 3, respectively.

Related studies describe antimicrobial effects of cinnamon leaf and oregano oils against multi-drug resistant *Salmonella* (Moore-Neibel et al. 2013, Todd et al. 2013). The results from another study indicate that carvacrol, cinnamaldehyde and citral can serve as effective natural antimicrobials against *E. coli* O157:H7 for washing organic leafy greens before they are bagged and shipped for sale (Denton et al. 2015).

Antimicrobial edible fruit and vegetable films

Edible coatings made from fruits and vegetables are being developed to extend the shelf-life and quality of foods (Du et al. 2008a,b, 2009b, McHugh and Senesi 2000). These films with added antimicrobials were evaluated for activity against pathogenic bacteria on meat and poultry products (Ravishankar et al. 2009). *S. enterica* or *E. coli* O157:H7 (10^7 CFU/g) cultures were surface inoculated on chicken breasts and *L. monocytogenes* (10^6 CFU/g) on ham. The inoculated

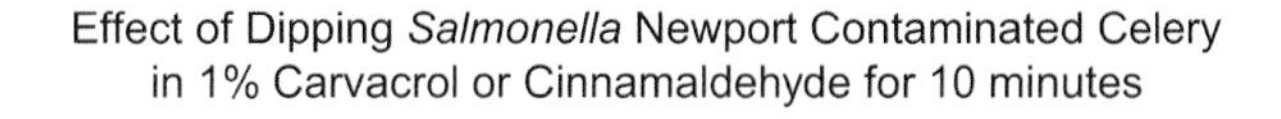

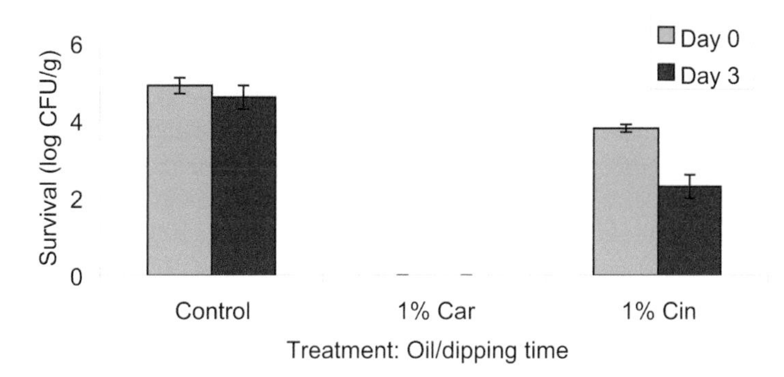

Figure 2.12. Inactivation of antibiotic-resistant *Salmonella* Newport on contaminated celery by carvacrol and cinnamaldehyde. Modified from Ravishankar et al. (2010).

products were then wrapped with edible films containing three concentrations (0.5, 1.5, and 3%) of cinnamaldehyde or carvacrol. Following incubation at either 23°C or 4°C for 72 hours, samples were stomached in buffered peptone water, diluted and plated for enumeration of survivors. The films exhibited antimicrobial concentration-dependent activities against the pathogens tested. The results showed that (a) carvacrol was a stronger antimicrobial agent against both *Salmonella* and *E. coli* O157:H7 than cinnamaldehyde on the chicken breast at 4°C; (b) at 23°C, *S. enterica* population reductions were similar for both carvacrol and cinnamaldehyde, but higher for carvacrol against of *E. coli* O157:H7; (c) carvacrol was also a stronger antimicrobial agent against *L. monocytogenes* than cinnamaldehyde on ham at 4°C and 23°C; (d) the apple films also inactivated the natural microflora on raw chicken breast; and (e) edible films containing carvacrol or cinnamaldehyde also inactivated *L. monocytogenes* on bologna and ham. The results suggest that the food industry and consumers could use these films as wrappings to control surface contamination by foodborne pathogens.

Vapors

In this laboratory, we incorporated EOs into films then tested the antimicrobial effect of the films on bacteria without any direct contact (Du et al. 2009a). *E. coli*, *S. enterica*, and *L. monocytogenes* were all inhibited by oregano oil vapors. In related studies, (a) López et al. (2007) demonstrated the efficacy of antimicrobial vapor-phase effects of cinnamon, thyme and oregano EOs against multiple gram-negative and gram-positive bacteria; (b) Mejía-Garibay et al. (2015) showed that vapor-phase contact of black mustard oil inhibited the growth of *Aspergillus* and *Penicillium* molds on contaminated bread; and (c) Petretto et al. (2013) developed a device that can be used to evaluate antimicrobial effects of volatile EOs by vapor-phase contact on foods. We think that utilizing the antimicrobial properties of the vapors of EOs holds much promise, for example, for use in bags of leafy greens, meats and poultry products.

Research Needs

In addition to research needs mentioned in the text, future studies should address the following aspects of natural antimicrobials:

- Screen different food categories with several antimicrobials listed in the tables to optimize antibiotic effects and minimize adverse organoleptic properties.

- Determine any reactivity with and longevity in the food system.
- Define additive and synergistic effects of mixtures of antimicrobials. Combinations that act synergistically will lessen amount needed for effective formulations. They will be safer and affect flavor and taste less compared to the use of individual compounds.
- Evaluate effectiveness of antimicrobials against additional antibiotic-resistant foodborne pathogens in human foods and animal feeds.
- Develop edible antimicrobial films and coatings from plants that contain natural antimicrobials and other bioactive compounds to protect foods against contamination and to concurrently benefit health.
- Determine whether molecular modeling/simulations of antimicrobial-pathogen cell membrane interactions can be used to predict antibiotic activities (Sirk et al. 2008, 2009).
- Develop methods to concurrently reduce both pathogens and other food hazards such carcinogenic heterocyclic amines and mycotoxins in processed meat and poultry products.
- Determine whether EOs and EO components can concurrently inhibit foodborne pathogens such as *E. coli* O157:H7, *S. aureus* and *Clostridium difficile* and the virulent toxins and endotoxins produced by these pathogens (Friedman and Rasooly 2013, Quiñones et al. 2009, Rasooly et al. 2010a,b).
- Explore antimicrobial potencies of EOs in combination with antimicrobial tea compounds (Friedman et al. 2007b), winery byproducts (Friedman 2014a), and mushroom polysaccharides (Friedman 2016).
- Determine the connection between consumption of contaminated foods treated with antimicrobials and lower risk of infection in humans.
- Determine using if EOs can protect rodents and humans against infections through the stimulation of the immune system, as described in detail for other food-compatible formulations (Kim et al. 2012a,b, 2014).

Conclusions

The general objective of our studies was to screen naturally occurring and food-compatible plant-derived compounds for their antimicrobial activities against bacterial foodborne pathogen. The following summary shows that edible plant compounds are a useful source of antimicrobials:

- EOs inactivated susceptible strains of *B. cereus*, *C. jejuni*, *Cl. perfringens*, *E. coli* O157:H7, *L. monocytogenes*, *Mycobacterium avium* subspecies *paratuberculosis*, *Salmonella* and *S. aureus* and resistant strains of *C. jejuni*, *Salmonella*, *E. coli*, *Mycobacterium luteus* and *S. aureus*.
- EO compounds rapidly inactivated foodborne pathogens in apple juice, tomato juice, wines, in meat and poultry products, and on oysters and leafy greens.
- EO compounds simultaneously inhibited the growth of *E. coli* and the formation of carcinogenic heterocyclic amines in grilled beef patties.
- Edible antimicrobial fruit and vegetable films used as wrappings on raw chicken breasts and inoculated with *S. enterica* and *E. coli* O157:H7 induced multilog reductions of the pathogens and of non-pathogenic microflora. The films also inactivated *L. monocytogenes* on ham.
- Disruption of bacterial cell membranes appear to largely govern the mechanism of antimicrobial effects of plant EO compounds.

Acknowledgment

I thank Carol E. Levin for assistance with the preparation of this chapter and colleagues whose names appear on the cited publications for excellent scientific collaboration designed to enhance microbial food safety.

References

Bevilacqua, A., Corbo, M.R. and Sinigaglia, M. 2010. Combining eugenol and cinnamaldehyde to control the growth of *Alicyclobacillus acidoterrestris*. Food Control 21: 172–177. doi:10.1016/j.foodcont.2009.05.002.

Burt, S. 2004. Essential oils: their antibacterial properties and potential applications in foods—a review. Int. J. Food Microbiol. 94: 223–253. doi:10.1016/j.ijfoodmicro.2004.03.022.

Centers for Disease Control and Prevention. 1996. Outbreak of *Escherichia coli* O157:H7 infections associated with drinking unpasteurized commercial apple juice—British Columbia, California, Colorado, and Washington, October 1996. MMWR Morb. Mortal. Wkly. Rep. 45: 975–975.

Cox, S.D., Mann, C.M., Markham, J.L., Bell, H.C., Gustafson, J.E., Warmington, J.R. and Wyllie, S.G. 2000. The mode of antimicrobial action of the essential oil of *Melaleuca alternifolia* (Tea tree oil). J. Appl. Microbiol. 88: 170–175. doi:10.1046/j.1365-2672.2000.00943.x.

Cristani, M., D'Arrigo, M., Mandalari, G., Castelli, F., Sarpietro, M.G., Micieli, D., Venuti, V., Bisignano, G., Saija, A. and Trombetta, D. 2007. Interaction of four monoterpenes contained in essential oils with model membranes: Implications for their antibacterial activity. J. Agric. Food Chem. 55: 6300–6308. doi:10.1021/jf070094x.

Davis, J.R. and Lederberg, J. (eds.). 2001. Emerging Infectious Diseases from the Global to Local Perspective: Workshop Summary. Washington DC, USA: National Academy of Sciences.

Denton, J.J., Ravishankar, S., Friedman, M. and Jaroni, D. 2015. Efficacy of plant derived compounds against *Escherichia coli* O157:H7 during flume-washing and storage of organic leafy greens. J. Food Process. Preserv. 39: 2728–2737. doi:10.1111/jfpp.12523.

Di Pasqua, R., Betts, G., Hoskins, N., Edwards, M., Ercolini, D. and Mauriello, G. 2007. Membrane toxicity of antimicrobial compounds from essential oils. J. Agric. Food Chem. 55: 4863–4870. doi:10.1021/jf0636465.

Du, W.X., Olsen, C.W., Avena-Bustillos, R.J., McHugh, T.H., Levin, C.E. and Friedman, M. 2008a. Antibacterial activity against *E. coli* O157:H7, physical properties, and storage stability of novel carvacrol-containing edible tomato films. J. Food Sci. 73: M378–383. doi:10.1111/j.1750-3841.2008.00892.x.

Du, W.X., Olsen, C.W., Avena-Bustillos, R.J., McHugh, T.H., Levin, C.E. and Friedman, M. 2008b. Storage stability and antibacterial activity against *Escherichia coli* O157:H7 of carvacrol in edible apple films made by two different casting methods. J. Agric. Food Chem. 56: 3082–3088. doi:10.1021/jf703629s.

Du, W.-X., Olsen, C.W., Avena-Bustillos, R.J., McHugh, T.H., Levin, C.E., Mandrell, R. and Friedman, M. 2009a. Antibacterial effects of allspice, garlic, and oregano essential oils in tomato films determined by overlay and vapor-phase methods. J. Food Sci. 74: M390–M397. doi:10.1111/j.1750-3841.2009.01289.x.

Du, W.X., Olsen, C.W., Avena-Bustillos, R.J., McHugh, T.H., Levin, C.E. and Friedman, M. 2009b. Effects of allspice, cinnamon, and clove bud essential oils in edible apple films on physical properties and antimicrobial activities. J. Food Sci. 74: M372–M378. doi:10.1111/j.1750-3841.2009.01282.x.

Dudai, N., Larkov, O., Ravid, U., Putievsky, E. and Lewinsohn, E. 2001. Developmental control of monoterpene content and composition in *Micromeria fruticosa* (L.) Druce. Ann. Bot. 88: 349–354. doi:10.1006/anbo.2001.1466.

Friedman, M., Kozukue, N. and Harden, L.A. 2000. Cinnamaldehyde content in foods determined by gas chromatography-mass spectrometry. J. Agric. Food Chem. 48: 5702–5709. doi:10.1021/jf000585g.

Friedman, M., Henika, P.R. and Mandrell, R.E. 2002. Bactericidal activities of plant essential oils and some of their isolated constituents against *Campylobacter jejuni*, *Escherichia coli*, *Listeria monocytogenes*, and *Salmonella enterica*. J. Food Prot. 65: 1545–1560.

Friedman, M., Henika, P.R. and Mandrell, R.E. 2003. Antibacterial activities of phenolic benzaldehydes and benzoic acids against *Campylobacter jejuni*, *Escherichia coli*, *Listeria monocytogenes*, and *Salmonella enterica*. J. Food Prot. 66: 1811–1821.

Friedman, M., Buick, R. and Elliott, C.T. 2004a. Antibacterial activities of naturally occurring compounds against antibiotic-resistant *Bacillus cereus* vegetative cells and spores, *Escherichia coli*, and *Staphylococcus aureus*. J. Food Prot. 67: 1774–1778.

Friedman, M., Henika, P.R., Levin, C.E. and Mandrell, R.E. 2004b. Antibacterial activities of plant essential oils and their components against *Escherichia coli* O157:H7 and *Salmonella enterica* in apple juice. J. Agric. Food Chem. 52: 6042–6048. doi:10.1021/Jf0495340.

Friedman, M. 2006. Antibiotic activities of plant compounds against non-resistant and antibiotic-resistant foodborne human pathogens. ACS Symp. Ser. 931: 167–183. doi:10.1021/bk-2006-0931.ch012.

Friedman, M., Buick, R. and Elliott, C.T. 2006a. Antimicrobial activities of plant compounds against antibiotic-resistant *Micrococcus luteus*. Int. J. Antimicrob. Agents 28: 156–158. doi:10.1016/j.ijantimicag.2006.05.023.

Friedman, M., Henika, P.R., Levin, C.E. and Mandrell, R.E. 2006b. Antimicrobial wine formulations active against the foodborne pathogens *Escherichia coli* O157:H7 and *Salmonella enterica*. J. Food Sci. 71: M245–M251. doi:10.1111/j.1750-3841.2006.00127.x.

Friedman, M., Henika, P.R., Levin, C.E., Mandrell, R.E. and Kozukue, N. 2006c. Antimicrobial activities of tea catechins and theaflavins and tea extracts against *Bacillus cereus*. J. Food Prot. 69: 354–361.

Friedman, M. 2007. Overview of antibacterial, antitoxin, antiviral, and antifungal activities of tea flavonoids and teas. Mol. Nutr. Food Res. 51: 116–134. doi:10.1002/mnfr.200600173.

Friedman, M., Henika, P.R., Levin, C.E. and Mandrell, R.E. 2007a. Recipes for antimicrobial wine marinades against *Bacillus cereus*, *Escherichia coli* O157:H7, *Listeria monocytogenes*, and *Salmonella enterica*. J. Food Sci. 72: M207–M213. doi:10.1111/j.1750-3841.2007.00418.x.

Friedman, M., Mackey, B.E., Kim, H.-J., Lee, I.-S., Lee, K.-R., Lee, S.-U., Kozukue, E. and Kozukue, N. 2007b. Structure-activity relationships of tea compounds against human cancer cells. J. Agric. Food Chem. 55: 243–253. doi:10.1021/jf062276h.

Friedman, M., Zhu, L., Feinstein, Y. and Ravishankar, S. 2009. Carvacrol facilitates heat-induced inactivation of *Escherichia coli* O157:H7 and inhibits formation of heterocyclic amines in grilled ground beef patties. J. Agric. Food Chem. 57: 1848–1853. doi:10.1021/jf8022657.

Friedman, M. and Rasooly, R. 2013. Review of the inhibition of biological activities of food-related selected toxins by natural compounds. Toxins 5: 743–775. doi:10.3390/toxins5040743.

Friedman, M. 2014a. Antibacterial, antiviral, and antifungal properties of wines and winery byproducts in relation to their flavonoid content. J. Agric. Food Chem. 62: 6025–6042. doi:10.1021/jf501266s.

Friedman, M. 2014b. Chemistry and multi-beneficial bioactivities of carvacrol (4-isopropyl-2-methylphenol), a component of essential oils produced by aromatic plants and spices. J. Agric. Food Chem. 62: 7652–7670. doi:10.1021/jf5023862.

Friedman, M. 2015. Antibiotic-resistant bacteria: prevalence in food and inactivation by food-compatible compounds and plant extracts. J. Agric. Food Chem. 63: 3805–3822. doi:10.1021/acs.jafc.5b00778.

Friedman, M., Henika, P.R. and Levin, C.E. 2015. Antimicrobial activities of red wine-based formulations containing plant extracts against *Escherichia coli* O157:H7 and *Salmonella enterica* serovar Hadar. Food Control 50: 652–658. doi:10.1016/j.foodcont.2014.10.005.

Friedman, M. 2016. Mushroom polysaccharides: chemistry and antiobesity, antidiabetes, anticancer, and antibiotic properties in cells, rodents, and humans. Foods 5: 80. doi:10.3390/foods5040080.

Grant, I.R., Williams, A.G., Rowe, M.T. and Muir, D.D. 2005. Efficacy of various pasteurization time-temperature conditions in combination with homogenization on inactivation of *Mycobacterium avium* subsp. *paratuberculosis* in milk. Appl. Environ. Microbiol. 71: 2853–2861. doi:10.1128/Aem.71.6.2853–2861.2005.

Gustafson, J.E., Liew, Y.C., Chew, S., Markham, J., Bell, H.C., Wyllie, S.G. and Warmington, J.R. 1998. Effects of tea tree oil on *Escherichia coli*. Lett. Appl. Microbiol. 26: 194–198. doi:10.1046/j.1472-765X.1998.00317.x.

Helander, I.M., Alakomi, H.L., Latva-Kala, K., Mattila-Sandholm, T., Pol, I.E., Smid, E.J., Gorris, L.G.M. and von Wright, A. 1998. Characterization of the action of selected essential oil components on gram-negative bacteria. J. Agric. Food Chem. 46: 3590–3595. doi:10.1021/Jf980154m.

Juneja, V.K., Thippareddi, H. and Friedman, M. 2006. Control of *Clostridium perfringens* in cooked ground beef by carvacrol, cinnamaldehyde, thymol, or oregano oil during chilling. J. Food Prot. 69: 1546–1551.

Juneja, V.K. and Friedman, M. 2007. Carvacrol, cinnamaldehyde, oregano oil, and thymol inhibit *Clostridium perfringens* spore germination and outgrowth in ground turkey during chilling. J. Food Prot. 70: 218–222.

Juneja, V.K. and Friedman, M. 2008. Carvacrol and cinnamaldehyde facilitate thermal destruction of *Escherichia coli* O157:H7 in raw ground beef. J. Food Prot. 71: 1604–1611.

Juneja, V.K., Hwang, C.-A. and Friedman, M. 2010. Thermal inactivation and postthermal treatment growth during storage of multiple *Salmonella* serotypes in ground beef as affected by sodium lactate and oregano oil. J. Food Sci. 75: M1–M6. doi:10.1111/j.1750-3841.2009.01395.x.

Juneja, V.K., Yadav, A.S., Hwang, C.-A., Sheen, S., Mukhopadhyay, S. and Friedman, M. 2012. Kinetics of thermal destruction of *Salmonella* in ground chicken containing *trans*-cinnamaldehyde and carvacrol. J. Food Prot. 75: 289–296. doi:10.4315/0362-028X.JFP-11-307.

Juneja, V.K., Gonzales-Barron, U., Butler, F., Yadav, A.S. and Friedman, M. 2013. Predictive thermal inactivation model for the combined effect of temperature, cinnamaldehyde and carvacrol on starvation-stressed multiple *Salmonella* serotypes in ground chicken. Int. J. Food Microbiol. 165: 184–199. doi:10.1016/j.ijfoodmicro.2013.04.025.

Kim, S.P., Kang, M.Y., Park, J.C., Nam, S.H. and Friedman, M. 2012a. Rice hull smoke extract inactivates *Salmonella* Typhimurium in laboratory media and protects infected mice against mortality. J. Food Sci. 77: M80–M85. doi:10.1111/j.1750-3841.2011.02478.x.

Kim, S.P., Moon, E., Nam, S.H. and Friedman, M. 2012b. *Hericium erinaceus* mushroom extracts protect infected mice against *Salmonella* Typhimurium-induced liver damage and mortality by stimulation of innate immune cells. J. Agric. Food Chem. 60: 5590–5596. doi:10.1021/jf300897w.

Kim, S.P., Nam, S.H. and Friedman, M. 2014. Rice hull smoke extract protects mice against a *Salmonella* lipopolysaccharide-induced endotoxemia. J. Agric. Food Chem. 62: 7753–7759. doi:10.1021/jf501533s.

Kwiecinski, J., Eick, S. and Wójcik, K. 2009. Effects of tea tree (*Melaleuca alternifolia*) oil on *Staphylococcus aureus* in biofilms and stationary growth phase. Int. J. Antimicrob. Agents 33: 343–347. doi:10.1016/j.ijantimicag.2008.08.028.

Lambert, R.J.W. and Pearson, J. 2000. Susceptibility testing: Accurate and reproducible minimum inhibitory concentration (MIC) and non-inhibitory concentration (NIC) values. J. Appl. Microbiol. 88: 784–790. doi:10.1046/j.1365-2672.2000.01017.x.

Lambert, R.J.W., Skandamis, P.N., Coote, P.J. and Nychas, G.-J.E. 2001. A study of the minimum inhibitory concentration and mode of action of oregano essential oil, thymol and carvacrol. J. Appl. Microbiol. 91: 453–462. doi:10.1046/j.1365-2672.2001.01428.x.

Lis-Balchin, M., Ochocka, R.J., Deans, S.G., Asztemborska, M. and Hart, S. 1999. Differences in bioactivity between the enantiomers of α-pinene. J. Essent. Oil Res. 11: 393–397. doi:10.1080/10412905.1999.9701162.

López, P., Sánchez, C., Batlle, R. and Nerín, C. 2007. Vapor-phase activities of cinnamon, thyme, and oregano essential oils and key constituents against foodborne microorganisms. J. Agric. Food Chem. 55: 4348–4356.

McHugh, T.H. and Senesi, E. 2000. Apple wraps: a novel method to improve the quality and extend the shelf life of fresh-cut apples. J. Food Sci. 65: 480–485. doi:10.1111/j.1365-2621.2000.tb16032.x.

Mejía-Garibay, B., E. Palou, and A. López-Malo. 2015. Composition, diffusion, and antifungal activity of black mustard (Brassica nigra) essential oil when applied by direct addition or vapor phase contact. J. Food Prot. 78: 843–848. doi:10.4315/0362-028X.JFP-14-485.

Mild, R.M., Joens, L.A., Friedman, M., Olsen, C.W., McHugh, T.H., Law, B. and Ravishankar, S. 2011. Antimicrobial edible apple films inactivate antibiotic resistant and susceptible *Campylobacter jejuni* strains on chicken breast. J. Food Sci. 76: M163–M168. doi:10.1111/j.1750-3841.2011.02065.x.

Moore-Neibel, K., Gerber, C., Patel, J., Friedman, M., Jaroni, D. and Ravishankar, S. 2013. Antimicrobial activity of oregano oil against antibiotic-resistant *Salmonella enterica* on organic leafy greens at varying exposure times and storage temperatures. Food Microbiol. 34: 123–129. doi:10.1016/J.Fm.2012.12.001.

Nowotarska, S.W., Nowotarski, K.I., Friedman, M. and Situ, C. 2014. Effect of structure on the interactions between five natural antimicrobial compounds and phospholipids of bacterial cell membrane on model monolayers. Molecules 19: 7497–7515. doi:10.3390/molecules19067497.

Petretto, G.L., Foddai, M., Maldini, M.T., Chessa, M., Venditti, T., D'Hallewin, G. and Pintore, G. 2013. A novel device for the study of antimicrobial activity by vapor-contact of volatile substances on food products. Commun. Agric. Appl. Biol. Sci. 78: 65–72.

Quiñones, B., Massey, S., Friedman, M., Swimley, M.S. and Teter, K. 2009. Novel cell-based method to detect Shiga toxin 2 from *Escherichia coli* O157:H7 and inhibitors of toxin activity. Appl. Environ. Microbiol. 75: 1410–1416. doi:10.1128/AEM.02230-08.

Rasooly, R., Do, P.M. and Friedman, M. 2010a. Inhibition of biological activity of Staphylococcal Enterotoxin A (SEA) by apple juice and apple polyphenols. J. Agric. Food Chem. 58: 5421–5426. doi:10.1021/jf904021b.

Rasooly, R., Do, P.M., Levin, C.E. and Friedman, M. 2010b. Inhibition of Shiga toxin 2 (Stx2) in apple juices and its resistance to pasteurization. J. Food Sci. 75: M296–M301. doi:10.1111/j.1750-3841.2010.01615.x.

Ravid, U. 2008. Enantiomeric distribution of odorous oxygenated monoterpenes in aromatic plants. pp. 155–187. *In*: Ikan, Raphael (ed.). Selected Topics in the Chemistry of Natural Products.World Scientific Publishing Co. Pte. Ltd., Singapore.

Ravishankar, S., Zhu, L., Law, B., Joens, L. and Friedman, M. 2008. Plant-derived compounds inactivate antibiotic-resistant *Campylobacter jejuni* strains. J. Food Prot. 71: 1145–1149.

Ravishankar, S., Zhu, L., Olsen, C.W., McHugh, T.H. and Friedman, M. 2009. Edible apple film wraps containing plant antimicrobials inactivate foodborne pathogens on meat and poultry products. J. Food Sci. 74: M440–M445. doi:10.1111/j.1750-3841.2009.01320.x.

Ravishankar, S., Zhu, L., Reyna-Granados, J., Law, B., Joens, L. and Friedman, M. 2010. Carvacrol and cinnamaldehyde inactivate antibiotic-resistant *Salmonella enterica* in buffer and on celery and oysters. J. Food Prot. 73: 234–240.

Rounds, L., Havens, C.M., Feinstein, Y., Friedman, M. and Ravishankar, S. 2012. Plant extracts, spices, and essential oils inactivate *Escherichia coli* O157:H7 and reduce formation of potentially carcinogenic heterocyclic amines in cooked beef patties. J. Agric. Food Chem. 60: 3792–3799. doi:10.1021/jf204062p.

Rounds, L., Havens, C.M., Feinstein, Y., Friedman, M. and Ravishankar, S. 2013. Concentration-dependent inhibition of *Escherichia coli* O157:H7 and heterocyclic amines in heated ground beef patties by apple and olive extracts, onion powder and clove bud oil. Meat Sci. 94: 461–467. doi:10.1016/j.meatsci.2013.03.010.

Scanu, A.M., Bull, T.J., Cannas, S., Sanderson, J.D., Sechi, L.A., Dettori, G., Zanetti, S. and Hermon-Taylor, J. 2007. *Mycobacterium avium* subspecies *paratuberculosis* infection in cases of irritable bowel syndrome and comparison with Crohn's disease and Johne's disease: common neural and immune pathogenicities. J. Clin. Microbiol. 45: 3883–3890. doi:10.1128/JCM.01371-07.

Sechi, L.A., Paccagnini, D., Salza, S., Pacifico, A., Ahmed, N. and Zanetti, S. 2008. *Mycobacterium avium* subspecies *paratuberculosis* bacteremia in type 1 diabetes mellitus: an infectious trigger? Clin. Infect. Dis. 46: 148–149. doi:10.1086/524084.

Sikkema, J., De Bont, J.A.M. and Poolman, B. 1994. Interactions of cyclic hydrocarbons with biological membranes. J. Biol. Chem. 269: 8022–8028.

Sirk, T.W., Brown, E.F., Sum, A.K. and Friedman, M. 2008. Molecular dynamics study on the biophysical interactions of seven green tea catechins with lipid bilayers of cell membranes. J. Agric. Food Chem. 56: 7750–7758. doi:10.1021/jf8013298.

Sirk, T.W., Brown, E.F., Friedman, M. and Sum, A.K. 2009. Molecular binding of catechins to biomembranes: relationship to biological activity. J. Agric. Food Chem. 57: 6720–6728. doi:10.1021/jf900951w.

Tapsell, L.C., Hemphill, I., Cobiac, L., Patch, C.S., Sullivan, D.R., Fenech, M., Roodenrys, S., Keogh, J.B., Clifton, P.M., Williams, P.G., Fazio, V.A. and Inge, K.E. 2006. Health benefits of herbs and spices: the past, the present, the future. Med. J. Aust. 185: S4–24.

Todd, J., Friedman, M., Patel, J., Jaroni, D. and Ravishankar, S. 2013. The antimicrobial effects of cinnamon leaf oil against multi-drug resistant *Salmonella* Newport on organic leafy greens. Int. J. Food Microbiol. 166: 193–199. doi:10.1016/j.ijfoodmicro.2013.06.021.

Ultee, A., Bennik, M.H.J. and Moezelaar, R. 2002. The phenolic hydroxyl group of carvacrol is essential for action against the food-borne pathogen *Bacillus cereus*. Appl. Environ. Microbiol. 68: 1561–1568. doi:10.1128/Aem.68.4.1561-1568.2002.

Upadhyaya, I., Upadhyay, A., Kollanoor-Johny, A., Mooyottu, S., Baskaran, S.A., Yin, H.B., Schreiber, D.T., Khan, M.I., Darre, M.J., Curtis, P.A. and Venkitanarayanan, K. 2015. In-feed supplementation of trans-cinnamaldehyde reduces layer-chicken egg-borne transmission of *Salmonella enterica* serovar Enteritidis. Appl. Environ. Microbiol. 81: 2985–2994. doi:10.1128/aem.03809-14.

Walsh, C. 2003. Antibiotics: Actions, Origins, Resistance. Washington, D.C.: ASM Press.

Wong, S.Y.Y., Grant, I.R., Friedman, M., Elliott, C.T. and Situ, C. 2008. Antibacterial activities of naturally occurring compounds against *Mycobacterium avium* subsp. *paratuberculosis*. Appl. Environ. Microbiol. 74: 5986–5990. doi:10.1128/AEM.00981-08.

Xu, J., Zhou, F., Ji, B.-P., Pei, R.-S. and Xu, N. 2008. The antibacterial mechanism of carvacrol and thymol against *Escherichia coli*. Lett. Appl. Microbiol. 47: 174–179. doi:10.1111/j.1472-765X.2008.02407.x.

Essential Oils Constituted by Prop-1(2)-enylbenzene Derivatives Used for Treatment of Microbial Infections

Amner Muñoz-Acevedo,[1,*] *María del Carmen González,*[1] *Erika Amparo Torres,*[1] *Martha Cervantes-Díaz*[2] and *Elena E. Stashenko*[3]

Introduction

Some plants, belonging mainly to Apiaceae, Aristolochiaceae, Asteraceae, Euphorbiaceae, Illiaceae, Lamiaceae, Lauraceae, Myrtaceae, Piperaceae and Rutaceae families, produce Essential Oils (EO) constituted by prop-1(2)-enylbenzene derivatives (Ojewole 2002, de Heluani et al. 2005, Guerrini et al. 2009, Jana and Shekhawat 2010, Dan et al. 2010, Wang et al. 2011, Salleh et al. 2016b), which have been traditionally used by the communities in different countries as a therapeutic alternative for the treatment of diseases/ailments (Suárez et al. 2003, Verspohl et al. 2005, Maxia et al. 2012).

These compounds based on prop-1(2)-enylbenzene structures include ethers (anethole, estragole, safrole, elemicin, dillapiole, etc.) and phenols (chavicol, eugenol, etc.), and in a lesser proportion, esters (acetyleugenol), whose antimicrobial, anti-inflammatory, antioxidant, antiviral, antitumoral, analgesic, anesthetic, insecticidal, antiparasitic, repellent and narcotic properties have been determined and established (Leal-Cardoso et al. 2004, Romero et al. 2007, Muñoz-Acevedo et al. 2009, 2014, Bendiabdellah et al. 2014).

Essential oils, containing phenylprop-1(2)-enoid compounds (10–99%), have showed a wide-range growth inhibition properties against bacteria and fungus, e.g., *Ocimum gratissimum* and *Pimenta racemosa* EO (eugenol chemotypes) against *Staphylococcus aureus* (MIC (Minimum Inhibition Concentration) 0.5 μg/mL), and *Aspergillus niger* (MIC 0.04 μg/mL); *Agastache mexicana* EO (high content of methyleugenol and estragole) against *Asp. flavus* (MIC 0.3 μg/mL); *Laurelia sempervirens*

[1] Chemistry and Biology Research Group – Department of Chemistry and Biology – Universidad del Norte, Barranquilla, Colombia.
[2] Environmental Research Group for Sustainable Development – Environmental Chemistry Faculty – Universidad Santo Tomás, Bucaramanga, Colombia.
[3] Mass Spectrometry and Chromatography Centre - CIBIMOL - CENIVAM, Universidad Industrial de Santander, Bucaramanga, Colombia.
* Corresponding author: amnerm@uninorte.edu.co

EO (high content of safrole and isosafrole) against *Acinetobacter* sp. (MIC 7 µg/mL); *Foeniculum vulgare* EO (high content of *E*-anethole/estragole) against *Candida albicans* (MIC 0.3 µg/mL), *C. guillermondii* (MIC 0.3 µg/mL), and *C. krusei* (MIC 0.6 µg/mL); *Artemisia dracunculus* EO (estragole chemotype) against *C. tropicalis* (MIC 0.6 µg/mL), and *Bacillus subtilis* (MIC 0.3 µg/mL); *F. vulgare* EO (*E*-anethole chemotype) against *Escherichia coli* (MIC 0.06 µg/mL); *Piper aduncum* EO (high content of dillapiole) against *Micrococcus luteus* (MIC 2 µg/mL); *Acorus calamus* EO (high content of *Z*-asarone) against *Pseudomonas aeruginosa* (MIC 0.3 µg/mL); *Pimenta dioca* EO (high content of eugenol/methyleugenol) against *Paenibacillus larvae* (MIC 78 µg/mL), and *Salmonella* sp. (MIC 6 µg/mL); *Crithmum maritimum* EO (high content of dillapiole) against *Tricophyton mentagrophytes* (MIC 0.08 µg/mL), and *T. rubrum* (MIC 0.08 µg/mL), etc. (Nakamura et al. 1999, Curini et al. 2006, Marongiu et al. 2007, Radušiene et al. 2007, Höferl et al. 2009, Montenegro et al. 2012, Bisht et al. 2014, Piras et al. 2014, Zabka et al. 2014, Juárez et al. 2015, Ansari et al. 2016); which make them of great interest for diverse and potential applications.

These phenylprop-1(2)-enoids/essential oils have been incorporated as ingredients for food and beverages, oral hygiene/dental products, cosmetics, etc. (EC 2009, 21CFR172.510, 21CFR184.1282). However, the control agencies (U.S. Food and Drugs Administration/European Commission) have 'restricted' the application/consumption of some substances (safrole, estragole, methyleugenol, etc.) due to the possible risk associated with 'toxicity/damage'. As a result of these restrictions, other interesting applications have arisen for example as precursors in organic synthesis (Sudo et al. 2006, Wang et al. 2009, Pandey and Bani 2010, Chen et al. 2011). Despite the 'restriction' on using of phenylprop-1(2)-enoids and essential oils containing them has been possible to carry out some inventive activities (patents), like those to promote the nanotechnological developments, taking advantages of their antimicrobial properties.

This chapter deals with essential oils containing phenylprop-1(2)-enoids as main components, its antimicrobial activities, the plants of different regions containing them, and their application in nanotechnology; and is divided in: (1) text mining; (2) importance of phenylprop-1(2)-enoids; (3) traditional and therapeutic uses of the plants; (4) antimicrobial activities of essential oils produce by some medicinal plants; (5) patents related to phenylprop-1(2)-enoids/essential oils; (6) nanotechnology applied to essential oils with antibacterial properties; and, (7) other applications involving to phenylprop-1(2)-enoids.

Text Mining

The monitoring of the scientific activity related to the antimicrobial properties of essential oils was performed through a bibliometric analysis of science articles indexed in the Scopus database (Elsevier 2016) using the search equation: (*TITLE-ABS-KEY ("essential oil*") AND TITLE-ABS-KEY (antimicrobial)) AND DOCTYPE (ar) AND PUBYEAR > 1999*. As results of the analysis, 4680 records of articles associated to this subject were found, and in the timeline 2000–2016, the most productive years were 2014 and 2015 with 604 records and 558 records, respectively. Moreover, the countries with the highest number of records were Brazil (499 articles), Iran (441 articles), India (429 articles) and Turkey (385 articles).

According to the science literature analysis on the antimicrobial activity-phenylprop-1(2)-enoid structure relationship, the main compounds to which have been attributed these activities (antibacterial and antifungal) were anethole, estragole, safrole, eugenol, methyleugenol, methylisoeugenol, myristicin, dillapiole, apiole, elemicin, isoeugenol and *Z*-asarone. Therefore, with the aim to establish the scientific dynamic correlated to the antimicrobial activities of these compounds, the following search equation was used in the Scopus database *(TITLE-ABS-KEY (anethol or estragole OR safrole OR eugenol OR "methyleugenol" OR "methylisoeugenol" OR myristicin OR dillapiole OR apiole OR elemicin OR isoeugenol OR "β-asarone") AND TITLE-ABS-KEY (antimicrobial)) AND DOCTYPE (ar) AND PUBYEAR > 1999,* with which were found 703 records.

Figure 3.1 displays the article distribution per year interrelated with the antimicrobial activities of prop-1(2)-enylbenzene compounds. An increasing trend in the publication numbers until 2015 is shown in this figure, with 2014–2015 being the years with the largest number of registers (70–72 articles) reported together with 2010 (71 articles). *This could be interpreted that there is a growing interest of research to establish the antimicrobial potential of phenylprop-1(2)-enoid compounds.* Over the course of 2016, 68 articles have been published.

In this dynamic of publications by countries, India (118 articles), Brazil (80 articles), United States (75 articles), and Italy (49 articles) are highlighted for having the highest number of records. Other countries, e.g., Turkey, Spain, China, Germany and Iran, had a lower but approximately the same number of publications (~ 24 registers). Finally, the most important authors/researcher were G. Buchbauer (13 registers) along with E. Schmidt, L. Jirovetz, M. Davidson, and J. Weiss, each one with 11 registers.

The data refining and clustering/differentiation in accordance with the antibacterial and/or antifungal activities of the most abundant phenylprop-1(2)-enoid compounds are presented in Fig. 3.2. The result of this analysis showed that eugenol was the compound with the highest number of registers about the antibacterial (153 records) and antifungal (73 records) activities. The other constituents with an important number of records were anethole, estragole and methyleugenol, each of them with an average value of 24 records and 22 records, respectively, for each activity (antibacterial and antifungal).

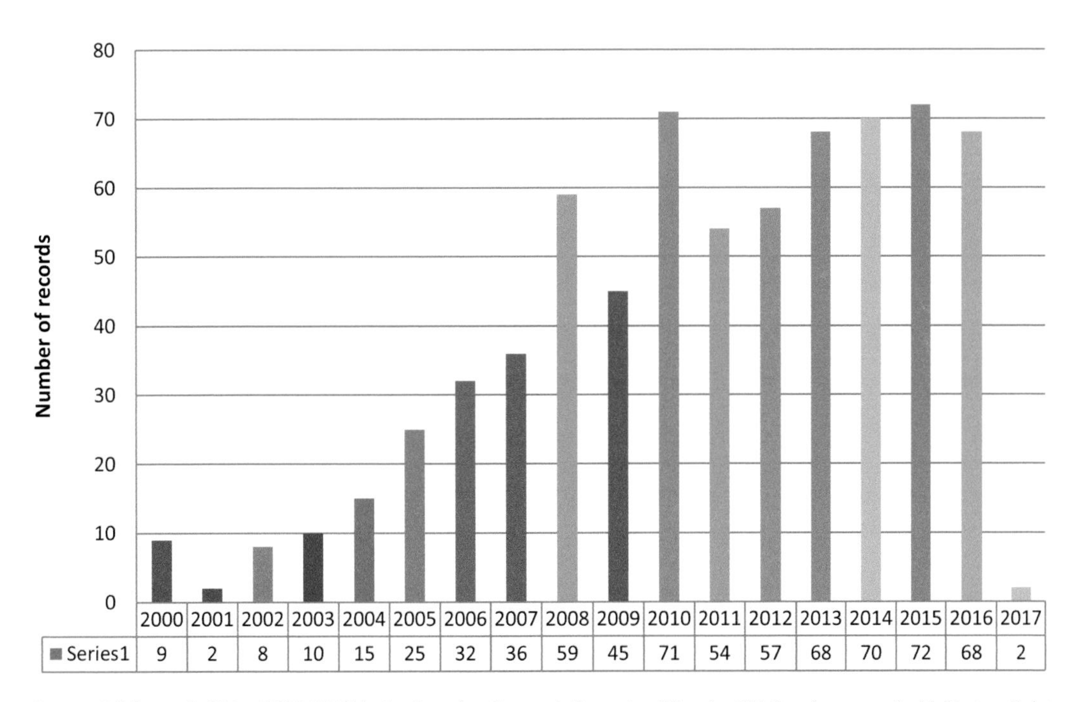

	2000	2001	2002	2003	2004	2005	2006	2007	2008	2009	2010	2011	2012	2013	2014	2015	2016	2017
Series1	9	2	8	10	15	25	32	36	59	45	71	54	57	68	70	72	68	2

Source: Bibliometric Unit - USTABUCA, data based on Scopus information (Elsevier 2016) and processed with VantagePoint software (Version 9.0, *Search Technology*).

Figure 3.1. Distribution of scientific articles per year related to the antimicrobial properties of prop-1(2)-enylbenzene compounds.

Importance of Prop-1(2)-enylbenzene Derivatives Compounds

Prop-1(2)-enylbenzene (C_6–C_3) compounds are secondary metabolites derived of *trans*-cinnamic acid, which come from shikimic acid mediated by L-phenylalanine (Solecka 1997, Seigler 1998,

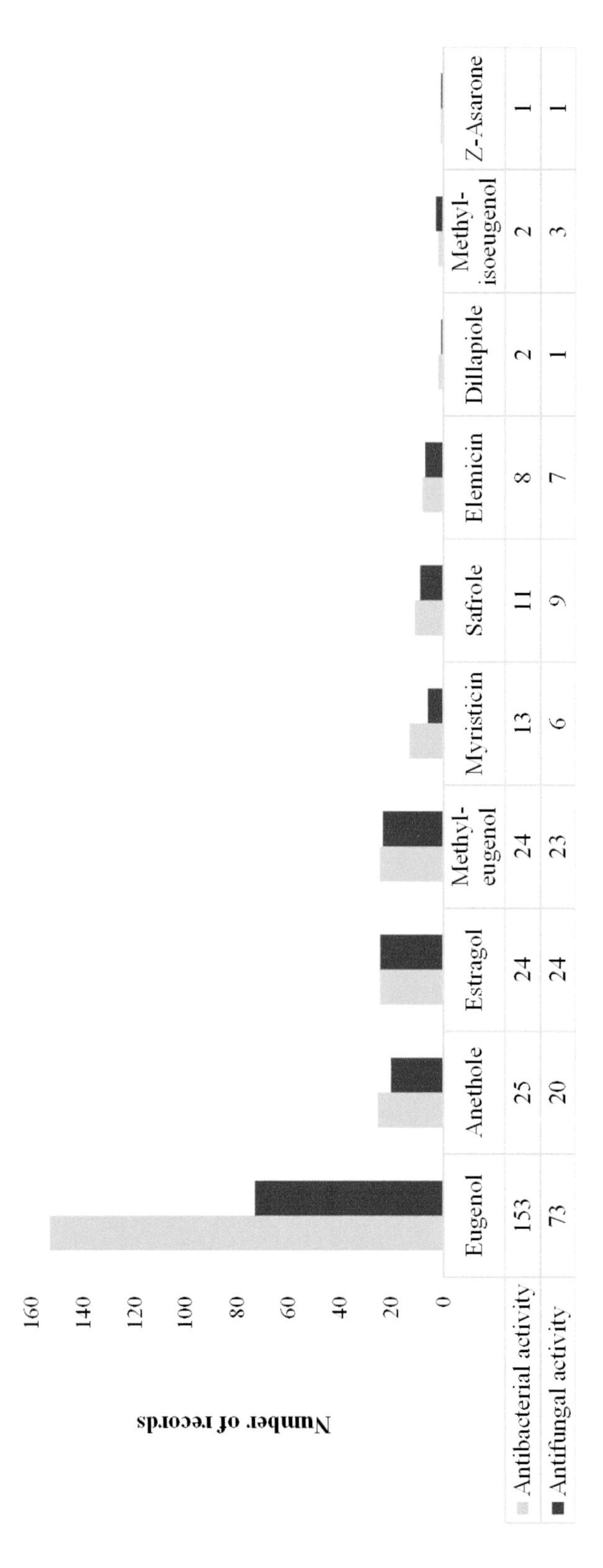

Number of records
160
140
120
100
80
60
40
20
0
Eugenol
Anethole
Estragol
Methyl-eugenol
Myristicin
Safrole
Elemicin
Dillapiole
Methyl-isoeugenol
Z-Asarone
Antibacterial activity
153
25
24
24
13
11
8
2
2
1
Antifungal activity
73
20
24
23
6
9
7
1
3
1

Source: Bibliometric Unit - USTABUCA, data based on Scopus information (Elsevier 2016) and processed with VantagePoint software (Version 9.0, *Search Technology*).

Figure 3.2. Distribution by number of registers of the antibacterial and antifungal activities of the most representative phenylprop-1(2)-enoids, in the timeline 2000–2017.

Vogt 2010). These types of molecules are closely related to some relatively volatile compounds found in the essential oils from plants, mainly. In addition, the prop-1(2)-enylbenzene derivatives (P1(2) ENB) have an ecological and economic importance because are probably involved in the animal/microorganism-plant interactions (e.g., kairomones, allomones and synomones), and are constituents of several commercially important spices and herbs (Seigler 1998). Most of the compounds/EO are incorporated as ingredients for flavor, food, cosmetic and perfumery industries (Seigler 1998, Burdock 2010).

Table 3.1 presents the general information (structures, IUPAC and common names, physicochemical and biological properties, applications/uses, plants containing to them) about P1(2) ENB compounds. According to this table and the scientific literature reviewed, the phenylprop-1(2)-enoids **1–5**, **7–10**, and **13** have been the compounds more frequently found in the essential oils from plants; whilst, the phenylprop-1(2)-enoids **11–12**, **14–17**, **19–20**, and **22–24** have been the less frequent compounds. These compounds have constituted between *ca.* 10–99% of the total essential oils.

Some plants that produce these types of compounds belong to different families and specific genera, e.g., Acoraceae (*Acorus*), Annonaceae (*Uvariodendron*, *Xylopia*), Apiaceae (*Ammi*, *Anethum*, *Anthricus*, *Apium*, *Athamanta*, *Bunium*, *Carum*, *Chaerophyllum*, *Crithmum*, *Daucus*, *Echinophora*, *Foeniculum*, *Heracleum*, *Petroselinum*, *Pimpinella*, *Pleurospermum*, *Ridolfia*, *Seseli*), Aristolochiaceae (*Asarum*), Asteraceae (*Artemisia*, *Psidia*, *Rhaponticum*, *Tagetes*), Atherospermataceae (*Laurelia*), Cupressaceae (*Juniperus*), Euphorbiaceae (*Croton*, *Phyllanthus*), Fabaceae (*Enterolobium*), Geranacieae (*Pelargonium*), Illiciaceae (*Illicium*), Lamiaceae (*Agastache*, *Cedronella*, *Ocimum*, *Platostoma*, *Pycnostachys*), Lauraceae (*Aniba*, *Beilschmiedia*, *Cinnamomum*, *Laurus*, *Lindera*, *Ocotea*, *Sassafras*), Myrtaceae (*Backhousia*, *Melaleuca*, *Pimenta*, *Syzygium*), Myristicaceae (*Myristica*), Nephrolepidaceae (*Nephrolepis*), Piperaceae (*Peperomia*, *Piper*), Poaceae (*Cymbopogon*), Rutaceae (*Clausena*, *Crowea*, *Empleurum*, *Zanthoxylum*, *Zieria*), Scrophulariaceae (*Limnophila*), Verbenaceae (*Lippia*), and Zingiberaceae (*Amomum*, *Etlingera*, *Hedychium*).

By analyzing in detail the information gathered, *Apiaceae is the family with the largest number of genera (ca. 18) and species (ca. 34, including varities; 6 Daucus* spp.*, and 5 Foeniculum* spp.*) containing phenylprop-1(2)-enoid compounds.* Nevertheless, other genera have greater numbers of species producing essential oils rich in the interest compounds, e.g., 14 *Ocimum* spp., 11 *Piper* spp., 6 *Clausena* spp., 6 *Pimenta* spp., and 5 *Cinnamomum* spp.

Data about the 50% letal dosage (LD_{50}) and 50% toxic dosage (TD_{50}) values of phenylpro-1(2)-enoids, which have allowed to define the restrictions in its uses and/or applications, are included in Table 3.1. Compounds **1–24** have LD_{50} values between ~0.05 g/kg – ~4.3 g/kg (oral/ingestion), ~0.3 g/kg –~1.3 g/kg (intraperitoneal), ~0.1 g/kg – ~0.3 g/kg (intravenous). *It should be noted that the compounds **2**, **3**, **5**, and **8** have TD_{50} values ranging between ~0.003 g/kg/day–0.4 g/kg/day, being safrole (0.003 g/kg/day, human) and myristicin (0.006 g/kg/day, human) who have the lowest values, respectively.*

The phenylpropenoids presenting allyl- and methoxy-groups (**2**, **3**, etc.) have been 'classified' as carcinogenic compounds, e.g., the safrole's carcinogenicity was evidenced in 1960–1961 when this compound caused preneoplastic changes (adenomas) in the rat's liver (Woo and Lai 1986, NTP, NTP TR 551, IPCS/WHO).

The metabolic pathway (Scheme 3.1) of safrole (**3**), related to their carcinogenic effect, involves oxidation reactions: (**a**) 1'-hydroxylation, and (**b**) 2',3'-epoxidation of prop-2-enyl chain; and, (**c**) demethylenation of methylenedioxy group (Klungsøyr and Scheline 1982, Wiseman et al. 1987, Luo and Guenther 1996, Gardner et al. 1997, IPCS/WHO).

The corresponding products are 1'-hydroxysafrole (**25**) (main metabolite), 2',3'-epoxidesafrole (**26**), and 2-hydroxychavicol (**27**); while 1'-sulfooxysafrole (**28**), 1'-oxosafrole (**29**), and 1'-hydroxysafrole-2',3'-oxide (**30**) are considered as the ultimate carcinogenic compounds. Compound **28** (produced by **25** with PAPS (sulfated coenzyme)) is probably responsible for generating reactive

Table 3.1. General information about prop-1(2)-enylbenzene derivatives compounds.

Structures	Physicochemical properties	Biological properties	Applications/Uses	Some plants containing P1(2)ENB	References
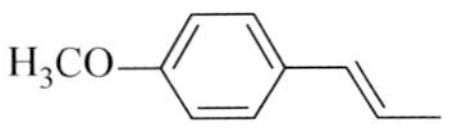 *E*-1-Methoxy-4-(prop-1-enyl)benzene *E*-Anethole (**1**)—CAS No. 4180-23-8	MW: 148.2 g/mol Bp: 234°C; Mp: 21°C P_{vap}: 5.45 Pa (294 K) LogP: 2.94 LD_{50} (g/kg) Oral—Rat: 2.1; Mouse: 3 Intraperitoneal—Mouse: 0.6 TD_{50} (g/kg/day): NR	Repellent, Fungicide, Insecticide, Inhibition of oviposition, Antimicrobial Anesthetic, Antioxidant Insecticidal activity against Mexican bean weevils	Diastereoselective synthesis, Flavoring for food and alcoholic drinks, Oral hygiene products	*C. heptaphylla, C. agasthyamalayana, P. anisum, I. verum, B. anisata, C. anisum-olens, C. grewioides, C. zehtneri, F. vulgare, L. rugosa, T. filifolia*	Newberne et al. 1999, Blewitt and Southwell 2000, Molino 2000, De et al. 2001, Ghelardini et al. 2001, de Lima et al. 2006, Romero et al. 2007, Silva et al. 2008, Camarillo et al. 2009, Muñoz-Acevedo et al. 2009, Shalaby et al. 2011, de Lima et al. 2013, Santhosh Kumar et al. 2014, Verma et al. 2014
 1-Allyl-4-methoxybenzene Estragole (**2**)—CAS No. 140-67-0	MW: 148.2 g/mol Bp: 216°C; Mp: NR P_{vap}: 0.21 mmHg (25°C) LogP: 3.23, pKa: –4.8 LD_{50} (g/kg) Intraperitoneal—Mouse: 1.3; Rat: 1 Ingestion—Mouse: 1.2 TD_{50} (g/kg/day) Mouse: 0.05	Local anesthetic, Antibacterial, Relaxing cavernosum in rats, Antiplasmodial, Antifungal, Antiviral	Perfumery, Flavoring, Anticonvulsant, Anesthetic, Antiinflamatorio, Vasoactive, Antioxidant, Antimicrobial, Antimelanogenic, Fumigant against grain pests	*A. foeniculum, F. vulgare, A. lingüiforme, C. dunniana, O. basilicum, O. americanum, T. filifolia, T. lucida, A. gramineus, A. cerefolium, A. dracunculus, C. canariensis* ssp. *anisata*	Mazza and Kiehn 1992, Engel et al. 1995, Wang et al. 1998, Leal-Cardoso et al. 2004, Ortuño 2006, da Costa et al. 2008, Muñoz-Acevedo et al. 2009, Ebadollahi et al. 2010a, 2010b, 2011, Chizzola 2011, Huang et al. 2012, Ponte et al. 2012, Cabral et al. 2014, Yamani et al. 2014
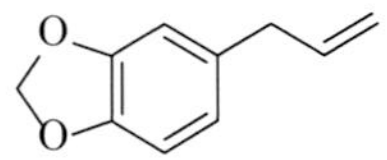 5-Allylbenzo[*d*][1,3]dioxole Safrole (**3**)—CAS No. 94-59-7	MW: 162.18 g/mol Bp: 234°C; Mp: 11°C P_{vap}: 0.071 mmHg (25°C) LogP: 2.23, pKa: –4.7 LD_{50} (g/kg) Oral—Mouse: 2.4; Rat: 2 TD_{50} (g/kg/day) Mouse: 0.05; Rat: 0.4 Human (60 kg): 0.003	Hepatotoxic, Genotoxic, Antibacterial, P450 enzyme inhibitor, Antifungal, Mosquito larvicidal	Precursor for MDMA and piperonal synthesis, Perfumery, Fragrances, Manufacture of soaps	*J. virginiana, C. micranthum, S. albidum, C. parthenoxylon, C. excavata, P. auritum, P. mikanianum, P. xylosteoides, Z. ovalifolium*	Brophy et al. 1995, 1997, Dũng et al. 1995, Lima et al. 2000, de Abreu et al. 2002, Ueng et al. 2005, Cheng et al. 2009, Leyva et al. 2009, Dognini et al. 2012, Khayyat 2013, Khayyat and Al-Zahrani 2014

H_3CO HO 4-Allyl-2-methoxyphenol Eugenol (**4**)—CAS No. 97-53-0	MW: 164.20 g/mol Bp: 225°C; Mp: –9°C P_{vap}: 0.022 mmHg (25°C) LogP: 2.66, pKa: 9.94 (Acidic), –4.9 (Basic) LD_{50} (g/kg) Oral—Mouse: 3; Guinea pig: 2.1; Rat: 1.9 TD_{50} (g/kg/day): NR	Thromboxane antiformation and antiplatelet in human blood, Microbial, Antifungal, Antioxidant Antipyretic, Repellency against *Anopheles gambiae*, Cytotoxicity in human lymphocites and zebrafish	Pharmacological properties, Stomatology, Preservative, Odontological applications, Precursor for synthesis	*P. dioica, C. zeylanicum, L. nobilis, O. americanum, O. sanctum, S. aromaticum, U. gorgonis*	González 2002, Schmidt et al. 2006, 2007, Jantan et al. 2008, Höferl et al. 2009, Muñoz-Acevedo et al. 2011, 2014, Anthony et al. 2012, Khayyat 2013, Innocent and Hassanali 2015, Xie et al. 2015
H_3CO H_3CO 4-Allyl-1,2-dimethoxybenzene Methyleugenol (**5**)—CAS No. 93-15-2	MW: 178.2 g/mol Bp: 255°C; Mp: –4°C P_{vap} : 0.012 mmHg (25°C) LogP: 2.99, pKa: –4.6 (Basic) LD_{50} (g/kg) Oral—Mouse: 0.5; Rat: 0.8 Intravenous—Mouse: 0.1 TD_{50} (g/kg/day) Mouse: 0.02; Rat: 0.02	Insecticide, Repellent, Attractant effect, Antinociceptive effect and acute toxicity, Cytotoxicity in human lymphocites and zebrafish	Perfumery, Detergents, Food additives (sweets, drinks, baked goods), Aromatherapy	*A. gramineus, M. bracteata, M. leucadendra, E. fragrans, A. rugosa* ssp. *methyleugenolifera, A. heterotropoides* var. *mandshuricum, B. myrtifolia, C. subavenium, C. malambo, O. canum*	Fujita and Fujita 1973, Brophy et al. 1995, Wang et al. 1998, Hee and Tan 2006, Ho et al. 2008, Wang et al. 2014, Xu et al. 2015
H_3CO HO *E*-2-Methoxy-4-(prop-1-enyl)phenol *E*-Isoeugenol (**6**)—CAS No. 97-54-1	MW: 164.2 g/mol Bp: 266°C; Mp: –10°C P_{vap}: 0.02 mmHg (77°F) LogP: 2.79, pKa: 10.01 (Acidic), –4.9 (Basic) LD_{50} (g/kg) Oral—Guinea pig: 1.4; Rat: 1.6; Birds: 0.3 TD_{50} (g/kg/day): NR	Antioxidant	Diastereoselective synthesis, Perfumery, Cosmetic products, Allergen, Dental materials, Foods	*H. coronarium, E. cevuga, O. basilicum*	Matsumoto et al. 1993, Rajakumar and Rao 1993, Camarillo et al. 2009, Yahya et al. 2010, Chudasama and Thaker 2012
H_3CO H_3CO *E*-1,2-Dimethoxy-4-(prop-1-enyl)benzene *E*-Methylisoeugenol (**7**)—CAS No. 6379-72-2	MW: 178.2 g/mol Bp: 271°C; Mp: 33°C P_{vap}: 0.011 mmHg (25°C) LogP: 2.78–3.47, pKa: –4.6 (Basic) LD_{50} (g/kg) Oral—Rat: 2.5 Intravenous—Mouse: 0.2 TD_{50} (g/kg/day): NR	Oviposition stimulant, Phagostimulant, Antibacterial, Anxiolytic, Antidepressant	Food additive, Treatment of mood disorder	*M. squamophloia, M. leucadendra, C. septentrionale, A. calamus, A. littoralis, B. myrtifolia*	Camps 1988, Brophy et al. 1995, 1999, Rossi et al. 2007, Wong et al. 2010, Fajemiroye et al. 2014

Table 3.1 contd....

... *Table 3.1 contd.*

Structures	Physicochemical properties	Biological properties	Applications/Uses	Some plants containing P1(2)ENB	References
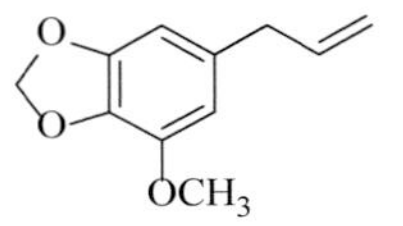 6-Allyl-4-methoxybenzo[*d*][1,3]dioxole Myristicin (**8**)—CAS No. 607-91-0	MW: 192.2 g/mol Bp: 276°C; Mp: –20°C P_{vap}: 6.46x10^{-3} mmHg (25°C) LogP: 2.47–2.54, pKa: –4.5 (Basic) LD_{50} (g/kg) Oral—Cat: 0.4; Guinea pig: 0.002; Rat: 4.3; Rabbit: 0.9 TD_{50} (g/kg/day) Humans: 0.006–0.007	Insecticide, Hepatoprotective	Rheumatism treatment, Anticholera, Digestive disorders	*L. neesiana, B. cylindricum, P. crispum, A. petelotii, A. sicula, C. anisum-olens, C. indica, D. carota* ssp. *hispanicus*	Srivastava et al. 2001, Morita et al. 2003, Camarda and Di Stefano 2003, Su et al. 2011, López et al. 2012, Thai et al. 2013a, Bendiabdellah et al. 2014
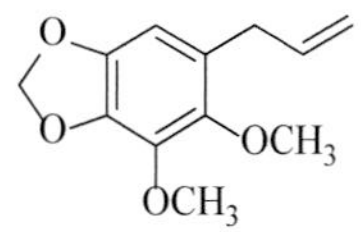 6-Allyl-4,5-dimethoxybenzo[*d*][1,3]dioxole Dillapiole (**9**)—CAS No. 484-31-1	MW: 222.2 g/mol Bp: 285°C; Mp: 28°C P_{vap}: 0.6 mmHg (25°C) LogP: 2.33–2.38, pKa: –4.4 (Basic) LD_{50} (g/kg) Oral—Mouse: 1 TD_{50} (g/kg/day): NR	Anti inflammatory, Insecticide, Antileishmanial, Antioxidant, Antifungical Larvicidal activity against *A. aegypti*	Synthesis of derivatives	*F. vulgare, A. sowa, A. graveolens, A. nodiflorum, C. maritimum, P. aduncum*	Bernard et al. 1995, Pino et al. 2004, Marongiu et al. 2007, de Almeida et al. 2009, Leyva et al. 2009, Parise-Filho et al. 2011, 2012, Maxia et al. 2012, Peerakam et al. 2014, Abdelkader and Lockwood 2016
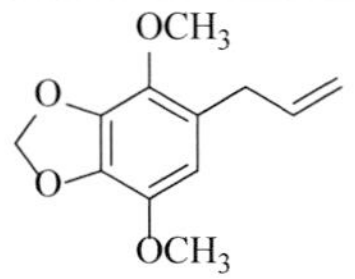 5-Allyl-4,7-dimethoxybenzo[*d*][1,3]dioxole Apiole (**10**)—CAS No. 523-80-8	MW: 222.2 g/mol Bp: 295°C; Mp: 29°C P_{vap}: 0.6 mmHg (25°C) LogP: 2.34–2.38, pKa: –4.4 (Basic) LD_{50} (g/kg) Oral—Mouse: 0.05 TD_{50} (g/kg/day): NR	Antitumoral, Abortive, Acaricide	Synthesis of derivatives, Dysmenorrhea and amenorrhea	*P. hortense, D. carota* ssp. *hispanicus*	Quinn et al. 1958, Song et al. 2011, Wei et al. 2012, Bendiabdellah et al. 2014
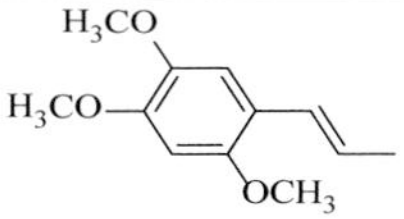 *E*-1,2,4-Trimethoxy-5-(prop-1-enyl)benzene *E*-Asarone (α-Asarone) (**11**)—CAS No. 2883-98-9	MW: 208.2 g/mol Bp: 296°C; Mp: 63°C P_{vap}: 1.5x10^{-3} mmHg (25°C) LogP: 2.62–3.35, pKa: –4.4 (Basic) LD_{50} (g/kg) Rat: 0.4 Intraabdominal-Rat: 0.3 TD_{50} (g/kg/day): NR	Repellent, Insecticide, Antirust, Antihyperlipemic, Anti inflammatory	Synthesis of derivatives, Treatment for epilepsy, cough, bronchitis and asthma	*P. marginatum, A. calamus, A. europaeum, A. forbesii, P. pseudocaryophyllus*	Mazza 1985, Park et al. 2003, Manikandan and Devi 2005, Zhang et al. 2005, Wilczewska et al. 2008, Autran et al. 2009, Limón et al. 2009, Marques et al. 2010, Shin et al. 2014

Z-1,2,4-Trimethoxy-5-(prop-1-enyl)benzene *Z*-Asarone (β-Asarone) (**12**)—CAS No. 5273-86-9	MW: 208.2 g/mol Bp: 267°C; Mp: 62°C P_{vap}: 0.003 mmHg (25°C) LogP: 3.4, pKa: NR LD_{50} (g/kg) Mice: 0.2; Rat: 1.0 TD_{50} (g/kg/day): NR	Antitumoral in Colon Cancer, Cardiovascular protector, Insecticide, Neuroprotection, Antioxidant, Antifungal, Antibacterial	Treatment for Alzheimer's disease, Flavoring for solid snacks, drinks and condiments	*A. calamus,* *A. gramineus,* *A. dracunculus*	Wang et al. 1998, Ciccia et al. 2000, Cho et al. 2002, Kordali et al. 2005, Liu et al. 2010, Qiu et al. 2011, Lee et al. 2011, Li et al. 2012, Padalia et al. 2014
5-Allyl-1,2,3-trimethoxybenzene Elemicin (**13**)—CAS No. 487-11-6	MW: 208.2 g/mol Bp: 314°C; Mp: 81°C P_{vap}: 0.007 mmHg (25°C) LogP: 2.6–3.03, pKa: –4.4 (Basic) LD_{50} (g/kg) Mouse: 0.3 Oral—Rabbit: 3.2; Rat: 1 TD_{50} (g/kg/day): NR	Antibacterial, Herbicide, Genotoxic	Synthesis of derivatives, Flavoring for foods	*A. dracunculus,* *C. microstachys,* *B. myrtifolia,* *A. cordifolium,* *A. balancae,* *M. squamophloia,* *C. glaucescens,* *C. austroindica*	Hasheminejad and Caldwell 1994, Brophy et al. 1995, 1999, Baruah and Nath 2006, Rossi et al. 2007, Thai et al. 2013a, 2013b
5-Allyl-4,6,7-trimethoxybenzo[*d*][1,3]dioxole Nothoapiole (**14**)—CAS No. 22934-74-3	MW: 252.3 g/mol Bp: 404°C; Mp: 192°C P_{vap}: 0.00103 mmHg (25°C) LogP: 2.17, pKa: NR LD_{50} (g/kg): NR TD_{50} (g/kg/day): NR	It does not attribute to any specific biological property	Have not been reported any application/uses	*P. angelicoides,* *C. montanum*	Laouer et al. 2009, Mathela et al. 2015
E-1-(3-Methyl-2-butenoxy)-4-(prop-1-enyl)benzene *E*-Foeniculin (**15**)—CAS No. 78259-41-3	MW: 202.3 g/mol Bp: 308°C; Mp: 11°C P_{vap}: 0.001 mmHg (25°C) LogP: 4.3–4.76 pKa: –4.9 (Basic) LD_{50} (g/kg): NR TD_{50} (g/kg/day): NR	Antioxidant, Inhibitory effect on TNF α, Anti-inflammatory	Chemical Reference Substances (CRS)	*C. Anisata,* *F. vulgare*	Garneau et al. 2000, Yang et al. 2015

Table 3.1 contd....

... *Table 3.1 contd.*

Structures	Physicochemical properties	Biological properties	Applications/Uses	Some plants containing P1(2) ENB	References
H_3CO, HO, H_3CO 4-Allyl-2,6-dimethoxyphenol 4-Allylsyringol (**16**)—CAS No. 6627-88-9	MW: 194.2 g/mol Bp: 169°C; Mp: 146°C P_{vap}: 0.001 mmHg (25°C) LogP: 2.23–2.45, pKa: 9.34 (Acidic), –4.6 (Basic) LD_{50} (g/kg): NR TD_{50} (g/kg/day): NR	Antioxidant	Flavoring for food, condiments, and sauces	*C. anisum-olens*	Ogata et al. 1997, Su and Huang 2011
HO 4-Allylphenol Chavicol (*p*-Allyphenol) (**17**)—CAS No. 501-92-8	MW: 134.2 g/mol Bp: 236°C; Mp: 15°C P_{vap}: 0.028 mmHg (25°C) LogP: 2.46–2.77, pKa: 9.5 (Acidic), –6 (Basic) LD_{50} (g/kg): NR TD_{50} (g/kg/day): NR	Antilarvicidal inhibitor, Anti inflammatory	Synthesis, Derivatization	*P. racemosa,* *O. gratissimum,* *P. racemosa* var. *racemosa*	Ohigashi et al. 1976, Williams et al. 2002
H_3CO, HO, OCH_3 *E*-2-Methoxy-4-(3-methoxyprop-1-enyl)phenol *E*-4-Methoxyisoeugenol (**18**)—CAS No. 63644-71-3	MW: 194.2 g/mol Bp: 302°C; Mp: 85°C P_{vap}: 0.00045 mmHg (25°C) LogP: 3.115, pKa: NR LD_{50} (g/kg): NR TD_{50} (g/kg/day): NR	It does not attribute to any specific biological property	Have not been reported any application/uses	*P. racemosa* var. *terebinthina,* *P. racemosa* var. *grisea*	García et al. 2002
H_3CO, H_3CO, H_3CO *E*-1,2,3-Trimethoxy-5-(prop-1-enyl)benzene *E*-Isoelemicin (**19**)—CAS No. 5273-85-8	MW: 208.2 g/mol Bp: 307°C; Mp: 66°C P_{vap}: 0.028 mmHg (25°C) LogP: 2.62–3.32, pKa: –4.4 (Basic) LD_{50} (g/kg): NR TD_{50} (g/kg/day): NR	It does not attribute to any specific biological property	Have not been reported any application/uses	*P. muellerianus*	Brusotti et al. 2012

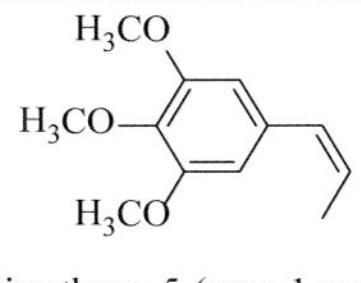 *Z*-1,2,3-Trimethoxy-5-(prop-1-enyl)benzene *Z*-Isoelemicin (**20**)—CAS No. 5273-84-7	MW: 208.2 g/mol Bp: 307°C; Mp: 66°C P_{vap}: 0.001 mmHg (25°C) LogP: 2.812, pKa: NR LD_{50} (g/kg): NR TD_{50} (g/kg/day): NR	It does not attribute to any specific biological property	Precursor synthesis Megaphone, 3,4,5-trimethoxy-benzaldehyde	*P. mikanianum*	Buechi and Chu 1981, Leal et al. 2005
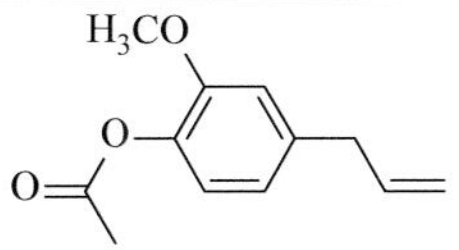 4-Allyl-2-methoxyphenyl acetate Acetyleugenol (**21**)—CAS No. 93-23-7	MW: 206.2 g/mol Bp: 281°C; Mp: 30°C P_{vap}: 3.19x10^{-3} mmHg (25°C) LogP: 2.52–3, pKa: –4.9 (Basic) LD_{50} (g/kg): Oral—Rat: 1.7 Intraperitoneal—Rat: 0.5 Dermal—Rabbit: > 5 TD_{50} (g/kg/day): NR	Antioxidant	Flavoring for food, Perfumery	*S. aromaticum*	Rovio et al. 1999, El-Mesallamy et al. 2012
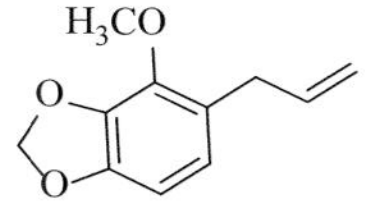 5-Allyl-4-methoxybenzo[*d*][1,3]dioxole Croweacin (**22**)—CAS No. 484-34-4	MW: 192.2 g/mol Bp: 304°C; Mp: 100°C LogP: 2.8, pKa: NR LD_{50} (g/kg): NR TD_{50} (g/kg/day): NR	Insecticidal	Synthesis	*C. saligna*, *C. angustifolia* var. *angustifolia*, *C. saligna* x *exalata*	Penfold et al. 1938, Brownell and Arthur 1951, Dinan et al. 2001
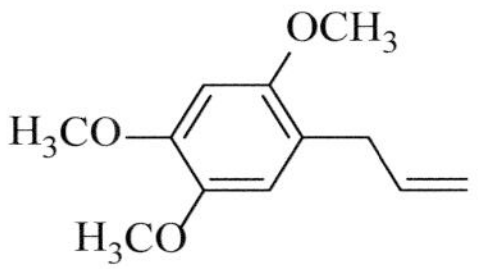 1-Allyl-2,4,5-trimethoxybenzene Euasarone (γ-Asarone) (**23**)—CAS No. 5353-15-1	MW: 208.2 g/mol Bp: 306°C; Mp: 81°C LogP: 3.1, pKa: –4.4 LD_{50} (g/kg): NR TD_{50} (g/kg/day): NR	Antibacterial	Have not been reported any application/uses	*A. calamus*, *A. hostmanniana*, *C. angustifolia* var. *angustifolia*, *S. libanotis*	Brophy et al. 1997, Ozturk and Ercisli 2006

Table 3.1 contd....

... *Table 3.1 contd.*

Structures	Physicochemical properties	Biological properties	Applications/Uses	Some plants containing P1(2)ENB	References
E-5-(Prop-1-en-1-yl)benzo[*d*][1,3]dioxole *E*-Isosafrole (**24**)—CAS No. 120-58-1	MW: 162.2 g/mol Bp: 253°C; Mp: 8°C P_{Vap}: 2.45x10^{-2} mmHg (25°C) LogP: 3.37, pKa: NR LD_{50} (g/kg) Oral—Mouse: 2.5; Rat: 1.3 Intraperitoneal—Mouse: 0.3 Intravenous—Rabbit: 0.3 Subcutaneous Mouse: 1.0 TD_{50} (g/kg/día): NR	Inhibitor and inducer of cytochrome P450 Hepatocarcinogenic	Synthesis	*I. religiosum*	Dickins et al. 1978

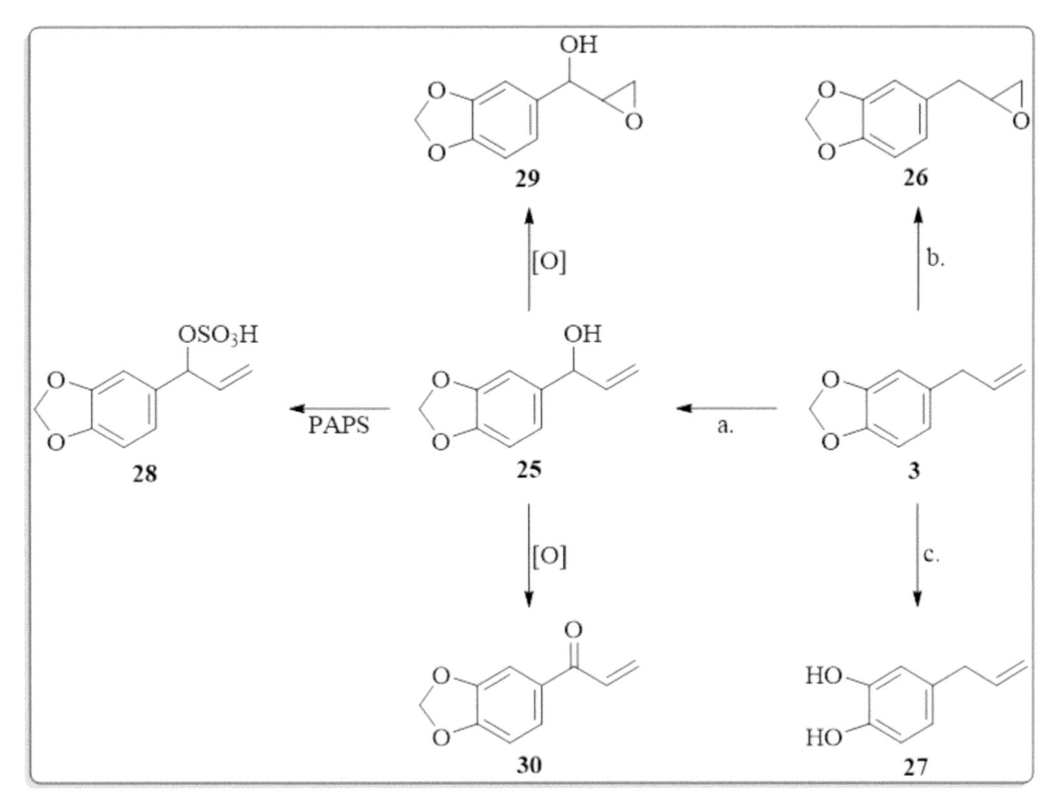

Scheme 3.1. Metabolic pathway of safrole (**3**) associated with their carcinogenicity.

intermediates which are covalently linked to nucleic acids (genotoxic mechanism). Hypothetically, all hydroxilated and carboxilated metabolites could be conjugate respectively, with glucuronic acid and glycine.

Other possible mechanism, unique for methylenedioxybenzene compounds (MDBC), is epigenetic nature (e.g., complexation with protein fragments from cytochromes); MDBC have showed two effects on mixed function oxidase enzymes (MFOE): (1) enzyme inhibition by interacting of the MDBC with cytochrome P-450; and, (2) increase in the enzyme activity by inducing the production of new cytochromes P-448 or P450. *These effects have been the basis for the development of new MDBC as synergists for pesticides* (Woo and Lai 1986, Gardner et al. 1997, NTP, NTP TR 491, EC 2001a, Ueng et al. 2005).

The metabolic pathway of estragole (**2**) involves the same oxidation reactions undergone by safrole; nevertheless, the demethylenation stage is substituted by O-demethylation stage. Main metabolites of estragole (**2**) related to carcinogenicity are shown in Scheme 3.2. Metabolite **34** (1'-sulfooxyestragole) is probably responsible for the DNA binding and carcinogenicity of estragole (Solheim and Scheline 1973, Bristol 2011, EC 2001b).

The relationship between the carcinogenicity and the phenylprop-1(2)-enoid structures is based on: (1) having at least one (or two as maximum number) methoxy substituent, which should be in *para*-position respect to prop-1(2)-enyl group. This *p*-methoxy group could stabilize (through electron delocalization) to the metabolically generated electrophilic intermediate and thus, would contribute to carcinogenicity. A larger number of methoxy substituents (e.g., **8**, **9**) along with O-demethylation of *p*-methoxy (e.g., **4**) inactive to the compounds probably due to steric hindrance and less stabilization by electron delocalization, respectively; (2) prop-1-enylbenzene congeners (e.g., **1**, **24**) have lower carcinogenic effects than their analogous prop-2-enylbenzene congeners (e.g., **2**, **3**), this is possibly

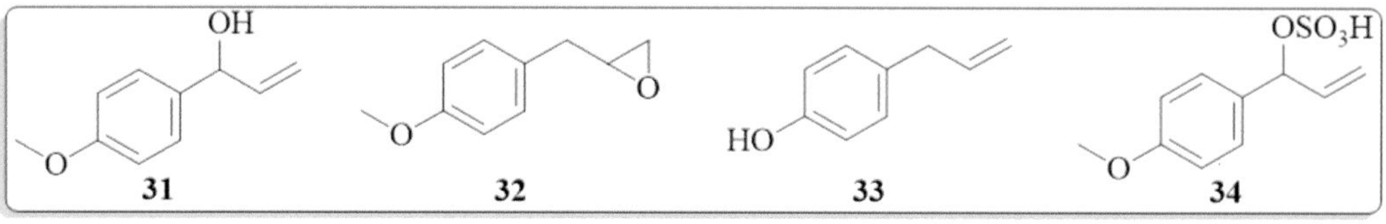

Scheme 3.2. Main metabolites of estragole (**2**) associated with their carcinogenicity.

due to the tendency to be oxidized to cinnamic derivatives; (3) the most of 1'-oxidized (e.g., **1'-OH**) prop-2-enylbenzene derivatives are more carcinogenic than their prop-2-enylbenzene precursors. 1'-Hydroxilation followed by esterification (with sulfate) is the main metabolic activation pathway of **3** and related compounds. It is worth noting some important biological effects related to phenylprop-1(2)-enoids: (1) the saturated congeners are less toxic than the corresponding unsaturated compounds; (2) one of the geometrical isomers (*E* and *Z*) of phenylprop-1-enoids is less toxic than the other (e.g., **11** <<< **12**); (3) the Central Nervous System (CNS) is mainly affected by the acute intoxication of these compounds, by showing the CNS depression or psychoactive effects (e.g., 'suspicion' of hallucinogenic effects of **8** and **13**); (4) according to subacute and chronic toxicity studies, the liver is the organ affected (Woo and Lai 1986, Kar 2003).

Based on Code of Federal Regulations Title 21 Vol. 3 of the U.S. FDA, essential oils, oleoresins, and extracts from *Artemisia dracunculus*, *Anethum graveolens*, *Anethum sowa*, *Cinnamomum cassia*, *Daucus carota*, *Foeniculum vulgare*, *Illicium verum*, *Laurus nobilis*, *Melaleuca leucadendron*, *Myristica fragrans*, *Ocimum basilicum*, *Petroselinum crispum*, *Pimenta dioica*, *Pimenta racemosa*, *Pimpinella anisum*, *Syzygium aromaticum*, and *Tagetes lucida*, along with *E*-anethole (**1**), estragole (**2**), eugenol (**4**), methyleugenol (**5**), isoeugenol (**6**), and eugenyl acetate (**21**), are substances generally recognized as safe for using as food (spices, natural seasonings and flavorings) for human consumption. Whereas, the use of safrole (**3**), EO and extracts from *Acorus calamus* (containing asarones) and *Sassafras albidum* (containing safrole) are prohibited for applications as flavoring agents of beverage and as food additives (21CFR189). For European Commission, essential oils, oleoresins, and extracts from *A. calamus*, *A. dracunculus*, *A. graveolens*, *A. sowa*, *Cinnamomum* spp., *D. carota*, *F. vulgare*, *I. verum*, *L. nobilis*, *Melaleuca* spp., *M. fragrans*, *O. basilicum*, *P. crispum*, *P. dioica*, *P. racemosa*, *P. anisum*, *S. aromaticum*, and *Tagetes* spp., along with *E*-anethole (**1**), estragole (**2**), eugenol (**4**), methyleugenol (**5**), isoeugenol (**6**), methylisoeugenol (**7**), and eugenyl acetate (**21**), are substances approved for using in cosmetic products (perfuming). Even though the use of safrole (**3**) is 'restricted' for applications, EC has permitted to incorporate the natural oils/extracts (e.g., *S. albidum*) containing safrole as long as the amount used in the cosmetic and oral hygiene/dental products do not exceed 100 μg/mL and 50 μg/mL, respectively (Committee of experts on cosmetic products 2008).

Finally, in accord with EC, the European legislation establishes the possible use of prop-1(2)-enylbenzenes (e.g., estragole (**2**), safrole (**3**), methyleugenol (**5**), and *Z*-asarone (**12**)) as food additives, if the maximum permissible levels are fulfilled for each one of compounds in the specific use as flavoring; namely, *E*-anethole (**1**) 0–2 mg/kg; estragole (**2**) 10–50 mg/kg; safrole (**3**) 1–15 mg/kg; eugenol (**4**) 0.5–221 mg/kg; methyleugenol (**5**) 1–60 mg/kg; isoeugenol (**6**) 30–40 mg/kg; myristicin (**8**) 1–2 mg/kg; *Z*-asarone (**12**) 2–115 mg/kg; and, eugenyl acetate (**21**) 0.5–10 mg/kg (Burdock 2010).

Traditional and Therapeutic Uses of Plants Containing Rich-phenylprop-1(2)-enoid Essential Oils

The majority of plants producing essential oils with high content of phenylprop-1(2)-enoid compounds have some similar ethnobotanical and therapeutic uses, which in a general way will be described below, for the most representative plants.

Aerial parts (seeds/fruits/leaves) or roots of the species (e.g., *Anethum* spp., *Apium* spp., *Carum* spp., *Crithmum* spp., *Daucus* spp., *Echinophora* spp., *Foeniculum* spp., *Petroselinum* spp., *Pimpinella* spp., *Illicium* spp.) belonging to Apiaceae and Illiciaceae families are mainly used to relieve stomach discomforts and digestive disorders. Similarly, the plants are galactogogues, emenagogues, carminatives, sedatives, expectorants, diuretics, anti-inflammatories, antibacterials and are beneficial to treat cough and vomiting. These species are also utilized as condiments or dressings and food additives (Ruberto et al. 2000, Tomar and Dureja 2001, Masoudi et al. 2005, Durmaz et al. 2006, 2015, Bullitta et al. 2007, Imamu et al. 2007, Marongiu et al. 2007, Rossi et al. 2007, Tavares et al. 2008, Jabrane et al. 2009, Maxia et al. 2009, 2012, Jana and Shekhawat 2010, Marzouki et al. 2010, Wang et al. 2011, Bendiabdellah et al. 2014, El Kolli et al. 2016).

In the ethnobotanical practices with *Ocimum* spp., *Agastache* spp., together with *Artermisia* spp., *Tagetes* spp., and other species of Lamiaceae and Asteraceae, respectively, the aerial parts (e.g., leaves) of plants have been used to treat eating and intestinal/digestive disorders. In the same way, the plants are febrifuges, carminatives, emenagogues, sedatives, digestives, diuretics, diaphoretics, antimicrobials, anti-inflammatories, anthelmintics and are helpful to mitigate headache, coughs, diarrhoea, constipation, etc. The plants have also been widely incorporated as flavor/food ingredients, used as spices (seasoning) and their essential oils, as additives for the food, pharmaceutical and cosmetic industries (De Feo et al. 1998, Wright 2003, Boussaada et al. 2008, Santillán-Ramírez et al. 2008, Yamani et al. 2014).

Aerial parts (leaves/fruits) of *Piper* spp. are frequently used for its astringent, digestive, sedative, carminative, diuretic, analgesic, antibacterial, and antiparasitic properties. Besides, these vegetable species are used against diarrhoea, dysentery, hemorrhoids and gonorrhea. For its spicy taste, these *Piper* spp. have been exploited as spices (Dasgupta and De 2004, Guerrini et al. 2009, Noriega Rivera et al. 2015).

The leaves of *Clausena* spp. and *Zieria* spp. (Rutaceae), in natural medicine, have commonly been used as effective remedies against febril conditions, pains (body, headache abdominal), some heart complaints (heart failure, hypertension), inflammatory conditions (arthritis, rheumatism), impotence/infertility issues, and have showed anticonvulsive, hypoglycaemic and antiparasitic effects. The fruits are exploited as food (Ojewole 2002, Govindarajan 2010).

Aerial parts (leaves/seeds/buds) of the species (*Pimenta* spp., *Syzygium* spp.) belonging to Myrtaceae are used in herbal medicine to treat respiratory diseases (e.g., pneumonia, influenza) and as a pain reliever (toothache). These plants have stimulant, febrifuge, anesthesic, carminative, antifungal, antibacterial, antioxidant and antidiarrhoeic actions. They have also been incorporated as flavor/food ingredients, used as spices (seasoning) for its preservative properties, and as additives for the food and cosmetic industries (Macía 1998, García et al. 2004, Sáenz et al. 2004).

Aerial parts (leaves/buds/stem) of *Cinnamomum* spp., *Laurus* spp., *Ocotea* spp., *Sassafras* spp. (Lauraceae), have been useful in ethnomedicine as cure for pulmonary (flu) and intestinal/digestive (stomaches, diarrhoea) disorders, and as appetite stimulants. Plants have presented antimicrobial, antimalarial, antitumoral, anti-inflammatory, antioxidant, anticholinesterase and antituberculous activities; and they are helpful as spices (seasoning) and as flavor (Mastura et al. 1999, Jantan et al. 2005, Verspohl et al. 2005, Ooi et al. 2006, Salleh et al. 2016b).

The barks of *Croton* spp. (Euphorbiaceae) and fruits/seeds of *Myristica* spp. (Myristicaceae) are widely used in ethnobotany as therapies for treatment of digestive/intestinal disorders (biles, diarrhoeas, colics, flatulence), diabetes, gastric ulcers, as pain relievers (neuralgia, rheumatism), and have promising effects as anti-inflammatories, analgesics, antiplatelets, antiulcerogenics, hypolipidaemisc, antitumorals and antimicrobials. *Myristica* spp. are used as spices (condiments and dressing) (Stäuble 1986, Suárez et al. 2003, Aiyeloja and Bello 2006, Chung et al. 2006).

Finally, in the ethnomedical practices with the *Acorus calamus* (Acoraceae) rhizomes, these are used as expectorant, laxant, diuretic, stomachic, sedative, analgesic, to relieve chest pain, bronchitis, and diarrhoea. With regard to the therapeutic uses these plants are antispasmodic, antibacterial, useful for the treatment of dysentery, gout and rheumatism (Motley 1994).

Antimicrobial Activities of the Essential Oils, Containing Phenylprop-1(2)-enoid Compounds

Plants containing these types of essential oils comprise grasses, shrubs and trees, and, its different parts such as flowers, seeds, buds, fruits, leaves, barks, stems, roots and rhizomes. General information (origin, used parts, yield, and microbial activities determined) related to some medicinal plants producing essential oils with high content of P1(2)ENB compounds is shown in Table 3.2. According to this table, the EO were mainly evaluated against 33 microorganisms (bacteria and fungus), and the MIC (μg/mL), Minimum Bactericidal Concentration (MBC, μg/mL), and/or diameters of inhibition zones (øIZ, mm) values were also determined.

Thus, for the case of Gram-positive bacteria, the values were: *Bacillus cereus*—MIC 0.6–2091 μg/mL, MBC 1.2–2091 μg/mL, øIZ 4–24 mm; *B. subtilis*—MIC 0.3–7500 μg/mL, MBC 2.1–1500 μg/mL, øIZ 4–44 mm; *Enterococcus faecalis*—MIC 1.2–100000 μg/mL, MBC 120–> 10000 μg/mL, øIZ 6–30 mm; *Listeria monocytogenes*—MIC 0.6–40 μg/mL, MBC 2.5–6000 μg/mL, øIZ 6–15 mm; *M. luteus*—MIC 2.1–5240 μg/mL, MBC 33–> 800 μg/mL, øIZ 4–16 mm; *Paenibacillus larvae*—MIC 78–1890 μg/mL, MBC 162–2111 μg/mL, øIZ 1.4–20 mm; *S. aureus*—MIC 0.5–100000 μg/mL, MBC 1.2–10000 μg/mL, øIZ 1–46 mm; *S. epidermidis*—MIC 0.2–15000 μg/mL, MBC 100–>800 μg/mL, øIZ 7–25 mm.

Likewise, Gram-negative bacteria showed the following values: *Acinetobacter* sp.—MIC 7–16000 μg/mL, øIZ 10 mm; *E. aerogenes*—MIC 0.6–500 μg/mL, MBC 2.2 μg/mL, øIZ 6–30 mm; *E. coli*—MIC 0.06–100000 μg/mL, MBC 2.1–5000 μg/mL, øIZ 5–95 mm; *Helicobacter pylori*—MIC 12 μg/mL; *Kliebsella pneumoniae*—MIC 0.2–>500 μg/mL, MBC 2.1–5000 μg/mL, øIZ 3–30 mm; *Proteus vulgaris*—MIC 0.6–8000 μg/mL, MBC >100 μg/mL, øIZ 6–24 mm; *P. aeruginosa*—MIC 0.3–100000 μg/mL, MBC 100–8364 μg/mL, øIZ 1–28 mm; *Salmonella* sp.—MIC 6–72 μg/mL, øIZ 1–27 mm; *Serratia marcescens*—MIC 0.3–>500 μg/mL, MBC 5000 μg/mL, øIZ 3–30 mm; *S. typhimurium*—MIC 0.2–>800 μg/mL, MBC 0.7–>5000 μg/mL, øIZ 6–18 mm.

Values for the filamentous fungus were: *A. flavus*—MIC 0.3–1000 μg/mL, MBC 0.5–2600 μg/mL, øIZ 4–84 mm; *A. fumigatus*—MIC 0.6–3750 μg/mL, MBC > 20 μg/mL, øIZ 2–30 mm; *A. niger*—MIC 0.04–500 μg/mL, MBC 15–12000 μg/mL, øIZ 4–100 mm; *A. parasiticus*—MIC 1–2 μg/mL, øIZ 5–80 mm; *A. terreus*—MIC 1–25 μg/mL, MBC 30 μg/mL.

Finally, the values for the yeast-like fungus were: *C. albicans*—MIC 0.3–100000 μg/mL, MBC 0.3–5000 μg/mL, øIZ 1–60 mm; *C. glabrata*—MIC 0.6–1250 μg/mL, MBC 625–2500 μg/mL, øIZ 26–30 mm; *C. guilliermondii*—MIC 0.3–2.5 μg/mL, MBC 0.3–2500 μg/mL; *C. krusei*—MIC 0.6–>20 μg/mL, MBC 0.6–1250 μg/mL, øIZ 18–22 mm; *C. neoformans*—MIC 0.3–10 μg/mL, MBC 0.3–1250 μg/mL, øIZ 1.4–22 mm; *C. parapsilosis*—MIC 1.2–2500 μg/mL, MBC 2.5–2500 μg/mL, øIZ 19 mm; *C. tropicalis*—MIC 0.6–2500 μg/mL, MBC 0.6–5000 μg/mL, øIZ 2–28 mm; *Microsporum canis*—MIC 0.08–> 5000 μg/mL, MBC 0.3–2500 μg/mL, øIZ 10–22 mm; *Trichophyton mentagrophytes*—MIC 0.08–250 μg/mL, MBC 0.3–1250 μg/mL; *T. rubrum*—MIC 0.08–150 μg/mL, MBC 0.3–1250 μg/mL, øIZ 20 mm.

From the 33 microorganisms mentioned above, those strains responsible for the infections (with highest recurrence) affecting humans were selected for discussion; scilicet, *L. monocytogenes*, *S. aureus (and MRSA)*, *E. coli*, *Acinetobacter* sp., *H. pylori*, *K. pneumoniae*, *P. aeruginosa*, *Salmonella* sp., *S. typhimurium*, *A. fumigatus*, *C. albicans* and *C. neoformans*.

Thus, EO from *A. dracunculus* (*E*-anethole chemotype), *D. carota* subsp. *hispanicus* (myristicin chemotype), *E. platyloba* (containing asarone, anethole and eugenol), *L. citriodora* (high content of *E*-anethole) were able to effectively inhibit (MIC 0.6 μg/mL–40 μg/mL) to *L. monocytogenes*. *L. citriodora EO achieved the lowest MIC value (0.6 μg/mL)* (Fayyaz et al. 2015, Abdollahnejad et al. 2016, Moosavi-Nasab et al. 2016).

Plant EO which evidenced a significative effects (MIC 0.5 μg/mL–500 μg/mL) on *S. aureus* were *A. dracunculus* (*E*-anethole chemotype), *E. supraxillaris* (high content of eugenol/methyleugenol),

Table 3.2. General information (origin, used parts, yield and microbial activities determined) on some medicinal plants containing essential oils with prop-1(2)-enylbenzene derivatives.

Species	Origin	Used parts	Yield,%	Prop-1(2)-enylbenzene found (relative amount, %)	Microbial activities	References
Acorus calamus	India	Rhizomes	NR	*Z*-Asarone (81.1–92.4)	Antifungal activity	Bisht et al. 2011
			3.5	*Z*-Asarone (74.8)	Antibacterial activity	Joshi et al. 2012
			0.5	*Z*-Asarone (72.7)	Antifungal activity	Shukla et al. 2013
	Korea		NR	Methylisoeugenol (41.5)	Antimicrobial activity	Kim et al. 2011
	Lithuania	Leaves	0.2–0.6	*Z*-Asarone (15.7–29.0) and *E*-asarone (1.1–7.7)	Antimicrobial effect	Radušiene et al. 2007
Agastache mexicana subsp. *xolocotziana*	Mexico	Aerial parts	1.0	Methyleugenol (36.4) and estragole (27.9)	Antifungal activity	Juárez et al. 2015
A. rugosa	Korea		NR	Estragole (49.4)	Antifungal activity	Shin and Kang 2003, Shin 2004
		Leaves	0.2–0.8	Estragole (77.5–94.9)	Antimicrobial testing on four microorganism	Kim 2008
Ammi visnaga	Algeria	Aerial part	1.3	Croweacin (12.2)	Antibacterial activities	Khalfallah et al. 2011
Amomum brevilabrum	Borneo	Rhizomes	0.3	Elemicin (35.6) and methyl isoeugenol (19.2)	Antibacterial activities	Vairappan et al. 2012
A. punicea	Thailand		NR	Estragole (95.7)	Antimicrobial activity	Tadtong et al. 2009
Anethum graveolens	Algeria	Seeds	2.1	Apiole (32.8)	Antifungal activity	Khaldi et al. 2015
	China		3.5	Apiole (16.8)		Tian et al. 2011, 2012, Chen et al. 2014
	Estonia	Aerial plants	0.6	Myristicin (1.7–28.2)	Antibacterial activity	Vokk et al. 2011
	India	Seeds	2.6 3.8	Dillapiole (14.4) Dillapiole (43.2) and *E*-anethole (11.0)	Antimicrobial potential	Singh et al. 2005
			1.3	Apiole (12.8)	Antimicrobial activity	Sintim et al. 2015
	India (commercial sample)	NR	NR	Apiole (30.8)	Antibacterial activity	Ansari et al. 2016

Table 3.2 contd. ...

... *Table 3.2 contd.*

Species	Origin	Used parts	Yield,%	Prop-1(2)-enylbenzene found (relative amount, %)	Microbial activities	References
Anethum graveolens	Iran	Aerial parts	2.5	Eugenol (49.6)	Anti-quorum sensing effect	Makhfian et al. 2015
		Seeds	NR	Dillapiole (16.8)	Antimicrobial activity	Mahmoodi et al. 2012, Roomiani et al. 2013
			0.7–1.8	Dillapiole (5.7–31.9)	Antibacterial activity	Salehiarjmand et al. 2014
	Thailand		1.5	Dillapiole (19.6)		Peerakam et al. 2014
A. sowa	India (commercial samples)		NR	Dillapiole (36.6)	Inhibitory activity on bacteria and fungi	Chao et al. 2000
Apium nodiflorum	Italy	NR	NR	Myristicin (18.5)	Antibacterial activity	Menghini et al. 2010
		Aerial parts	1.1	Dillapiole (57.5–70.8)		Maxia et al. 2012
	Portugal		1.0	Myristicin (29.1) and dillapiole (22.5)		
Artemisia dracunculus	Austria		1.5	Estragole (75.0–84.1)	Antifungal activity	Obistioiu et al. 2014
	Canada		0.4	Methyleugenol (35.8) and estragole (16.2)	Antimicrobial activity	Lopes-Lutz et al. 2008
	EE.UU.		NR	Methyleugenol (25.1)	Antifungal activity	Meepagala et al. 2002
	France (commercial samples)	Whole plant	NR	Estragole (77.6)	Inhibitory activity on bacteria and fungi	Chao et al. 2000
	Iran	Aerial parts	0.01 0.04	*E*-Anethole (49.0) *E*-Anethole (53.7)	Antimicrobial activity	Abdollahnejad et al. 2016
	Italy		0.5	*E*-Anethole (53.4)		Curini et al. 2006
	Turkey		1.0	*Z*-Anethole (81.0)	Antifungal and antibacterial activities	Kordali et al. 2005
Asarum heterotropoides subsp. *mandshuricum*	China	Whole plant	1.2	Methyleugenol (59.4)	Antimicrobial activity	Dan et al. 2010
Athamanta turbith subsp. *hungarica*	Serbia	Rhizomes	0.6	Myristicin (54.2) and apiole (32.6)		Tomić et al. 2009
A. turbith subsp. *haynaldii*			0.3	Apiole (48.5) and myristicin (22.3)		
Beilschmiedia pulverulenta	Malaysia	Aerial parts	2.8	Eugenol (45.3)		Salleh et al. 2016b
Bunium cylindricum	Iran		0.09	Myristicin (43.1)	Antibacterial activity	Masoudi et al. 2005
Carum montanum	France	Aerial parts Roots	1.1 0.7	Nothoapiole (62.8)	Antibacterial and antifungal activities	Laouer et al. 2009

Chaerophyllum macropodum	Iran	Flower Leaves Stem	NR	Myristicin (42.5) Myristicin (15.7) Myristicin (22.4)	Antibacterial activity	Shafaghat 2009
Cinnamomum cassia	Brazil (commercial sample)	NR	NR	Eugenol (72.1)	Antimicrobial activity	Andrade et al. 2014
C. rhyncophyllum	Malaysia	Bark	1.5	Safrole (41.5)	Antifungal activity	Jantan et al. 2008
C. subavenium	Taiwan	Leaves	0.7–0.8	Methyleugenol (75.9–78.7)	Antimicrobial activity	Ho et al. 2008
C. tamala	India	Leaves: Fresh Dried	0.5	Eugenol (57.0) Eugenol (56.9–60.4)	Antibacterial potential	Padalia et al. 2012
		Leaves	2.2	Eugenol (66.1)	Antimicrobial activity	Kapoor et al. 2009
C. zeylanicum	Brazil		NR	Eugenol (75.5)	Antifungal activity	Trajano et al. 2012
	Italy (commercial samples)	NR	NR	Eugenol (64.8)	Antibacterial activity	Fratini et al. 2014
	Mauritus	Leaves	0.5	Eugenol (58.1)	Antimicrobial, antibiotic potentiating activity	Aumeeruddy-Elalfi et al. 2015
	Malaysia		5.5	Eugenol (90.2)	Antifungal activity	Jantan et al. 2008
	NR (commercial sample)		NR	Eugenol (60.4) and eugenyl acetate (18.3)	Antifungal activity	Roselló et al. 2015
	Sri Lanka		NR	Eugenol (74.9)	Antibacterial and antifungal activities	Höferl et al. 2009, Schmidt et al. 2007
	Taiwan		4.2	Eugenol (93.5)	Antipathogenic activity	Cheng et al. 2011
Clausena anisata	Ghana		NR	Estragole (98.8) *E*-Anethole (95.5) *E*-Foeniculin (97.0) *E*-Anethole (57.1–67.5) and *E*-foeniculin (29.2–38.2)	Antimicrobial screening	Osei-Safo et al. 2010
C. anisum-olens	China	Nutlets		Myristicin (47.1)), elemicin (8.2) and 4-allylsyringol (7.2)	Antimicrobial activity	Su and Huang 2011
C. austroindica	India	Leaves		Elemicin (66.6) and myristicin (19.1)	Antibacterial activities	Gopan et al. 2004
C. indica	Vietnam	Branches and leaves		Myristicin (35.3)	Antimicrobial activity	Diep et al. 2009

Table 3.2 contd. ...

... Table 3.2 contd.

Species	Origin	Used parts	Yield,%	Prop-1(2)-enylbenzene found (relative amount,%)	Microbial activities	References
C. pentaphylla	India	Leaves	0.8	Methyleugenol (38.1) and safrole (6.7)	Fungitoxicant activity	Pandey et al. 2012, 2013
C. suffruticosa	Bangladesh		NR	Estragole (58.2) and *E*-anethole (33.2)	Antimicrobial activity	Rahman et al. 2012
Crithmum maritimum	Italy			Dillapiole (32.9–64.2)	Antifungal activity	Marongiu et al. 2007
		Leaves and stems	0.5	Dillapiole (9.5)	Antimicrobial activity	Flamini et al. 1999
Croton hieronymi	Argentina	Roots	0.07	Euasarone (25.7) and *E*-asarone (11.4)		de Heluani et al. 2005
C. malambo	Venezuela	Leaves	NR	Methyleugenol (94.2)	Antibacterial activities	Suárez et al. 2008
C. nepetaefolius	Brazil			Methyleugenol (15.7)	Antifungal activity	Fontenelle et al. 2008
C. zehtneri		Bark Leaves		Estragole (72.6) and *E*-anethole (23.3) Estragole (93.6)	Antimicrobial activity	Donati et al. 2015
		Leaves		Estragole (72.9) and *E*-anethole (14.3)	Antifungal activity	Fontenelle et al. 2008
Daucus carota	Corsica	Aerial parts		Elemicin (11.4–16.3) and *E*-methyl-isoeugenol (21.8–33.0)	Antimicrobial activity	Rossi et al. 2007
D. carota sativa	Uzbekistan	Seeds	2.2	*E*-Asarone (8.8)		Imamu et al. 2007
D. carota subsp. *carota*	Italy	Umbel with seeds	1.0	*E*-methylisoeugenol (10.0)	Antifungal activity	Maxia et al. 2009
	Tunisia	Flowering and Umbels with seeds	NR	Elemicin (31.5–35.3)	Antimicrobial activity	Marzouki et al. 2010
D. carota subsp. *halophilus*	Portugal	Ripe umbels	0.6–1.0	Elemicin (26.0–31.0)	Antifungal activity	Tavares et al. 2008
D. carota subsp. *hispanicus*	Algeria	Roots Aerial parts Stems Leaves Flowers	0.4 2.2 0.1 1.6 2.1	Apiole (80.3) and myristicin (16.6) Myristicin (73.2) Myristicin (66.9) Myristicin (80.2) Myristicin (83.8)	Fungitoxic activity	Bendiabdellah et al. 2014
		Roots Aerial parts	0.1–1.6 0.4–3.1	Myristicin (15.6–83.4) and apiole (13.2–81.3) Myristicin (62.9–86.2)		

D. carota subsp. *maritimus*	Tunisia	Roots	0.07	Dillapiole (46.6) and myristicin (29.7)	Antibacterial activity	Jabrane et al. 2009
D. gracilis	Algeria	Aerial parts	0.7	Elemicin (35.3)	Antimicrobial activity	El Kolli et al. 2016
Echinophora platyloba	Iran		NR	Asarone (10.2), anethole (7.4) and eugenol (6.7)		Fayyaz et al. 2015
E. tenuifolia	Turkey	Leaves	1.6	*m*-Eugenol (29.5)		Aridoğan et al. 2002
Enterolobium contortisiliquum	Egypt	Seeds	~ 0.3	Estragole (20.7)		Shahat et al. 2008
Eugenia supraxillaris		Fruits	2.7	Eugenol (35.5) and methyleugenol (32.8)		Aboutabl et al. 2011
Foeniculum vulgare	Argentina		0.5–1.2	*E*-Anethole (92.7)		Gende et al. 2009
	Brazil (commercial sample)	NR	NR	*E*-Anethole (95.7)		Andrade et al. 2014
	China	Whole plant	NR	*E*-Anethole (88.9)		Qiu et al. 2012
		Seeds	1.7	*E*-Anethole (68.5) and estragole (10.4)	Antibacterial activity	Diao et al. 2014
	Egypt		2.0	*E*-Anethole (56.4)	Antimicrobial activity	Roby et al. 2013
	India		NR	*E*-Anethole (50.4) and estragole (22.4)	Antibacterial activity	Bisht et al. 2014
	Iran		2.6	*E*-Anethole (71.2)	Antifungal activity	Sharifi et al. 2008
			3.5	*E*-Anethole (75.2)		Mohammadi et al. 2014
		Fruits	NR	*E*-Anethole (64.4)		Abdolahi et al. 2010
	Italy (commercial samples)	NR	NR	*E*-Anethole (54.5)	Antibacterial activity	Fratini et al. 2014
	Pakistan	Fruits: Immature intermediate mature	2.8 3.2 3.5	*E*-Anethole (65.2) *E*-Anethole (69.7) *E*-Anethole (72.6)	Antimicrobial activity	Anwar et al. 2009
	Spain (commercial sample)	Plants	NR	*E*-Anethole (23.5)		Marín et al. 2016

Table 3.2 contd. ...

... Table 3.2 contd.

Species	Origin	Used parts	Yield,%	Prop-1(2)-enylbenzene found (relative amount, %)	Microbial activities	References
Foeniculum vulgare	Turkey	Seeds	NR	*E*-Anethole (82.8)		Soylu et al. 2006, Türkölmez and Soylu 2014
			4.5	*E*-Anethole (76.4)		Aridoğan et al. 2002
		Seeds (commercial sample)	NR	*E*-Anethole (80.4)	Antibacterial effect	Evrendilek 2015
	Yugoslavia	Seeds	2.8–3.4	*E*-Anethole (72.3–74.2)	Antifungal activity	Mimica-Dukić et al. 2003
F. vulgare var. *azoricum*	Iran		NR	*E*-Anethole (61.1)	Antimicrobial activity	Chang et al. 2016
F. vulgare var. *dulce*	Bulgaria	Fruits	1.4	*E*-Anethole (72.3)		Damianova et al. 2004, Damyanova and Stoyanova 2007
	Iran	Seeds	NR	*E*-Anethole (46.3)		Chang et al. 2016
F. vulgare subsp. *piperitum*	Turkey	Flowers Unripe fruits Ripe fruits	2.1 6.0 4.4	Estragole (53.1) Estragole (56.1) Estragole (61.1)	Antifungal activity	Özcan et al. 2006b
F. vulgare var. *vulgare*	Iran	Seeds	NR	Estragole (57.9)	Antimicrobial activity	Chang et al. 2016
	Italy	Aerial parts	NR	Estragole (34.9–42.6) and *E*-anethole (24.6–43.4)	Antifungal activity	Piras et al. 2014
F. vulgare var. *vulgare* var. *azoricum*		Leaves	0.02–0.04 0.03–0.04	*E*-Anethole (59.8–66.0) *E*-Anethole (66.3–90.4)	Antibacterial activity	Senatore et al. 2013
Heracleum anisactis	Iran	Aerial parts Roots	0.3 1.3	Myristicin (93.5) Myristicin (95.2)		Torbati et al. 2013, 2014
H. ternatum	Belgrade	Leaves	0.1	*Z*-isoelemicin (35.1), elemicin (12.6) and methyleugenol (10.7)	Antimicrobial activity	Ušjak et al. 2016
H. transcaucasicum	Iran	Aerial parts Roots	0.2 1.7	Myristicin (70.1) Myristicin (96.9)	Antibacterial activity	Torbati et al. 2013, 2014

Illicium verum	Brazil (commercial sample)	NR	NR	*E*-Anethole (90.4)	Antimicrobial potential	Freire et al. 2011
	China	Fruits	7.5	*E*-Anethole (89.5)	Antifungal activity	Huang et al. 2010
	Egypt (commercial sample)	NR	NR	*E*-Anethole (82.7)	Antifungal activity	Aly et al. 2016
	India	Fruits	2.6	*E*-Anethole (94.4)	Antimicrobial potential	Singh et al. 2006a
	India (commercial sample)	NR	NR	*E*-Anethole (89.5)	Antibacterial activity	Ansari et al. 2016
Laurelia sempervirens	Chile	Bark	1.0	Safrole (65.0) and isosafrole (11.9)	Antimicrobial activity	Montenegro et al. 2012
Laurus nobilis	Algeria	Leaves	0.5–1.3	Methyleugenol (10.6–11.0)		Marzouki et al. 2009a
	India (commercial sample)	NR	NR	Eugenol (50.4)	Antipathogenic activity	Chudasama and Thaker 2012, 2014
	Mauritus	Leaves	0.5	Eugenol (13.3) and methyleugenol (12.9)	Antimicrobial, antibiotic potentiating activity	Aumeeruddy-Elalfi et al. 2015
	Tunisia		1.6–2.1	Methyleugenol (10.2–10.6)	Antimicrobial activity	Marzouki et al. 2009a
		Bud Flowers Leaves Stem	2.5 0.3–0.8 0.6–1.5 0.2–0.5	Methyleugenol (1.0–16.8) Methyleugenol (3.9–14.3) Methyleugenol (13.1–33.6) Methyleugenol (6.6–17.8)	Antibacterial activity	Marzouki et al. 2009b
Limnophila rugosa	Vietnam	Aerial parts	0.14–0.27	Estragole (64.2–71.0) and *E*-anethole (25.0–30.4)	Antimicrobial activity	Linh and Thac 2011
Lippia citriodora	Iran	Leaves	NR	*E*-Anethole (68.2)	Antibacterial activity	Moosavi-Nasab et al. 2016
Melaleuca bracteata var. *revolution gold*	South Africa		0.2–0.8	Methyleugenol (82.5–84.6)		Oyedeji et al. 2014
M. bracteata var. *revolution green*			0.1–0.2	Methyleugenol (89.7) Methyleugenol (72.5) and isoeugenol (8.3)		
M. ericifolia	Egypt		0.4–0.8	Methyleugenol (96.8)	Antimicrobial activity	Farag et al. 2004
Myristica fragrans	Brazil (commercial sample)	NR	NR	Myristicin (17.6)		Andrade et al. 2014
Nephrolepis cordifolia	Egypt	Aerial parts	0.6	Eugenol (7.2) and anethole (4.6)		El-Tantawy et al. 2016

Table 3.2 contd. ...

... *Table 3.2 contd.*

Species	Origin	Used parts	Yield,%	Prop-1(2)-enylbenzene found (relative amount, %)	Microbial activities	References
Ocimum basilicum	China	Leaves	NR	Estragole (35.1)	Effects on mycelial growth and morphogenesis of some phytopathogens	Lu et al. 2013
	Germany (commercial sample)	NR		Estragole (86.1)	Antimicrobial activity	Baratta et al. 1998
	India	Leaves		Methyleugenol (42.3)	Preservative against fungal and aflatoxin contamination	Kumar et al. 2011
	India (commercial sample)	NR		*iso*-Eugenol (47.2)	Antipathogenic activity	Chudasama and Thaker 2012
	Iran	Aerial parts	0.4	Estragole (62.5)	Antibacterial activity	Yavari et al. 2011
			NR	Estragole (24.7)		Azizkhani and Parsaeimehr 2015
	Italy	NR		Estragole (91.1)	Antimicrobial activity	Scazzocchio et al. 2016
		Aerial parts	0.2	Estragole (85.7)	Antibacterial activity	Senatore et al. 2003
	United Kingdom	NR	NR	Estragole (76.1) Eugenol (12.4)	Antifungal activity	Oxenham et al. 2005
O. basilicum var. *album*	France			Estragole (83.5)	Antimicrobial activity againts *Helicobacter pylori*	Ohno et al. 2003
O. basilicum var. *difforme*	Croatia	Leaves and flowers		Estragole (47.5)	Antifungal activity	Carović-Stanko et al. 2013
O. basilicum var. *purpurascens*				Estragole (94.6)		
O. canum	Cameroon	Aerial parts	0.3	Eugenol (17.7)	Antimicrobial activity	Vyry Wouatsa et al. 2014
O. ciliatum	Iran	Aerial parts (flowering stage)	1.3	Estragole (87.6)	Antibacterial activity	Moghaddam et al. 2014
O. forskolei	Oman	Flowers Leaves	0.6 0.5	Estragole (59.4–65.2) Estragole (42.1–78.4)	Antimicrobial activity	Fatope et al. 2008
O. gratissimum	Brazil	Leaves	0.2	Eugenol (67.0)	Antibacterial activity	Nakamura et al. 1999
			NR	Eugenol (91.5)		Sutili et al. 2015
			0.4	Eugenol (65.3)		Aguiar et al. 2015
O. micranthum			0.6	Eugenol (64.1)	Antifungal activity	Vieira et al. 2014

O. sanctum	India	Aerial parts	1.8	Eugenol (46.7)	Antagonistc activity on grown and mycotoxin production	Kalagatur et al. 2015
	Togo		NR	Methyleugenol (74.4)	Antimicrobial activity	Chaumont et al. 2001
O. selloi	Brazil	Leaves	0.2	Anethole (52.2)	Antifungal activity	Vieira et al. 2014
O. tenuiflorum	Australia	Leaves and flowers	0.6	Eugenol (13.8)	Antimicrobial activity	Yamani et al. 2016
	Croatia		NR	Estragole (21.6)	Antifungal activity	Carović-Stanko et al. 2013
	India (commercial sample)	NR		Eugenol (75.3)	Antibacterial activity	Ansari et al. 2016
O. trichodon	Rwanda	Plants	0.4–0.8	Eugenol (44.0–81.0)	Antimicrobial activity	Ntezurubanza et al. 1986, Janssen et al. 1998
O. urticifolium			NR	*E*-Methylisoeugenol (35.6–49.7) Methyleugenol (73.6–86.6) Eugenol (37.6–71.1)		Janssen et al. 1998
Ocotea puchury-major	Brazil	Leaves	1.5	Safrole (39.4)		Leporatti et al. 2014
Pelargonium odoratissimum	United Kingdom		NR	Methyleugenol (31.2–79.8)		Lis-Balchin and Roth 2000
P. x *fragrans*				Methyleugenol (5.4–13.3)		
Peperomia inaequalifolia	Ecuador		0.2	Safrole (32.1), myristicin (13.3) and elemicin (10.1)		Noriega Rivera et al. 2015
Petroselinum crispum	Algeria	Aerial plants	0.08–0.1	Dillapiole (47.5)		Nawel et al. 2014
	Estonia		0.2–0.3	Myristicin (30.7–42.6)	Antibacterial activity	Vokk et al. 2011
	Peru	Leaves	0.02	Apiole (8.4) and myristicin (7.4)		Vivanco et al. 2012
	Spain (commercial sample)	Plants	NR	Myristicin (36.2) and apiole (21.0)	Antimicrobial activity	Marín et al. 2016
Phyllanthus muellerianus	Cameroon	Bark	0.06	*E*-Isoelemicin (36.4)	Antimicrobial activity	Brusotti et al. 2012
Pimenta dioca	Jamaica	Leaves	NR	Eugenol (76.0)	Antibacterial and antifungal activities	Höferl et al. 2009
		NR		Eugenol (64.3) and methyleugenol (20.6)	Antifungal activity	Zabka et al. 2009
	India (commercial sample)			Eugenol (62.1) and methyleugenol (22.9)	Antibacterial activity	Ansari et al. 2016
	Mauritus	Leaves	0.7	Eugenol (79.9) and methyleugenol (9.3)	Antimicrobial, antibiotic potentiating activity	Aumeeruddy-Elalfi et al. 2015

Table 3.2 contd. ...

... *Table 3.2 contd.*

Species	Origin	Used parts	Yield,%	Prop-1(2)-enylbenzene found (relative amount, %)	Microbial activities	References
P. racemosa	France (commercial sample)	NR	NR	Eugenol (60.0) and chavicol (10.4)	Antifungal activity	Delespaul et al. 2000
	Jamaica	Leaves	NR	Eugenol (45.6)	Antibacterial and antifungal activities	Höferl et al. 2009
			1.9	Eugenol (64.4) and chavicol (7.7)	Antifungal activity	Zabka et al. 2014
	Korea	Commercial sample	NR	Eugenol (46.2) and chavicol (10.5)	Fumigant activity against phytopathogenic fungi	Kim et al. 2008
P. racemosa var. *grisea*	Dominican Republic	Leaves	0.2	4-Methoxy isoeugenol (75.2) and 4-methoxy eugenol (4.5)	Antibacterial activity	García et al. 2002, Sáenz et al. 2004
P. racemosa var. *racemosa*	Guadeloupe		NR	Estragole (32.0) and methyleugenol (48.1) Chavicol (17.1) and eugenol (56.1)	Antibacterial and antifungal activities	Aurore et al. 1998
P. racemosa var. *terebinthina*	Dominican Republic		0.8	4-Methoxyeugenol (12.6)	Antibacterial activity	García et al. 2002, Sáenz et al. 2004
Pimpinella anisum	Argentina	Fruits	0.5–2.0	*E*-Anethole (96.3)	Antimicrobial activity	Gende et al. 2009
	Egypt (commercial sample)	NR	NR	*E*-Anethole (61.7)		Dawidar et al. 2008
	India	Seeds		*E*-Anethole (90.0)	Antifungal and phytotoxic effect	Sharma et al. 2009
	India (commercial sample)	NR		*E*-Anethole (NR)	Antifungal activity	Kamble and Patil 2008
	Iran	Seeds		*E*-Anethole (90.7)		Fatemi et al. 2013
			3.3	*E*-Anethole (92.9)		Sharifi et al. 2008
		Plants	NR	*E*-Anethole (37.8) and isoeugenol (11.2)	Antimicrobial effect on *E. coli*	Ehsani and Mahmoudi 2012
	Turkey	Fruits (commercial sample)	2.6	*E*-Anethole (94.6)	Alternative antibiotic	Özkırım et al. 2012
		Seeds (commercial sample)	NR	*E*-Anethole (84.0) and estragole (9.1)	Antibacterial effect	Evrendilek 2015
		Fruits	1.9	*E*-Anethole (98.3)	Antifungal activity	Özcan and Chalchat 2006

Piper aduncum	Brazil	Aerial parts	3.0	Dillapiole (76.5)	Antibacterial activity	Brazão et al. 2014
			2.8	Dillapiole (79.9–86.9)	Antifungal activity	de Almeida et al. 2009
			3.0	Dillapiole (85.9)	Antimicrobial activity	Ferreira et al. 2016
	Ecuador		0.8	Dillapiole (45.9)		Guerrini et al. 2009
P. arborescens	Malaysia	Stem	0.2	Methyleugenol (11.0)		Salleh et al. 2016a
P. auritum	Colombia	Aerial parts	0.1	Safrole (69.5)	Antifungal activity	Pineda et al. 2012
	Cuba	Leaves	0.4	Safrole (84.4)	Antibacterial activity	Sánchez et al. 2013
P. betle	India		0.3	Safrole (25.7) and eugenol (18.3)	Antimicrobial activity	Sugumaran et al. 2011
	Mauritus		0.7	Safrole (49.0) and eugenol (14.8)	Antimicrobial, antibiotic potentiating activity	Aumeeruddy-Elalfi et al. 2015
P. caninum	Malaysia	Leaves Stem	0.5 0.3	Safrole (17.1) Safrole (25.5)	Antimicrobial activity	Salleh et al. 2011
P. divaricatum	Brazil	NR	NR	Methyleugenol (63.8) and eugenol (23.6)	Antifungal activity	da Silva et al. 2010
		Aerial parts	3.0	Methyleugenol (77.1) and eugenol (7.9)		da Silva et al. 2014
P. holtonii	Colombia		0.1	Apiole (64.2)		Pineda et al. 2012
P. obliquum	Ecuador		0.2	Safrole (45.9)	Antimicrobial activity	Guerrini et al. 2009
P. xylosteoides	Brazil	Leaves	2.0–3.0	Safrole (75.8–84.1)	Antibacterial activity	Dognini et al. 2012
Platostoma africanum	Togo	Aerial parts	NR	Methyleugenol (14.2) and eugenol (10.8)	Antimicrobial activity	Chaumont et al. 2001
Psiadia arguta	Mauritus	Leaves	0.5	*iso*-Eugenol (50.3), methyleugenol (11.2)	Antimicrobial, antibiotic potentiating activity	Aumeeruddy-Elalfi et al. 2015
P. terebinthina			0.7	Acetyl eugenol (10.9)		Aumeeruddy-Elalfi et al. 2015
Pycnostachys abyssinica	Ethiopia	Stem	0.01	Estragole (70.4)	Antimicrobial activity	Hussien et al. 2010
Rhaponticum acaule	Tunisia	Capitula and aerial parts	0.2 0.04	Methyleugenol (11.4) Methyleugenol (10.6)		Boussaada et al. 2008
Ridolfia segetum		Flowers	0.01–1.4	Dillapiole (29.5–85.4) and myristicin (13.1–31.5)	Antibacterial activity	Jannet and Mighri 2007
		Roots	0.02	Dillapiole (47.4) and myristicin (19.2)		Jabrane et al. 2010

Table 3.2 contd. ...

... *Table 3.2 contd.*

Species	Origin	Used parts	Yield,%	Prop-1(2)-enylbenzene found (relative amount, %)	Microbial activities	References
Sassafras albidum	Serbia (commercial sample)	NR	NR	Safrole (82.0)	Antifungal activity	Simić et al. 2004
Seseli libanotis	Turkey	Aerial parts	0.4	Euasarone (10.7)	Antibacterial activity	Ozturk and Ercisli 2006
Syzygium aromaticum	Brazil (commercial sample)	NR	NR	Eugenol (83.6)	Antimicrobial activity	Andrade et al. 2014
	Egypt	Buds	9.5	Eugenol (75.8) and eugenyl acetate (18.9)		El-Mesallamy et al. 2012
	France (commercial sample)	NR	NR	Eugenol (82.0) and eugenyl acetate (14.0)	Antifungal activity	Delespaul et al. 2000
				Eugenol (90.0)	Antibacterial activity	Rhayour et al. 2003
	India (commercial sample)			Eugenol (NR)	Antifungal activity	Kamble and Patil 2008
				Eugenol (47.6)	Antipathogenic activity	Chudasama and Thaker 2012, 2014
		Buds		Eugenol (68.9) and eugenol acetate (12.4)	Antimicrobial activity	Mahboubi and Mahboubi 2015
		NR		Eugenol (72.5)	Antifungal activity	Ahmad et al. 2005
	Indonesia	Buds	5.1–16.2	Eugenol (56.1–71.8) and eugenyl acetate (4.7–10.6)	Antimicrobial activity	Kennouche et al. 2015
	Madagascar	Leaves	NR	Eugenol (76.8)	Antibacterial and antifungal activities	Höferl et al. 2009
	Madagascar (commercial sample)	NR		Eugenol (89.5)	Antifungal activity	Ahmad et al. 2005
	NR (commercial sample)	Leaves		Eugenol (89.8)		Roselló et al. 2015
	Tunisia	Buds (commercial sample)		Eugenol (88.6)		Chaieb et al. 2007, Miladi et al. 2010
	Turkey	Seeds (commercial sample)		Eugenol (67.3) and eugenyl acetate (13.0)	Antibacterial effect	Evrendilek 2015

Tagetes filifolia	Argentina	Leaves		*E*-Anethole (71.3) and estragole (20.5)	Antifungal activity	Zygadlo et al. 1994
T. lucida	Cuba		0.8	Estragole (96.8)	Screening of antibacterial and antifungal activities	Regalado et al. 2011
Xylopia aethiopica	Nigeria	Fruits	1.3	Eugenol (12.2) and acetyleugenol (7.0)	Antibacterial activity	Usman et al. 2016
Zieria arborescens	Australia	Branches and leaves	NR	Methyleugenol (58.9)	Antimicrobial activity	Griffin et al. 1998
Z. hindii				Safrole (64.0)		
Z. prostata				Methyleugenol (21.4)		
Z. smithii subsp. *smithii*				Methyleugenol (49.5) Elemicin (50.6) and methyleugenol (19.7) Safrole (73.4)		

C. cassia (high content of eugenol), *M. fragrans* (containing myristicin), *F. vulgare* (high content of *E*-anethole), *C. zeylanicum* (high content of eugenol), *P. dioca* (high content of eugenol/methyleugenol), *P. betle* (high content of safrole/eugenol), *P. arguta* (high content of isoeugenol/methyleugenol), *P. terebinthina* (containing acetyleugenol), *O. basilicum* (high content of estragole), *D. carota* subsp. *hispanicus* (myristicin chemotype), *R. acaule* (containing methyleugenol), *P. aduncum* (high content of dillapiole), *E. platyloba*, *Z. smithii* subsp. *smithii* (elemicin and safrole chemotypes), *P. racemosa* (high content of eugenol), *D. carota sativa* (high content of *E*-asarone), *D. carota* subsp. *maritimus* (high content of dillapiole/myristicin), *A. calamus* (high content of *Z*-asarone), *A. visnaga* (containing croweacin), *L. nobilis* (high content of eugenol), *L. sempervirens* (high content of safrole/isosafrole), *L. citriodora* (high content of *E*-anethole), *O. gratissimum* (high content of eugenol, MBC 1.2 µg/mL), *P. inaequalifolia* (high content of safrole and myristicin), *M. bracteata* var. *revolution gold* (high content of methyleugenol), *M. bracteata* var. *revolution green* (containing methyleugenol), *A. graveolens* (dillapiole chemotype), *P. racemosa* var. *grisea* (high content of 4-methoxyisoeugenol), *P. racemosa* var. *terebinthina* (containing 4-methoxyeugenol), *B. pulverulenta* (high content of eugenol), *P. arborescens* (containing methyleugenol), *A. turbith* subsp. *hungarica* (high content of myristicin/apiole), *A. turbith* subsp. *haynaldii* (high content of apiole and myristicin), *A. brevilabrum* (high content of elemicin/methylisoeugenol), *O. canum* (high content of eugenol), and *O. tenuiflorum* (containing eugenol). *The EO that showed the lowest MIC values (0.5–0.8 µg/mL) were obtained from C. zeylanicum, P. betle, P. arguta, Z. smithii* subsp. *smithii, and O. gratissimum, which had a high content of phenols (eugenol and isoeugenol)* (Griffin et al. 1998, Nakamura et al. 1999, Senatore et al. 2003, Sáenz et al. 2004, Singh et al. 2005, Imamu et al. 2007, Boussaada et al. 2008, Shahat et al. 2008, Anwar et al. 2009, Höferl et al. 2009, Jabrane et al. 2009, 2010, Tomić et al. 2009, Aboutabl et al. 2011, Khalfallah et al. 2011, Su et al. 2011, Joshi et al. 2012, Montenegro et al. 2012, Vairappan et al. 2012, Andrade et al. 2014, Bendiabdellah et al. 2014, Brazão et al. 2014, Oyedeji et al. 2014, Peerakam et al. 2014, Vyry Wouatsa et al. 2014, Aumeeruddy-Elalfi et al. 2015, Azizkhani and Parsaeimehr 2015, Fayyaz et al. 2015, Noriega Rivera et al. 2015, Abdollahnejad et al. 2016, Moosavi-Nasab et al. 2016, Salleh et al. 2016a, 2016b, Yamani et al. 2016).

An interesting case is the high growth inhibition (MIC values 2.5 µg/mL–40 µg/mL) of the strain **methicillin-resistant *S. aureus* (MRSA)** by EO from *A. dracunculus* (*E*-anethole chemotype), *C. zeylanicum* (high content of eugenol), *P. dioca* (high content of eugenol/methyleugenol), *P. betle* (high content of safrole/eugenol), *P. terebinthina* (containing acetyleugenol), *D. carota* subsp. *maritimus* (high content of dillapiole/myristicin), and *R. segetum* (high content of dillapiole/myristicin). *EO of R. segetum (MIC 2.5 µg/mL), and D. carota* subsp. *maritimus (MIC 2.5–5 µg/mL) reached the lowest MIC values, constituted mainly by methylenedioxybenzene compounds* (Jabrane et al. 2009, 2010, Aumeeruddy-Elalfi et al. 2015, Abdollahnejad et al. 2016, Yamani et al. 2016).

The EO from plants which effectively inhibited (MIC 0.06 µg/mL–60 µg/mL) to ***E. coli*** were *A. dracunculus* (*E*-anethole chemotype), *E. supraxillaris* (high content of eugenol/methyleugenol), *C. cassia* (high content of eugenol), *M. fragrans* (containing myristicin), *S. aromaticum* (high content of eugenol), *F. vulgare* (*E*-anethole chemotype; *E*-anethole/estragole chemotype), *C. zeylanicum* (high content of eugenol), *P. dioca* (high content of eugenol/methyleugenol), *P. betle* (high content of safrole/eugenol), *P. arguta* (high content of isoeugenol/methyleugenol), *P. terebinthina* (containing acetyleugenol), *C. hieronymi* (high content of euasarone/*E*-asarone), *E. platyloba* (containing asarone, anethole and eugenol), *Z. smithii* subsp. *smithii* (elemicin and safrole chemotypes), *P. dioca* (eugenol chemotype), *P. racemosa* (eugenol chemotype), *D. carota* subsp. *maritimus* (high content of dillapiole/myristicin), *R. segetum* (high content of dillapiole/myristicin), *A. calamus* (high content of *Z*-asarone), *A. visnaga* (containing croweacin), *S. aromaticum* (high content of eugenol/acetyleugenol), *L. citriodora* (high content of *E*-anethole), *O. gratissimum* (high content of eugenol), *P. inaequalifolia* (containing safrole, myristicin and elemicin), *M. bracteata* var. *revolution gold* (high content of methyleugenol), *M. bracteata* var. *revolution green* (high content of methyleugenol), *A. graveolens* (dillapiole chemotype), *O. basilicum* (estragole chemotype), *F. vulgare* var. *vulgare* var. *azoricum* (high content of *E*-anethole), *C. anisum-olens* (high content of myristicin), *A. turbith* subsp. *haynaldii* (high content of apiole/myristicin), and *O. tenuiflorum* (containing eugenol). *EO that showed the lowest MIC values were isolated from F. vulgare (E-anethole/estragole (50%/22%) chemotype, MIC 0.06 μg/mL), C. anisum-olens (MIC 0.09 μg/mL), F. vulgare (E-anethole/estragole (68%/10%) chemotype, MIC 0.2 μg/mL), Z. smithii* subsp. *smithii (MIC 0.3 μg/mL), A. dracunculus (MIC 0.3 μg/mL),* and *P. inaequalifolia (MIC 0.8 μg/mL), which were mainly constituted by aromatic ethers (estragole, anethole, methylenedioxybenzene compounds)* (Griffin et al. 1998, Nakamura et al. 1999, Senatore et al. 2003, de Heluani et al. 2005, Singh et al. 2005, Curini et al. 2006, Höferl et al. 2009, Jabrane et al. 2009, 2010, Tomić et al. 2009, Khalfallah et al. 2011, Su et al. 2011, Joshi et al. 2012, Diao et al. 2014, Oyedeji et al. 2014, Peerakam et al. 2014, Bisht et al. 2014, Aumeeruddy-Elalfi et al. 2015, Fayyaz et al. 2015, Mahboubi and Mahboubi 2015, Noriega Rivera et al. 2015, Moosavi-Nasab et al. 2016, Yamani et al. 2016).

Vegetable species with EO which showed a marked inhibition (MIC 7 μg/mL–600 μg/mL) against ***Acinetobacter* sp.** were *O. sanctum* (methyleugenol chemotype), *P. africanum* (containing methyleugenol and eugenol), and *L. sempervirens* (high content of safrole/isosafrole). *EO which showed the lowest MIC (7 μg/mL) value was isolated from L. sempervirens, which had a high content of methylenedioxybenzene compounds* (Chaumont et al. 2001, Montenegro et al. 2012).

On the other hand, EO from *A. nodiflorum* (myristicin chemotype), and *O. basilicum* var. *album* (high content of estragole) substantially inhibited to bacterium ***H. pylori***. *The lowest MIC value (12 μg/mL) was for EO from A. nodiflorum* (Ohno et al. 2003, Menghini et al. 2010).

The bacterium ***K. pneumoniae*** presented a high susceptibility (MIC 0.2 μg/mL–50 μg/mL) to the EO from *A. dracunculus* (*E*-anethole chemotype), *C. zeylanicum* (high content of eugenol), *P. dioca* (high content of eugenol/methyleugenol), *P. betle* (high content of safrole/eugenol), *P. terebinthina* (containing acetyleugenol), *C. hieronymi* (high content of euasarone/*E*-asarone), *N. cordifolia* (containing eugenol and *E*-anethole), *P. dioca* (high content of eugenol), *P. racemosa* (high content of eugenol), *S. aromaticum* (high content of eugenol), *D. carota* subsp. *maritimus* (high content of dillapiole/myristicin), *R. segetum* (high content of dillapiole/myristicin), *A. visnaga* (containing croweacin), *O. gratissimum* (high content of eugenol), *M. bracteata* var. *revolution gold* (high content of methyleugenol), *M. bracteata* var. *revolution green* (high content of methyleugenol/isoeugenol), *A. turbith* subsp. *hungarica* (high content of myristicin/apiole), *A. turbith* subsp. *haynaldii* (high content of apiole/myristicin), and *O. canum* (high content of eugenol). *The lowest MIC values were achieved from EO of N. cordifolia (MIC 0.2 μg/mL), M. bracteata* var. *revolution green (MIC 0.6–1.2 μg/mL), S. aromaticum (MIC 1 μg/mL), M. bracteata* var. *revolution gold (MIC 1.2 μg/mL), and O. canum (MIC 1.7 μg/mL). These EO were mainly represented by phenol or aromatic ether compounds* (Nakamura et al. 1999, de Heluani et al. 2005, Höferl et al. 2009, Jabrane et al. 2009, 2010, Tomić et al. 2009, Khalfallah et al. 2011, Oyedeji et al. 2014, Vyry Wouatsa et al. 2014, Aumeeruddy-Elalfi et al. 2015, Mahboubi and Mahboubi 2015, Abdollahnejad et al. 2016, El-Tantawy et al. 2016).

EO isolated from *A. dracunculus* (*E*-anethole chemotype), *E. supraxillaris* (high content of eugenol/methyleugenol), *P. arguta* (high content of eugenol/methyleugenol), *N. cordifolia* (containing eugenol and *E*-anethole), *P. aduncum* (high content of dillapiole), *P. dioca* (high content of eugenol), *P. racemosa* (high content of eugenol), *D. carota* subsp. *maritimus* (high content of dillapiole/myristicin), *R. segetum* (high content of dillapiole/myristicin), *A. visnaga* (containing croweacin), *L. sempervirens* (high content of safrole/isosafrole), *P. inaequalifolia* (high content of safrole and myristicin/elemicin), *M. bracteata* var. *revolution gold* (high content of methyleugenol), *M. bracteata* var. *revolution green* (high content of methyleugenol/isoeugenol), *A. calamus* (high content of *Z*-asarone/*E*-asarone), *P. racemosa* var. *grisea* (high content of 4-methoxyisoeugenol), *B. pulverulenta* (high content of eugenol), and *P. arborescens* (containing methyleugenol) displayed a growth inhibitory activity (MIC 0.3 μg/mL–500 μg/mL) against ***P. aeruginosa***. *The lowest MIC values were reached by EO from A. calamus (MIC 0.3–0.6 μg/mL), M. bracteata* var. *revolution gold (MIC 1.2–1.5 μg/mL), M. bracteata* var. *revolution green (MIC 1.2–2.5 μg/mL), D. carota* subsp. *maritimus (MIC 2.5 μg/mL), R. segetum (MIC 2.5 μg/mL), and P. inaequalifolia (MIC 3 μg/mL). These EO were characterized by aromatic ether and methylenedioxybenzene compounds* (Sáenz et al. 2004, Singh et al. 2005, Radušiene et al. 2007, Guerrini et al. 2009, Höferl et al. 2009, Jabrane et al. 2009, 2010, Aboutabl et al. 2011, Khalfallah et al. 2011, Montenegro et al. 2012, Oyedeji et al. 2014, Aumeeruddy-Elalfi et al. 2015, Noriega Rivera et al. 2015, Abdollahnejad et al. 2016, El-Tantawy et al. 2016, Salleh et al. 2016a, 2016b).

EO with high content of eugenol from four plants (*C. zeylanicum*, *P. dioca*, *P. racemosa*, and *S. aromaticum*) and one (*A. dracunculus*) with high content of *E*-anethole were able to inhibit the growth (MIC 6 µg/mL–40 µg/mL) of ***Salmonella* sp.** *EO from C. zeylanicum, P. dioca, P. racemosa, and S. aromaticum showed the lowest MIC values, each one of them with 6 µg/mL* (Höferl et al. 2009, Abdollahnejad et al. 2016). Growth inhibition (MIC 0.2 µg/mL–500 µg/mL) on ***S. typhimurium*** were produced by EO from *A. dracunculus* (*E*-anethole chemotype), *C. hieronymi* (high content of euasarone and *E*-asarone), *F. vulgare* (high content of *E*-anethole/estragole), *N. cordifolia* (containing eugenol and *E*-anethole), *D. carota* subsp. *maritimus* (high content of dillapiole/myristicin), *R. segetum* (high content of dillapiole/myristicin), *S. aromaticum* (high content of eugenol/acetyleugenol), *L. citriodora* (high content of *E*-anethole), *P. racemosa* var. *grisea* (high content of 4-methoxyisoeugenol), *P. racemosa* var. *terebinthina* (containing 4-methoxyeugenol), and *O. canum* (high content of eugenol). *EO that showed the lowest MIC values were secluded from F. vulgare (MIC 0.2 µg/mL), O. canum (MIC 0.4 µg/mL), and S. aromaticum (MIC 2 µg/mL)* (Sáenz et al. 2004, de Heluani et al. 2005, Jabrane et al. 2009, 2010, Diao et al. 2014, Vyry Wouatsa et al. 2014, Mahboubi and Mahboubi 2015, Abdollahnejad et al. 2016, El-Tantawy et al. 2016, Moosavi-Nasab et al. 2016).

From elsewhere, the MIC values determined from EO of *O. basilicum* (methyleugenol chemotype), *D. carota* subsp. *carota* (containing *E*-methylisoeugenol), *A. nodiflorum* (dillapiole chemotype), *D. carota* subsp. *halophilus* (high content of elemicin), and *P. aduncum* (high content of dillapiole) evaluated on ***A. fumigatus*** were found between 0.6 µg/mL–10 µg/mL. *EO with lowest MIC values were withdrawn from A. nodiflorum (0.6–10 µg/mL), and O. basilicum (1 µg/mL). These EO were mainly represented by aromatic ether compounds* (Tavares et al. 2008, Maxia et al. 2009, 2012, Kumar et al. 2011, Ferreira et al. 2016).

EO of plants which demonstrated a noticeable inhibition (MIC 0.3 µg/mL –620 µg/mL) on ***C. albicans*** were separated from *E. supraxillaris* (high content of eugenol/methyleugenol), *D. carota* subsp. *hispanicus* (myristicin chemotype), *O. sanctum* (high content of eugenol), *P. africanum* (containing methyleugenol/eugenol), *A. dracunculus* (*E*-anethole chemotype), *C. hieronymi* (high content of euasarone/*E*-asarone), *Z. smithii* subsp. *smithii* (elemicin and safrole chemotypes), *P. dioca* (high content of eugenol), *P. racemosa* (high content of eugenol), *S. aromaticum* (high content of eugenol), *D. carota sativa* (high content of *E*-asarone), *C. rhyncophyllum* (high content of safrole), *C. maritimum* (high content of dillapiole), *D. carota* subsp. *carota* (containing *E*-methylisoeugenol), *A. nodiflorum* (high content of dillapiole), *P. inaequalifolia* (high content of safrole/myristicin), *A. graveolens* (dillapiole chemotype), *F. vulgare* var. *vulgare* (estragole/*E*-anethole and *E*-anethole/estragole chemotypes), *A. calamus* (high content of *Z*-asarone), *B. pulverulenta* (high content of eugenol), *C. zeylanicum* (high content of eugenol), *A. rugosa* (high content of estragole), *D. carota* subsp. *halophilus* (high content of elemicin), *A. turbith* subsp. *hungarica* (high content of myristicin/apiole), and *A. turbith* subsp. *haynaldii* (high content of apiole/myristicin). *EO which showed the lowest MIC values (0.3 µg/mL, and 0.6 µg/mL) were found from F. vulgare* var. *vulgare, and A. calamus, respectively, which had a high content of aromatic ethers (estragole/E-anethole, and Z-asarone)* (Chaumont et al. 2001, Shin and Kang 2003, Shin 2004, de Heluani et al. 2005, Curini et al. 2006, Imamu et al. 2007, Marongiu et al. 2007, Radušiene et al. 2007, Schmidt et al. 2007, Jantan et al. 2008, Tavares et al. 2008, Höferl et al. 2009, Tomić et al. 2009, Aboutabl et al. 2011, Bendiabdellah et al. 2014, Peerakam et al. 2014, Piras et al. 2014, Noriega Rivera et al. 2015, Salleh et al. 2016b).

EO from *C. maritimum* (high content of dillapiole; MBC 320 µg/mL), *D. carota* subsp. *carota* (containing *E*-methylisoeugenol), *A. nodiflorum* (dillapiole and myristicin/dillapiole chemotypes; MBC 0.6 µg/mL), *F. vulgare* var. *vulgare* (estragole/*E*-anethole and *E*-anethole/estragole chemotypes), *A. rugosa* (high content of estragole), and *D. carota* subsp. *halophilus* (high content of elemicin) were able to inhibit to ***C. neoformans*** with MIC values between 0.2 µg/mL–10 µg/mL. *These compounds* *were chemically characterized by aromatic ethers (estragole, anethole, methylenedioxybenzene compounds)* (Shin and Kang 2003, Shin 2004, Marongiu et al. 2007, Tavares et al. 2008, Maxia et al. 2009, 2012, Piras et al. 2014).

To finish the analysis, only 31 EO with high content of one phenylprop-1(2)-enoid or in combination with other(s) (*E*-anethole, eugenol, myristicin, methyleugenol, dillapiole, *E*-methylisoeugenol, estragole, elemicin, safrole, *Z*-asarone safrole/eugenol, isoeugenol/methyleugenol, safrole/isosafrole, dillapiole/myristicin, estragole/*E*-anethole, myristicin/dillapiole, *E*-anethole/estragole and methyleugenol/isoeugenol), isolated of 27 plants (*L. citriodora*, *P. betle*, *P. arguta*, *O. gratissimum*, *A. dracunculus*, *L. sempervirens*, *C. anisum-olens*, *D. carota* subsp. *maritimus*, *R. segetum*, *P. dioca*, *P. racemosa*, *O. basilicum*, *F. vulgare* var. *vulgare*, *C. maritimum*, *D. carota* subsp. *carota*, *A. rugosa*, *D. carota* subsp. *halophilus*, *S. aromaticum*, *A. nodiflorum*, *F. vulgare*, *C. zeylanicum*, *Z. smithii* subsp. *smithii*, *P. inaequalifolia*, *M. bracteata* var. *revolution green*, *M. bracteata* var. *revolution gold*, *O. canum*, and *A. calamus*) from each one of them listed in Table 3.2, resulted highly effectives (lowest MIC values) against 12 micro-organisms (*L. monocytogenes*, *S. aureus*, *E. coli*, *Acinetobacter* sp., *H. pylori*, *K. pneumoniae*, *P. aeruginosa*, *Salmonella* sp., *S. typhimurium*, *A. fumigatus*, *C. albicans* and *C. neoformans*) which mainly cause some human infections/ailments. EO which effectively inhibited to a greater number of microorganism (3) were insulated from *A. nodiflorum* (myristicin and dillapiole chemotypes and myristicin/dillapiole chemotype against *H. pylori*, *A. fumigatus*, and *C. neoformans*, respectively), and *S. aromaticum* (high content of eugenol against *K. pneumoniae*, *Salmonella* sp., and *S. typhimurium*). Nevertheless, the most sensitive microorganisms (6) to the EO evaluated were *P. aeruginosa* (growth inhibited by 6 EO); *S. aureus* and *C. neoformans* (growth inhibited by 5 EO); *E. coli*, *K. pneumoniae* and *Salmonella* sp. (growth inhibited by 4 EO).

Inventive Capacity Related to Rich-phenylprop-1(2)-enoid Essential Oils or Phenylprop-1(2)-enoid Compounds

In order to establish the inventive activity involving the antimicrobial activities of prop-1(2)-enylbenzene derivatives, a search equation was proposed in the *Derwent Innovations* database (Thomson Reuters 2016), using the keywords: *(anethole or estragole OR safrole OR eugenol OR methyleugenol OR methylisoeugenol OR myristicin OR dillapiole OR apiole OR elemicin OR isoeugenol OR asarone) AND (antimicrobial)*. The result was 46 registers of patents indexed in this database. The distribution of numbers of patents per year is shown in the Fig. 3.3. In this case, 2011–2014 were the years with the largest number (7–9) of patents reported, in the timeline consulted (2000–2016).

Regarding the categorization of patents in accordance with the IPC (International Patent Classification), 28 records were found in the A61K section (human necessities—Fig. 3.4), which corresponds to preparations for medical, dental or toilet purposes, and includes the products demonstrating antimicrobial properties. Another important section labeled A01N covers the inventions (19) related to disinfectant, pesticide or herbicide properties. Companies such as Unilever with 8 patents, along with Conopco Inc. and Hindustan Ltd., each one with 6 registers, are found among the most important asignees.

Inventive capacity analysis based on the rich-prop-1(2)-enylbenzene essential oils or prop-1(2)-enylbenzene compounds is presented in Table 3.3, which includes the code, title, IPC, and Asignees of patents. This information was obtained by monitoring *Reaxys database* (2016). In agreement with this analysis were identified three main groups related to: (1) antimicrobial properties; (2) pest control; and, (3) other applications. A summary of the most important patents and the relationship with the phenylprop-1(2)-enoid compounds and its essential oils are mentioned below.

Group I—antimicrobial properties. Patent WO 2002012421 A1 deals with a nutritional (functional food/medicament) formulation incorporating a specific essential oil (e.g., clove, tarragon or combination

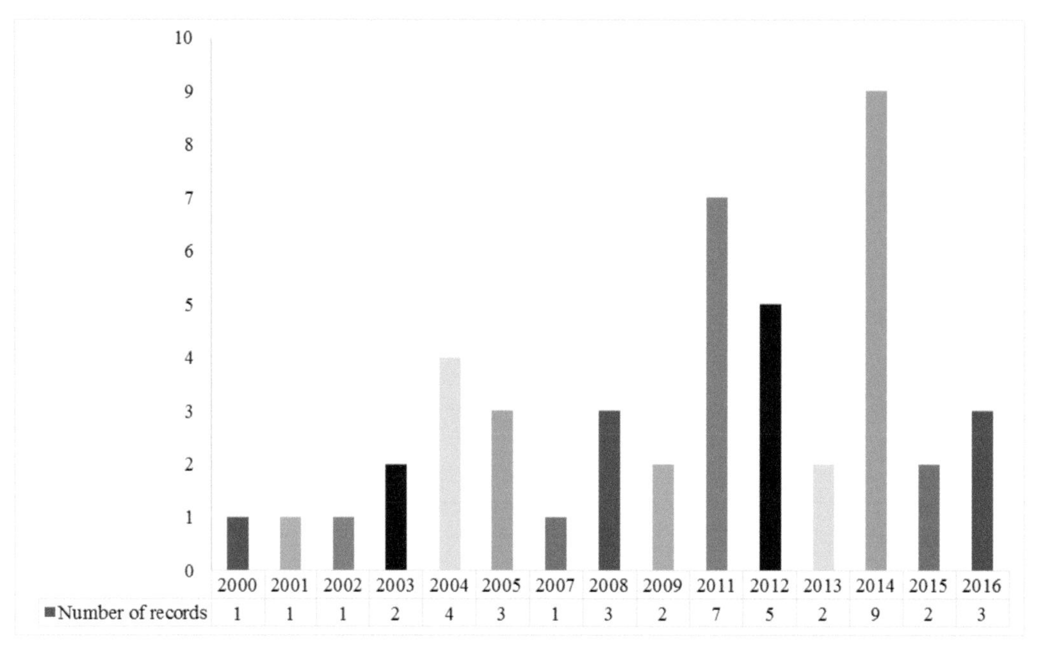

Source: Bibliometric Unit - USTABUCA, data based on *Derwent Innovations* information (Thomson Reuters 2016) and processed with VantagePoint software (Version 9.0, *Search Technology*).

Figure 3.3. Dynamic of the inventive activity (number of patents) per year related to the antimicrobial effects of prop-1(2)-enylbenzene compounds.

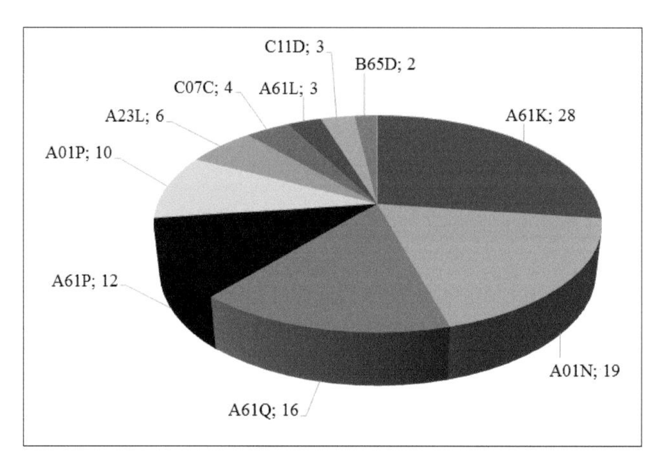

Source: Bibliometric Unit - USTABUCA, data based on Derwent Innovations information (Thomson Reuters 2016) and processed with VantagePoint software (Version 9.0, *Search Technology*).

Figure 3.4. Distribution of patents based on the International Patent Classification (IPC) (http://cip.oepm.es/ipcpub/#lang=es&menulang=ES&refresh=page).

thereof) and/or specific pure compound (e.g., eugenol (**4**), estragole (**2**), or combination thereof) isolated from the essential oil for prevention or treatment of infection by a gastric *Helicobacter*-like organism. Likewise, patent EP 2645862 A1 reports the manufacture and use of an emulsion composed by an antimicrobial essential oil (e.g., clove leaf 0.06%), acacia gum and water with the purpose of improving the antimicrobial effect of essential oils. Then, the emulsion is incorporated as an ingredient for products with aqueous composition (beverage or food product/supplement, etc.) and is effective against *A. niger*, *P. nalgiovense* and *Eurotium amstelodami* up to 12 months.

Table 3.3. Some patents related to rich-phenylprop-1(2)-enoid essential oils or phenylprop-1(2)-enoid compounds.

Patent code	Compound/OE	Title	IPC	Assignees	Date
US 20070004686 A1	*E*-Anethole (**1**)	Attractant for apple fruit moth and other insect pests of apple	A01N37/40, A01M, A01N35/04, A01N37/02, A01N27/00, A01N65/34	Bengtsson Marie, P. Witzgall, S. Kobro, G. Jaastad, J. Lofqvist, C. Lindhe	2007-01-04
EP 2572579 A1		Compositions for attracting Tortricidae (fruit moths)	A01N31/04, A01N35/02, A01N37/02, A01N43/40, A01P19/00	Bioforsk	2014-04-23
US 6340710 B1	*E*-Anethole (**1**), eugenol (**4**)	Non-Hazardous pest control	A01N61/00, A01N25/08, A01N31/04, A01N59/04	Ecosmart Technologies, Inc.	2002-01-22
WO 2002012421 A1	Estragole (**2**), eugenol (**4**) and isoeugenol (**6**) Tarragon and clove EO	Use of essential oils for combating gi tract infection by helicobacter-like organisms	A61P31/04, A61K31/22, C11B9/00, A23F3/16, A23L1/222, A23G3/48	Société des Produits Nestlé S.A.	2002-02-14
WO 2011092600 A2	Safrole (**4**), dillapiole (**9**), apiole (**10**) *Piper aduncum* EO	Methods and product for reducing the population size of *Papilio demoleus* L. (Papilionidae)	A01N27/00, A01N65/36, A01N25/00, A01P7/04	University of the West Indies	2011-08-04
US 20150086421 A1	Rich-eugenol EO (clove or cinnamon)	Antibacterial composition	A01N31/04, A01N31/08	Conopco, Inc., D/B/A Unilever	2015-03-26
CN 104212136 A	Clove and fennel EO	Polylactic acid anti-bacterial-activity packaging material and preparation method thereof	B29C47/92, B29C55/28, C08L67/04, C08L69/00, C08L91/00	Univ. Kunming Science & Tech.	2014-12-17
EP 2645862 A1	Clove EO	Delivery carrier for antimicrobial essential oils	A23L23/00, A01N65/00, A01N25/04, A01N65/28, A01N65/34, A23L3/3472	Nestec S.A.	2013-10-09
US 7871649 B2	Clove EO	Antimicrobial compositions containing synergistic combinations of quaternary ammonium compounds and essential oils and/or constituents thereof	A61K33/30, A01N65/00, A61K45/06, A61K36/752, A61K36/9068, A61K36/63	The Trustees of Columbia University in the City of New York	2011-01-18
US 20120276022 A1	Eugenol (**4**)	Disinfecting agent comprising eugenol, terpineol and thymol	A61K8/34, A01P1/00, A01N33/12, A61Q11/00	S. Venkata Medepalli, A. Chakrabortty, B. Cheviti	2012-11-01

Table 3.3 contd....

Table 3.3 contd....

Patent code	Compound/OE	Title	IPC	Assignees	Date
US 20080118461 A1	Methyeugenol (**5**)	Sprayable controlled-release, male annihilation technique (MAT) formulation and insect control	A01P7/04, A01N25/00	Dow Agrosciences Llc, Isca Technologies Inc.	2008-05-22
WO 03105794 A1	Eugenol (**4**), *E*-isoeugenol (**6**), methyl isoeugenol (**7**)	Antibacterial flavor and fragrance composition and halitosis-inhibition flavor and fragrance composition and oral care composition containing the same	A61Q13/00, A61K8/33, A23G4/00, C11B9/00	Takasago International Corporation	2003-12-24
WO2008080980 A1	Myristicin (**8**)	Medicinal composition for treating animal skin comprising a wound healing agent and a deterrent	A61K45/06	Patrick Franke, Georg Roessling	2008-07-10
US 2005049230 A1	Apiole (**10**), myristicin (**8**), methyleugenol (**5**), safrole (**3**), eugenol (**4**), clove EO	Compounds that act to modulate insect growth and methods and systems for identifying such compounds	C12N15/11, A01N53/00, G01N33/68, C12Q1/68	Henrich Vincent, Weinberger Cary Alan	2005-03-03
WO 2013133723 A1	Eugenol (**4**), elemicin (**13**), isoelemicin (**19**)	Synthesis of elemicin and topical analgesic compositions	A61P29/00, A61P17/00, A61K31/09	La Grange Martin James	2013-09-12
US 6495172 B1	Elemicin (**13**), myristicin (**8**), isoelemicin (**19**) Nutmeg EO	Method of using steam ironing of fabrics as a way of causing reduction of physiological and/or subjective reactivity to stress in humans	A61K9/72, A61P25/18, A61K36/534, A61K36/84, A61K36/537	Procter & Gamble	2002-12-17
WO 2007082864 A2	Chavicol (**17**)	Use of chavicol as an antiseptic	A61P17/10, A61P17/14, A61P17/00, A61K31/05	Polichem SA, Carlo Ghisalberti	2007-07-26
WO 2006120567 A2	Eugenol (**4**)	Pharmaceutical composition comprising an anti-bacterial agent and an active ingredient selected from carveol, thymol, eugenol, borneol and carvacrol	A61P31/04, A61K45/06, A61K31/43, A61K31/05, A61K31/045	Advanced Scientific Developments	2006-11-16
US 9144544 B1	*Pimpinella anisum* L.	Synthesis of silver nanoparticles from *Pimpinella anisum* seeds	A61K9/14, A61K36/23, A61K33/38	King Saud University	2015-09-29
WO 2012114201 A1	Clove (buds/leaves) EO	Nanoencapsulation d'huiles essentielles pour la prévention et la guérison des maladies infectieuses seules ou associées à un antibiotique	A61K9/51, A61K9/127, A61K9/107, A61K31/00	Aroma Technologies, Institut National de La Sante et de La Recherche Medicale, Université d'Angers	2012-08-30

WO 2009029046 A1	Clove EO	Sugar-based surfactant microemulsions containing essential oils for cosmetic and pharmaceutical use	A61K8/00, A61K 8/34, A61K8/60, A61K8/97, A61K9/00, A61K9/107	Agency For Science, Technology and Research	2009-03-05
WO 2009043987 A1	Clove (buds/leaves) EO	Compositions antibiotiques a base d'huiles essentielles - prophylaxie et traitement d'infections nosocomiales	A61K36/00, A61K31/407, A61K31/496,A61K31/65, A61P 31/04, A61L101/56	Aroma Technologies	2009-04-09
WO 2015181084 A1	Eugenol (**4**)	Use of a nanoemulsion of cinnamaldehyde and/or a metabolite thereof, possibly in association with eugenol and/or carvacrol, to prevent resistance to antibiotics	A61P31/04, A61K9/107, A61K31/085, A61K31/11, A61K31/05	Eydo Pharma	2015-12-03

Invention US 20150086421 A1 is associated with a method for external surface disinfection (human/animal skin or hair) and an antibacterial composition, which in a synergistic combination of select antibacterial agents (e.g., 0.02%–1% essential oil containing eugenol (**4**)), specific polymers (chitosan or polyvinyl alcohol) and select hydrotropes (sodium salicylate, sodium acetate, etc.), can reduce the bacterial (e.g., *E. coli*) count in less than 5 minutes. Similarly, patent US 20120276022 A1 discloses an antimicrobial composition for surface disinfection (oral care, personal or hard surface cleaning) that has relatively fast action. Eugenol (**4**) (0.005–5%) and other constituents showed a synergic antimicrobial effects (against *E. coli*) when are included in the formulation.

A method to decrease the number of bacteria present on a surface together with a method for preventing/treating a bacterial infection (nosocomial infections) in humans are included in the invention WO 2009043987 A1. This patent relates to a novel antibiotic composition, drug combination, and formulations containing essential oils (e.g., clove bud/leaf (eugenol) EO 13–85% by weight), to use in prophylaxis/treatment of bacterial infections (in respiratory tract (pneumonia), in urinary tract, in skin/soft tissue, legionellosis (CNS), invasive aspergillosis, meningo-encephalitis, empyema, gastrointestinal, cardiopulmonary complications (endocarditis), bacteremia, site operative or generalized infection (sepsis)).

Another invention associated to this group has code WO 03105794 A1 and protects an antibacterial flavor/fragrance and anti-halitosis compositions (e.g., isoeugenol (**6**), eugenol (**4**), methyleugenol (**5**)) with inhibitory action against periodontal disease-causing bacteria (e.g., *Fusobacterium nucleatum*) and the halitosis.

Patent WO 2006120567 A2 deals with a pharmaceutical composition conformed by two therapeutically active compounds (terpenoids/phenols (e.g., eugenol) and antibiotic) one of which exerts a potentiating action on each other, and the use of this composition.

An antimicrobial composition for topical use, constituted by a quaternary ammonium compound (0.01–0.5% w/w), a polyhexamethyl biguanide along with an essential oils/individual constituents (e.g., isoeugenol (**6**), eugenol (**4**); 0.05–1.0% w/w), and a zinc salts, is specified in Patent US 7871649 B2. All constituents of the formulation are present in amounts which exhibit a synergistic antimicrobial activity (e.g., against *S. aureus*), and the formulation can be incorporated in lotions, gels, creams, soaps, for application to skin or mucous membranes.

Patent CN 104212136 A informs about the invention of a packaging material (with antibacterial activity against, e.g., *E. coli* and *S. aureus*) from polylactic acid and their preparation method. The material is prepared by blending (extruder) 100 parts by weight of the base resin and 2–10 parts by weight of the antimicrobial essential oils (e.g., clove, fennel, etc.). The film can be biodegradable and has elongation and tensile strength.

The use of chavicol (**17**) (0.1–10% w/w) for the manufacturing of a topically applicable antiseptic product (e.g., cream, lotion, mousse, spray, emulsion, shampoo or gel) for the treatment of skin and scalp infected by pathogenic microorganisms such as yeast/fungi or bacteria (e.g., *Malassezia* spp., *Pityrosporum* spp. and/or *Candida* spp.) is claimed in the patent WO 2007082864 A2.

Patent WO 2009029046 A1 reports the invention of microemulsions containing some essential oils (0.5–10% w/w, e.g., clove leaf EO) along with an aqueous phase (60–95% w/w) and a polyhydric alcohol (1–10% w/w), stabilized by biocompatible sugar-based surfactant (1–20% w/w, e.g., sucrose ester, an alkyl polyglucoside). The formulation is incorporated to personal care and pharmaceutical products for topical applications.

A nanoemulsion (oil-in-water, size of hydrophobic droplets 20–350 nm, zeta potential ranges 5–60 mV) produced from cinnamaldehyde and a metabolite thereof along with eugenol/carvacrol, for preventing the infection caused by a pathogen (e.g., *A. baumannii*, *A. calcoaceticus*, *S. aureus*, *S. epidermidis*, *E. coli*, *E. coli* BLSE, *E. hermannii*, *K. pneumonia*, *K. oxytoca*, *L. monocytogenes*, *S. enteritidis*, *P. aeruginosa*, *P. fluorescens*, *P. putida*, *P. stutzeri*) and resistance to antibiotics, is protected in patent WO 2015181084 A1.

Finally, patent WO 2012114201 A1 is an invention involving a nanoparticle (1–100 nm, lipidic nature) encapsulated (10–500 μL EO) formulation constituted by at least one essential oil (e.g., clove bud/leaf (eugenol ~ 63%) EO 1–75% by weight) or extract having a broad spectrum of biological activities, e.g., antibacterial (*S. aureus*, *E. coli*, *E. cloacae*, *P. aeroginosa*, *A. baumannii*, *K. oxyloca*), antifungal, antiparasitic, plant antipathogenic, etc. At least one antibiotic and a pharmaceutically acceptable carrier are included optionally in this formulation. Nano-encapsulated product can be administered orally, intravenously or through an aerosol or topically or through a diffuser.

Group II – pest control. Patents US 20070004686 A1 and EP 2572579 A1 were developed for controlling pests (e.g., *Argyresthia conjugella*, *Cydia pomonella*, *Hedya nubiflerana*, and *Pandemis heparama*) of fruits (e.g., *Sorbus aucuparia* and *Malus domestica*) through modifying the insect behavior by using volatile compounds (kairomones). For example, anethole in synergic action with 2-phenylethanol, is the phenylpropenoid that efficiently attract to females and males of the moth *A. conjugella*. In the case of patent US 6340710 B1, the phenylprop-1(2)-enoids anethole (**1**) or eugenol (**4**) were employed as octopamine receptor site affectors (pesticides) of insects, arachnids and larvae. Deposition of substances used on a surface provides a residual toxicity for up to 30 days and promotes antifeedant effects on insects or larvae.

Furthermore, patent WO 2011092600 A2 deals with the methods and products for reducing the population of *Papilio demoleus* (butterfly, destructive pest for citrus) by using a non-host plant (e.g., *P. aduncum*) containing natural toxins (e.g., essential oils richs in dillapiole (**9**), apiole (**10**), or safrole (**4**), etc.) and a kairomone applied. The non-host plants cause the death (by refusing to feed) or deformity on *P. demoleus*. Finally, the patents US 20080118461 A1 and US 2005049230 A1 include the use of methyleugenol (**5**) as a male-specific attractant (parapheromone) to control of populations of fruit fly species using male anhililation technique (MAT), and relate to compounds (myristicin (**8**) and methyleugenol (**5**)) that act as insecticide and/or modulator of insect growth, respectively.

Group III—other applications. This patent group is related to (1) the synthesis of elemicin and its isomeric forms (WO 2013133723 A1) for uses in analgesia; (2) synthesis of silver nanoparticles from *P. anisum* seeds (aqueous extract) (US 9144544 B1); (3) a method for decreasing the physiological/subjective stress (US 6495172 B1) in humans by employing the beneficial effects of aromatherapy (essential oils or individual components).

Nanotechnology Applied to Rich-phenylprop-1(2)-enoid Essential Oils/Phenylprop-1(2)-enoid Compounds with Antibacterial Properties

The scientific dynamic encompassing to theme 'nanotechnology applied to essential oils' was carried out using the search equation (*TITLE-ABS-KEY ("essential oil *") AND TITLE-ABS-KEY (nanotechnolog*)) AND DOCTYPE (ar) AND PUBYEAR> 1999* in the Scopus database (Elsevier 2016). 41 Registers (articles) were found (Fig. 3.5). The most productive year was 2015 with 13 registers; the countries with the highest number of records were Brazil (10 articles), India (8 articles), E.E.U.U. (4 articles), and Italy (4 articles); and, the most important application areas were industrial crops and products (4 registers) and experimental parasitology (3 registers). *The analysis showed that just in the last 6 years this area has been emerging and has become a new area for knowledge generation.*

The application of nanotechnology for treatment, monitoring, and control of biological systems has been incorporated to the medicine. According to this development, some methods emerged with distinctive advantages, e.g., the nanoparticles as an efficient system (*in vitro*/*in vivo*) for pharmaceutical delivery (carriers) of some types of drugs (anticancer, antibacterial, etc.) (Torchilin 2006).

There are different kinds of nanoparticles; however, in this chapter only three will be defined: (1) nanoemulsions (NE), which are an oil-in-water dispersions (from pure substances stabilized by surfactants) of small droplet size (range 20–200 nm, narrow distribution), translucent, of low viscosity, kinetically stable, with high stability against sedimentation and require a low surfactant concentration in order to form them (Torchilin 2006); (2) polymeric nanoparticles, which are

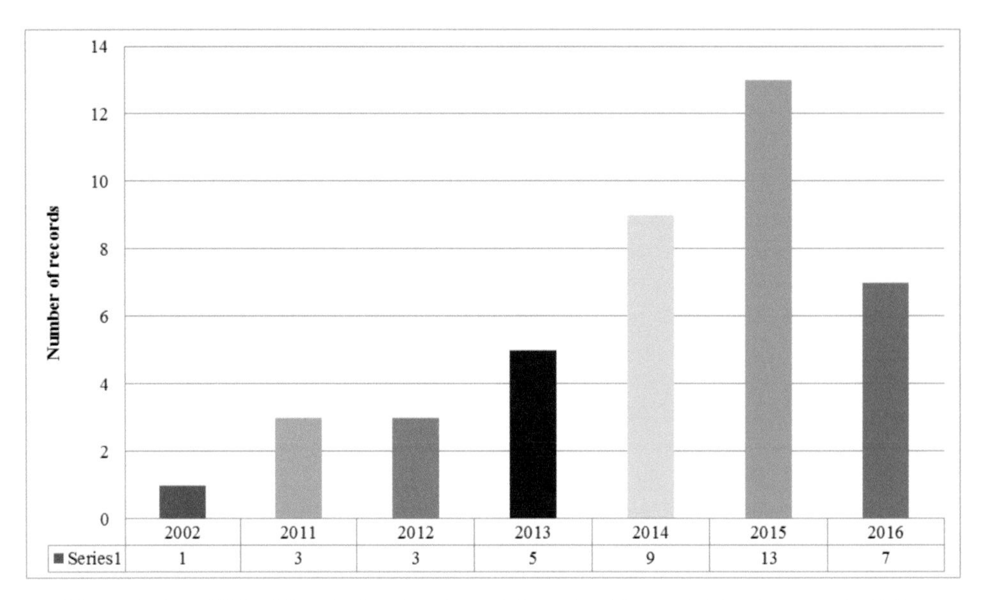

Source: Bibliometric Unit - USTABUCA, data based on Scopus information (Elsevier 2016) and processed with VantagePoint software (Version 9.0, *Search Technology*).

Figure 3.5. Distribution of scientific articles per year related to the nanotechnology applied to essential oils.

submicron size entities (range 10–1000 nm) produced from some biodegradable (e.g., chitosan) and non-biodegradable polymers; and, (3) solid-lipid nanoparticles, constituted by a solid lipid (e.g., lecithin), a emulsifier and water (D'Mello et al. 2009, Pathak 2009). The most important parameters determining the characteristic of nanoparticles are three: droplet size, polydispersity index, and zeta potential (Zambaux et al. 1998).

Nanotechnology applied to rich-prop-1(2)-enylbenzene essential oils/prop-1(2)-enylbenzene compounds are recorded in the Table 3.4, which includes types of nanoparticles, conditions or specifications of formulation, types of components encapsuled, and microorganisms affected by formulations. In agreement with this analysis three main types of technologies were identified: nanoemulsion, polymeric nanoparticles and solid-lipidic nanocapsules.

The prototype of nanoemulsion, from clove EO containing eugenol, reported by Shahavi et al. (2016) showed effectiveness against *E. coli* by means of a bactericidal action mechanism. The MIC/MBC values was 16 µg/mL for this bacterium; whereas for *B. cereus* these values were 32 µg/mL and 64 µg/mL for its MIC and MBC, correspondingly. Otherwise, nanoemulsion from basil EO containing estragole (Ghosh et al. 2013) was effective against *E. coli* in a way dose-dependent (pure nanoemulsion – 100% inhibition; nanoemulsion 10 fold diluted – 60% inhibition).

In the case of spherical nanoparticles from PLGA (poly(lactide-co-glycolide)) and eugenol (Gomes et al. 2011), these nanoparticles showed inhibition against *Salmonella* spp. and *Listeria* spp. with MIC values of 10000 µg/mL and 20000 µg/mL, respectively. According to Chen et al. (2009), nanoparticles from chitosan and eugenol aldehydes resulted most active than pure eugenol (MIC/MBC—900 µg/mL) against *E. coli* (MIC—250–500 µg/mL; MBC—2000 µg/mL) and *S. aureus* (MIC—500 µg/mL; MBC—1000 µg/mL).

Antibacterial efficacy of nanoparticles based on microemulsion from clove bud EO was determined through the agar disk diffusion method (Hamed et al. 2012). Diameters of inhibition zones for each bacterium from differents concentrations evaluated of microemulsion were: 8 ± 1 mm—12 ± 2 mm (*S. aureus*, 0.6–0.9% EO); 8.0 ± 0.1 mm (*E. coli*, 0.9% EO); 7.00 ± 0.01 mm—8 ± 1 mm (*B. cereus*, 0.6–0.9% EO); 9.6 ± 0.5 mm (*S. typhi*, 0.9% EO), 11 ± 1 mm (*P. aeruginosa*, 0.9% EO),

Table 3.4. Some nanoscale formulations applied to rich-phenylprop-1(2)-enoid essential oils or phenylprop-1(2)-enoid compounds.

Type of technology	Conditions/specifications of formulation	Type of sample encapsulated	Microorganism affected by formulations	References
Nanoemulsion	Ultrasonic emulsification Surfactant (SF): Tween® 80; Span® 80 Oil:SF (w/w): 3:1 Emulsification time: 8 min Droplet size: 32 ± 2 nm PDI: 0.38 ± 0.06 Zeta potential: –41 ± 2 mV	Clove EO	*E. coli*; *B. cereus*	Shahavi et al. 2016
Spherical PLGA nanoparticles	Ultrasonic/Emulsion evaporation method Emulsification time: 10 minutes Ultrafiltration Freeze-drying Surfactant: PVA Oil:SF (w/w): 1:5 Average size: 179 nm PDI: –0.33 Entrapment yield: 98.3%	Eugenol (**4**)	*Salmonella* spp.; *Listeria* spp.	Gomes et al. 2011
Chitosan nanoparticles	Ionic gelation method Hydrodinamic size: 235 ± 12 nm PDI: 0.318 Zeta potential: 28 ± 2 mV	Eugenol aldehydes	*E. coli*; *S. aureus*	Chen et al. 2009
Nanoparticles from microemulsion	Oil titration method Surfactant (SF): Tween® 20 Oil:SF (w/w): 1:5 Equilibrium time: 24 hours Microemulsion method Average particle diameter: 7.90 ± 0.05 nm	Cloves EO Eugenol (**4**)	*S. aureus*; *E. coli*; *B. cereus*; *S. typhi*; *P. aeruginosa*; *L. monocytogenes*	Hamed et al. 2012
Nanoemulsion	Ultrasonic emulsification Emulsification time: 5 minutes Surfactant: Tween® 80 Oil:SF (w/w): 1:1–1:3 Droplet size: 41.2 ± 0.4 nm –29.3 ± 0.2 nm PDI: 0.092 ± 0.001–0.236 ± 0.002 pH: 4.43 ± 0.01–6.02 ± 0.06 Zeta potential: –3.7 ± 0.4 mV (1:3 ratio, pH 5.53)	*Ocimum basilicum* (Estragole)	*E. coli*	Ghosh et al. 2013
Nanoemulsion	Surfactant: Poloxamer Droplet size: 30–250 nm PDI: NR; pH: 7.0 Zeta potential: –10 mV – (–40 mV)	Include eugenol (**4**)	*Acinetobacter* spp., *Staphylococcus* spp., *Escherichia* spp., *Klebsiella* spp., *Listeria* sp., *Salmonella* sp., *Pseudomonas* spp.	Patent WO 2015181084 A1
Lipidic (mixture of soybean lecithin) nanocapsules	Solvent-free and soft energy procedures Surfactant: PEG 660 and PEG 66-hydroxystearate Oil:SF (w/w): 3:1 Average size: 20–200 nm PDI: NR Zeta potential: –6 ± 1 – (–10 ± 1)	Clove (bud/leaf) EO (Eugenol)	*Lactobacillus* spp.; *E. coli*; *Enterobacteria* spp. *P. aeruginosa*; *A. baumannii*	Patent WO 2012114201 A1

and 9.6 ± 0.5 mm (*L. monocytogenes*, 0.9% EO). The most effective concentration of microemulsion was 0.9% EO against all bacteria tested.

Finally, solid-lipid nanoparticles from soybean lecithin with an essential oil (clove EO, 63–83% eugenol) showed greater effectiveness than essential oil corresponding against *Acinetobacter* spp. (MIC 500–750 μg/mL), *Staphylococcus* spp. (MIC 500–1000 μg/mL), *Escherichia* spp. (MIC 750–1000 μg/mL), *K. oxytoca* (MIC 750–1000 μg/mL), *S. enteritidis* (MIC 750–1000 μg/mL), and *Pseudomonas* spp. (MIC 750–2500 μg/mL).

Other Applications Involving to Phenylprop-1(2)-enoids and/or Essential Oils Containing to Them

Principal application involving phenylprop-1(2)-enoids and/or essential oils containing them includes the fine organic synthesis which are used as precursor reactives or to obtain biologically active molecules. The precursors, conditions of reactions, and products related to organic synthesis are summarized in Table 3.5.

Table 3.5. Application of some prop-1(2)-enylbenzene compounds and essential oils containing them in fine organic synthesis.

Precursor (EO/ Compound)	Conditions	Product	References
1	With 4-methoxyphenol, $(NH_4)_2(SO_4)_2$; $Ru(bpz)_3(PF_6)_2$ in ACN 27 h; Schlenk technique	5-Methoxy-2-(4-methoxyphenyl)-3-methyl-2,3-dihydrobenzofuran (**35**)	Blum et al. 2014
	With S_8 in DMF, 8 h (heating)	5-(4-Hydroxyphenyl)-3*H*-1,2-dithiole-3-thione (**36**)	Wallace et al. 2006
	With O_2, $N_2H_4.H_2O$ in n-Propanol, T:120°C; P: 15001.5 Torr; 0.5 h	1-Methoxy-4-propylbenzene (**37**)	Pieber et al. 2013
	With $[Ph(OCOCF_3)I]_2O$ in CH_2Cl_2 T: 0–20°C, 3 h	(2S,3S)-5-Methoxy-2-(4-methoxyphenyl)-3-methyl-1-tosylindoline (**38**)	Kita et al. 2015
2	With $(C_2H_5)_3N$, $(CH_3COO)_2Pd$, CyJohnPhos in DMF T: 90°C, 15 h	Methyl (*E*)-3-butoxy-4-(3-(4-methoxyphenyl)-prop-1-enyl)benzoate (**39**)	Boiteau et al. 2010

Table 3.5 contd....

...Table 3.5 contd.

Precursor (EO/ Compound)	Conditions	Product	References
	With H_2O, DDQ, $PdCl_2$ in $C_2H_4Cl_2$ T: 50°C, 2 h; stereoselective reaction	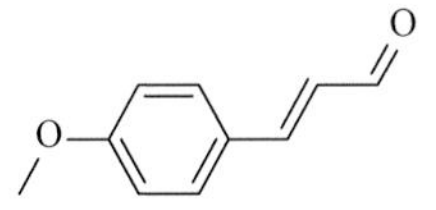 (*E*)-3-(4-Methoxyphenyl) acrylaldehyde (**40**)	Wang et al. 2009, Chen et al. 2011
3	With KOH in ButOH 3 h, T_{room}	E-Isosafrole (**24**)	Sudo et al. 2006
	With BCl_3, TBAI in CH_2Cl_2 T: –78°C, dealkylation, 1 h	2-Hydroxychavicol (**27**)	Pandey and Bani 2010
4	With 4,4'-di-tert-butylbiphenyl; Li_2; 2-propanol, $NiCl_2$ in THF; T: 20–76°C; inert atmosphere; chemoselective reaction	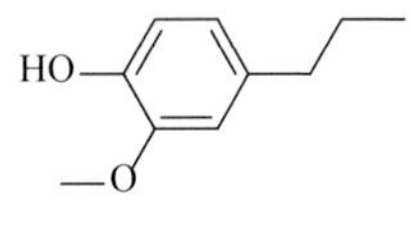 2-Methoxy-4-propylphenol (**41**)	Alonso et al. 2009
	With CH_3I, K_2CO_3 in acetone; T: 60°C; 24 h	**5**	Thota et al. 2010
6	With H_2O_2; HRP in CH_3OH T: 18°C; addition; oxidation; pH: 3; 1 hour With $Ce(NH_4)_2(NO_3)_6$ in THF T: 0°C; 1.5 hours; inert atmosphere; reagent/catalyst solvent	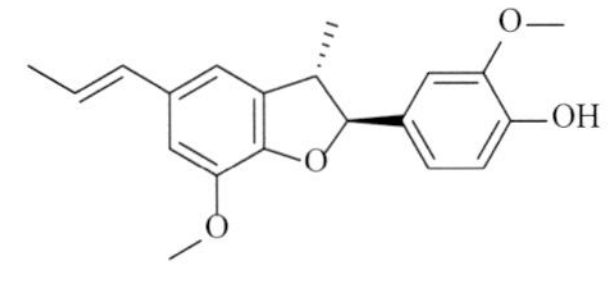 (*E*)-2-Methoxy-4-(7-methoxy-3-methyl-5-(prop-1-enyl)-2,3-dihydrobenzofuran-2-yl)phenol (**43**) or Licarin A	Nascimento et al. 2000 Chen et al. 2013, Néris et al. 2013
	Stage 1: 3,4-dimethoxy-benzaldehyde; benzo[1,3] dioxolo-5-ylamine in CH_3CN T: 23°C; 0.3 hours; inert atmosphere Stage 2: 6 with BF_3OEt_2 in ACN; T: 60°C; inert atmosphere	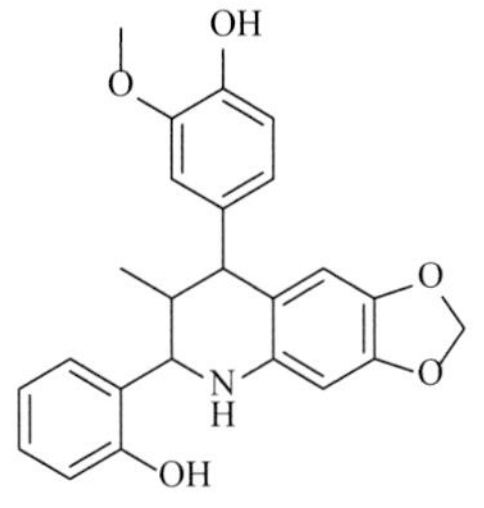 2-(2-Hydroxyphenyl)-4-(4-hydroxy-3-methoxy-phenyl)-3-methyl-6,7-methylenedioxy-1,2,3,4-tetrahydroquinoline (**44**)	Kouznetsov et al. 2016
7	With $KMnO_4$; $CuSO_4$ $5H_2O$ 0.125 h; neat (no solvent) microwave irradiation	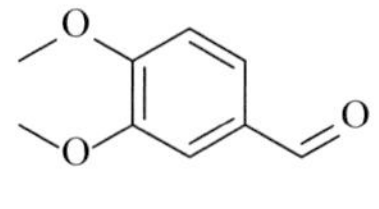 3,4-Dimethoxybenzaldehyde (**45**)	Luu et al. 2009, Faria et al. 2011

Table 3.5 contd....

...Table 3.5 contd.

Precursor (EO/ Compound)	Conditions	Product	References
8	With 4-methoxy-N-oxidanyl-benzenecarboximidoyl chloride, Et_3N in CH_2Cl_2 T: 0–20°C; 17 h	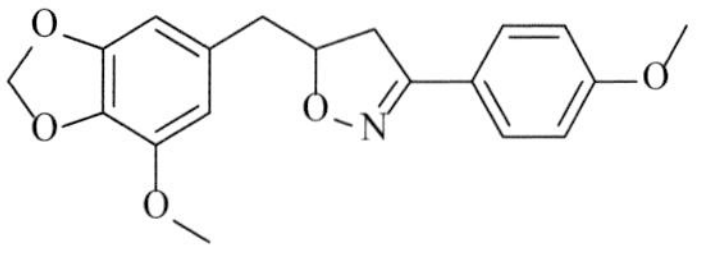 5-((7-Methoxybenzo[*d*][1,3]dioxol-5-yl)methyl)-3-(4-methoxyphenyl)-4,5-dihydroisoxazole (**46**)	Tsyganov et al. 2016
9	1. **9** with C_4H_9Li in THF T: –78°C –0°C; 1 hour 2. Benzophenone in THF T: –78°C; 0.08 hours	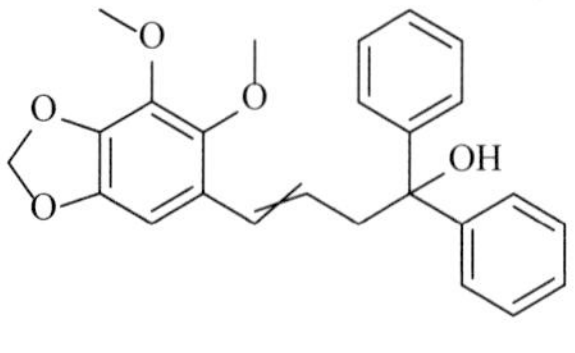 4-(6,7-Dimethoxybenzo[*d*][1,3]dioxol-5-yl)-1,1-diphenylbut-3-en-1-ol (**47**)	Arnason et al. 2013
Ac. calumus oil **12**	1. OsO_4 (cat), $NaIO_4$ 2. DDQ/Dioxane	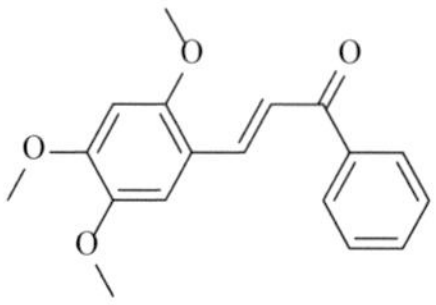 (*E*)-1-Phenyl-3-(2,4,5-trimethoxy-phenyl)prop-2-en-1-one (**48**)	Kumar et al. 2010
13	Multi-step reaction with two steps 1. TBAB; KOH, 0.67 h, 100°C neat (no solvent) 2. pyridine; O_3/CH_3OH; $CHCl_3$, –15°C	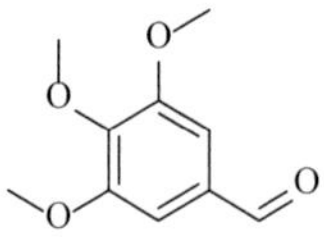 3,4,5-Trimethoxybenzaldehyde (**49**)	Titov et al. 2011
17	With BBr_3 in CH_2Cl_2 T: 0°C; 1 hour; inert atmosphere	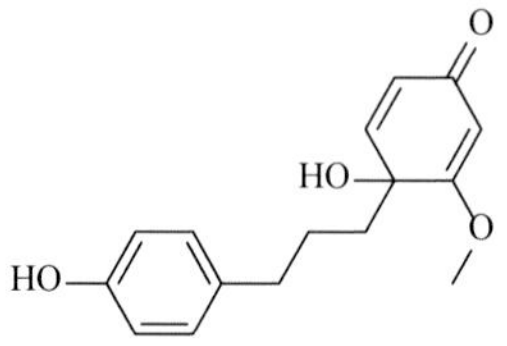 4-Hydroxy-4-(3-(4-hydroxyphenyl)propyl)-3-methoxycyclohexa-2,5-dien-1-one (**50**) or Broussonone A	Ahn et al. 2012, Jo et al. 2015
Ill. religiosum **24**	With $NaIO_4$; OsO_4; BTMAC in THF; water; microwave irradiation	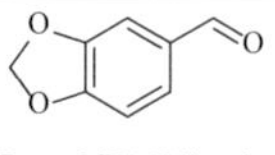 Benzo[*d*][1,3]dioxole-5-carbaldehyde (**51**)	Beckford et al. 2011, Sharma et al. 2013
Anise EO **1**	C_6H_5CHO and o-NO_2ArNH_2, BF_3OEt_2 in ACN T: 70°C, imino Diels-Alder reaction, 10 hours	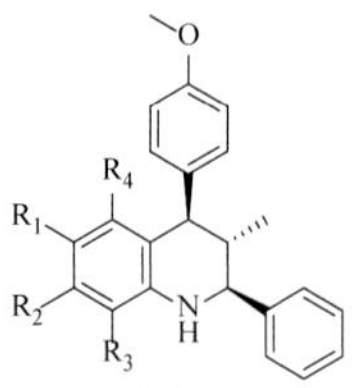 (**52a-h**)	Kouznetsov et al. 2007, Sheikhhosseini et al. 2012

Table 3.5 contd....

...Table 3.5 contd.

Precursor (EO/ Compound)	Conditions	Product	References
16	With quinone, BF_3OEt_2 in different solvent(s) (PEG-400) T: 0–20°C, 14 hours	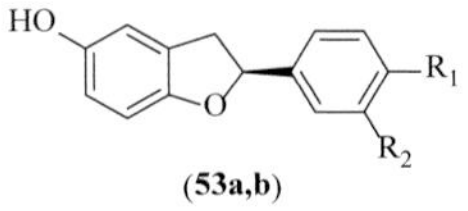 HO R_1 R_2 **(53a,b)**	Kouznetsov et al. 2008, Sheikhhosseini et al. 2012
	With $ArNH_2$, PhCHO in MeCN and in PEG-400 T: 0–20°C, 14 hours	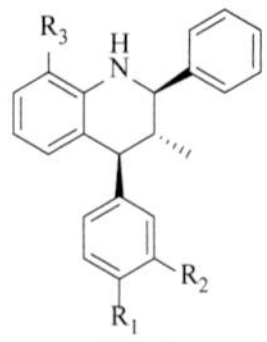 R_3 H N R_2 R_1 **(54a-f)**	Kouznetsov et al. 2008, Sheikhhosseini et al. 2012

Accordingly, compound **36** is a precursor for obtaining of 4(5)-aminosalicylic acid derivatives. Meanwhile, **38** is a promoter for oxidating of coupling-phenolic styrenes required to obtain 2-aryldihydrobenzofuran derivatives. Other interesting compounds are **39** and **49**, which are precursors for the synthesis of 3-phenyl acrylic acids and trimethoprim (antibacterial) and are incorporated into some cosmetic/pharmaceutical products for uses in human/veterinary medicine. Respectively, the compounds **24** and **27** are a precursor for preparing salts with positive contractibility of cardiac muscle (treatment for heart failure), and a protective agent on induced Alzheimer's disease (in rat). Compound **51** is a precursor for biosynthesis of mixed-ligand-piperonal thiosemicarbazone diimine complexes of ruthenium (II) for evaluation as anticancer and antibacterial agent.

A group of compounds are biologically active molecules, e.g.: **37**—antifeedant agent (against tobacco armyworm); **40**—antiviral agent (against human respiratory syncytial virus, in larynx carcinoma cell line); **43** (Licarin A)—antiprotozoal agent (*Leishmania* promastigotes); **44**, **52**, **54** (tetrahydroquinoline derivatives)—antitumoral, cytotoxic and antiprotozoal agents; **45**—antifungal agent (*Candida* and *Cryptococcus*); **46** (pycnanthulignene D derivative)—antiproliferative agent (sea urchin embryo); **47** (dillapiole derivative)—synergic agent (with pesticides or as pharmaco-enhancer); **48** (methoxylated chalcone derivative)—antiparasitic agent (*Plasmodium falciparum*); **50** (Broussonone A)—enzyme inhibitory agent (pancreatic lipase).

Finally, only one compound (**41**) has showed dual functions as, e.g., catalytic agent (with nickel nanoparticles) or kairomone (attractive male of *Bactrocera latifrons*—Diptera:Tephritidae).

Conclusions

Based on the analysis of the scientific literature consulted, we could conclude that:

The prop-1(2)-enylbenzene compounds and the EO that contain them are one kind of natural products of great importance due to the roles that they fulfill in nature by allowing the balance and preservation of several species and the occurrence of them is not restricted only for a specific family/genus/species of plants. Additionally, humans have been 'exploiting' these plants/EO/phenylprop-1(2)-enoids for commercial purposes (e.g., spices and herbs, flavors, food, cosmetic and perfumery industries). However, some EO/oleoresins/extracts containing specific phenylprop-1(2)-enoids have been 'regulated' by control agencies (FDA/EC) in certain type of applications (e.g., human consumption products). Even so, the same agencies have established permissible applications/levels generating opportunities to 'profit' of these natural resources.

Regarding the antimicrobial potential of these EO and its phenylprop-1(2)-enoids, it was found that only 31 EO belonging to certain families/species of plants (27) showed the highest effectiveness against 12 microorganisms that cause human infections/ailments, with the lowest values of MIC and MBC, and the highest value of øIZ. These EO were rich in particular phenylprop-1(2)-enoids *with specific structural characteristics,* e.g., *prop-1(2)-enylphenols and/or prop-1(2)-enyl-methoxy-(methylenedioxy)-benzenes*.

Finally, the nanotechnology field that uses to EO has become an emerging area of knowledge in the last six years. Consequently is expected that in the future (*ca.* one decade) the strengthening (background/basing) of this technology will be successful in boosting its exploitation in all areas of knowledge. Nonetheless, the major developments using nanotechnology for the treatment of diseases caused by microorganisms, employing phenylprop-1(2)-enoids and EO containing them, were only focused on eugenol and some essential oils that contain it, mainly. There is still, a wide variety of other active essential oils (rich in phenylprop-1(2)-enoids) against microbial strains (affecting humans/animals). To this extent, it will be necessary to determine the effective/actual biological potentials of the EO (enhancement of activity/synergism) when they are incorporated in the nanoparticle formulations, thus allowing the expansion of the frontier of knowledge about the treatment of microbial infections applying nanotechnology to essential oils with high content of phenylprop-1(2)-enoids.

Acknowledgements

The authors would like to thank: Universidad del Norte, for its financial support through Strategic Area 'Biodiversidad, Servicios Ecosistémicos y Bienestar Humano'; Universidad Santo Tomás de Aquino (Bucaramanga) by using of the Bibliometric Unit. Finally, Colciencias ('Patrimonio Autónomo Fondo Nacional de Financiamiento para la Ciencia, la Tecnología y la Innovación, Francisco José de Caldas', Contract RC-0572–2012) for its support by means of Bio-Red-Co-CENIVAM, and Program 'Jóvenes Investigadores e Innovadores 2014/2015'.

References

21CFR172.510—Code of Federal Regulations. Title 21—Foods and Drugs. Chapter I—Food and drug administration. Department of health and human services. Subchapter B—Food for human consumption. Part 172—Food additives permitted for direct addition to food to human consumption. Subpart F—Flavoring agents and related substances. Sec. 172.510—Natural flavoring substances and natural substances used in conjunction with flavors. 2016. U.S. Food and Drugs Administration (FDA). Available from: https://www.accessdata.fda.gov/scripts/cdrh/cfdocs/cfCFR/CFRSearch.cfm?fr=172.510. (Consulted online: May 2016).

21CFR184.1282—Code of Federal Regulations. Title 21—Foods and Drugs. Chapter I—Food and drug administration. Department of health and human services. Subchapter B—Food for human consumption. Part 184—Direct food substances affirmed as generally recognized as safe. Subpart B—Listing of specific substances affirmed as GRAS. Sec. 184.1282—Dill and its derivates. 2016. U.S. Food and Drugs Administration (FDA). Available from: http://www.accessdata.fda.gov/scripts/cdrh/cfdocs/cfcfr/CFRSearch.cfm?fr=184.1282. (Consulted online: May 2016).

21CFR189—Code of Federal Regulations. Title 21—Foods and Drugs. Chapter I—Food and drug administration. Department of health and human services. Subchapter B—Food for human consumption. Part 189—Substances prohibited from use in human food. Subpart C—Substances generally prohibited from direct addition or use as human food. Sec. 189.110—Calamus and its derivatives; Sec. 189.180—Safrole. 2016. U.S. Food and Drugs Administration (FDA). Available from: http://www.accessdata.fda.gov/scripts/cdrh/cfdocs/cfcfr/CFRSearch.cfm?CFRPart=189. (Consulted online: May 2016).

Abdelkader, M.S.A. and Lockwood, G.B. 2016. Essential oils from the plant, hairy root cultures and shoot cultures of Egyptian *Anethum graveolens* (dill). J. Essent. Oil Res. 28: 104–112.

Abdolahi, A., Hassani, A., Ghuosta, Y., Bernousi, I. and Meshkatalsadat, M.H. 2010. *In vitro* efficacy of four plant essential oils against *Botrytis cinerea* Pers.:Fr. and *Mucor piriformis* A. Fischer. J. Essent. Oil Bear. Pl. 13: 97–107.

Abdollahnejad, F., Kobarfard, F., Kamalinejad, M., Mehrgan, H. and Babaeian, M. 2016. Yield, chemical composition and antibacterial activity of *Artemisia dracunculus* L. essential oils obtained by two different methods. J. Essent Oil Bear. Pl. 19: 574–81.

Aboutabl, E.A., Meselhy, K.M., Elkhreisy, E.M., Nassar, M.I. and Fawzi, R. 2011. Composition and bioactivity of essential oils from leaves and fruits of *Myrtus communis* and *Eugenia supraxillaris* (Myrtaceae) grown in Egypt. J. Essent. Oil Bear. Pl. 14: 192–200.

Aguiar, J.J.S., Sousa, C.P.B., Araruna, M.K.A., Silva, M.K.N., Portelo, A.C., Lopes, J.C., Carvalho, V.R.A., Figueredo, F.G., Bitu, V.C.N., Coutinho, H.D.M., Miranda, T.A.S. and Matias, E.F.F. 2015. Antibacterial and modifying-antibiotic activities of the essential oils of *Ocimum gratissimum* L. and *Plectranthus amboinicus* L. Eur. J. Integr. Med. 7: 151–56.

Ahmad, N., Alam, M.K., Shehbaz, A., Khan, A., Mannan, A., Hakim, S.R., Bisht, D. and Owais, M. 2005. Antimicrobial activity of clove oil and its potential in the treatment of *Vaginal candidiasis*. J. Drug Target. 13: 555–61.

Ahn, J.H., Liu, Q., Lee, C., Ahn, M.-J., Yoo, H.-S., Hwang, B.Y. and Lee, M.K. 2012. A new pancreatic lipase inhibitor from *Broussonetia kanzinoki*. Bioorg. Med. Chem. Lett. 22: 2760–63.

Aiyeloja, A. and Bello, O. 2006. Ethnobotanical potentials of common herbs in Nigeria: A case study of Enugu state. Educ. Res. Rev. 1: 16–22.

Alonso, F., Riente, P. and Yus, M. 2009. Transfer hydrogenation of olefins catalysed by nickel nanoparticles. Tetrahedron. 65: 10637–43.

Aly, S.E., Sabry, B.A., Shaheen, M.S. and Hathout, A.S. 2016. Assessment of antimycotoxigenic and antioxidant activity of star anise (*Illicium verum*) *in vitro*. J. Saudi Soc. Agr. Sci. 15: 20–27.

Andrade, B.F.M.T., Barbosa, L.N., Probst, I.S. and Junior, A.F. 2014. Antimicrobial activity of essential oils. J. Essent. Oil Res. 26: 34–40.

Ansari, M.J., Al-Ghamdi, A., Usmani, S., Al-Waili, N., Nuru, A., Sharma, D., Khan, K.A., Kaur, M. and Omer, M. 2016. *In vitro* evaluation of the effects of some plant essential oils on *Paenibacillus larvae*, the causative agent of American foulbrood. Biotechnol. Biotec. Eq. 30: 49–55.

Anthony, K.P., Deolu-Sobogun, S.A. and Saleh, M.A. 2012. Comprehensive assessment of antioxidant activity of essential oils. J. Food Sci. 77: C839–43.

Anwar, F., Hussain, A.I., Sherazi, S.T.H. and Bhanger, M.I. 2009. Changes in composition and antioxidant and antimicrobial activities of essential oil of fennel *Foeniculum vulgare* Mill. fruit at different stages of maturity. J. Herbs Spic. Med. Plants 15: 187–202.

Aridoğan, B.C., Baydar, H., Kaya, S., Demirci, M., Özbaşar, D. and Mumcu, E. 2002. Antimicrobial activity and chemical composition of some essential oils. Arch. Pharm. Res. 25: 860–64.

Arnason, J.T., Durst, T. and Foster, B. 2013. Derivatives of dillapiol and related monolignans and use thereof. (Patent US20130012477 A1).

Aumeeruddy-Elalfi, Z., Gurib-Fakim, A. and Mahomoodally, F. 2015. Antimicrobial, antibiotic potentiating activity and phytochemical profile of essential oils from exotic and endemic medicinal plants of Mauritius. Ind. Crop. Prod. 71: 197–204.

Aurore, G.S., Abaul, J., Bourgeois, P. and Luc, J. 1998. Antibacterial and antifungal activities of the essential oils of *Pimenta racemosa* var. *racemosa* P. Miller (J.W. Moore) (Myrtaceae). J. Essent. Oil Res. 10: 161–64.

Autran, E.S., Neves, I.A., da Silva, C.S.B., Santos, G.K.N., da Câmara, C.A.G. and Navarro, D.M.A.F. 2009. Chemical composition, oviposition deterrent and larvicidal activities against *Aedes aegypti* of essential oils from *Piper marginatum* Jacq. (Piperaceae). Bioresource Technol. 100: 2284–88.

Azizkhani, M. and Parsaeimehr, M. 2015. Effects of *Cinnamomum zeylanicum* and *Ocimum basilicum* essential oils on the growth of *Staphylococcus aureus* ATCC 29213 and gene expression of enterotoxins A, C and E. J. Essent. Oil Res. 27: 506–13.

Baratta, M.T., Dorman, H.J.D., Deans, S.G., Figueiredo, A.C., Barroso, J.G. and Ruberto, G. 1998. Antimicrobial and antioxidant properties of some commercial essential oils. Flavour Frag. J. 13: 235–44.

Baruah, A. and Nath, S.C. 2006. Leaf essential oils of *Cinnamomum glanduliferum* (Wall) Meissn and *Cinnamomum glaucescens* (Nees) Meissn. J. Essent. Oil Res. 18: 200–02.

Beckford, F.A., Thessing, J., Shaloski, M., Mbarushimana, P.C., Brock, A., Didion, J., Woods, J., Gonzalez-Sarrías, A. and Seeram, N.P. 2011. Synthesis and characterization of mixed-ligand diimine-piperonal thiosemicarbazone complexes of ruthenium (II): Biophysical investigations and biological evaluation as anticancer and antibacterial agents. J. Mol. Struct. 992: 39–47.

Bendiabdellah, A., El Amine Dib, M., Djabou, N., Hassani, F., Paolini, J., Tabti, B., Costa, J. and Muselli, A. 2014. *Daucus carota* ssp. *hispanicus* Gouan. essential oils: chemical variability and fungitoxic activity. J. Essent. Oil Res. 26: 427–40.

Bernard, C.B., Krishinamurty, H.G., Chauret, D., Durst, T., Philogene, B.J., Sanchés-Vindas, P., Hasbun, C., Poveda, L., San Román, L. and Arnason, J.T. 1995. Insecticidal defenses of Piperaceae from the neotropics. J. Chem. Ecol. 21: 801–14.

Bisht, D.S., Menon, K.R.K. and Kumar-Singhal, M. 2014. Comparative antimicrobial activity of essential oils of *Cuminum cyminum* L. and *Foeniculum vulgare* Mill. seeds against *Salmonella typhimurium* and *Escherichia coli*. J. Essent. Oil Bear. Pl. 17: 617–22.

Bisht, D., Pal, A., Chanotiya, C.S., Mishra, D. and Pandey, K.N. 2011. Terpenoid composition and antifungal activity of three commercially important essential oils against *Aspergillus flavus* and *Aspergillus niger*. Nat. Prod. Res. 25: 1993–98.

Blewitt, M. and Southwell, I.A. 2000. *Backhousia anisata* Vickery, an alternative source of *E*-anethole. J. Essent. Oil Res. 12: 445–54.

Blum, T.R., Zhu, Y., Nordeen, S.A. and Yoon, T.P. 2014. Photocatalytic synthesis of dihydrobenzofurans by oxidative [3+2] cycloaddition of phenols. Angew. Chem. Int. Ed. 53: 11056–59.

Boiteau, J.-G., Clary, L., Pascal, J.-C. and Chambon, S. 2010. Novel 3-phenyl acrylic acid compound activators of type ppar receptors and pharmaceutical/cosmetic compositions comprised thereof. (Patent US8404836 B2).

Boussaada, O., Ammar, S., Saidana, D., Chriaa, J., Chraif, I., Daami, M., Helal, A.N. and Mighri, Z. 2008. Chemical composition and antimicrobial activity of volatile components from capitula and aerial parts of *Rhaponticum acaule* DC growing wild in Tunisia. Microbiol. Res. 163: 87–95.

Brazão, M.A.B., Brazão, F.V., Maia, J.G.S. and Monteiro, M.C. 2014. Antibacterial activity of the *Piper aduncum* oil and dillapiole, its main constituent, against multidrug-resistant strains. Bol. Latinoam. Caribe 13: 517–26.

Bristol, D. 2011. NTP 3-month toxicity studies of estragole (CAS No. 140-67-0) administered by gavage to F344/N rats and B6C3F1 mice. Toxic. Rep. Ser. 82: 1–111.

Brophy, J.J., Goldsack, R.J., Fookes, C.J.R. and Forster, P.I. 1995. Leaf oils of the genus *Backhousia* Myrtaceae. J. Essent. Oil Res. 7: 237–54.

Brophy, J.J., Goldsack, R.J., Doran, J.C., Craven, L.A. and Lepschi, B.J. 1999. A comparison of the leaf oils of *Melaleuca squamophloia* with those of its close relatives, *M. styphelioides* and *M. bracteata*. J. Essent. Oil Res. 11: 327–32.

Brophy, J.J., Goldsack, R.J., Punruckvong, A., Forster, P.I. and Fookes, C.J.R. 1997. Essential oils of the genus *Crowea* Rutaceae. J. Essent. Oil Res. 9: 401–09.

Brownell, W.B. and Weston, A.W. 1951. A new synthesis of croweacin aldehyde. J. Am. Chem. Soc. 73: 4971–72.

Brusotti, G., Cesari, I., Gilardoni, G., Tosi, S., Grisoli, P., Picco, A.M. and Caccialanza, G. 2012. Chemical composition and antimicrobial activity of *Phyllanthus muellerianus* Kuntze Excel essential oil. J. Ethnopharmacol. 142: 657–62.

Buechi, G. and Chu, P.-S. 1981. Synthesis of megaphone. J. Am. Chem. Soc. 103: 2718–21.

Bullitta, S., Piluzza, G. and Viegi, L. 2007. Plant resources used for traditional ethnoveterinary phytotherapy in Sardinia (Italy). Genet. Resour. Crop. Ev. 54: 1447–64.

Burdock, G.A. 2010. Fenaroli's handbook of flavor ingredients. 6th Ed. Boca Raton: CRC Press (Taylor and Francis Group). 2136 pp.

Cabral, P.H.B., Campos, R.M., Fonteles, M.C., Santos, C.F., Cardoso, J.H.L. and do Nascimento, N.R.F. 2014. Effects of the essential oil of *Croton zehntneri* and its major components, anethole and estragole, on the rat corpora cavernosa. Life Sci. 112: 74–81.

Camarda, L. and Di Stefano, V. 2003. Essential oil of leaves and fruits of *Athamanta sicula* L. Apiaceae. J. Essent. Oil Res. 15: 133–34.

Camarillo, G., Ortega, L.D., Serrato, M.A. and Rodríguez, C. 2009. Biological activity of *Tagetes filifolia* (Asteraceae) on *Trialeurodes vaporariorum* (Hemiptera: Aleyrodidae). Rev. Colomb. Entomol. 35: 177–84.

Camps, F. 1988. Relación planta-insecto. Insecticida de origen vegetal. pp. 69–85. *In*: Insecticidas biorracionales. Editado por Xavier Bellés. Consejo Superior de Investigaciones Cientificas, Madrid.

Carović-Stanko, K., Fruk, G., Satovic, Z., Ivić, D., Politeo, O., Sever, Z., Grdiša, M., Strikić, F. and Jemrić, T. 2013. Effects of *Ocimum* spp. essential oil on *Monilinia laxa in vitro*. J. Essent. Oil Res. 25: 143–48.

Chaieb, K., Hajlaoui, H., Zmantar, T., Kahla-Nakbi, A.B., Rouabhia, M., Mahdouani, K. and Bakhrouf, A. 2007. The chemical composition and biological activity of clove essential oil, *Eugenia caryophyllata* (*Syzigium aromaticum* L. Myrtaceae): a short review. Phytother. Res. 21: 501–06.

Chang, S., Nafchi, A.M. and Karim, A.A. 2016. Chemical composition, antioxidant activity and antimicrobial properties of three selected varieties of Iranian fennel seeds. J. Essent. Oil Res. 28: 357–63.

Chao, S.C., Young, D.G. and Oberg, C.J. 2000. Screening for inhibitory activity of essential oils on selected bacteria, fungi and viruses. J. Essent. Oil Res. 12: 639–49.

Chaumont, J.-P., Mandin, D., Sanda, K., Koba, K. and de Souza, C.A. 2001. Activités antimicrobiennes *in vitro* de cinq huiles essentielles de Lamiacées togolaises vis-à-vis de germes représentatifs de la microflore cutanée. Acta. Bot. Gallica 148: 93–101.

Chen, F., Shi, Z., Neoh, K.G. and Kang, E.T. 2009. Antioxidant and antibacterial activities of eugenol and carvacrol-grafted chitosan nanoparticles. Biotechnol. Bioeng. 104: 30–39.

Chen, H., Jiang, H., Cai, C., Dong, J. and Fu, W. 2011. Facile synthesis of (*E*)-alkenyl aldehydes from allyl arenes or alkenes via Pd(II)-catalyzed direct oxygenation of allylic C-H bond. Org. Lett. 13: 992–94.

Chen, P.-Y., Wu, Y.-H., Hsu, M.-H., Wang, T.-P. and Wang, E.-C. 2013. Cerium ammonium nitrate-mediated the oxidative dimerization of *p*-alkenylphenols: a new synthesis of substituted (±)-*trans*-dihydrobenzofurans. Tetrahedron 69: 653–57.

Chen, Y., Zeng, H., Tian, J., Ban, X., Ma, B. and Wang, Y. 2014. Dill *Anethum graveolens* L. seed essential oil induces *Candida albicans* apoptosis in a metacaspase-dependent manner. Fungal Biol. 118: 394–401.

Cheng, S.-S., Chang, H.-T., Lin, C.-Y., Chen, P.-S., Huang, C.-G., Chen, W.-J. and Chang, S.-T. 2009. Insecticidal activities of leaf and twig essential oils from *Clausena excavata* against *Aedes aegypti* and *Aedes albopictus* larvae. Pest. Manag. Sci. 65: 339–43.

Cheng, S.-S., Chung, M.-J., Chen, Y.-J. and Chang, S.-T. 2011. Antipathogenic activities and chemical composition of *Cinnamomum osmophloeum* and *Cinnamomum zeylanicum* leaf essential oils. J. Wood Chem. Technol. 31: 73–87.

Chizzola, R. 2011. Composition of the essential oils from *Anthriscus cerefolium* var. *trichocarpa* and *A. caucalis* growing wild in the urban area of Vienna Austria. Nat. Prod. Commun. 6: 1147–50.

Cho, J., Kim, Y.H., Kong, J.Y., Yang, C.H. and Park, C.G. 2002. Protection of cultured rat cortical neurons from excitotoxicity by asarone, a major essential oil component in the rhizomes of *Acorus gramineus*. Life Sci. 71: 591–99.

Chudasama, K.S. and Thaker, V.S. 2012. Screening of potential antimicrobial compounds against *Xanthomonas campestris* from 100 essential oils of aromatic plants used in India: an ecofriendly approach. Arch. Phytopathol. Pfl. 45: 783–95.

Chudasama, K.S. and Thaker, V.S. 2014. Biological control of phytopathogenic bacteria *Pantoea agglomerans* and *Erwinia chrysanthemi* using 100 essential oils. Arch. Phytopathol. Pfl. 47: 2221–32.

Chung, J.Y., Choo, J.H., Lee, M.H. and Hwang, J.K. 2006. Anticariogenic activity of macelignan isolated from *Myristica fragrans* nutmeg against *Streptococcus mutans*. Phytomedicine 13: 261–66.

Ciccia, G., Coussio, J. and Mongelli, E. 2000. Insecticidal activity against *Aedes aegypti* larvae of some medicinal South American plants. J. Ethnopharmacol. 72: 185–89.

Committee of experts on cosmetic products. 2008. Active ingredients used in cosmetics: safety survey. Strasbourg: Council of Europe Publishing. 445 pp.

Curini, M., Epifano, F., Genovese, S., Tammaro, F. and Menghini, L. 2006. Composition and antimicrobial activity of the essential oil of *Artemisia dracunculus* Piemontese from Italy. Chem. Nat. Compd. 42: 738–39.

D'Mello, S.R., Das, S.K. and Das, N.G. 2009. Polymeric nanoparticles for small-molecule drugs: biodegradation of polymers and fabrication of nanoparticles. pp. 16–34. *In*: Pathak, Y. and Thassu, D. (eds.). Drug Delivery Nanoparticles Formulation and Characterization. CRC Press, Boca Raton.

da Costa, J.G.M., Rodrigues, F.F.G., Angélico, E.C., Pereira, C.K.B., de Souza, E.O., Caldas, G.F.R., Silva, M.R., Santos, N.K.A., Mota, M.L. and dos Santos, P.F. 2008. Composição química e avaliação da atividade antibacteriana e toxicidade do óleo essencial de *Croton zehntneri* variedade estragol. Braz. J. Pharmacog. 18: 583–86.

da Silva, J.K., Andrade, E.H., Guimarães, E.F. and Maia, J.G. 2010. Essential oil composition, antioxidant capacity and antifungal activity of *Piper divaricatum*. Nat. Prod. Commun. 5: 477–80.

da Silva, J.K.R., Silva, J.R.A., Nascimento, S.B., da Luz, S.F.M., Meireles, E.N., Alves, C.N., Ramos, A.R. and Maia, J.G.S. 2014. Antifungal activity and computational study of constituents from *Piper divaricatum* essential oil against *Fusarium* infection in black pepper. Molecules 19: 17926–42.

Damianova, S., Stoyanova, A., Konakchiev, A. and Djurdjev, I. 2004. Supercritical carbon dioxide extracts of spices. 2. Fennel (*Foeniculum vulgare* Mill. var. *dulce* Mill.). J. Essent. Oil Bear. Pl. 7: 247–49.

Damyanova, S. and Stoyanova, A. 2007. Antimicrobial activity of aromatic products. 14 extracts from fruits of sweet fennel (*Foeniculum vulgare* Mill. var. *dulce* Mill.) and coriander (*Coriandrum sativum* L.). J. Essent. Oil Bear. Pl. 10: 440–45.

Dan, Y., Liu, H.Y., Gao, W.W. and Chen, S.L. 2010. Activities of essential oils from *Asarum heterotropoides* var. *mandshuricum* against five phytopathogens. Crop. Prot. 29: 295–99.

Dasgupta, N. and De, B. 2004. Antioxidant activity of *Piper betle* L. leaf extract *in vitro*. Food Chem. 88: 219–24.

Dawidar, A.M., Abdel Mogib, M., El-Ghorab, A.H., Mahfouz, M., Elsaid, F.G. and Hussien, Kh. 2008. Chemical composition and effect of photo-oxygenation on biological activities of Egyptian commercial anise and fennel essential oils. J. Essent. Oil Bear. Pl. 11: 124–36.

de Abreu, A.M., Sevegnani, L., Machicado, A.R., Zimermann, D. and Rebelo, R.A. 2002. *Piper mikanianum* Kunth Steudel from Santa Catarina, Brazil—a new source of safrole. J. Essent. Oil Res. 14: 361–63.

de Almeida, R.R.P., Souto, R.N.P., Bastos, C.N., da Silva, M.H.L. and Maia, J.G.S. 2009. Chemical variation in *Piper aduncum* and biological properties of its dillapiole-rich essential oil. Chem. Biodivers. 6: 1427–34.

De Feo, V., Della Porta, G., Urrunaga Soria, E., Urrunaga Soria, R. and Senatore, F. 1998. Composition of the essential oil of *Tagetes filifolia* Lag. Flavour Frag. J. 13: 145–47.

de Heluani, C.S., de Lampasona, M.P., Vega, M.I. and Catalan, C.A.N. 2005. Antimicrobial activity and chemical composition of the leaf and root oils from *Croton hieronymi* Griseb. J. Essent. Oil Res. 17: 351–53.

de Lima, G.P.G., de Souza, T.M., Freire, G.P., Farias, D.F., Cunha, A.P., Ricardo, N.M.P.S., de Morais, S.M. and Carvalho, A.F.U. 2013. Further insecticidal activities of essential oils from *Lippia sidoides* and *Croton*species against *Aedes aegypti* L. Parasitol. Res. 112: 1953–58.

de Lima, M.G.A., Maia, I.C.C., de Sousa, B.D., de Morais, S.M. and Freitas, S.M. 2006. Effect of stalk and leaf extracts from Euphorbiaceae species on *Aedes aegypti* (Diptera, Culicidae) Larvae. Rev. Inst. Med. Trop. S. Paulo 48: 211–14.

De, M., De, A.K., Mukhopadhyay, R., Miró, M. and Anerjee, A.B. 2001. Actividad antimicrobiana de *Illicium verum* Hook. f. Ars. Pharm. 42: 209–220.

Delespaul, Q., de Billerbeck, V.G., Roques, C.G., Michel, G., Marquier-Viñuales, C. and Bessière, J.M. 2000. The antifungal activity of essential oils as determined by different screening methods. J. Essent. Oil Res. 12: 256–66.

Diao, W.-R., Hu, Q.-P., Zhang, H. and Xu, J.-G. 2014. Chemical composition, antibacterial activity and mechanism of action of essential oil from seeds of fennel *Foeniculum vulgare* Mill. Food Control 35: 109–16.

Dickins, M., Bridges, J.W., Elcombe, C.R. and Netter, K.J. 1978. A novel haemoprotein induced by isosafrole pretreatment in the rat. Biochem. Biophys. Res. Commun. 80: 89–96.

Diep, P.T.M., Pawloska, A.M., Cioni, P.L., Minh, C.V., Huong, L.M. and Braca, A. 2009. Chemical composition and antimicrobial activity of *Clausena indica* (Dalz) Oliv. (Rutaceae) essential oil from Vietnam. Nat. Prod. Commun. 4: 869–72. (Abstract).

Dinan, L., Bourne, P.C., Meng, Y., Sarker, S.D., Tolentino, R.B. and Whiting, P. 2001. Assessment of natural products in the *Drosophila melanogaster* BII cell bioassay for ecdysteroid agonist and antagonist activities. Cell. Mol. Life Sci. 58: 321–42.

Dognini, J., Meneghetti, E.K., Teske, M.N., Begnini, I.M., Rebelo, R.A., Dalmarco, E.M., Verdi, M. and de Gasper, A.L. 2012. Antibacterial activity of high safrole contain essential oils from *Piper xylosteoides* (Kunth) Steudel. J. Essent. Oil Res. 24: 241–44.

Donati, M., Mondin, A., Chen, Z., Miranda, F.M., do Nascimento Jr, B.B., Schirato, G., Pastore, P. and Froldi, G. 2015. Radical scavenging and antimicrobial activities of *Croton zehntneri*, *Pterodon emarginatus* and *Schinopsis brasiliensis* essential oils and their major constituents: estragole, *trans*-anethole, β-caryophyllene and myrcene. Nat. Prod. Res. 29: 939–46.

Dũng, N.X., Mõi, L.D., Hung, N.D. and Leclercq, P.A. 1995. Constituents of the essential oils of *Cinnamomum parthenoxylon* Jack Nees from Vietnam. J. Essent. Oil Res. 7: 53–56.

Durmaz, H., Sagun, E., Tarakci, Z. and Ozgokce, F. 2006. Antibacterial activities of *Allium vineale*, *Chaerophyllum macropodum* and *Prangos ferulacea*. Afr. J. Biotechnol. 5: 1795–98.

Durmaz, H., Aygun, O., Sancak, H. and Celik, H. 2015. Oxidant/antioxidant status of herbs *Allium vineale* and *Chaerophyllum macropodum* used for manufacture of van herby cheese. Int. J. Sci. Technol. Res. 1: 288–96.
Ebadollahi, A., Safaralizadeh, M.H., Hoseini, S.A., Ashouri, S. and Sharifian, I. 2010a. Insecticidal activity of essential oil of *Agastache foeniculum* against *Ephestia kuehniella* and *Plodia interpunctella* (Lepidoptera: Pyralidae). Mun. Ent. Zool. 5: 785–91.
Ebadollahi, A., Safaralizadeh, M.H., Pourmirza, A.A. and Gheibi, S.A. 2010b. Toxicity of essential oil of *Agastache foeniculum* (Pursh) Kuntze to *Oryzaephilus surinamensis* L. and *Lasioderma serricorne* F. J. Plant Protect. Res. 50: 215–19.
Ebadollahi, A. 2011. Chemical constituents and toxicity of *Agastache foeniculum* (Pursh) Kuntze essential oil against two stored-product insect pests. Chilean J. Agric. Res. 71: 212–17.
EC—European Commission. Health & Consumer Protection Directorate. 2001a. General. Scientific committee on food. Opinion of the scientific committee on food on estragole (1-Allyl-4-methoxybenzene). SCF/CS/FLAV/FLAVOUR/6 ADD2 FINAL. Available from: http://ec.europa.eu/food/fs/sc/scf/out104_en.pdf.
EC—European Commission. Health & Consumer Protection Directorate. 2001b. General. Scientific committee on food. Opinion of the committee on food on the presence of methyleugenol in flavourings and other food ingredients with flavouring properties. SCF/CS/FLAV/FLAVOUR/4 ADD1 Final. Available from: http://ec.europa.eu/food/fs/sc/scf/out102_en.pdf.
EC—European Parliament and Council of the European Union. 2009. Regulation (EC) No 1223/2009 of the European Parliament and of the Council of 30 november 2009 on cosmetic products. Official Journal of the European Union (22.12.2009). L 342/59. http://eur-lex.europa.eu/lexuriserv/LexUriServ/LexUriServ.do?uri=OJ:L:2009:342:0059:0209:en:pdf.
Ehsani, A. and Mahmoudi, R. 2012. Phytochemical properties and hygienic effects of *Allium ascalonicum* and *Pimpinella anisum* essential oils in Iranian white brined cheese. J. Essent. Oil Bear. Pl. 15: 1013–20.
El Kolli, M., Laouer, H., El Kolli, H., Akkal, S. and Sahli, F. 2016. Chemical analysis, antimicrobial and anti-oxidative properties of *Daucus gracilis* essential oil and its mechanism of action. Asian Pac. J. Trop. Biomed. 6: 8–15.
El-Mesallamy, A.M.D., El-Gerby, M., Abd El Azim, M.H.M. and Awad, A. 2012. Antioxidant, antimicrobial activities and volatile constituents of clove flower buds oil. J. Essent. Oil Bear. Pl. 15: 900–07.
El-Tantawy, M.E., Shams, M.M. and Afifi, M.S. 2016. Chemical composition and biological evaluation of the volatile constituents from the aerial parts of *Nephrolepis exaltata* L. and *Nephrolepis cordifolia* L. C. Presl grown in Egypt. Nat. Prod. Res. 30: 1197–201.
Elsevier. 2016. Scopus database. Avalaible: url=http://www.scopus.com/home.url/ (consulted online: January–July 2016).
Engel, R., Nahrstedt, A. and Hammerschmidt, F.J. 1995. Composition of the essential oils of *Cedronella canariensis* (L.) Webb et Berth, ssp. *canariensis* and ssp. *anisata* f. *glabra* and *f. pubescens*. J. Essent. Oil Res. 7: 473–87.
Evrendilek, G.A. 2015. Empirical prediction and validation of antibacterial inhibitory effects of various plant essential oils on common pathogenic bacteria. Int. J. Food Microbiol. 202: 35–41.
Fajemiroye, J.O., Galdino, P.M., De Paula, J.A., Rocha, F.F., Akanmu, M.A., Vanderlinde, F.A., Zjawionye, J.K. and Costa, E.A. 2014. Anxiolytic and antidepressant like effects of natural food flavour E-methyl isoeugenol. Food Funct. 5: 1819–28.
Farag, R.S., Shalaby, A.S., El-Baroty, G.A., Ibrahim, N.A., Ali, M.A. and Hassan, E.M. 2004. Chemical and biological evaluation of the essential oils of different *Melaleuca* species. Phytother. Res. 18: 30–35.
Faria, N.C.G., Kim, J.H., Gonçalves, L.A.P., Martins, M.L., Chan, K.L. and Campbell, B.C. 2011. Enhanced activity of antifungal drugs using natural phenolics against yeast strains of *Candida* and *Cryptococcus*. Lett. Appl. Microbiol. 52: 506–13.
Fatemi, H., Aminifard, M.H. and Mohammadi, S. 2013. Efficacy of plant essential oils on post-harvest control of rot caused by *Botrytis cinerea* on kiwi fruits. Arch. Phytopathol. Pfl. 46: 536–47.
Fatope, M.O., Marwah, R.G., Al Hadhrami, N.M., Onifade, A.K. and Williams, J.R. 2008. Identification of the chemotypes of *Ocimum forskolei* and *Ocimum basilicum* by NMR spectroscopy. Chem. Biodivers. 5: 2457–63.
Fayyaz, N., Sani, A.M. and Najafi, M.N. 2015. Antimicrobial activity and composition of essential oil from *Echinophora platyloba*. J. Essent. Oil Bear. Pl. 18: 1157–64.
Ferreira, R.G., Monteiro, M.C., da Silva, J.K.R. and Maia, J.G.S. 2016. Antifungal action of the dillapiole-rich oil of *Piper aduncum* against dermatomycoses caused by filamentous fungi. Brit. J. Medic. Med. Res. 15: 1–10.
Flamini, G., Mastrorilli, E., Cioni, P.L., Morelli, I. and Panizzi, L. 1999. Essential oil from *Crithmum maritimum* grown in Liguria (Italy): seasonal variation and antimicrobial activity. J. Essent. Oil Res. 11: 788–92.
Fontenelle, R.O.S., Morais, S.M., Brito, E.H.S., Brilhante, R.S.N., Cordeiro, R.A., Nascimento, N.R.F., Kerntopf, M.R., Sidrim, J.J.C. and Rocha, M.F.G. 2008. Antifungal activity of essential oils of *Croton* species from the Brazilian Caatinga Biome. J. Appl. Microbiol. 104: 1383–90.
Fratini, F., Casella, S., Leonardi, M., Pisseri, F., Ebani, V.V., Pistelli, L. and Pistelli, L. 2014. Antibacterial activity of essential oils, their blends and mixtures of their main constituents against some strains supporting livestock mastitis. Fitoterapia 96: 1–7.
Freire, J.M., Cardoso, M.G., Batista, L.R. and Andrade, M.A. 2011. Essential oil of *Origanum majorana* L., *Illicium verum* Hook. f. and *Cinnamomum zeylanicum* Blume: chemical and antimicrobial characterization. Rev. Bras. Pl. Med. 13: 209–14.
Fujita, Y. and Fujita, S. 1973. Miscellaneous contributions to the essential oils of plants from various territories XXXIII. essential oil of *Agastache rugosa* O. Kuntze. Yakugaku Zasshi. 93: 1679–81.
García, D., Álvarez, A., Tornos, P., Fernández, A. and Sáenz, T. 2002. Gas chromatographic-mass spectrometry study of the essential oils of *Pimenta racemosa* var. *terebinthina* and *P. racemosa* var. *grisea*. Z. Naturforsch. C. 57: 449–51.
García, M.D., Fernández, M.A., Álvarez, A. and Sáenz, M.T. 2004. Antinociceptive and anti-inflammatory effect of the aqueous extract from leaves of *Pimenta racemosa* var. *ozua* Mirtaceae. J. Ethnopharmacol. 91: 69–73.

Gardner, I., Wakazono, H., Bergin, P., de Waziers, I., Beaune, P., Kenna, J.G. and Caldwell, J. 1997. Cytochrome P450 mediated bioactivation of methyleugenol to 1'-hydroxymethyleugenol in Fischer 344 rat and human liver microsomes. Carcinogenesis 18: 1775–83.

Garneau, F.-X., Pichette, A., Gagnon, H., Jean, F.-I., Addae-Mensah, I., Osei-Safu, D., Asomaning, W.A., Oteng-Yeboah, A., Moudachirou, M. and Koumaglo, K.H. 2000. (*E*)- and (*Z*)-Foeniculin, constituents of the leaf oil of a new chemovariety of *Clausena anisata*. J. Essent. Oil Res. 12: 757–62.

Gende, L.B., Maggi, M.D., Fritz, R., Eguaras, M.J., Bailac, P.N. and Ponzi, M.I. 2009. Antimicrobial activity of *Pimpinella anisum* and *Foeniculum vulgare* essential oils against *Paenibacillus larvae*. J. Essent. Oil Res. 21: 91–93.

Ghelardini, C., Galeotti, N. and Mazzanti, G. 2001. Local anaesthetic activity of monoterpenes and phenylpropanes of essential oils. Planta Med. 67: 564–66.

Ghosh, V., Mukherjee, A. and Chandrasekaran, N. 2013. Ultrasonic emulsification of food-grade nanoemulsion formulation and evaluation of its bactericidal activity. Ultrason. Sonochem. 20: 338–44.

Gomes, C., Moreira, R.G. and Castell-Perez, E. 2011. Poly(DL-lactide-co-glycolide) (PLGA) nanoparticles with entrapped *trans*-cinnamaldehyde and eugenol for antimicrobial delivery applications. J. Food Sci. 76: N16–24.

González, R. 2002. Eugenol: propiedades farmacológicas y toxicológicas. Ventajas y desventajas de su uso. Rev. Cubana Estomatol. 39: 139–56.

Gopan R., Shiburaj, S., Sethuraman, M.G. and George, V. 2004. Chemical composition and antibacterial activity of the leaf oil from *Clausena austroindica*. J. Trop. Med. Plants 5: 233–35.

Govindarajan, M. 2010. Chemical composition and larvicidal activity of leaf essential oil from *Clausena anisata* Willd. Hook. f. ex Benth (Rutaceae) against three mosquito species. Asian Pac. J. Trop. Med. 3: 874–77.

Griffin, S.G., Leach, D.N., Markham, J.L. and Johnstone, R. 1998. Antimicrobial activity of essential oils from *Zieria*. J. Essent. Oil Res. 10: 165–74.

Guerrini, A., Sacchetti, G., Rossi, D., Paganetto, G., Muzzoli, M., Andreotti, E., Tognolini, M., Maldonado, M.E. and Bruni, R. 2009. Bioactivities of *Piper aduncum* L. and *Piper obliquum* Ruiz & Pavon Piperaceae essential oils from Eastern Ecuador. Environ. Toxicol. Phar. 27: 39–48.

Hamed, S.F., Sadek, Z. and Edris, A. 2012. Antioxidant and antimicrobial activities of clove bud essential oil and eugenol nanoparticles in alcohol-free microemulsion. J. Oleo Sci. 61: 641–48.

Hasheminejad, G. and Caldwell, J. 1994. Genotoxicity of the alkenylbenzenes α- and β-asarone, myristicin and elemicin as determined by the UDS assay in cultured rat hepatocytes. Food Chem. Toxicol. 32: 223–31.

Hee, A.K.-W. and Tan, K.-H. 2006. Transport of methyl eugenol-derived sex pheromonal components in the male fruit fly, *Bactrocera dorsalis*. Comp. Biochem. Phys. C 143: 422–28.

Ho, C.-L., Wang, E.I.-C., Wei, X.-T., Lu, S.-Y. and Su, Y.-C. 2008. Composition and bioactivities of the leaf essential oils of *Cinnamomum subavenium* Miq. from Taiwan. J. Essent. Oil Res. 20: 328–34.

Höferl, M., Buchbauer, G., Jirovetz, L., Schmidt, E., Stoyanova, A., Denkova, Z., Slavchev, A. and Geissler, M. 2009. Correlation of antimicrobial activities of various essential oils and their main aromatic volatile constituents. J. Essent. Oil Res. 21: 459–63.

Huang, H.-C., Wang, H.-F., Yih, K.-H., Chang, L.-Z. and Chang, T.-M. 2012. The dual antimelanogenic and antioxidant activities of the essential oil extracted from the leaves of *Acorus macrospadiceus* (Yamamoto) F.N. Wei et Y.K. Li. Evid-Based. Compl. Alt. 2012: 781280 (1-10).

Huang, Y., Zhao, J., Zhou, L., Wang, J., Gong, Y., Chen, X., Guo, Z., Wang, Q. and Jiang, W. 2010. Antifungal activity of the essential oil of *Illicium verum* fruit and its main component *trans*-anethole. Molecules 15: 7558–69.

Hussien, J., Hymete, A. and Rohloff, J. 2010. Volatile constituents and biological activities of *Pycnostachys abyssinica* and *Pycnostachys eminii* extracts. Pharm. Biol. 48: 1384–91.

Imamu, X., Yili, A., Aisa, H.A., Maksimov, V.V., Veshkurova, O.N. and Salikhov, S.I. 2007. Chemical composition and antimicrobial activity of essential oil from *Daucus carota sativa* seeds. Chem. Nat. Compd. 43: 495–96.

Innocent, E. and Hassanali, A. 2015. Constituents of essential oils from three plant species used in traditional medicine and insect control in Tanzania. J. Herbs Spic. Med. Plants 21: 219–29.

IPCS/WHO—International Programme on Chemical Safety/World Heath Organization. Estragole (WHO Food Additives Series 14). 472. Available: http://www.inchem.org/documents/jecfa/jecmono/v14je08.htm (consulted online: May 2016).

Jabrane, A., Jannet, H.B., Harzallah-Skhiri, F., Mastouri, M., Casanova, J. and Mighri, Z. 2009. Flower and root oils of the Tunisian *Daucus carota* L. ssp. *maritimus* (Apiaceae): integrated analyses by GC, GC/MS, and ^{13}C-NMR spectroscopy, and *in vitro* antibacterial activity. Chem. Biodivers. 6: 881–89.

Jabrane, A., Jannet, H.B., Mastouri, M., Mighri, Z. and Casanova, J. 2010. Chemical composition and *in vitro* evaluation of antioxidant and antibacterial activities of the root oil of *Ridolfia segetum* L. Moris from Tunisia. Nat. Prod. Res. 24: 491–99.

Jana, S. and Shekhawat, G.S. 2010. *Anethum graveolens*: An Indian traditional medicinal herb and spice. Pharmacogn. Rev. 4: 179–84.

Jannet, H.B. and Mighri, Z. 2007. Hydrodistillation kinetic and antibacterial effect studies of the flower essential oil from the Tunisian *Ridolfia segetum* (L.). J. Essent. Oil Res. 19: 258–61.

Janssen, A.M., Scheffer, J.J.C., Ntezurubanza, L. and Baerheim Svendesen, A. 1989. Antimicrobial activities of some *Ocimum* species grown in Rwanda. J. Ethnopharmacol. 26: 57–63.

Jantan, I.B., Yalvema, M.F., Ahmad, N.W. and Jamal, J.A. 2005. Insecticidal activities of the leaf oils of eight *Cinnamomum* species against *Aedes aegypti* and *Aedes albopictus*. Pharm. Biol. 43: 526–32.

Jantan, I.B., Moharam, B.A.K., Santhanam, J. and Jamal, J.A. 2008. Correlation between chemical composition and antifungal activity of the essential oils of eight *Cinnamomum* species. Pharm. Biol. 46: 406–12.

Jo, H., Choi, M., Viji, M., Lee, Y.H., Kwak, Y.-S., Lee, K., Choi, N.S., Lee, Y.-J., Lee, H., Hong, J.T., Lee, M.K. and Jung, J.-K. 2015. Concise synthesis of Broussonone A. Molecules 20: 15966–75.

Joshi, N., Prakash, O. and Pant, A.K. 2012. Essential oil composition and *in vitro* antibacterial activity of rhizome essential oil and β-asarone from *Acorus calamus* L. collected from Lower Himalayan Region of Uttarakhand. J. Essent. Oil Bear. Pl. 15: 32–37.

Juárez, Z.N., Hernández, L.R., Bach, H., Sánchez-Arreola, E. and Bach, H. 2015. Antifungal activity of essential oils extracted from *Agastache mexicana* ssp. *xolocotziana* and *Porophyllum linaria* against post-harvest pathogens. Ind. Crop. Prod. 74: 178–82.

Kalagatur, N.K., Mudili, V., Siddaiah, C., Gupta, V.K., Natarajan, G., Sreepathi, M.H., Vardhan, B.H. and Putcha, V.L. 2015. Antagonistic activity of *Ocimum sanctum* L. essential oil on growth and zearalenone production by *Fusarium graminearum* in maize grains. Front Microbiol. 6: 892 (1-11).

Kamble, V.A. and Patil, S.D. 2008. Spice-derived essential oils: effective antifungal and possible therapeutic agents. J. Herbs Spic. Med. Plants 14: 129–43.

Kapoor, I.P.S., Singh, B., Singh, G., Isidorov, V. and Szczepaniak, L. 2009. Chemistry, antimicrobial and antioxidant potentials of *Cinnamomum tamala* Nees and Eberm. (Tejpat) essential oil and oleoresins. Nat. Prod. Rad. 8: 106–16.

Kar, A. 2003. Pharmacognosy and pharmacobiotechnology. New Delhi: New Age International Limited Publisher. 900 pp.

Kennouche, A., Benkaci-Ali, F., Scholl, G. and Eppe, G. 2015. Chemical composition and antimicrobial activity of the essential oil of *Eugenia caryophyllata* cloves extracted by conventional and microwave techniques. J. Biol. Act. Prod. Nat. 5: 1–11.

Khaldi, A., Meddah, B., Moussaoui, A., Sonnet, P. and Akermy, M.M. 2015. Chemical composition and antifungal activity of essential oil of *Anethum graveolens* L. from South-Western Algeria (Bechar). J. Chem. Pharm. Res. 7: 615–20.

Khalfallah, A., Labed, A., Semra, Z., Ai Kaki, B., Kabouche, A., Touzani, R. and Kabouche, Z. 2011. Antibacterial activity and chemical compositon of the essential oil of *Ammi visnaga* L. Apiaceae from Constatine, Algeria. Int. J. Med. Arom. Plants 1: 302–05.

Khayyat, S.A. 2013. Photosynthesis of dimeric cinnamaldehyde, eugenol, and safrole as antimicrobial agents. J. Saudi Chem. Soc. 17: 61–65.

Khayyat, S.A. and Al-Zahrani, S.H. 2014. Thermal, photosynthesis and antibacterial studies of bioactive safrole derivative as precursor for natural flavor and fragrance. Arab. J. Chem. 7: 800–04.

Kim, J.J. 2008. Phytotoxic and antimicrobial activities and chemical analysis of leaf essential oil from *Agastache rugosa*. J. Plant Biol. 51: 276–83.

Kim, J., Lee, Y.-S., Lee, S.-G., Shin, S.-C. and Park, I.-K. 2008. Fumigant antifungal activity of plant essential oils and components from West Indian bay *Pimenta racemosa* and thyme *Thymus vulgaris* oils against two phytopathogenic fungi. Flavour Frag. J. 23: 272–77.

Kim, W.J., Hwang, K.H., Park, D.G., Kim, T.J., Kim, D.W., Choi, D.K., Moon, W.K. and Lee, K.H. 2011. Major constituents and antimicrobial activity of Korean herb *Acorus calamus*. Nat. Prod. Res. 25: 1278–81.

Kita, Y., Dohi, T., Nakae, T., Toyoda, Y., Koseki, D., Kubo, H. and Kamitanaka, T. 2015. Phenol and aniline oxidative coupling with alkenes by using hypervalent iodine dimer for the rapid access to dihydrobenzofurans and indolines. Heterocycles 90: 631–44.

Klungsøyr, J. and Scheline, R.R. 1982. Metabolism of isosafrole and dihydrosafrole in the rat. Biol. Mass. Spectrom. 9: 323–29.

Kordali, S., Kotan, R., Mavi, A., Cakir, A., Ala, A. and Yildirim, A. 2005. Determination of the chemical composition and antioxidant activity of the essential oil of *Artemisia dracunculus* and of the antifungal and antibacterial activities of turkish *Artemisia absinthium*, *A. dracunculus*, *Artemisia santonicum*, and *Artemisia spicigera* essential oils. J. Agric. Food Chem. 53: 9452–58.

Kouznetsov, V.V., Romero, A.R. and Stashenko, E.E. 2007. Three-component imino Diels–Alder reaction with essential oil and seeds of anise: generation of new tetrahydroquinolines. Tetrahedron Lett. 48: 8855–60.

Kouznetsov, V.V., Merchan, D.R. and Romero, A.R. 2008. PEG-400 as green reaction medium for Lewis acid-promoted cycloaddition reactions with isoeugenol and anethole. Tetrahedron Lett. 49: 3097–100.

Kouznetsov, V.V., Merchan-Arenas, D.R., Tangarife-Castaño, V., Correa-Royero, J. and Betancur-Galvis, L. 2016. Synthesis and cytotoxic evaluation of novel 2-aryl-4-(4-hydroxy-3-methoxyphenyl)-3-methyl-6,7-methylendioxy-1,2,3,4-tetrahydroquinolines, podophyllotoxin-like molecules. Med. Chem. Res. 25: 429–37.

Kumar, A., Shukla, R., Singh, P., Prakash, B. and Dubey, N.K. 2011. Chemical composition of *Ocimum basilicum* L. essential oil and its efficacy as a preservative against fungal and aflatoxin contamination of dry fruits. Int. J. Food Sci. Tech. 46: 1840–46.

Kumar, R., Mohanakrishnan, D., Sharma, A., Kaushik, N.K., Kalia, K., Sinha, A.K. and Sahal, D. 2010. Reinvestigation of structure-activity relationship of methoxylated chalcones as antimalarials: synthesis and evaluation of 2,4,5-trimethoxy substituted patterns as lead candidates derived from abundantly available natural β-asarone. Eur. J. Med. Chem. 45: 5292–301.

Laouer, H., Meriem, El-K., Prado, S. and Baldovini, N. 2009. An antibacterial and antifungal phenylpropanoid from *Carum montanum* (Coss. et Dur.) Benth. et Hook. Phytother. Res. 23: 1726–30.

Leal, L.F., Miguel, O.G., Silva, R.Z., Yunes, R.A., Santos, A.S. and Cechinel-Filho, V. 2005. Chemical composition of *Piper mikanianum* essential oil. J. Essent. Oil Res. 17: 316–17.

Leal-Cardoso, J.H., Matos-Brito, B.G., Lopes-Junior, J.E.G., Viana-Cardoso, K.V., Sampaio-Freitas, A.B., Brasil, R.O., Coelho-de-Souza, A.N. and Albuquerque, A.A.C. 2004. Effects of estragole on the compound action potential of the rat sciatic nerve. Braz. J. Med. Biol. Res. 37: 1193–98.

Lee, M.-H., Chen, Y.-Y., Tsai, J.-W., Wang, S.-C., Watanabe, T. and Tsai, Y.-C. 2011. Inhibitory effect of β-asarone, a component of *Acorus calamus* essential oil, on inhibition of adipogenesis in 3T3-L1 cells. Food Chem. 126: 1–7.

Leporatti, M.L., Pintore, G., Foddai, M., Chessa, M., Piana, A., Petretto, G.L., Masia, M.D., Mangano, G. and Nicoletti, M. 2014. Chemical, biological, morphoanatomical and antimicrobial study of *Ocotea puchury*-major Mart. Nat. Prod. Res. 28: 294–300.

Leyva, M., Marquetti, M.C., Tacoronte, J.E., Scull, R., Tiomno, O., Mesa, A. and Montada, D. 2009. Actividad larvicida de aceites esenciales de plantas contra *Aedes aegypti* L. Diptera: Culicidae. Rev. Biomed. 20: 5–13.

Li, Z., Zhao, G., Qian, Z., Sanqi, Y., Zijun, C., Chen, J., Cai, C., Liang, X. and Guo, J. 2012. Cerebrovascular protection of β-asarone in Alzheimer's disease rats: A behavioral, cerebral blood flow, biochemical and genic study. J. Ethnopharmacol. 144: 305–12.

Lima, P.C., Lima, L.M., da Silva, K.C.M., Léda, P.H.O., de Miranda, A.L.P., Fraga, C.A.M. and Barreiro, E.J. 2000. Synthesis and analgesic activity of novel N-acylarylhydrazones and isosters, derived from natural safrole. Eur. J. Med. Chem. 35: 187–203.

Limón, I.D., Mendieta, L., Díaz, A., Chamorro, G., Espinosa, B., Zenteno, E. and Guevara, J. 2009. Neuroprotective effect of alpha-asarone on spatial memory and nitric oxide levels in rats injected with amyloid-β 25–35. Neurosci. Lett. 453: 98–103.

Linh, N.T. and Thach, L.N. 2011. Study of the essential oil of *Limnophila rugosa* (Roth.) Merr. in the South of Vietnam. J. Essent. Oil Bear. Pl. 14: 366–72.

Lis-Balchin, M. and Roth, G. 2000. Composition of the essential oils of *Pelargonium odoratissimum*, *P. exstipulatum*, and *P. x fragrans* (Geraniaceae) and their bioactivity. Flavour Frag. J. 15: 391–94.

Liu, Z.-B., Niu, W.-M., Yang, X.-H., Wang, Y. and Wang, W.-G. 2010. Study on perfume stimulating olfaction with volatile oil of *Acorus gramineus* for treatment of the Alzheimer's disease rat. J. Trad. Chin. Med. 30: 283–87.

Lopes-Lutz, D., Alviano, D.S., Alviano, C.S. and Kolodziejczyk, P.P. 2008. Screening of chemical composition, antimicrobial and antioxidant activities of *Artemisia* essential oils. Phytochemistry 69: 1732–38.

López, A., Gil, A.G. and Bello, J. 2012. Alimentos con sustancias toxicas de origen natural: plantas superiores alimenticias. pp. 191–209. *In*: Cameán, A.M. and Repetto, M. (eds.). Toxicología alimentaria. Diaz de Santos, Madrid.

Lu, M., Han, Z., Xu, Y. and Yao, L. 2013. Effects of essential oils from Chinese indigenous aromatic plants on mycelial growth and morphogenesis of three phytopathogens. Flavour Frag. J. 28: 84–92.

Luo, G. and Guenthner, T.M. 1996. Covalent binding to DNA *in vitro* of 2',3'-oxides derived from allylbenzene analogs. Drug. Metab. Dispos. 24: 1020–27.

Luu, T.X.T., Lam, T.T., Le, T.N. and Duus, F. 2009. Fast and green microwave-assisted conversion of essential oil allylbenzenes into the corresponding aldehydes *via* alkene isomerization and subsequent potassium permanganate promoted oxidative alkene group cleavage. Molecules 14: 3411–24.

Macía, M.J. 1998. La pimienta de Jamaica [*Pimenta dioica* (L.) Merrill, Myrtaceae] en la Sierra Norte de Puebla México. Anales Jard. Bot. Madrid 56: 337–49.

Mahboubi, M. and Mahboubi, M. 2015. Chemical composition, antimicrobial and antioxidant activities of *Eugenia caryophyllata* essential oil. J. Essent. Oil Bear. Pl. 18: 967–75.

Mahmoodi, A., Soltani, M., Basti, A.A., Roomiani, L., Kamali, A. and Taheri, S. 2012. Chemical composition and antibacterial activity of essential oils and extracts from *Rosmarinus officinalis*, *Zataria multiflora*, *Anethum graveolens* and *Eucalyptus globulus*. Global Veterin. 9: 73–79.

Makhfian, M., Hassanzadeh, N., Mahmoudi, E. and Zandyavari, N. 2015. Anti-quorum sensing effects of ethanolic crude extract of *Anethum graveolens* L. J. Essent. Oil Bear. Pl. 18: 687–96.

Manikandan, S. and Devi, R.S. 2005. Antioxidant property of alpha-asarone against noise-stress-induced changes in different regions of rat brain. Pharmacol. Res. 52: 467–74.

Marín, I., Sayas-Barberá, E., Viuda-Martos, M., Navarro, C. and Sendra, E. 2016. Chemical composition, antioxidant and antimicrobial activity of essential oils from organic fennel, parsley, and lavender from Spain. Foods 5: 18 (1-10).

Marongiu, B., Maxia, A., Piras, A., Porcedda, S., Tuveri, E., Gonçalves, M.J., Cavaleiro, C. and Salgueiro, L. 2007. Isolation of *Crithmum maritimum* L. volatile oil by supercritical carbon dioxide extraction and biological assays. Nat. Prod. Res. 21: 1145–50.

Marques, F.A., Wendler, E.P., Baroni, A.C.M., de Oliveira, P.R., Sasaki, B.S. and Guerrero Jr., P.G. 2010. Leaf essential oil composition of *Pimenta pseudocaryophyllus* (Gomes) L. R. Landrum native from Brazil. J. Essent. Oil Res. 22: 150–52.

Marzouki, H., Khaldi, A., Chamli, R., Bouzid, S., Piras, A., Falconieri, D. and Marongiu, B. 2009a. Biological activity evaluation of the oils from *Laurus nobilis* of Tunisia and Algeria extracted by supercritical carbon dioxide. Nat. Prod. Res. 23: 230–37.

Marzouki, H., Piras, A., Salah, K.B., Medini, H., Pivetta, T., Bouzid, S., Marongiu, B. and Falconieri, D. 2009b. Essential oil composition and variability of *Laurus nobilis* L. growing in Tunisia, comparison and chemometric investigation of different plant organs. Nat. Prod. Res. 23: 343–54.

Marzouki, H., Khaldi, A., Falconieri, D., Piras, A., Marongiu, B., Molicotti, P. and Zanetti, S. 2010. Essential oils of *Daucus* carota subsp. *carota* of Tunisia obtained by supercritical carbon dioxide extraction. Nat. Prod. Commun. 5: 1955–58.

Masoudi, S., Monfared, A., Rustaiyan, A. and Chalabian, F. 2005. Composition and antibacterial activity of the essential oils of *Semenovia dichotoma* (Boiss.) Manden., *Johreniopsis seseloides* (C.A.Mey) M. Pimen and *Bunium cylindricum* (Boiss. et Hohen.) Drude., three umbelliferae herbs growing wild in Iran. J. Essent. Oil Res. 17: 691–94.

Mastura, M., Nor Azah, M.A., Khozirah, S., Mawardi, R. and Manaf, A.A. 1999. Anticandidal and antidermatophytic activity of *Cinnamomum* species essential oils. Cytobios. 98: 17–23.

Mathela, C.S., Joshi, R.K., Bisht, B.S. and Joshi, S.C. 2015. Nothoapiole and α-asarone rich essential oils from himalayan *Pleurospermum angelicoides* Benth. Rec. Nat. Prod. 9: 546–52.

Matsumoto, F., Idetsuki, H., Harada, K., Nohara, I. and Toyoda, T. 1993. Volatile components of *Hedychium coronarium* Koenig flowers. J. Essent. Oil Res. 5: 123–33.

Maxia, A., Marongiu, B., Piras, A., Porcedda, S., Tuveri, E., Gonçalves, M.J., Cavaleiro, C. and Salgueiro, L. 2009. Chemical characterization and biological activity of essential oils from *Daucus carota* L. subsp. *carota* growing wild on the Mediterranean Coast and on the Atlantic Coast. Fitoterapia 80: 57–61.

Maxia, A., Falconieri, D., Piras, A., Porcedda, S., Marongiu, B., Frau, M.A., Gonçalves, M.J., Cabral, C., Cavaleiro, C. and Salgueiro, L. 2012. Chemical composition and antifungal activity of essential oils and supercritical CO_2 extracts of *Apium nodiflorum* L. Lag. Mycopathologia 174: 61–67.

Mazza, G. 1985. Gas chromatographic and mass spectrometric studies of the constituents of the rhizome of *Calamus*: I. The volatile constituents of the essential oil. J. Chromatogr. 328: 179–94.

Mazza, G. and Kiehn, F.A. 1992. Essential oil of *Agastache foeniculum*, a potential source of methyl chavicol. J. Essent. Oil Res. 4: 295–99.

Meepagala, K.M., Sturtz, G. and Wedge, D.E. 2002. Antifungal constituents of the essential oil fraction of *Artemisia dracunculus* L. var. *dracunculus*. J. Agric. Food Chem. 50: 6989–92.

Menghini, L., Leporini, L., Tirillini, B., Epifano, F. and Genovese, S. 2010. Chemical composition and inhibitory activity against *Helicobacter pylori* of the essential oil of *Apium nodiflorum* Apiaceae. J. Med. Food 13: 228–30.

Miladi, H., Chaieb, K., Ammar, E. and Bakhrouf, A. 2010. Inhibitory effect of clove oil (*Syzygium aromaticum*) against *Listeria monocytgenes* cells incubated in fresh-cut salmon. J. Food Safety 30: 432–42.

Mimica-Dukić, N., Kujundžić, S., Soković, M. and Couladis, M. 2003. Essential oil composition and antifungal activity of *Foeniculum vulgare* Mill. obtained by different distillation conditions. Phytother. Res. 17: 368–71.

Moghaddam, M., Reza Alymanesh, M., Mehdizadeh, L., Mirzaei, H. and Ghasemi Pirbalouti, A. 2014. Chemical composition and antibacterial activity of essential oil of *Ocimum ciliatum*, as a new source of methyl chavicol, against ten phytopathogens. Ind. Crop. Prod. 59: 144–48.

Mohammadi, S., Aroiee, H., Hossein Aminifard, M., Tehranifar, A. and Jahanbakhsh, V. 2014. Effect of fungicidal essential oils against *Botrytis cinerea* and *Rhizopus stolonifer* rot fungus *in vitro* conditions. Arch. Phytopathol. Pfl. 47: 1603–10.

Molino, J.-F. 2000. The inheritance of leaf oil composition in *Clausena anisum-olens* (Blanco) Merr. J. Essent. Oil Res. 12: 135–39.

Montenegro, I., Madrid, A., Zaror, L., Martínez, R., Werner, E., Carrasco-Altamirano, H., Cuellar, M. and Palma-Flemming, H. 2012. Antimicrobial activity of ethyl acetate extract and essential oil from bark of *Laurelia sempervirens* against multiresistant bacteria. Bol. Latinoam. Caribe 11: 306–15.

Moosavi-Nasab, M., Saharkhiz, M.J., Ziaee, E., Moayedi, F., Koshani, R. and Azizi, R. 2016. Chemical compositions and antibacterial activities of five selected aromatic plants essential oils against food-borne pathogens and spoilage bacteria. J. Essent. Oil Res. 28: 241–51.

Morita, T., Jinno, K., Kawagishi, H., Arimoto, Y., Suganuma, H., Inakuma, T. and Sugiyama, K. 2003. Hepatoprotective effect of myristicin from nutmeg (*Myristica fragrans*) on lipopolysaccharide/D-galactosamine-induced liver injury. J. Agric. Food Chem. 51: 1560–65.

Motley, T.J. 1994. The ethnobotany of sweet flag, *Acorus calamus* Araceae. Econ. Bot. 48: 397–412.

Muñoz-Acevedo, A., Kouznetsov, V.V. and Stashenko, E.E. 2009. Composition and *in vitro* antioxidant capacity of essential oils rich in thymol, carvacrol, *trans*-anethole or estragole. Rev. Univ. Ind. Santander 41: 287–94.

Muñoz-Acevedo, A., Stashenko, E.E., Kouznetsov, V.V. and Martínez, J.R. 2011. Differentiation of leaf and flower extracts of Basil (*Ocimum* sp.) varieties grown in Colombia. J. Essent. Oil Bear. Pl. 14: 387–95.

Muñoz-Acevedo, A., Puerto, C.E., Rodríguez, J.D., Aristizabal-Córdoba, S. and Kouznetsov, V.V. 2014. Chemical-biological study of the essential oils from *Croton malambo* H. Karst. and their major component, methyleugenol. Bol. Latinoam. Caribe 13: 336–43.

Nakamura, C.V., Ueda-Nakamura, T., Bando, E., Fernandes, A., Aparício, D. and Prado, B. 1999. Antibacterial activity of *Ocimum gratissimum* L. essential oil. Mem. Inst. Oswaldo Cruz 94: 675–78.

Nascimento, I.R., Lopes, L.M.X., Davin, L.B. and Lewis, N.G. 2000. Stereoselective synthesis of 8,9-licarinediols. Tetrahedron 56: 9181–93.

Nawel, O., Ahmed, H. and Douniazad, E.A. 2014. Phytochemical analysis and antimicrobial bioactivity of the Algerian parsley essential oil (*Petroselinum crispum*). African J. Microbiol. Res. 8: 1157–69.

Néris, P.L., Caldas, J.P., Rodrigues, Y.K., Amorim, F.M., Leite, J.A., Rodrigues-Mascarenhas, S., Barbosa-Filho, J.M., Rodrigues, L.C. and Oliveira, M.R. 2013. Neolignan Licarin A presents effect against Leishmania (Leishmania) major associated with immunomodulation *in vitro*. Exp. Parasitol. 135: 307–13.

Newberne, P., Smith, R.L., Doull, J., Goodman, J.I., Munro, I.C., Portoghese, P.S., Wagner, B.M., Weil, C.S., Woods, L.A., Adams, T.B., Lucas, C.D. and Ford, R.A. 1999. The FEMA GRAS assessment of *trans*-anethole used as a flavouring substance. Food Chem. Toxicol. 37: 789–811.

Noriega Rivera, P., Mosquera, T., Baldisserotto, A., Abad, J., Aillon, C., Cabezas, D., Piedra, J., Coronel, I. and Manfredini, S. 2015. Chemical composition and *in vitro* biological activities of the essential oil from leaves of *Peperomia inaequalifolia* Ruiz & Pav. Am. J. Essent. Oils Nat. Prod. 2: 29–31.
Ntezurubanza, L., Scheffer, J.J.C. and Baerheim Svendsen, A. 1986. Composition of the essential oil of *Ocimum trichodon* grown in Rwanda. J. Nat. Prod. 49: 945–47.
NTP—National Toxicology Program. U.S. Department of Health and Human Services. Public Health Service. 2000. Report on carcinogens background document for methyleugenol. Available: https://ntp.niehs.nih.gov/ntp/newhomeroc/roc10/me_no_appendices_508.pdf.
NTP TR 491—National Toxicology Program. 2000. Toxicology and carcinogenesis studies of methyleugenol (CAS No. 93-15-2) in F344/N rats and B6C3F1 mice (gavage studies). Natl. Toxicol. Program. Tech. Rep. Ser. 491: 1–412.
NTP TR 551—National Toxicology Program. 2010. NTP Toxicology and carcinogenesis studies of isoeugenol (CAS No. 97-54-1) in F344/N rats and B6C3F1 mice (gavage studies). Natl. Toxicol. Program. Tech. Rep. Ser. 551: 1–178.
Obistioiu, D., Cristina, R.T., Schmerold, I., Chizzola, R., Stolze, K., Nichita, I. and Chiurciu, V. 2014. Chemical characterization by GC-MS and *in vitro* activity against *Candida albicans* of volatile fractions prepared from *Artemisia dracunculus*, *Artemisia abrotanum*, *Artemisia absinthium* and *Artemisia vulgaris*. Chem. Cent. J. 8: 6 (1-11).
Ogata, M., Hoshi, M., Shimotohno, K., Urano, S. and Endo, T. 1997. Antioxidant activity of magnolol, honokiol, and related phenolic compounds. J. Am. Oil Chem. Soc. 74: 557–72.
Ohigashi, H. and Koshimizu, K. 1976. Chavicol, as a larva-growth inhibitor, from *Viburnum japonicum* Spreng. Agr. Biol. Chem. 40: 2283–87.
Ohno, T., Kita, M., Yamaoka, Y., Imamura, S., Yamamoto, T., Mitsufuji, S., Kodama, T., Kashima, K. and Imanishi, J. 2003. Antimicrobial activity of essential oils against *Helicobacter pylori*. Helicobacter 8: 207–15.
Ojewole, J.A.O. 2002. Hypoglycaemic effect of *Clausena anisata* (Willd) Hook methanolic root extract in rats. J. Ethnopharmacol. 81: 231–37.
Ooi, L.S.M., Li, Y., Kam, S.L., Wang, H., Wong, E.Y.L. and Ooi, V.E.C. 2006. Antimicrobial activities of cinnamon oil and cinnamaldehyde from the Chinese medicinal herb *Cinnamomum cassia* Blume. Am. J. Chin. Med. 34: 511–22.
Ortuño, M.F. 2006. Manual práctico de aceites esenciales, aromas y perfumes. España: Aiyana Ediciones. 276 pp.
Osei-Safo, D., Addae-Mensah, I., Garneau, F.-X. and Koumaglo, H.K. 2010. A comparative study of the antimicrobial activity of the leaf essential oils of chemo-varieties of *Clausena anisata* (Willd.) Hook. f. ex Benth. Ind. Crop. Prod. 32: 634–38.
Oxenham, S.K., Svoboda, K.P. and Walters, D.R. 2005. Antifungal activity of the essential oil of basil (*Ocimum basilicum*). J. Phytopathol. 153: 174–80.
Oyedeji, O.O., Oyedeji, A.O. and Shode, F.O. 2014. Compositional variations and antibacterial activities of the essential oils of three *Melaleuca* species from South Africa. J. Essent. Oil Bear. Pl. 17: 265–76.
Özcan, M.M. and Chalchat, J.C. 2006a. Chemical composition and antifungal effect of anise (*Pimpinella anisum* L.) fruit oil at ripening stage. Ann. Microbiol. 56: 353–58.
Özcan, M.M., Chalchat, J.C., Arslan, D., Ateş, A. and Ünver, A. 2006b. Comparative essential oil composition and antifungal effect of bitter fennel (*Foeniculum vulgare* ssp. *piperitum*) fruit oils obtained during different vegetation. J. Med. Food 9: 552–61.
Özkirim, A., Keskin, N., Kürkçüoğlu, M. and Can Başer, K.H. 2012. Evaluation of some essential oils as alternative antibiotics against American foulbrood agent *Paenibacillus larvae* on honey bees *Apis mellifera* L. J. Essent. Oil Res. 24: 465–70.
Ozturk, S. and Ercisli, S. 2006. Chemical composition and *in vitro* antibacterial activity of *Seseli libanotis*. World J. Microbiol. Biotechnol. 22: 261–65.
Padalia, R.C., Verma, R.S., Sah, A., Karki, N., Chauhan, A., Sakia, D. and Krishna, B. 2012. Study on chemotypic variations in essential oil of *Cinnamomum tamala* (Buch-Ham.) Nees et Eberm. and their antibacterial and antioxidant potential. J. Essent. Oil Bear. Pl. 15: 800–08.
Padalia, R.C., Chauhan, A., Verma, R.S., Bisht, M., Thul, S. and Sundaresan, V. 2014. Variability in rhizome volatile constituents of *Acorus calamus* L. from Western Himalaya. J. Essent. Oil Bear. Pl. 17: 32–41.
Pandey, A. and Bani, S. 2010. Hydroxychavicol inhibits immune responses to mitigate cognitive dysfunction in rats. J. Neuroimmunol. 226: 48–58.
Pandey, A.K., Singh, P., Mohan, M. and Tripathi, N.N. 2012. New report on the chemical composition of the essential oil from leaves of *Clausena pentaphylla* from India. Chem. Nat. Compd. 48: 896–97.
Pandey, A.K., Palni, U.T. and Tripathi, N.N. 2013. Evaluation of *Clausena pentaphylla* Roxb. DC oil as a fungitoxicant against storage mycoflora of pigeon pea seeds. J. Sci. Food Agric. 93: 1680–86.
Parise-Filho, R., Pastrello, M., Pereira Camerlingo, C.E., Silva, G.J., Agostinho, L.A., de Souza, T., Motter Magri, F.M., Ribeiro, R.R., Brandt, C.A. and Polli, M.C. 2011. The anti-inflammatory activity of dillapiole and some semisynthetic analogues. Pharm. Biol. 49: 1173–79.
Parise-Filho, R., Pasqualoto, K.F., Magri, F.M., Ferreira, A.K., da Silva, B.A., Damião, M.C., Tavares, M.T., Azevedo, R.A., Auada, A.V., Polli, M.C. and Brandt, C.A. 2012. Dillapiole as antileishmanial agent: discovery, cytotoxic activity and preliminary SAR studies of dillapiole analogues. Arch. Pharm. Chem. Life Sci. 345: 934–44.
Park, C., Kim, S.-I. and Ahn, Y.-J. 2003. Insecticidal activity of asarones identified in *Acorus gramineus* rhizome against three coleopteran stored-product insects. J. Stored Prod. Res. 39: 333–42.
Patent CN 104212136 A. 2014. Polylactic acid anti-bacterial-activity packaging material and preparation method thereof.
Patent EP 2572579 A1. 2014. Compositions for attracting Tortricidae (fruit moths).
Patent EP 2645862 A1. 2013. Delivery carrier for antimicrobial essential oils.

Patent US 2005049230 A1. 2005. Compounds that act to modulate insect growth and methods and systems for identifying such compounds.
Patent US 20070004686 A1. 2007. Attractant for apple fruit moth and other insect pests of apple.
Patent US 20080118461 A1. 2008. Sprayable controlled-release, male annihilation technique (MAT) formulation and insect control.
Patent US 20120276022 A1. 2012. Disinfecting agent comprising eugenol, terpineol and thymol.
Patent US 20150086421 A1. 2015. Antibacterial composition.
Patent US 6340710 B1. 2002. Non-Hazardous pest control.
Patent US 6495172 B1. 2002. Method of using steam ironing of fabrics as a way of causing reduction of physiological and/or subjective reactivity to stress in humans.
Patent US 7871649 B2. 2011. Antimicrobial compositions containing synergistic combinations of quaternary ammonium compounds and essential oils and/or constituents thereof.
Patent US 9144544 B1. 2015. Synthesis of silver nanoparticles from *Pimpinella anisum* seeds.
Patent WO 03105794 A1. 2003. Antibacterial flavor and fragrance composition and halitosis-inhibition flavor and fragrance composition and oral care composition containing the same.
Patent WO 2002012421 A1. 2002. Use of essential oils for combating gi tract infection by *helicobacter-like* organisms.
Patent WO 2006120567 A2. 2006. Pharmaceutical composition comprising an anti-bacterial agent and an active ingredient selected from carveol, thymol, eugenol, borneol and carvacrol.
Patent WO 2007082864 A2. 2007. Use of chavicol as an antiseptic.
Patent WO 2009029046 A1. 2009. Sugar-based surfactant microemulsions containing essential oils for cosmetic and pharmaceutical use.
Patent WO 2009043987 A1. 2009. Compositions antibiotiques a base d'huiles essentielles - prophylaxie et traitement d'infections nosocomiales.
Patent WO 2011092600 A2. 2011. Methods and product for reducing the population size of *Papilio demoleus* L. (Papilionidae).
Patent WO 2013133723 A1. 2013. Synthesis of elemicin and topical analgesic compositions.
Patent WO 2015181084 A1. 2015. Use of a nanoemulsion of cinnamaldehyde and/or a metabolite thereof, possibly in association with eugenol and/or carvacrol, to prevent resistance to antibiotics.
Patent WO2008080980 A1. 2008. Medicinal composition for treating animal skin comprising a wound healing agent and a deterrent.
Patent WO 2012114201 A1. 2012. Nanoencapsulation d'huiles essentielles pour la prévention et la guérison des maladies infectieuses seules ou associées à un antibiotique (Nanocapsulation of essential oils for preventing or curing infectious diseases alone or with an antibiotic).
Pathak, Y. 2009. Chapter 1: Recent developments in nanoparticulate drug delivery systems. pp. 1–15. *In*: Pathak, Y. and Thassu, D. (eds.). Drug Delivery Nanoparticles Formulation and Characterization. CRC Press, Boca Raton.
Peerakam, N., Wattanathorn, J., Punjaisee, S., Buamongkol, S., Sirisa-ard, P. and Chansakaow, S. 2014. Chemical profiling of essential oil composition and biological evaluation of *Anethum graveolens* L. seed grown in Thailand. J. Nat. Sci. Res. 4: 34–41.
Penfold, A.R., Ramage, G.R. and Simonsen, J.L. 1938. 142. The constitution of croweacin. J. Chem. Soc. 756–58.
Pieber, B., Martínez, S.T., Cantillo, D. and Kappe, C.O. 2013. *In situ* generation of diimide from hydrazine and oxygen: continuous-flow transfer hydrogenation of olefins. Angew. Chem. Int. Ed. 52: 10241–44.
Pineda, R., Vizcaíno, S., García, C.M., Gil, J.H. and Durango, D.L. 2012. Chemical composition and antifungal activity of *Piper auritum* Kunth and *Piper holtonii* C. DC. against phytopathogenic fungi. Chilean J. Agric. Res. 72: 507–15.
Pino, J.A., Marbot, R., Bello, A. and Urquiola, A. 2004. Essential oils of *Piper peltata* (L.) Miq. and *Piper aduncum* L. from Cuba. J. Essent. Oil Res. 16: 124–26.
Piras, A., Falconieri, D., Porcedda, S., Marongiu, B., Gonçalves, M.J., Cavaleiro, C. and Salgueiro, L. 2014. Supercritical CO_2 extraction of volatile oils from sardinian *Foeniculum vulgare* ssp. *vulgare* Apiaceae: chemical composition and biological activity. Nat. Prod. Res. 28: 1819–25.
Ponte, E.L., Sousa, P.L., Rocha, M.V., Soares, P.M., Coelho-de-Souza, A.N. and Leal-Cardoso, J.H. and Assreuy, A.M. 2012. Comparative study of the anti-edematogenic effects of anethole and estragole. Pharmacol. Rep. 64: 984–990.
Qiu, D., Hou, L., Chen, Y., Zhou, X., Yuan, W., Rong, W., Zhu, L. and Wang, J. 2011. Beta-asarone inhibits synaptic inputs to airway preganglionic parasympathetic motoneurons. Respir. Physiol. Neurobiol. 177: 313–19.
Qiu, J., Li, H., Su, Dong, J., Luo, M., Wang, J., Leng, B., Deng, Y., Liu, J. and Deng, X. 2012. Chemical composition of fennel essential oil and its impact on *Staphylococcus aureus* exotoxin production. World J. Microbiol. Biotechnol. 28: 1399–405.
Quinn, L.J., Harris, C. and Joron, G.E. 1958. Apiol poisoning. Can. Med. Assoc. J. 78: 635–36.
Radušiene, J., Judžentiene, A., Pečiulyte, D. and Janulis, V. 2007. Essential oil composition and antimicrobial assay of *Acorus calamus* leaves from different wild populations. Plant Genet. Resour. 5: 37–44.
Rahman, A., Chakma, J.S., Bhuiyan, N.I. and Islam, S. 2012. Composition of the essential oil of *Clausena suffruticosa* leaf and evaluation of its antimicrobial and cytotoxic activities. Trop. J. Pharm. Res. 11: 739–46.
Rajakumar, D.V. and Rao, M.N. 1993. Dehydrozingerone and isoeugenol as inhibitors of lipid peroxidation and as free radical scavengers. Biochem. Pharmacol. 46: 2067–72.
Regalado, E.L., Fernández, M.D., Pino, J.A., Mendiola, J. and Echemendia, O.A. 2011. Chemical composition and biological properties of the leaf essential oil of *Tagetes lucida* Cav. from Cuba. J. Essent. Oil Res. 23: 63–67.

Rhayour, K., Bouchikhi, T., Tantaoui-Elaraki, A., Sendide, K. and Remmal, A. 2003. The mechanism of bactericidal action of oregano and clove essential oils and of their phenolic major components on *Escherichia coli* and *Bacillus subtilis*. J. Essent. Oil Res. 15: 356–62.

Roby, M.H.H., Sarhana, M.A., Selim, K.A.-H. and Khalel, K.I. 2013. Antioxidant and antimicrobial activities of essential oil and extracts of fennel (*Foeniculum vulgare* L.) and chamomile (*Matricaria chamomilla* L.). Ind. Crop. Prod. 44: 437–45.

Romero, A.R., Merchan, D.R. and Kouznetsov, V.V. 2007. Reacción de imino diels-alder de tres componentes con precursores de origen natural. Generación de nuevas tetrahidroquinolinas 2,4-diaril disustituidas. Sci. Et Technica. 7: 91–95.

Roomiani, L., Soltani, M., Akhondzadeh Basti, A., Mahmoodi, A.T., Mirghaed, A. and Yadollahi, F. 2013. Evaluation of the chemical composition and *in vitro* antimicrobial activity of *Rosmarinus officinalis*, *Zataria multiflora*, *Anethum graveolens* and *Eucalyptus globulus* against *Streptococcus iniae*; the cause of zoonotic disease in farmed fish. Iranian J. Fish. Sci. 12: 702–16.

Roselló, J., Sempere, F., Sanz-Berzosa, I., Chiralt, A. and Santamarina, M.P. 2015. Antifungal activity and potential use of essential oils against *Fusarium culmorum* and *Fusarium verticillioides*. J. Essent. Oil Bear. Pl. 18: 359–67.

Rossi, P.G., Bao, L., Luciani, A., Panighi, J., Desjobert, J.M., Costa, J., Casanova, J., Bolla, J.M. and Berti, L. 2007. *E*-Methylisoeugenol and elemicin: antibacterial components of *Daucus carota* L. essential oil against *Campylobacter jejuni*. J. Agric. Food Chem. 55: 7332–36.

Rovio, S., Hartonen, K., Holm, Y., Hiltunen, R. and Riekkola, M.-L. 1999. Extraction of clove using pressurized hot water. Flavour Frag. J. 14: 399–404.

Ruberto, G., Baratta, M.T., Deans, S.G. and Dorman, H.J.D. 2000. Antioxidant and antimicrobial activity of *Foeniculum vulgare* and *Crithmum maritimum* essential oils. Plant. Med. J. 66: 687–93.

Sáenz, M.T., Tornos, M.P., Álvarez, A., Fernández, M.A. and García, M.D. 2004. Antibacterial activity of essential oils of *Pimenta racemosa* var. *terebinthina* and *Pimenta racemosa* var. *grisea*. Fitoterapia 75: 599–602.

Salehiarjmand, H., Ebrahimi, S.N., Hadian, J. and Ghorbanpour, M. 2014. Essential oils main constituents and antibacterial activity of seeds from Iranian local landraces of dill (*Anethum graveolens* L.). J. Hortic. Forest. Biotechn. 18: 1–9.

Salleh, W., Ahmad, F. and Khong, H.Y. 2016a. Essential oil compositions and antimicrobial activity of *Piper arborescens* Roxb. *Marmara*. Pharm. J. 20: 111–15.

Salleh, W., Ahmad, F., Yen, K.H. and Zulkifli, R. 2016b. Chemical composition and biological activities of essential oil of *Beilschmiedia pulverulenta*. Pharm. Biol. 54: 322–30.

Salleh, W.M.N.H.W., Ahmad, F., Yen, K.H. and Sirat, H.M. 2011. Chemical compositions, antioxidant and antimicrobial activities of essential oils of *Piper caninum* Blume. Int. J. Mol. Sci. 12: 7720–31.

Sánchez, Y., Correa, T.M., Abreu, Y. and Pino, O. 2013. Efecto del aceite esencial de *Piper auritum* Kunth y sus componentes sobre *Xanthomonas albilineans* (Ashby) Dowson y *Xanthomonas campestris* pv. *campestris* (Pammel) Dowson. Rev. Protección Veg. 28: 204–10.

Santillán-Ramírez, M.A., López-Villafranco, M.E., Aguilar-Rodríguez, S. and Aguilar Contreras, A. 2008. Etnobotany, leaf architecture, and vegetative anatomy of *Agastache mexicana* ssp. *mexicana* and *A. mexicana* ssp. *xolocotziana*. Rev. Mex. Biodivers. 79: 513–24.

Santhosh Kumar, E.S., Shareef, S.M., Roy, P.E. and Veldkamp, J.F. 2014. *Clausena agasthyamalayana* sp. nov. Rutaceae from Kerala, India. Nord. J. Bot. 33: 151–54.

Scazzocchio, F., Garzoli, S., Conti, C., Leone, C., Renaioli, C., Pepi, F. and Angiolella, L. 2016. Properties and limits of some essential oils: chemical characterisation, antimicrobial activity, interaction with antibiotics and cytotoxicity. Nat. Prod. Res. 30: 1909–18.

Schmidt, E., Jirovetz, L., Buchbauer, G., Eller, G.A., Stoilova, I., Krastanov, A., Stoyanova, A. and Geissler, M. 2006. Composition and antioxidant activities of the essential oil of Cinnamon (*Cinnamomum zeylanicum* Blume) leaves from Sri Lanka. J. Essent. Oil Bear. Pl. 9: 170–82.

Schmidt, E., Jirovetz, L., Wlcek, K., Buchbauer, G., Gochev, V., Girova, T., Stoyanova, A. and Geissler, M. 2007. Antifungal activity of eugenol and various eugenol-containing essential oils against 38 clinical isolates of *Candida albicans*. J. Essent. Oil Bear. Pl. 10: 421–29.

Seigler, D.S. 1998. Plant secondary metabolism. New York: Springer Science + Business Media. Pag. 106–114.

Senatore, F., De Fusco, R., Grassia, A., Moro, C.O., Rigano, D. and Napolitano, F. 2003. Chemical composition and antibacterial activity of essential oils from five culinary herbs of the Lamiaceae family growing in Campania, Southern Italy. J. Essent. Oil Bear. Pl. 6: 166–73.

Senatore, F., Oliviero, F., Scandolera, E., Taglialatela-Scafati, O., Roscigno, G., Zaccardelli, M. and De Falco, E. 2013. Chemical composition, antimicrobial and antioxidant activities of anethole-rich oil from leaves of selected varieties of fennel [*Foeniculum vulgare* Mill. ssp. *vulgare* var. *azoricum* (Mill.) Thell]. Fitoterapia 90: 214–19.

Shafaghat, A. 2009. Antibacterial activity and composition of essential oils from flower, leaf and stem of *Chaerophyllum macropodum* Boiss. from Iran. Nat. Prod. Commun. 4: 861–64.

Shahat, A.A., El-Barouty, G., Hassan, R.A., Hammouda, F.M., Abdel-Rahman, F.H. and Saleh, M.A. 2008. Chemical composition and antimicrobial activities of the essential oil from the seeds of *Enterolobium contortisiliquum* leguminosae. J. Environ. Sci. Health B. 43: 519–25.

Shahavi, M.H., Hosseini, M., Jahanshahi, M., Meyer, R.L. and Darzi, G.N. 2016. Clove oil nanoemulsion as an effective antibacterial agent: Taguchi optimization method. Desalin. Water Treat. 57: 18379–90.

Shalaby, A.S., Hendawy, S.F. and Khalil, M.Y. 2011. Evaluation of some types of fennel (*Foeniculum vulgare* Mill.) newly introduced and adapted in Egypt. J. Essent. Oil Res. 23: 35–42.

Sharifi, R., Kiani, H., Farzaneh, M. and Ahmadzadeh, M. 2008. Chemical composition of essential oils of Iranian *Pimpinella anisum* L. and *Foeniculum vulgare* Miller and their antifungal activity against postharvest pathogens. J. Essent. Oil Bear. Pl. 11: 514–22.

Sharma, P.K., Raina, A.P. and Dureja, P. 2009. Evaluation of the antifungal and phytotoxic effects of various essential oils against *Sclerotium rolfsii* (Sacc) and *Rhizoctonia bataticola* (Taub). Arch. Phytopathol. Pfl. 42: 65–72.

Sharma, U.K., Sood, S., Sharma, N., Rahi, P., Kumar, R., Sinha, A.K. and Gulati, A. 2013. Synthesis and SAR investigation of natural phenylpropene-derived methoxylated cinnamaldehydes and their novel Schiff bases as potent antimicrobial and antioxidant agents. Med. Chem. Res. 22: 5129–40.

Sheikhhosseini, E., Farrokhi, E. and Bigdeli, M.A. 2012. Synthesis of novel tetrahydroquinoline derivatives from α,α′-bis(substituted-benzylidene) cycloalkanones. J. Saudi Chem. Soc. 20: S227–30.

Shin, J.-W., Cheong, Y.-J., Koo, Y.-M., Kim, S., Noh, C.-K., Son, Kang, C. and Sohn, N.-W. 2014. α-Asarone ameliorates memory deficit in lipopolysaccharide-treated mice via suppression of pro-inflammatory cytokines and microglial activation. Biomol. Ther. 22: 17–26.

Shin, S. and Kang, C.A. 2003. Antifungal activity of the essential oil of *Agastache rugosa* Kuntze and and its synergism with ketoconazole. Lett. Appl. Microbiol. 36: 111–15.

Shin, S. 2004. Essential oil compounds from *Agastache rugosa* as antifungal agents against *Trichophyton* species. Arch. Pharm. Res. 27: 295–99.

Shukla, R., Singh, P., Prakash, B. and N.K. Dubey. 2013. Efficacy of *Acorus calamus* L. essential oil as a safe plant-based antioxidant, Aflatoxin B1 suppressor and broad spectrum antimicrobial against food-infesting fungi. Int. J. Food Sci. Tech. 48: 128–35.

Silva, C.G.V., Zago, H.B., Júnior, H.J.G.S., da Camara, C.A.G., de Oliveira, J.V., Barros, R., Schwartz, M.O.E. and Lucena, M.F.A. 2008. Composition and insecticidal activity of the essential oil of *Croton grewioides* Baill. against Mexican Bean *Weevil zabrotes* subfasciatus *Boheman*. J. Essent. Oil Res. 20: 179–82.

Simić, A., Soković, M.D., Ristić, M., Grujić-Jovanović, S., Vukojević, J. and Marin, P.D. 2004. The chemical composition of some Lauraceae essential oils and their antifungal activities. Phytother. Res. 18: 713–17.

Singh, G., Maurya, de Lampasona, M.P. and Catalan, C. 2005. Chemical constituents, antimicrobial investigations, and antioxidative potentials of *Anethum graveolens* L. essential oil and acetone extract: Part 52. J. Food Sci. 70: M208–15.

Singh, G., Maurya, S., de Lampasona, M.P. and Catalan, C. 2006a. Chemical constituents, antimicrobial investigations and antioxidative potential of volatile oil and acetone extract of star anise fruits. J. Sci. Food Agr. 86: 111–21.

Singh, G., Maurya, S., de Lampasona, M.P. and Catalan, C. 2006b. Chemical constituents, antifungal and antioxidative potential of *Foeniculum vulgare* volatile oil and its acetone extract. Food Control 17: 745–52.

Sintim, H.Y., Burkhardt, A., Gawde, A., Cantrell, C.L., Astatkie, T., Obour, A.E., Zheljazkov, V.D. and Schlegel, V. 2015. Hydrodistillation time affects dill seed essential oil yield, composition, and bioactivity. Ind. Crop. Prod. 63: 190–96.

Solecka, D. 1997. Role of phenylpropanoid compounds in plant responses to different stress factors. Acta. Physiol. Plant. 7: 257–68.

Solheim, E. and Scheline, R.R. 1973. Metabolism of alkenebenzene derivatives in the rat. I. p-Methoxyallylbenzene(estragole) and p-methoxypropenylbenzene (anethole). Xenobiotica 3: 493–510.

Song, H.Y., Yang, J.Y., Suh, J.W. and Lee, H.S. 2011. Acaricidal activities of apiol and its derivatives from *Petroselinum sativum* seeds against *Dermatophagoides pteronyssinus*, *Dermatophagoides farinae*, and *Tyrophagus putrescentiae*. J. Agric. Food Chem. 59: 7759–64.

Soylu, E.M., Soylu, S. and Kurt, S. 2006. Antimicrobial activities of the essential oils of various plants against tomato late blight disease agent *Phytophthora infestans*. Mycopathologia 161: 119–28.

Srivastava, S., Gupta, M.M., Prajapati, V., Tripathi, A.K. and Kumar, S. 2001. Insecticidal activity of myristicin from *Piper mullesua*. Pharm. Biol. 39: 226–29.

Stäuble, N. 1986. Ethnobotany of Euphorbiaceae of West Africa. J. Ethnopharmacol. 16: 23–103.

Su, X.-F., Huang, L.-J. and Feng, P.-Z. 2011. Chemical composition analysis and antimicrobial activity of volatile oil from the nutlets of *Clausena anisum-olens*. Food Sci. 32: 30–32.

Suárez, A.I., Compagnone, R.S., Salazar-Bookaman, M.M., Tillett, S., Delle Monache, F., Di Giulio, C. and Bruges, G. 2003. Antinociceptive and anti-inflammatory effects of *Croton malambo* bark aqueous extract. J. Ethnopharmacol. 88: 11–14.

Suárez, A.I., Vásquez, L.J., Taddei, A., Arvelo, F. and Compagnone, R.S. 2008. Antibacterial and cytotoxic activity of leaf essential oil of *Croton malambo*. J. Essent. Oil Bear. Pl. 11: 208–13.

Sudo, R.T., Albuquerque, E.X., de Barreiro, E.J., Aracava, Y., Cintra, W.M., Melo, P.A., Noel, F.G., Sudo, G.Z., da Silva, C.L.M., de Castro, N.G., Fernandes, P.D., Fraga, C.A.M. and de Miranda, A.L.P. 2006. Thienylhydrazon with digitalis-like properties (positive inotropic effects). (Patent US7091238 B1).

Sugumaran, M., Suresh Gandhi, M., Sankarnarayanan, K., Yokesh, M., Poornima, M. and Sreerama, R. 2011. Chemical composition and antimicrobial activity of vellaikodi variety of *Piper betle* Linn leaf oil against dental pathogens. Int. J. Pharm. Tech. Res. 3: 2135–39.

Sutili, F.J., de LimaSilva, L., Gressler, L.T., Gressler, L.T., Battisti, E.K., Heinzmann, B.M., de Vargas, A.C. and Baldisserotto, B. 2015. Plant essential oils against *Aeromonas hydrophila*: *in vitro* activity and their use in experimentally infected fish. J. Appl. Microbiol. 119: 47–54.

Tadtong, S., Wannakhot, P., Poolsawat, W., Athikomkulchai, S. and Ruangrungsi, N. 2009. Antimicrobial activities of essential oil from *Etlingera punicea* rhizome. J. Health Res. 23: 77–79.

Tavares, A.C., Gonçalves, M.J., Cavaleiro, C., Cruz, M.T., Lopes, M.C., Canhoto, J. and Salgueiro, L.R. 2008. Essential oil of *Daucus carota* subsp. *halophilus*. Composition, antifungal activity and cytotoxicity. J. Ethnopharmacol. 119: 129–34.
Thai, T.H., Bazzali, O., Hoi, T.M., Tuan, N.A., Tomi, F., Casanova, J. and Bighelli, A. 2013a. Chemical composition of the essential oils from two Vietnamese *Asarum* species: *A. glabrum* and *A. cordifolium*. Nat. Prod. Commun. 8: 235–38.
Thai, T.H., Hien, N.T., Hoi, T.M., Tuan, N.A., Dat, N.T. and Hai, N.T. 2013b. Chemical composition of essential oil from some species of *Asarum* L. genus in Vietnam. Tap. Chi. Sinh. Hoc. 35: 55–60. (Abstract).
Thomson Reuters. 2016. Web of science database. Available: url=http: //www.webofscience.com (consulted online: January—July 2016).
Thota, N., Reddy, M.V., Kumar, A., Khan, I.A., Sangwan, P.L., Kalia, N.P., Koul, J.L. and Koul, S. 2010. Substituted dihydronaphthalenes as efflux pump inhibitors of *Staphylococcus aureus*. Eur. J. Med. Chem. 45: 3607–16.
Tian, J., Ban, X., Zeng, H., Huang, B., He, J. and Wang, Y. 2011. *In vitro* and *in vivo* activity of essential oil from dill (*Anethum graveolens* L.) against fungal spoilage of cherry tomatoes. Food Control 22: 1992–99.
Tian, J., Ban, X., Zeng, H., He, J., Chen, Y. and Wang, Y. 2012. The mechanism of antifungal action of essential oil from dill (*Anethum graveolens* L.) on *Aspergillus flavus*. Plos One 7: e30147.
Titov, I.Y., Sagamanova, I.K., Gritsenko, R.T., Karmanova, I.B., Atamanenko, O.P., Semenova, M.N. and Semenov, V.V. 2011. Application of plant allylpolyalkoxybenzenes in synthesis of antimitotic phenstatin analogues. Bioorg. Med. Chem. Lett. 21: 1578–81.
Tomar, S.S. and Dureja, P. 2001. New minor constituents from *Anethum sowa*. Fitoterapia 72: 76–77.
Tomić, A., Petrović, S., Pavlović, M., Tzakou, O., Couladis, M., Milenković, M., Vučićević, D. and Lakušić, B. 2009. Composition and antimicrobial activity of the rhizome essential oils of two *Athamanta turbith* subspecies. J. Essent. Oil Res. 21: 276–79.
Torbati, M., Nazemiyeh, H., Lotfipour, F., Asnaashari, S., Nemati, M. and Fathiazad, F. 2013. Composition and antibacterial activity of *Heracleum transcaucasicum* and *Heracleum anisactis* aerial parts essential oil. Adv. Pharm. Bull. 3: 415–18.
Torbati, M., Nazemiyeh, H., Lotfipour, F., Nemati, M., Asnaashari, S. and Fathiazad, F. 2014. Chemical composition and *in vitro* antioxidant and antibacterial activity of *Heracleum transcaucasicum* and *Heracleum anisactis* roots essential oil. Bioimpacts 4: 69–74.
Torchilin, V. 2006. Micellar nanocarriers: pharmaceutical perspectives. Pharm. Res. 24: 1–16.
Trajano, V.N., Lima, E.O. and de Souza, F.S. 2012. Antifungal activity of the essential oil of *Cinnamomum zeylanicum* Blume and eugenol on *Aspergillus flavus*. J. Essent. Oil Bear. Pl. 15: 785–93.
Tsyganov, D.V., Krayushkin, M.M., Konyushkin, L.D., Strelenko, Y.A., Semenova, M.N. and Semenov, V.V. 2016. Facile synthesis of natural alkoxynaphthalene analogues from plant alkoxybenzenes. J. Nat. Prod. 79: 923–28.
Türkölmez, S. and Soylu, E.M. 2014. Antifungal efficacies of plant essential oils and main constituents against soil-borne fungal disease agents of bean. J. Essent. Oil Bear. Pl. 17: 203–11.
Ueng, Y.F., Hsieh, C.H. and Don, M.J. 2005. Inhibition of human cytochrome P450 enzymes by the natural hepatotoxin safrole. Food Chem. Toxicol. 43: 707–12.
Ušjak, L.J., Petrović, S.D., Drobac, M.M., Soković, M.D., Stanojković, T.P., Ćirić, A.D., Grozdanić, N.Đ. and Niketić, M.S. 2016. Chemical composition, antimicrobial and cytotoxic activity of *Heracleum verticillatum* Pančić and *H. ternatum* Velen. (Apiaceae) essential oils. Chem. Biodivers. 13: 466–76.
Usman, L.A., Akolade, J.O., Odebisi, B.O. and Olanipekun, B. 2016. Chemical composition and antibacterial activity of fruit essential oil of *Xylopia aethiopica* D. grown in Nigeria. J. Essent. Oil Bear. Pl. 19: 648–55.
Vairappan, C.S., Nagappan, T. and Palaniveloo, K. 2012. Essential oil composition, cytotoxic and antibacterial activities of five *Etlingera* species from Borneo. Nat. Prod. Commun. 7: 239–42.
Verma, R.S., Padalia, R.C. and Chauhan, A. 2014. Geographical impact on essential oil composition of *Limnophila rugosa* Roth. Merr. J. Essent. Oil Res. 26: 338–41.
Verspohl, E.J., Bauer, K. and Neddermann, E. 2005. Antidiabetic effect of *Cinnamomum cassia* and *Cinnamomum zeylanicum in vivo* and *in vitro*. Phytother. Res. 19: 203–06.
Vieira, P.R.N., de Morais, S.M., Bezerra, F.H.Q., Ferreira, P.A.T., Oliveira, I.R. and Silva, M.G.V. 2014. Chemical composition and antifungal activity of essential oils from *Ocimum* species. Ind. Crop. Prod. 55: 267–71.
Vivanco, R., León, E., Castro, A. and Ramos, N.J. 2012. Composición química del aceite esencial de *Petroselinum crispum* (Mill) Nyman ex A.W. Hill "perejil" y determinación de su actividad antibacteriana. Cienc. Investig. 15: 78–83.
Vogt, T. 2010. Phenylpropanoid biosynthesis. Mol. Plant 3: 2–20.
Vokk, R., Lõugas, T., Mets, K. and Kravets, M. 2011. Dill *Anethum graveolens* L. and parsley *Petroselinum crispum* Mill. Fuss from Estonia: seasonal differences in essential oil composition. Agron. Res. 9: 515–20.
Vyry Wouatsa, N.A., Misra, L. and Venkatesh Kumar, R. 2014. Antibacterial activity of essential oils of edible spices, *Ocimum canum* and *Xylopia aethiopica*. J. Food Sci. 79: M972–77.
Wallace, J.L., Cirino, G., Caliendo, G., Sparatore, A., Santagada, V. and Fiorucci, S. 2006. Derivatives of 4- or 5-aminosalicylic acid (Patent US7910568 B2).
Wang, B., Qi, W., Wang, L., Kong, D., Kano, Y., Li, J. and Yuan, D. 2014. Comparative study of chemical composition, antinociceptive effect and acute toxicity of the essential oils of three *Asarum* drugs. J. Chin. Pharm. Sci. 23: 480–89.
Wang, G.-W., Hu, W.T., Huang, B.K. and Qin, L.P. 2011. *Illicium verum*: A review on its botany, traditional use, chemistry and pharmacology. J. Ethnopharmacol. 136: 10–20.
Wang, H.-Z., Chen, Y.-G. and Fan, C.-S. 1998. Review of studies on chemical constituents and pharmacology of genus *Acorus* in China. Acta. Bot. Yunnanica Suppl. X: 96–100.

Wang, K.C., Chang, J.S., Chiang, L.C. and Lin, C.C. 2009. 4-Methoxycinnamaldehyde inhibited human respiratory syncytial virus in a human larynx carcinoma cell line. Phytomedicine 16: 882–86.

Wei, P.-L., Tu, S.-H., Lien, H.-M., Chen, L.-C., Chen, C.-S., Wu, C.-H., Huang, C.-S., Chang, H.-W., Chang, C.-H., Tseng, H. and Ho, Y.-S. 2012. The *in vivo* antitumor effects on human COLO 205 cancer cells of the 4,7-dimethoxy-5-(2-propen-1-yl)-1,3-benzodioxole (apiole) derivative of 5-substituted 4,7-dimethoxy-5-methyl-l,3-benzodioxole (SY-1) isolated from the fruiting body of *Antrodia camphorate*. J. Canc. Res. Ther. 8: 532–36.

Wilczewska, A.Z., Ulman, M., Chilmończyk, Z., Maj, J., Koprowicz, T., Tomczyk, M. and Tomczykowa, M. 2008. Comparison of volatile constituents of *Acorus calamus* and *Asarum europaeum* obtained by different techniques. J. Essent. Oil Res. 20: 390–95.

Williams, L.A.D., Vasquez, E.A., Milan, P.P., Zebitz, C. and Kraus, W. 2002. *In vitro* anti-inflammatory and antimicrobial activities of phenylpropanoids from *Piper betle* L. (Piperaceae). pp. 221–27. *In*: Rauter, A.P., Brito Palma, F., Justino, J., Araújo, M.E. and dos Santo, S.P. (eds.). Natural Products in the New Millennium: Prospects and Industrial Application. Springer-Science + Business Media. Dordrecht.

Wiseman, R.W., Miller, E.C., Miller, J.A. and Liem, A. 1987. Structure-activity studies of the hepatocarcinogenicities of alkenylbenzene derivatives related to estragole and safrole on administration to preweanling Male C57BL/6J × C3H/HeJ F1 Mice. Canc. Res. 47: 2275–83.

Wong, K.C., Sivasothy, Y., Boey, P.L., Osman, H. and Sulaiman, B. 2010. Essential oils of *Etlingera elatior* Jack R. M. Smith and *Etlingera littoralis* Koenig Giseke. J. Essent. Oil Res. 22: 461–66.

Woo, Y.T. and Lai, D.Y. 1986. Safrole and its alkenylbenzene congeners. Safrole, estragole, and related compounds. Carcinogenicity and structure activity relationships. Other biological properties. Metabolism. Environmental significance. U.S. EPA Current Awareness Document. National Service Center for Environmental Publications. 82 p.

Wright, C.W. 2003. Artemisia. Boca Raton: CRC Press. 344 pp.

Xie, Y., Yang, Z., Cao, D., Rong, F., Ding, H. and Zhang, D. 2015. Antitermitic and antifungal activities of eugenol and its congeners from the flower buds of *Syzgium aromaticum* clove. Ind. Crop. Prod. 77: 780–86.

Xu, H.-X., Zheng, X.-S., Yang, Y.-J., Tian, J.-C., Lu, Y.-H., Tan, K.-H., Heong, K.-L. and Lu, Z.-X. 2015. Methyl eugenol bioactivities as a new potential botanical insecticide against major insect pests and their natural enemies on rice *Oriza sativa*. Crop Prot. 72: 144–49.

Yahya, M.A.A., Yaacob, W.A., Din, L.B. and Nazlina, I. 2010. Analysis of essential oils of *Etlingera sphaerocephala* var. *grandiflora* by two-dimensional gas chromatography with time-of-flight mass spectrometry. Malaysian J. Anal. Sci. 14: 32–40.

Yamani, H., Mantri, N., Morrison, P.D. and Pang, E. 2014. Analysis of the volatile organic compounds from leaves, flower spikes, and nectar of Australian grown *Agastache rugosa*. BMC Compl. Altern. Med. 14: 495(1-6).

Yamani, H.A., Pang, E.C., Mantri, N. and Deighton, M.A. 2016. Antimicrobial activity of Tulsi (*Ocimum tenuiflorum*) essential oil and their major constituents against three species of bacteria. Front Microbiol. 7: 681(1-10).

Yang, I.J., Lee, D.U. and Shin, H.M. 2015. Anti-inflammatory and antioxidant effects of coumarins isolated from *Foeniculum vulgare* in lipopolysaccharide-stimulated macrophages and 12-O-tetradecanoylphorbol-13-acetate-stimulated mice. Immunopharmacol. Immunotoxicol. 37: 308–17.

Yavari, M., Mirdamadi, S., Masoudi, S., Tabatabaei-Anaraki, M., Larijani, K. and Rustaiyan, A. 2011. Composition and antibacterial activity of the essential oil of a green type and a purple type of *Ocimum basilicum* L. from Iran. J. Essent. Oil Res. 23: 1–4.

Zabka, M., Pavela, R. and Slezakova, L. 2009. Antifungal effect of *Pimenta dioica* essential oil against dangerous pathogenic and toxinogenic fungi. Ind. Crop. Prod. 30: 250–53.

Zabka, M., Pavela, R. and Prokinova, E. 2014. Antifungal activity and chemical composition of twenty essential oils against significant indoor and outdoor toxigenic and aeroallergenic fungi. Chemosphere 112: 443–48.

Zambaux, M.F., Bonneaux, F., Gref, R., Maincent, P., Dellacherie, E., Alonso, M.J., Labrude, P. and Vigneron, C. 1998. Influence of experimental parameters on the characteristics of poly(lactic acid) nanoparticles prepared by a double emulsion method. J. Control. Release 50: 31–40.

Zhang, F., Xu, Q., Fu, S., Ma, X., Xiao, H. and Liang, X. 2005. Chemical constituents of the essential oil of *Asarum forbesii* Maxim (Aristolochiaceae). Flavour Frag. J. 20: 318–20.

Zygadlo, J.A., Guzman, C.A. and Grosso, N.R. 1994. Antifungal properties of the leaf oils of *Tagetes minuta* L. and *T. filifolia* Lag. J. Essent. Oil Res. 6: 617–21.

4

Antibacterial and Anti-biofilm Activities of Essential Oils and Their Components Including Modes of Action

Julio A. Zygadlo, María P. Zunino, Romina P. Pizzolitto, Carolina Merlo, Alejandra Omarini* and *José S. Dambolena*

Introduction

Essential Oils (EOs) are mixtures of monoterpenes (MT), sesquiterpenes (ST) and/or phenylpropanoids (PhP) (Zygadlo 2011, Raut and Karuppayil 2014), with MT and ST being biosynthesized from mevalonate and/or deoxyxylulose (methyl D-erythritol 4-phosphate) pathways, while PhP come from a biosynthetic route of sikimic acid. The chemical diversity of EOs is high with aliphatic, cyclic and bicyclic structures, and is increased with the presence of different functional groups such as hydroxyls, carbonyls, carboxyls or esteres (Sell 2010, Zygadlo 2011). The great variety of bioactivities and the many mechanisms of action or targets are associated with the large molecular diversity of EOs (Koroch et al. 2007, Franz 2010, Sadgrove et al. 2015). In fact, several botanical families produce EOs, for examples *Zingiberaceae, Rutaceae, Poaceae, Apiaceae, Asteraceae, Geraniaceae, Lamiaceae, Lauraceae, Myrtaceae, Pinaceae, Piperaceae, Verbenaceae* and *Santalaceae,* with the EOs being biosynthesized and stored in special structures, such as glandular hairs or esquizogene channels (Franz 2010, Zygadlo 2011, Raut and Karuppayil 2014).

The physician Arnald de Villanova was the first to obtain EOs by hydrodistillation. After EOs could be isolated, they had an impact on the medicine of the 13th century, and from this time are specified in various European pharmacopoeia (Guenther 1948, Sadgrove et al. 2015). However, the first studies on EO composition were not carried out until the 19th century by the French chemist M.J. Dumas (Kubeczka 2010). The chemical composition of EOs depend on climatic conditions, type soil (Hussain et al. 2008, Oliva et al. 2010, Lopez et al. 2012, Ben El Hadj Ali et al. 2015), genetic material, chemotype (Jordán et al. 2013), culture techniques (Rioba et al. 2015), and the parts of the plants (leaves, fruits, flowers) where the EOs are obtained (Saïdana et al. 2008, Maggi et al. 2009, de Almeida et al. 2013, Alipour et al. 2014, Popovic et al. 2015, Villa-Ruano et al. 2015).

Universidad Nacional de Córdoba, Instituto Multidisciplinario de Biologia Vegetal, CONICET, Córdoba, Argentina, Avenida Velez Sarsfiled 1611.
* Corresponding author: jzygadlo@unc.edu.ar

Hydrodistillation, steam distillation, solvent extraction, water microwave assisted hydrodistillation, and more recently, supercritical fluid extraction with CO_2, are the techniques most frequently used to obtain EOs (Zygadlo 2011), but each method used has produced differences in the chemical composition of the final product (Wenqiang et al. 2007, Okoh et al. 2010).

Nowadays, there is a wealth of information available related to the biological activities of EOs and their components (Koroch et al. 2007, Lang and Buchbauer 2012, Nazzaro et al. 2013, Raut et al. 2014, Hassanien et al. 2015, Sadgrove et al. 2015), which is the starting point for the development of many commercial products (Brud 2010). At present, there is great interest in using EOs in the industrial development of hygiene products, particularly due to their antimicrobial properties (Lang and Buchbauer 2012, Dreger and Wielgus 2013, Calo et al. 2015, Prakash et al. 2015). In addition, since 2008, there has been an increase in bacterial diseases worldwide (Raut et al. 2014), with a rise of above 30% in the consumption of antibiotics, with monolactams, cephalosporins and fluoroquinolones being the antibiotics most in demand (Van Boeckel et al. 2014, Gelband et al. 2015, Price et al. 2015). This increase in the consumption of antibiotics has been accompanied by a corresponding growth in the bacterial population that is β-lactamase positive and not susceptible to third generation cephalosporins (Guzman-Blanco et al. 2014). In the context of this global problem MT and ST have shown the ability to synergize and optimize the efficiency of antibiotics, for examples, geraniol increased the performance of β-lactams, quinolones and chloramphenicol (Lorenzi et al. 2009), while vancomycin was synergized by aromadendrene and 1,8-cineole on vancomycin-resistant enterococci. The susceptibility of *Staphylococcus aureus* to several antibiotics (ciprofloxacin, clindamycin, erythromycin, polymycin, gentamicin, tetracycline, oxacillin, methacillin and vancomycin) was also improved by EOs of *Cassia*, Peru balsam, red thyme (Kavanaugh and Ribbeck 2012), nerolidol, farnesol and bisabolol (Wolska et al. 2012). Furthermore, cinnamaldehyde increased the potency of clindamycin (Abreu et al. 2012), while carvacrol, thymol and eugenol enhanced the activity of multiple antibiotics (Langeveld et al. 2014). Hence, EOs and their components have been shown to be synergist agents of several antibiotics (Brud 2010, Abreu et al. 2012, Wolska et al. 2012, Langeveld et al. 2014, Price et al. 2015).

The antibacterial activity of EOs, however, is not always attributed to its main component (Bakkali et al. 2008), because a combination of different constituents may exhibit antagonistic, additive or synergistic effects. Concerning this, the concentration values of phenols (thymol+carvacrol) in some *Thymus* sp. did not show a direct correlation with the values of antibacterial activity (Nikolić et al. 2014). Thus, the toxic effects of phenols included in an EO may not be directly related to their concentrations, because there are compounds in EOs with inhibitory activity. Therefore, it would be interesting to be able to identify the compounds that generate opposing effects and thereby reduce their presence in the EOs. While it seems evident that the presence of phenols, such as eugenol or thymol, in an EO ensure a significant degree of antibacterial activity, a high concentration of hydrocarbons in an EO is an indicator of poor antimicrobial activity (Tables 4.1 and 4.2). Moreover, aldehydes and ketones are electrophiles that produce adducts with biological nucleophiles, hence strong toxic activity was also shown for ketones and aldehydes such as menthone (Soković et al. 2010), thymoquinone (Harzallah et al. 2011), p-anisaldehyde, cinnamaldehyde and cuminaldehyde (Andrade-Ochoa et al. 2015), the magnitude of antibacterial activity of carbonyl compounds was related to both partition coefficient (Log P) and electrophilicity (Schultz and Yarbrough 2004, Lopachin and Gavin 2014). The antibacterial capacity of EOs is also conditioned by the strain used in bioassays, with in general, Gram positive bacteria being more sensitive than Gram negative bacteria for the same EOs.

We collected information concerning the antibacterial activity of EOs and their components since 2011 (Tables 4.1 and 4.2), with compounds or EOs having MIC values less than 30 μg/mL being categorized as showing excellent antibacterial activity. Those with MIC values between 30 and 200 μg/mL had good antibacterial activity, while EOs with MIC values above 200 μg/L were considered to be inactive.

Table 4.1. Essential oils of aromatic plants with antimicrobial activity against Gram positive bacteria.

Aromatic plants	Main components from EOs (> 9%)	Bacteria[a] (MIC)[b] MIC μg/mL > 200	References
Nigella sativa	α-thujene, p-cymene		1) Harzallah et al. 2011
Cymbopogon citratus	neral, geranial	*B. bifidum (4)*	2) Bassole et al. 2011
C. giganteus	limonene, cis- and trans-p-mentha-2,8-dien-1-ol	*B. cereus (7, 13, 24, 27, 28)*	
Origanum vulgare	p-cymene, γ-terpinene, carvacrol	*B. longum (4)*	3) Lv et al. 2011
Ocimum basilicum	eugenol, caryophyllene	*B. subtilis (3, 7, 9, 10, 18, 22, 27, 28, 30)*	
Citrus bergamia	limonene, linalool, bergamol		
Perilla arguta	2-pentanoylfuran, benzene, 1-methoxy-4-(1-methyl propyl)	*B. infantis (4)*	
Paeonia lactiflora	myrtenal, myrtenol, paeonol	*Bifidobacterium adolescentes (4)*	4) Ngan et al. 2012
Pelargonium graveolens	citronellol, geraniol	*C. butyricum (4)*	5) Bigos et al. 2012
Teucrium marum	caryophyllene oxide	*C. difficile (4)*	6) Djabou et al. 2013
T. polium ssp. *capitatum*	α-pinene	*C. paraputrificum (4)*	
T. flavum ssp. *glaucum*	α-pinene, β-pinene, limonene	*C. perfringens (4)*	
T. scorodonia ssp. *scorodonia*	α-caryophyllene	*C. xerosis (17)*	
T. chamaedrys	α-caryophyllene, germacrene D	*E. durans (9)*	
T. massiliense	6-methyl-3-heptyl acetate	*E. faecalis (1, 2, 11, 14, 15, 20, 38)*	
Callistemom viminalis	α-pinene, 1,8-cineole		7) Salem et al. 2013
Satureja horvatii	p-cymene, thymol, thymol methyl ether	*E. hirae (9)*	8) Bukvicki et al. 2014
Citrus aurantifolia	β-pinene, limonene, γ-terpinene	*G. haemolysans (1)*	9) Costa et al. 2014
Foeniculum vulgare	estragole, *trans*-anethole	*G. morbillorum (1)*	10) Diao et al. 2014
Rosa alba	geraniol, heneicosane, nonadecane, citronellol	*L. acidophilus (4)*	11) Mileva et al. 2014
Toona sinensis	β-caryophyllene, α-caryophyllene	*L. casei (4)*	12) Wu et al. 2014
Pinus nigra sp. *nigra*	α-pinene, germacrene D	*L. monocytogenes (2, 9, 29)*	13) Sarac et al. 2014
P. nigra sp. *pallasiana*		*Lactobacillus acidophilus (14)*	
P. nigra sp. *banatica*		*M. luteus (7)*	
Mentha piperita	menthone, menthol		14) Nikolic et al. 2014
Mentha pulegium	pulegone, piperitone	*M. luteus (17)*	
Salvia lavandulifolia	1,8-cineole, camphor	*P. acnes (20)*	
Cymbopogon giganteus	limonene, cis- and trans-p-mentha-2,8-dien-1-ol, 1,3,8-p-menthatriene	*S. albus (10)*	15) Ahmad and Viljoen 2015
C. flexuosus	neral, geranial	*S. anginosus (1)*	
C. nardus	geraniol	*S. aureus (2, 3, 4, 5, 6, 7, 9, 10, 12, 13, 14, 15, 16, 18, 20, 21, 22, 23, 24, 25, 26, 28, 30, 31, 41)*	
Lavandula officinalis	linalool		16) Martucci et al. 2015
Origanum vulgare	carvacrol, p-cymene, γ-terpinene		
Origanum vulgare	γ-terpinene, terpinen-4-ol, carvacrol	*S. constellatus (1)*	17) Suzuki et al. 2015
Agathis cammara	β-myrcene, limonene, β-bisabolene		18) Chen et al. 2015
Eucalyptus globulus	isovaleraldehyde, 1,8-cineole	*S. epidermidis (6, 9, 17, 20, 21, 23)*	19) Harkat-Madouri et al. 2015
Cinnamomum zeylanicum	eugenol, cinnamaldehyde	*S. lutea (7, 13)* *S. mitis (1)*	20) Aumeeruddy-Elalfi et al. 2015
Laurus nobilis	camphene, linalool, methyl eugenol, eugenol	*S. mutans (1, 11, 14, 19, 23)*	
Piper bette	β-caryophyllene, eugenol, safrole	*S. oralis (1)*	
Pimenta dioica	eugenol	*S. peroris (20)*	
Rosmarinus officinalis	α-pinene, 1,8-cineole		
Salvia officinalis	1,8-cineole, camphor, thujone, aromadrendene	*S. pneumoniae (30)* *S. pyogenes (14)*	
Schinus terebinthifolius	α-pinene, 3-carene, α-phellandrene	*S. salivarius (1, 14, 38)*	
Psiada arguta	caryophyllene oxide, methyl eugenol, isoeugenol	*S. sanguinis (14)* *S. sanguis (1)*	
Psiadia terebinthina	α-curcumene, acetyl eugenol	*S. sobrinus (19)*	

Table 4.1 contd....

...Table 4.1 contd.

Aromatic plants	**Main components from EOs (> 9%)**	**Bacteria[a] (MIC)[b]** **MIC μg/mL > 200**	**References**
Geniosporum rotundifolium	spathulenol		21) Ngassapa et al. 2016
Haumaniastrum villosum	caryophyllene oxide, humulene epoxide II		
Plectranthus rugosus	α-pinene, β-caryophyllene, germacrene D		22) Gani et al. 2015
Agathis robusta	β-selinene, a-selinene, spathulenol, caryophyllene oxide		23) Verma et al. 2016
Aloysia gratissima	1,8-cineole, β-caryophyllene, germacrene D, β-pinene		24) Santos et al. 2015
Forsythia koreana	cis-3-hexenol, linalool, trans-phytol		25) Yang et al. 2015b
Coriandrum sativum	limonene, linalool, linalyl acetate		26) Bogavac et al. 2015
Thymus vulgaris	p-cymene, thymol, carvacrol		
Lippia origanoides	p-cymene, carvacrol, thymol		27) Sarrazin et al. 2015
Mentha spicata	limonene, carvone		28) Shahbazi 2015a
Origanum vulgare	thymol, p-cymene		29) de Medeiros Barbosa et al. 2016
Rosmarinus officinalis	1,8-cineole, camphor, limonene		30) Miguel et al. 2016
Cordia globosa	α-pinene, camphene		
		MIC μg/mL 20–200	
T. maroccanus	carvacrol	*B. cereus (31)*	31) Fadli et al. 2012
T. broussonetti	borneol, carvacrol	*B. subtilis (31, 32, 33)*	
Skimmia laureola	linalool, linalyl acetate	*E. faecalis (35, 36)*	32) Shah et al. 2012
Ferula assafoetida	α-pinene, 1,2-dithionine	*L. acidophilus (34, 35)* *L. casei (38)*	33) Kavoosi et al. 2013
Thymus algeriensis	thymol, carvacrol	*L. monocytogenes (37)*	34) Nikolic et al. 2014
Thymus vulgaris	p-cymene, thymol	*M. luteus (31)*	
Lavandula angustifolia	linalool, borneol, linalyl acetate	*S. aureus (8, 32, 33, 34, 35, 36, 40, 41)*	35) Nikolic et al. 2014
Satureja montana	p-cymene, thymol		
C. flexuosus	neral, geranial	*S. epidermidis (32)*	36) Ahmad and Viljoen 2015
C. martinii	geraniol	*S. mitis (38)*	
C. citratus	neral, geranial	*S. mutans (34, 35)*	
C. winterianus	citronellal, citronellol, geraniol	*S. pneumoniae (39)*	
Salvia sclareoides	β-trans-ocimene, linalool, (E)-caryophyllene	*S. pyogenes (34, 35, 39, 40)* *S. salivarius (34, 35)*	37) Sepahvand et al. 2014
Plectranthus neochilus	α-pinene, (E)-caryophyllene, caryophyllene oxide	*S. sanguinis (34, 35, 38)* *S. sobrinus (38)*	38) Crevelin et al. 2015
Artemisia capillaris	β-pinene, capillin		39)Yang et al. 2015b
Thymus transcaucasicus	thymol, carvacrol		40) Bektas et al. 2016
		MIC μg/mL < 19	
Alpinia purpurata red ginger	β-pinene, linalool, β-caryophyllene	*A. flavus (42)* *B. cereus (42, 46)*	41) Santos et al. 2012
Gnaphlium affine	linalool, eugenol, (E)-caryophyllene	*B. laubach (42)* *B. subtilis (40, 43, 44)*	42) Zeng et al. 2011
Juglans regia	α-pinene, β-pinene, germacrene D	*E. faecalis (45)*	43) Rather et al. 2012
Carum copticum	p-cymene, γ-terpinene, thymol	*L. acidophilus (45)* *M. flavus (46)*	44) Kavoosi et al. 2013
Thymus serpyllum	thymol	*M. luteus (46)*	45) Nikolic et al. 2014
Ruta graveolens	2-nonanone, 2-undecanone	*S. aureus (41, 42, 43,44,45,46,47)* *S. epidermidis (41,43)* *S. mutans (38,45)* *S. pyogenes (45)* *S. salivarius (45)* *S. sanguis (45)*	46) Franca Orlanda and Nascimento 2015
Salvia sclareoides	β-trans-ocimene, linalool, (E)-caryophyllene		47) Sepahvand et al. 2014

[a] In parenthesis besides the microorganism are the references numbered in the last column of the table.

[b] For an easier comparison, the MICs of the references were transformed to μg/mL, in the case of μL/mL it was considered equal to water density.

Table 4.2. Essential oils of aromatic plants with antimicrobial activity against Gram negative bacteria.

Aromatic plants	**Mains components from EOs (> 9%)**	**Bacteria[a] (MIC)[b]**	**References**
		MIC µg/mL > 200	
Cymbopogon citratus	neral, geranial	*A. baumanii (18, 25, 33)* *A. actinomycetemcomita (21)* *B. thetaiotaomicron (3)* *Bacteroides fragilis (3)* *C. freundi (14)* *C. jejuni (6)* *E. aerogenes (1, 6)* *E. cloacae (5, 7, 14, 22, 34)* *E. coli (1, 2, 3, 4, 5, 7, 9, 11, 12, 14, 17, 18, 19, 22, 23, 24, 25, 26, 28, 29, 30, 31, 32, 33)* *E. faecalis (1, 8, 13, 17, 18)* *E. sakazakii (14)* *F. nucleatum (21)* *K. pneumonia (3, 5, 12, 18, 22, 23, 24, 33)* *Klebsiella* sp. *(4)* *L. innocua (6)* *L. monocytogenes (1, 7, 26, 29, 32, 35)* *M. catarrhalis (17)* *P. aeruginosa (1, 4, 5, 7, 8, 9, 10, 11, 14, 15, 18, 22, 23, 24, 25, 31, 33)* *P. gingivalis (21)* *P. mirabilis (7, 9, 12, 28)* *P. vulgaris (15, 18, 20, 23)* *Proteus* sp. *(4, 14)* *S. dysenteria (1, 11)* *S. entérica (1)* *S. enteritidis (32)* *S. marcescens (7, 15)* *S. typhimurium (1, 3, 11, 24, 27, 29)* *Salmonella* sp. *(4, 5, 26)* *Shigella* sp. *(4)*	1) Bassole et al. 2011
C. giganteus	limonene, *cis*- and *trans*-p-mentha-2,8-dien-1-ol		
Origanum vulgare	p-cymene, γ-terpinene, carvacrol		2) Lv et al. 2011
Ocimum basilicum	eugenol, caryophyllene		
Citrus bergamia	limonene, linalool, bergamol		
Perilla arguta	2-pentanoylfuran, benzene, 1-methoxy-4-(1-methylpropyl)		
Paeonia lactiflora	myrtenal, myrtenol, paeonol		3) Ngan et al. 2012
Alpinia purpurata red ginger	β-pinene, linalool, β-caryophyllene		4) Santos et al. 2012
Ferula assafoetida	α-pinene, 1,2-dithionine		
T. maroccanus	carvacrol		5) Fadli et al. 2012
T. broussonetti	carvacrol, borneol		
Teucrium marum	caryophyllene oxide		6) Djabou et al. 2013
T. polium ssp. *capitatum*	α-pinene		
T. flavum ssp. *glaucum*	α-pinene, β-pinene, limonene		
T. scorodonia ssp. *scorodonia*	α-caryophyllene		
T. chamaedrys	α-caryophyllene, germacrene D		
T. massiliense	6-methyl-3-heptyl acetate		
Citrus aurantifolia	β-pinene, limonene, γ-terpinene		7) Costa et al. 2014
Mentha piperita	menthone, menthol		8) Nikolic et al. 2014
Mentha pulegium	pulegone, piperitone		
Lavandula angustifolia	linalool, borneol, linalyl acetate		
Salvia lavandulifolia	1,8-cineole, camphor		
Pinus nigra sp. *nigra; P. nigra* sp. *pallasiana; P. nigra* sp. *banatica*	α-pinene, germacrene D		9) Sarac et al. 2014
Skimmia laureola	linalool, linalyl acetate		10) Shah et al. 2012
Foeniculum vulgare	estragole, *trans*-anethole		11) Diao et al. 2014
Azorella trifurcata	limonene, spathulenol		12) Lopez et al. 2014
Senecio pogonias	α-pinene, α-phellandrene		
Senecio oreophyton	α-pinene, p-mentha,1-(7),8-diene		
Rosa alba	geraniol, heneicosane, nonadecane, citronellol		13) Mileva et al. 2014
Pelargonium graveolens	citronellol, geraniol, nerol		14) Sienkiewicz et al. 2014
Callistemom viminalis	α-pinene, 1,8-cineole		15) Salem et al. 2013
Artemisia capillaris	β-pinene, capillin		16) Yang et al. 2015a
Cymbopogon giganteus	limonene, *cis*- and *trans*-p-mentha-2,8-dien-1-ol, 1,3,8-p-menthatriene		17) Ahmad and Viljoen 2015
C. winterianus	citronellal, citronellol, geraniol		
C. flexuosus	neral, geranial		
C. nardus	geraniol		
C. citratus	neral, geranial		
Cinnamomum zeylanicum	eugenol, cinnamaldehyde		18) Aumeeruddy-Elalfi et al. 2015
Laurus nobilis	camphene, linalool, methyl eugenol, eugenol		

Table 4.2 contd....

...Table 4.2 contd.

Aromatic plants	Mains components from EOs (> 9%)	Bacteria[a] (MIC)[b]	References
		MIC μg/mL > 200	
Psiada arguta	caryophyllene oxide, methyl eugenol, isoeugenol		
Piper bette	β-caryophyllene, eugenol, safrole		
Pimenta dioica	eugenol		
Psiadia terebinthina	α-curcumene, acetyl eugenol		
Rosmarinus officinalis	α-pinene, 1,8-cineole		
Salvia officinalis	1,8-cineole, camphor, thujone, aromadrendene		
Schinus terebinthifolius	α-pinene, 3-carene, α-phellandrene		
Lavandula officinalis	linalool		19) Martucci et al. 2015
Origanum vulgare	carvacrol, p-cymene, γ-terpinene		
Origanum vulgare	γ-terpinene, terpinen-4-ol, carvacrol		20) Suzuki et al. 2015
Eucalyptus globulus	isovaleraldheyde, 1,8-cineole		21) Harkat-Madouri et al. 2015
Geniosporum rotundifolium	spathulenol		22) Ngassapa et al. 2016
Haumaniastrum villosum	caryophyllene oxide, humulene epoxide II		
Plectranthus rugosus	α-pinene, β-caryophyllene, germacrene D		23) Gani et al. 2015
Agathis robusta	β-selinene, α-selinene, spathulenol, caryophyllene oxide		24) Verma et al. 2016
Aloysia gratissima	1,8-cineole, β-caryophyllene, germacrene D, β-pinene		25) Santos et al. 2015
Forsythia koreana	cis-3-hexenol, linalool, trans-phytol		26) Yang et al. 2015b
Lippia origanoides	p-cymene, carvacrol, thymol		27) Sarrazin et al. 2015
Coriandrum sativum	limonene, linalool, linalyl acetate		28) Bogavac et al. 2015
Thymus vulgaris	p-cymene, thymol, carvacrol		
Mentha spicata	limonene, carvone		29) Shahbazi 2015
Thymus transcaucasicus	thymol, carvacrol		30) Bektas et al. 2016
Cordia globosa	α-pinene, camphene		31) Miguel et al. 2016
Origanum vulgare	thymol, p-cymene		32) de Medeiros Barbosa et al. 2016
Rosmarinus officinalis	1,8-cineole, camphor, limonene		
Eucalyptus radiata	limonene, α-terpineol		33) Luis et al. 2016
E. globulus	1,8-cineole, α-pinene		
Pentacalia ledifolia	α-pinene, 4-terpineol		34) Carrillo-Hormaza et al. 2015
Baccharis antioquensis	β-pinene, β-caryophyllene, α-caryophyllene, δ-cadinene		
Baccharis tricuneata	limonene, cyclosativene		
Diplostephium antioquense	β-pinene, p-mentha-2,8-dien-1-ol, β-copaene, δ-cadinene		
Diplostephium rosmarinifolius	α-pinene		
Ageratina tinifolia	α-caryophyllene, β-sesquiphellandrene		
Baccharis brachylaenoides	α-pinene, β-pinene, (Z)-β–ocimene		

Table 4.2 contd....

...Table 4.2 contd.

Aromatic plants	**Mains components from EOs (> 9%)**	**Bacteria[a] (MIC)[b]**	**References**
Pentacalia trianae	β-caryophyllene, (Z)-nerolidol	**MIC µg/mL 30–200**	
Juglans regia	α-pinene, β-pinene, germacrene D	*E. coli (35, 36, 39)* *E. cloacae (30)* *E. faecalis (30, 38)* *K. pneumoniae (35)* *M. catarrhalis (39)* *Helicobacter pylori(40)* *P. aeruginosa (35, 38,41,46)* *P. vulgaris (35)* *S. dysenteriae (35,37)* *S. typhimurium (36)* *S. typhy (35)* *V. alginolyticus (43)* *V. cholerae (42)* *V. cincinnatiensis (43)* *V.diazotrophicus (43)* *V. fluvialis(43)* *V. mimicus (43)* *V. natrigens (43)* *V. splendis (43)* *V. tapetis (43)*	35) Rather et al. 2012
Satureja horvatii	p-cymene, thymol, thymol methyl ether		36) Bukvicki et al. 2014
Foeniculum vulgare	estragole, *trans*-anethole		37) Diao et al. 2014
Thymus algeriensis	thymol, carvacrol		38) Nikolic et al. 2014
Thymus vulgaris	p-cymene, thymol		
Cymbopogon martinii	geraniol		39) Ahmad & Viljoen 2015
C. nardus			
C. citratus	neral, geranial		
Satureja bachtiarica	thymol, carvacrol		40) Falsafi et al. 2015
Satureja montana	p-cymene, thymol		41) Nikolic et al. 2014
Cordia globosa	a-pinene, camphene		42) Miguel et al. 2016
Petroselinum crispum	apiole, myristicin, β-phellandrene, terpinolene, 1, 3, 8-p-menthatriene		43) Snoussi et al. 2016
Ocimum basilicum	linalool, (E)-methyl cinnamate		
		MIC µg/mL < 30	
Gnaphlium affine	linalool, eugenol, (E)-caryophyllene	*A. baumannii (30)* *A. hydrophila (47)* *E. aerogenes (46)* *E. coli (44, 46)* *E. fecalis (45)* *P. aeruginosa (45)* *P. vulgaris (30)* *S. typhi (46)* *S. typhimurium (44)* *V. alginolyticus (47)* *V. anguillarum (47)* *V. carhiaccae (47)* *V. cholera (47)* *V. cincinnatiensis (47)* *V. fluvialis (47)* *V. furnisii (47)* *V. harveyii (47)* *V. mimicus (47)* *V. natrigens (47)* *V. parahaemolyticus (47) V. pectenicidae (47)* *V. proteolyticus (47)* *V. splendidus (47)* *V. vulnificus (47)*	44) Zeng et al. 2011
Thymus serpyllum *Thymus algeriensis*	thymol		45) Nikolic et al. 2014
Ruta graveolens	2-nonanone, 2-undecanone		46) Franca Orlanda and Nascimento 2015
Petroselinum crispum	apiole, myristicin, β-phellandrene, terpinolene, 1, 3, 8-p-menthatriene		47) Snoussi et al. 2016
Ocimum basilicum	linalool, (E)-methyl cinnamate		

[a] In parenthesis besides the bacteria are the references numbered in the last column of the table.
[b] For an easier comparison, the MICs of the references were transformed to µg/mL, in the case of µL/mL it was considered equal to water density.

Anti-biofilm Activity of Essential Oils and Their Components

Bacteria live in either free or in aggregate forms, with these aggregations being referred to as biofilms or slime. However, bacteria living as a group or colony have many advantages over a life as a single cell, as they can access nutrients more easily and develop mechanisms to protect against desiccation, with toxic substances being optimized when bacteria live in a biofilm structure (Harshey 2003, Adukwu et al. 2012). Bacterial biofilms are created when bacteria aggregate and fix to a surface, where they then develop an extracellular matrix whose main components are polysaccharides and proteins, but in which it can also be found DNA, teichoic acids, peptideglycan and dead cells (Niu and Gilbert 2004, Branda et al. 2005, Chaignon et al. 2007, Amalaradjou et al. 2010). The physico-chemical characteristics of the surface define the ease of biofilm growth. Thus, hydrophobic surfaces have a better bacterial adherence than hydrophilic ones (Mah and O'Toole 2001, Cerca et al. 2005, Branda et al. 2005, Oliveira et al. 2010).

The chemical characteristics and physiological properties of extracellular matrices, such as those of the extracellular polysaccharide secreted post-induction cell/surface, are dependent on the nature of the resident bacteria (Harshey 2003). Cellulose was found in major polysaccharide biofilms assembled by *Cronobacter sakazakii*, *Enterobacter sakazakii*, *Escherichia coli* and *Salmonella typhimurium* (Grimm et al. 2008, Amalaradjou and Venkitanarayanan 2011). In contrast, *Pseudomonas aeruginosa* revealed alginate as the main component of its slime, while *Staphylococcus epidermidis, S. aureus* and some Gram negative bacteria showed the adhesin, poly-N-acetyl glucosamine polymers, as the fundamental constituent soft their biofilms (Branda et al. 2005).

Hygiene is the main factor that helps in protecting the health of people, especially in sensitive areas such as hospitals, potable water supplies and food preparation, which are all vulnerable to the development of biofilms. The inclusion of bacteria in biofilms increases its resistance to antibiotics, as well as to various cleaning techniques, when compared with their planktonic forms or cells in suspension (Niu and Gilbert 2004, Budzyńska et al. 2011, Adukwu et al. 2012, Bazargani and Rohloff 2016). The following techniques can to be used to prevent or eliminate biofilms: to minimize the adhesion of bacteria to surfaces, to break down the matrix, and to avoid the development of bacteria with in the biofilm. Related to this, EOs have shown excellent antibacterial and anti adhesion activities against the microorganisms contained in biofilms. Moreover, EOs with higher concentrations of MT or PhP with phenols or carbonyl groups have revealed stronger toxic properties to prevent bacterial biofilms. The first stages of biofilm development have demonstrated a low resistance to MT or PhP, because this involves primarily a planktonic population, but when the adherent cell population of the biofilm increases, so does the resistance to MT or PhP.

Carvacrol has shown toxic effects on biofilm (Szabo et al. 2010, Burt et al. 2014). In addition, the eradication of bacterial biofilms by multiple treatments with phenols was resisted, the fundamental reason was an increase in the thickness of slime, due to the high contribution of extracellular polysaccharide increasing the biofilm thickness and preventing antibacterial activity of the phenols (Knowles et al. 2005). As protease and elastase play a fundamental role in the first stage of colonization and adherence to surfaces in the development of biofilms, the loss of protease or elastase activities could have negative implications for the virulence of bacteria. Related to this, viridiflorol was able to decrease by 70% the elastase activity of *P. aeruginosa*, while triterpenes reduced elastase activity by 90%, without modifying bacterial biofilm development (Gilabert et al. 2015). Moreover, protease, chitinase and elastase activities were reduced by clove EO (Husain et al. 2013) and cinnamaldehyde (Brackman et al. 2008).

Bacteria first need to be mobilized in order to locate the substrate. Then, after fixing to the surface they begin to form an extracellular matrix. Therefore, a strategy for their control is to decrease the bacteria mobilization capacity. Cinnamaldehyde and methyl eugenol, were able to reduce the swimming motility of *E. coli* and *P. aeruginosa* (Niu and Gilbert 2004, Brackman et al. 2008, Packiavathy et al. 2012), with cinnamaldehyde also reducing the formation and function of flagella in *C. sakazakii* (Amalaradjou and Venkitanarayanan 2011). Furthermore, carvacrol inhibited the synthesis of flagella with a loss of cell motility (Faleiro 2011), while α-bisabolol showed an anti-swarming activity dependent on the concentration (Sethupathy et al. 2015).

Effects of EO on Quorum Sensing

The Quorum Sensing (QS) system is an intercellular communication used between bacteria to regulate many activities, such as biofilm formation, sporulation, virulence and the production of bacterial pheromones or auto inducer. Gram negative and Gram positive bacteria have revealed different auto inducers (Nazzaro et al. 2013) and their control could be used to regulate biofilm development. A high potential anti-QS activity has been shown by clove, cinnamom, peppermint and lavender oils (Khan et al. 2008), rose, geranium and rosemary oils (Szabo et al. 2010), clary sage, marjoram and juniper oils (Kerekes et al. 2013), with moderate anti-QS activity being observed by *Eucalyptus*, citrus, and chamomile oils (Szabo et al. 2010). Quorum sensing of *Chromobacterium violaceum* was reduced by *Piper bredemeyeri, P. brachypodom* and *P. bogotence* oils with IC_{50} values of 45.6 μg/mL, 93.1 μg/mL, and 513.8 μg/mL, respectively (Olivero et al. 2011). However, the QS of *C. violaceum* was not controlled by eugenol (Khan et al. 2009), both LasR (a bacterial auto inducer of *P. aeruginosa*) and biofilm growth were inhibited by methyl eugenol at concentrations of 10 ug/mL (Packiavathy et al. 2012). In addition, the system of long chain AHL-QS for plasmid pkR-C12 was inhibited by citral at a concentration of 250 mM, while α-pinene, carvone and geranyl acetate showed only moderate inhibition when were tested on *E. coli* containing plasmid pJBA132 of a short chain (Jaramillo-Colorado et al. 2012). Finally, the tertiary alcohols, linalool and terpinen-4-ol showed strong inhibitory properties on QS (Kerekes et al. 2013).

Monoterpenes and Phenylpropanoids: Mode of Action as Antibacterial Compounds

As EOs components have different functional groups and structural characteristics, the antibacterial activity of these compounds may be able to act on several bacterial targets.

Phospholipids

Phosphatidylglycerol (PG), phosphatidylethanolamine (PE) and cardiolipin (CL) are the most common bacterial membrane phospholipids, while phosphatidylcholine (PC) and phosphatidylinositol (PI) are present at a lower frequency in the formation of bacterial membranes (Epand et al. 2008, Hagi et al. 2015, Mingeot-Leclercq and Décout 2016, Sohlenkamp and Geiger 2016). Other components of bacterial membranes are lipopolysaccharides (LPS), lipoteichoic acids (LTA), hopanoides and pentacyclic triterpenoid lipids, which are analogues of the sterols present in eukaryotic cells (Denich et al. 2003, Mingeot-Leclercq and Décout 2016). The concentration of these compounds in the membrane fluctuates according to the different strains or due to changes in the environmental conditions, and consequently can easily produce variations in the membrane properties. The phospholipid composition is also important and discriminates against the toxic effects of MT and PhP (Trombetta et al. 2005, Anaya-López et al. 2012, Cox et al. 2014). Generally, PC, phosphatidylserine (PS) and PG have revealed a tubular shape, whereas CL and PG, have evidenced a cone shape (hexagonal type H_{II}) in the presence of cations (Denich et al. 2003), but these shapes can change when other hydrophobic compounds interact with them. The phospholipid aggregated with MT or PhP is elongated, with this effect expanding into the neighbouring phospholipid complex (Pham et al. 2015). The presence of thymol among phospholipid leads to a condensing effect, as a consequence of modification of the membrane thickness and a decrease in the head group area where the phospholipids occur, with similar effect being produced by sterols (Denich et al. 2003, Hung et al. 2007, Pham et al. 2015). As a result of a decrease in the membrane thickness physiological aspects of cells area affected (Murínová and Dercová 2014), so this imbalance could be important in the survival of the bacteria.

Changes of bacterial fatty acid composition of the cell membrane

The antimicrobial activity of EOs in common with other phytochemicals is mainly related to the disruption of lipid components of the cell bilayer of the bacterial membrane, and generates the death of microorganisms or works as a barrier thereby avoiding the access of toxic compounds (Epand et al. 2008, Wydro et al. 2012). Hence, the magnitude of the antimicrobial activity is partially dependent on the chemical composition of the membrane, which determines its structural and functional architecture (Epand et al. 2008). A stressful situation on bacterial membrane is generated by the presence of MT or PhP, and for this reason changes in the fatty acid composition are produced to affect the physiological stability and integrity of the membrane (Di Pasqua et al. 2007, Tabanelli et al. 2013, Siroli et al. 2015). In this way, the changes in the fatty acid composition can avoid damage to the membrane enzymes and proteins implicated in electron transport chains (Murínová and Dercová 2014). The adaptive strategies of microbial in their fatty acids composition due to the presence of MT or PhP include alteration of the branching position, development of cyclopropane rings, modification of the saturation degree, changes in chain length and geometrical isomerisation.

Treatments with citral, thymol, eugenol or carvacrol on *P. fluorescens, Salmonella enterica, L. monocytogenes, S. typhimurium* and *S. enteritidis* have revealed dissimilar results. *P. fluorescens* and *S. enterica* produced an increase in concentration of saturated fatty acids (SaFAs) (Di Pasqua et al. 2007, Dubois-Brissonnet et al. 2011, Nazzaro et al. 2013), with eugenol originating the highest concentrations of saturated fatty acids, followed by thymol, carvacrol and citral. This was highly linked to the dose of MT or PhP and also to the bacteria that entered the stationary phase (Dubois-Brissonnet et al. 2011). The other strains investigated, in general increased the concentration of unsaturated fatty acids (UnFAs) (Di Pasqua et al. 2006, 2007, Siroli et al. 2015).

Escherichia coli was only able to synthesize UnFAs by the insertion of double bonds during '*de novo*' biosynthesis of fatty acids. In addition, this bacterium was unable to introduce a double bond into pre-existing SaFAs as *Pseudomonas* (Altabe et al. 2013, Kim and Lee 2015). Hence, the ability of bacteria to change their membrane composition using slow mechanisms such as '*de novo*' acyl chain biosynthesis or the use of desaturase enzymes may explain how well bacteria can tolerate a certain environmental stress. Depending on the presence of MT or PhP in the culture media of *E. coli* the different UnFAs were increased, 16:1 cis acid by the cinnamaldehyde, 18.1 cis acid by the eugenol and 18:2 trans and 20:5 cis acids by the limonene (Di Pasqua et al. 2006).

The balance in the membrane fluidity of the bacteria under stress was investigated by transformation of the geometric isomers of *cis* to *trans* (Denich et al. 2003, Di Pasqua et al. 2007). Membrane fatty acid composition of some *Pseudomonas* strains, *E. coli* and *S. enteritidis* were modulated in response to volatile organic compound exposure, thereby increasing the *trans* isomers (Patrignani et al. 2008, Dubois-Brissonnet et al. 2011, Siroli et al. 2015). However, MT or PhP did not activate the *cis-trans* isomerases that manage the alterations of geometrical isomers (Nazzaro et al. 2013).

Changes in the physico-chemical properties of the membrane by the synthesis of branched fatty acids or cyclopropanes have only been induced by biosynthesis of a new membrane (Denich et al. 2003, Altabe et al. 2013). *Listeria monocytogenesis, Staphylococcus* spp., *Legionella* spp. and *Bacillus* spp. species revealed a cytoplasmic membrane highly enriched in branched-chain fatty acids, with this chemical composition of the membrane playing a key role in virulence regulation (Sun et al. 2012). Treatment with carvacrol decreased the concentration of 17-iso and 17-anteiso, while the concentrations of 15-iso, 15-anteiso and 16-iso were increased by thymol (Siroli et al. 2015). Hence, this suggests a differential effect of thymol and carvacrol on branched-chain-alpha-keto acid enzyme, which could be affecting the branched fatty acids biosynthesis (Vizcaino et al. 2014). The antimicrobial activity of (+)-α-pinene shown to be weak on *S. aureus* and *M. luteus,* with a high concentration of 15-anteiso and 17-anteiso fatty acids in their membranes, whereas the membrane composition of *E. coli* without anteiso fatty acids was more resistant to antimicrobial compounds (Dhar et al. 2014).

Fatty acids with a cyclopropane ring are commonly found in Gram negative bacteria, and are largely synthesized either by a cyclase enzyme or less frequently by conversion of *cis*-fatty acids (Denich et al. 2003, Poger and Mark 2015, Siroli et al. 2015), which are mainly produced in the stationary phase (Dubois-Brissonnet et al. 2011). The double bonds in FAs are mainly sensitive to oxidation, and it is hypothesized that cyclopropane FAs improve resistance to oxidative stress. Therefore, cyclopropane FAs may perform a better stabilizing function of the microbial membrane than UnFAs (Poger and Mark 2015). Treatment of *E. coli* and *S. enteritidis* with thymol increased the percentage of cyclopropanic FA, while carvacrol produced a strong reduction (Siroli et al. 2015). A significant decrease in the synthesis of FAs with cyclopropane rings by *S. enterica* and *Lactobacillus plantarum* being observed after treatments using carvacrol, thymol, eugenol, citral or acid phenolic compounds (Dubois-Brissonnet et al. 2011). Dubois-Brissonnet et al. (2011) hypothesized that the accumulation of MT or PhP may limit the accessibility of S-adenosilmethionine (the enzyme responsible for cyclopropane ring synthesis) between the *cis*-double bonds of fatty acids; thus the cyclopropane ring is not normally synthesized.

Previous studies have suggested that an increase in fatty acid length is another membrane adaptation to environmental stress (Di Pasqua et al. 2006). Related to this, although the acyl chain length increased when oregano and thyme EOs were used on *L. monocytogenes*, *E. coli* and *S. enteriditis* did not reveal any changes in the length of their FAs of phospholipids (Siroli et al. 2015).

Stereochemistry

The determination of the enantiomeric composition in EOs is important when using them as antibacterial compounds (Table 4.3). Differential changes in the membrane conformation can be produced by enantiomers of neomenthol, α-, β-pinene and carvone (Fig. 4.1) as a result of stereospecific interactions (Nandi 2003, Zunino et al. 2011, Tsuchiya and Mizogami 2012). In this way, MT stereoisomers have shown the ability to distinguish, the chirality associated with 'lipid domains' in the membrane (Nandi 2003, Zunino et al. 2011, Tsuchiya and Mizogami 2012), therefore, properties could be very important in the antimicrobial activity of MT and PhP. In fact, there are some contradictory results, several reports have shown that the enantiomers of MT have differential antibacterial activities, with (–)-enantiomers being more toxic than (+)-enantiomers (Aggarwal et al. 2002, Demirci et al. 2002, Carrillo-hormaza et al. 2015), while the results of other studies have demonstrated similar toxic effects for enantiomers (Ngan et al. 2012, Mun et al. 2014, Snoussi et al. 2015).

The antibacterial activity of enantiomers of pinene is, however, controversial. Enantiomers of β- and α-pinene did not show any differences in the antimicrobial activities of *Enterobacter cloacae* (Carrillo-hormaza et al. 2015), *S. mutans* or *S. sobrinus* (Choi et al. 2016), but the strains *S. aureus* methicillin resistant (MRSA), *E. coli* and *M. luteus* were inhibited by (+)-β-pinene and (+)-α-pinene, while the enantiomers (–)-β- and (–)-α-pinene did not show antibacterial effects (Rivas da Silva et al. 2012, Dhar et al. 2014). In another study, (+)-α-pinene enantiomer inhibited the phospholipase and esterase activities of MRSA more than (+)-β-pinene (Rivas da Silva et al. 2012). It was also observed that changes in membrane dynamics were caused by pinene isomers, with the membrane showing a stereoselective activity. The (+)-α-pinene had a stronger effect on the anisotropy than its enantiomer (–)-α-pinene, whereas the opposite was detected for β-pinene enantiomers (Zunino et al. 2011).

In contrast, carvone enantiomers decreased the anisotropy values and revealed dissimilar breaks of the biophysical structure of bilayers (Zunino et al. 2011). From the bibliography data, the constitutional isomers α-pinene have mostly been reported to be more active than β-pinene against various bacteria (Leite et al. 2007, Rivas da Silva et al. 2012, Yang et al. 2013, Yang et al. 2015a, Ngassapa et al. 2016). It should be noted that the Molar Volume of β-pinene is less than that of α-pinene, with it being speculated that α-pinene would consequently produce a greater separation between the achyl chain of phospholipids than β-pinene and result is an increase in toxicity. The Molar Volume descriptor has been included in QSAR models of antifungal (Dambolena et al. 2012) and antibacterial compounds (Andrade-Ochoa et al. 2015) and used to illustrate a mechanism of action of EO components related

Table 4.3. Terpenes and phenylpropanoids with antimicrobial activity.

Terpenes and Phenylpropanoids	**Bacteria**[a] **(MIC**[b] **- μg/mL)**	**References**
	MIC μg/mL > 200	
Thymoquinone	*A. actimomycetemcomitans (6)* *B. adolescentes (3)* *B. bifidum (3)* *B. breve (3)* *B. fragilis (3)* *B. infantis (3)* *B. longum (3)* *B. subtilis (2)* *B. thetaiotaomicron (3)* *C. butyricum (3)* *C. coli (3, 11)* *C. difficile (3)* *C. jejuni (11)* *C. paraputrificum (3)* *C. perfringens (3)* *E. cloacae (13)* *E. coli (2, 8, 9)* *E. faecalis (1, 6, 7, 9)* *E. faecium (1)* *F. nucleatum (4)* *K. pneumonia (2, 3, 9)* *L. acidophilus (3)* *L. casei (3, 7)* *M. luteus (9)* *P. aeruginosa (4, 9)* *P. gingivalis (4)* *P. intermedia (4)* *P. vulgaris (2)* *S. aureus (3, 8, 9, 10, 12)* *S. dysenteriae (2)* *S. epidermidis (4, 9)* *S. mitis (1, 7)* *S. mutans (4, 5, 6, 7)* *S. pyogenes (4)* *S. salivarius (7)* *S. sanguinis (4, 5, 7)* *S. sobrinus (4, 5, 7)* *S. typhimurium (3, 4)*	1) Harzallah et al. 2011
Myrcene, (Z)-β-Ocimene, (E)-β–Ocimene		2) Rather et al. 2012
Linalool, Cuminaldehyde, (S)-(–)-Perillaldehyde, (Z)-Verbenol, (1S)-(–)-Verbenone, Paeonol, (E)-Anethole, (R)-(+)-Citronellal, (S)-(–)-Citronellal,1,8-Cineole, Thymol, (S)-(–)-β-Citronellol, (R)-(+)-β-citronellol Lavandulol, Nerol, (1R)-(–)-Myrtenol (–)-Borneol, (1S, 2S, 5S)-(–)-Myrtanol Geraniol, (–)-Perilla alcohol, (1R)-(–)-Myrtenal, α-Terpinolene.		3) Ngan et al. 2012
Terpinen-4-ol, α-Terpineol, 1,8-Cineole, α-Pinene, Sabinene α-Terpinene, γ-Terpinene, Aromadendrene, p-Cymene, Limonene		4) Lee et al. 2013
α-Pinene, β-Pinene, Myrcene, Limonene, α-Terpinolene, 1,8-Cineole, Linalool, Isopulegol, α-Terpineol, β-Caryophyllene, Caryophyllene oxide, Geranyl acetate, Citronellal, Citral, Citronellol, Nerol, Geraniol		5) Yang et al. 2013
Nerol, Citral, Eugenol, Citronellol, Methyleugenol, Geraniol		6) Mileva et al. 2014
α-Pinene, β-Pinene, (E)-Caryophyllene, Caryophyllene oxide		7) Crevelin et al. 2015
Thymol, Carvacrol, Linalool		8) Martucci et al. 2015
Thymol		9) Popovic et al. 2015
α-Pinene, β-Pinene, Limonene, 1,8-Cineole, Piperitone, β-Caryophyllene, Capillin		10) Yang et al. 2015a
Linalool		11) Duarte et al. 2016
(+)-α-Pinene, (+)-β-Pinene		12) Rivas et al. 2012
Cinnamaldehyde, Carvacrol,Thymol, Linalool, β–(+)Pinene, β–(–)Pinene 3-Carene, (S)-(+)-Citronellal, α-Bisabolol, (+)-α-Pinene, (–)-α-Pinene, γ-Terpinene, p-Cymene, α–Phellandrene		13) Carrillo-Hormaza et al. 2015

Table 4.3 contd....

...Table 4.3 contd.

<table>
<tr><th>Terpenes and Phenylpropanoids</th><th>Bacteria[a] (MIC[b] - µg/mL)</th><th>References</th></tr>
<tr><td></td><td>MIC µg/mL 30–200</td><td></td></tr>
<tr><td>Thymoquinone</td><td rowspan="8">A. actimomycetemcomitans (21)
B. breve (16)
B. fragilis (16)
B. subtilis (15)
B. thetaiotaomicron (16)
C. paraputrificum (16)
C. perfringens (16)
E. coli (15, 19)
F. nucleatum (17)
G. haemolysans (14)
G. morbillorum (14)
H. influenzae (19)
H. pylori (20)
K. pneumoniae (15, 16, 19)
M. bovis (18)
M. tuberculosis (18)
P. aeruginosa (15)
P. gingivalis (17)
P. intermedia (17)
P. vulgaris (15)
S. anginosus (14)
S. aureus (15, 16, 19)
S. constellatus (14)
S. criceti (17)
S. dysenteriae (15)
S. epidermidis (15)
S. gordonii (17)
S. mitis (14)
S. mutans (14, 21)
S. oralis (14)
S. pneumoniae (19)
S. pyogenes (17, 19)
S. salivarius (14)
S. typhi (15)
S. typhimurium (16)</td><td>14) Harzallah et al. 2011</td></tr>
<tr><td>α-Pinene, β-Pinene, Myrcene, Limonene, (Z)-β-Ocimene, (E)-β–Ocimene, β-Caryophyllene, Germacrene D</td><td>15) Rather et al. 2012</td></tr>
<tr><td>(R)-(+)-β-citronellol; (S)-(–)-β-Citronellol, Thymol, Nerol, (1R)-(–)-Myrtenol, (–)-Borneol, (1S, 2S, 5S)-(–)-Myrtanol, Geraniol, (–)-Perilla alcohol, (1R)-(–)-Myrtenal, α-Terpinolene</td><td>16) Ngan et al. 2012</td></tr>
<tr><td>Terpinen-4-ol, α-Terpineol, 1,8-Cineole, α-Pinene, Sabinene
α-Terpinene, γ-Terpinene, aromadendrene, p-Cymene, Limonene</td><td>17) Lee et al. 2013</td></tr>
<tr><td>Camphor, p-Anisaldehyde, Cinnamaldehyde, Cuminaldehyde
(+)-Carvone, Carvacrol, Eugenol
Thymol, β-Citronellol, Geraniol, Linalool, Menthol, t-Anethole, Estragole, Eucalyptol, (+)-Limonene
β-Pinene, 3-Carene, Myrcene, Sabinene, β-Caryophyllene, α-Terpinene, Terpinolene, p-Cymene</td><td>18) Andrade-Ochoa et al. 2015</td></tr>
<tr><td>α-Pinene, β-Pinene, Limonene, 1,8-Cineole, Piperitone, β-Caryophyllene, Capillin</td><td>19) Yang et al. 2015a</td></tr>
<tr><td>Thymol</td><td>20) Falsafi et al. 2015</td></tr>
<tr><td>Eugenol, citronellol</td><td>21) Mileva et al. 2014</td></tr>
</table>

Table 4.3 contd....

...Table 4.3 contd.

Terpenes and Phenylpropanoids	Bacteria[a] (MIC[b] - μg/mL)	References
	MIC < 30 μg/mL	
Thymoquinone	*H. pylori (24)* *M. bovis (23)* *M. tuberculosis (23)* *S. constellatus (22)* *S. mutans (22)* *S. oralis (22)* *S. salivarius (22)* *S. sanguis (22)*	22) Harzallah et al. 2011
Cinnamic acid, Camphor, p-Anisaldehyde, Cinnamaldehyde, Cuminaldehyde (+)-Carvone, Carvacrol, Eugenol Thymol, β-Citronellol, Geraniol Linalool, Menthol, *trans*-Anethole, Estragole, β–Pinene		23) Andrade-Ochoa et al. 2015
Carvacrol		24) Falsafi et al. 2015

[a] In parenthesis besides the microorganism are the references numbered in the last column of the table.
[b] For an easier comparison, the MICs of the references were transformed to μg/mL, in the case of μL/mL it was considered equal to water density.

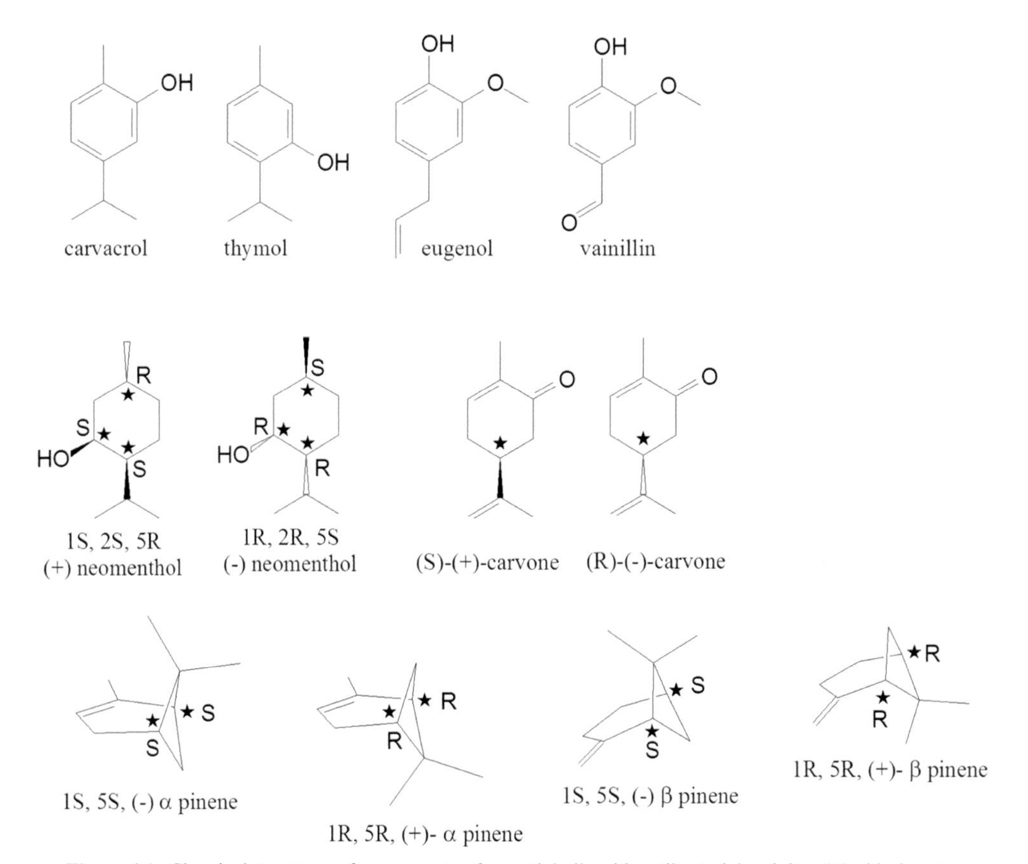

Figure 4.1. Chemical structures of components of essential oils with antibacterial activity. (★) chiral center.

to bacterial membrane damage. Hence, the Molar Volume value appears to an important factor in the expansion of the membrane and therefore its disruption.

Effects on enzymes and proteins

When bovine albumin is added to the culture medium, the antimicrobial activity of carvacrol or thymol decreases (Juven et al. 1994). This suggests that the hydroxyl group of phenols (carvacrol, etc.), alcohols (menthol, etc.), and acid (geranic acid, etc.), may form hydrogen bonds with proteins, and that this action can inhibit enzymes such as coagulase, lipases (de Barros et al. 2009, Souza et al. 2013), α-glucosidases (Tan et al. 2016), monoamine oxidase (Devi et al. 2010) and ATPase (Gill and Holley 2006b, Nazzaro et al. 2013). However, the inhibition of enzyme activity by hydroxylated MT arises in many situations at the same concentration in which the membrane was broken, indicating that enzyme inactivation is sometimes a secondary consequence of cell death (Gill and Holley 2006b).

Glycolysis enzymes were reported to be inhibited by carvacrol, inflicting a change in the metabolism of the bacteria which was accompanied from respiration (Krebs cycle) to fermentation by a decrease in formate, citrate and succinate (Picone et al. 2013). This inhibition of cellular respiration in the bacteria was also produced by eugenol, although the effects on glycolysis or the Krebs cycle are unknown (Devi et al. 2010).

Bacteria have shown an increase in the concentration of heat shock proteins (HSPs), which are chaperones and proteases that contribute to the conservation and rebuilding of denatured polypeptides after a stressful situation. Among, the heat shock genes are DnaK and GroEL, with the former being involved in the repair and maintenance of the structure of the membrane and also assisting in denatured protein refolding (Cuaron et al. 2013), and the latter, producing protein folding when spontaneous folding did not take place (Lin and Rye 2006). Although treatment with MT (carvacrol, thymol and p-cymene) creates a stressful situation for bacteria HSP were not affected (Burt et al. 2007, Faleiro 2011, Nazzaro et al. 2013). However, (–)-α-pinene had a negative effect on the production of HSPs (Kovac et al. 2015).

The cell division of bacteria is controlled by FtsZ, a prokaryotic homologue of tubulin, cinnamaldehyde has shown an inhibitory effect on FtsZ (Domadia et al. 2007), while *in silico* studies was observed that the germacrene D-4-ol interacts with FtsZ (Šarac et al. 2014).

Effects on glucose and cytosolic Ca^{2+}

Glucose has a very important function in the homeostasis of the pH of the bacterial cells, as it acts as a power source for proton H^+ -ATPase-pumping (Lambert et al. 2001), among other functions. Bacterial cells have demonstrated an inability to metabolize internal glucose and have evidenced a loss of viability when treated with carvacrol (Picone et al. 2013). However, the presence of glucose in the culture medium improved the resistance of the bacteria to the toxic effects of phenols (carvacrol, thymol and eugenol) (Gill and Holley 2006a, Li 2011, Zhang et al. 2012). Cinnamaldehyde inhibited glucose uptake (Gill and Holley 2004, Amalaradjou et al. 2010), affecting the proton-pumping H^+ -ATPase (Lambert et al. 2001, Gill and Holley 2004), but did not produce a loss in ATP or inhibit ATPase (Gill and Holley 2006a). However, this aldehyde inhibited enzymes of glycolysis but did not reveal changes in fermentation (Gill and Holley 2004).

It has been hypothesized that an increase of the concentration of aromatic amino acids is produced by carvacrol, although other amino acids did not show changes in their concentrations (Picone et al. 2013). This effect could be linked with production of 3-deoxy-D-arabino-heptulosonate-7-phosphate, the intermediary between glycolysis and the shikimate pathway, and with the release of allosteric control affected by the aromatic amino acids. Moreover, carvacrol has also shown an ability to inhibit chorismate mutase enzyme, which is very important as a target for the development of antibacterial compounds, since vertebrates do not have the shikimate pathway. Parent compounds of carvacrol have revealed a lower inhibitory activity than carvacrol, with the structural characteristics that reduced the activity of the parent compounds being the hydroxyl group position relative to the alkyl substituents, replacement of the hydroxyl group by an amino group, the presence of halogenated groups and the absence of an aromatic ring, ether or ester formation (Alokam et al. 2014).

In the eukaryotic cell carvacrol and thymol produced increases in cytosolic Ca^{2+}, and the antimicrobial activity which were links to the time and grade of the cytosolic Ca^{2+} concentration (Zhang et al. 2012). Eugenol did not show any changes in cytosolic Ca^{2+} when used at antibacterial doses suggesting that it did not work in a similar manner to that of carvacrol (Hyldgaard et al. 2012, Nazzaro et al. 2013). Bacterial cells such as eukaryotic cells are equipped with biological mechanisms to control Ca^{2+} homeostasis. However, in recent years only a few studies have been carried out into, the role of bacteria calcium in the toxic mechanism of MT phenols (Dominguez et al. 2015).

Effects on membrane potential

Different ion concentrations inside (negative) and outside the bacterial cell produce an electrochemical potential (or membrane potential), which show a primary performance in the control of the exchange of solutes. A measure of the health of bacteria is the magnitude of the membrane potential, which varies from 100 to 200 mV (Páez et al. 2013). Also, the chemical composition of the membrane is related to the values of the membrane potential (Epand et al. 2008, Magalhaes and Glogauer 2010, Machado de Araújo et al. 2011, Anaya-López et al. 2012, Cox et al. 2014). Hence, the degree of interaction between the MT or PhP with phospholipids is shown in change in membrane potential (Epand et al. 2008, Machado de Araújo et al. 2011, Nowotarska et al. 2014). This results in the inhibition of the binding or entry of polarized molecules (Anaya-López et al. 2012, Cox et al. 2014), changes in thickness of the lipid bilayer, movement and adhesion of ions to the surface of the cell (Khandelia et al. 2010, Rodrigues et al. 2013), and the formation of new 'lipid domains' (Nowotarska et al. 2014), which generate a loss of membrane potential as an event previous to membrane permeabilization (Turina et al. 2006, Wang et al. 2008, Silva et al. 2011, Nowotarska et al. 2014).

The strong hydrophobic attraction between the aromatic region of cinnamaldehyde with chains of phospholipids cause a strong reduction of surface potential (Nowotarska et al. 2014). However, cinnamaldehyde does not induce changes in the polarity or permeability of the membrane at MIC values (Hammer and Heel 2012).

Changes in membrane potential values were evidenced when phenols were included in the membrane of the bacteria, and as a consequence, the orientation of the phospholipids changed (Korchowiec et al. 2015). In fact, the membrane potential is more affected by carvacrol than by hydrocarbons (Ultee et al. 2002, Saad et al. 2013), because hydrocarbons release potassium ions without affecting the pH, whereas phenols release protons, that cause alterations to the pH (Machado de Araújo et al. 2011). *Escherichia coli* and *S. typhi* treated with EOs revealed reductions in their intracellular pH values from 6.2 to 5.2 (Faleiro 2011). When bacteria are exposed to EO components, the protein synthesis and enzymatic activity depend on an appropriate maintenance of the intracellular pH levels. For example, the membrane potential of *B. cereus*, *P. aeruginosa* and *S. aureus* was reduced by carvacrol (Ben Arfa et al. 2006, Saad et al. 2013) when phenol increased the pH gradient (Ultee et al. 1999). Terpenes also showed a depolarizing effect of the membrane as a result of their accumulation within the membrane, which were able to increase the ionic permeability and the trans membrane ion gradient (Hammer and Heel 2012). The utilization of carveol or citronellol was reported to result in an increase in the bacteria surface hydrophilic character (Lopez-Romero et al. 2015). At physiological conditions the surface charges of *E. coli* and *S. aureus* before treatment with monoterpenes were negative due to the presence of anionic groups. However, after exposure to citronellol and carveol, these surface charges had less negative values (Lopez-Romero et al. 2015). Terpinen-4-ol and 1,8-cineole also produced depolarization at a smaller magnitude than phenols (Hammer and Heel 2012).

The MT, (+)-menthyl acetate, (–)-limonene and 1,8-cineole when applied to eukaryotic cells did not produce any effects on the membrane potential, whereas (+)-menthofuran, (+)-pulegone, (+)-neomenthol, (–)-menthol and (–)-menthone improved the ability to depolarize the membrane, and consequently changes occurred in the flow of K^+, Ca^{2+} and H^+ across the membrane (Maffei et al. 2001). It is interesting to recall that at low concentrations neomenthols displayed an enantioselective interaction with the membrane, with this effect influencing primarily the anisotropy (Zunino et al. 2011), and also possibly the membrane potential. These alterations in membrane potential were related to an increase in toxicity and modification of the transduction pathways (Maffei et al. 2001).

As the net surface charge of the bacterial membranes is dependent on the phospholipidic composition, the net surface charge of the membrane is neutral for PC, positive for PC/SA and negative for PC/PS (Cristani et al. 2007). When the membrane is composed of PC or PC/stearylamine (SA) the order of the membrane damage was as follows: thymol > (+)-menthol > linalyl acetate. When the composition of the membrane is PC/PS the order was: (+) menthol > thymol > linalyl acetate, whereas for a PG/CL membrane composition the order was (+)-menthol > linalyl acetate > thymol (Cristani et al. 2007). Thus, it is suggested that membrane damage by MT or PhP could be partially affected by the net surface charge of the membrane.

Antibacterial effects of substituents on aromatic rings of phenols

Although the presence of the hydroxyl group is critical for antibacterial activity, there are varying magnitudes of activity among different phenols (Fig. 4.1). An approach using a ranking of the phenols isolated from EOs has been utilized by several authors in order of decreasing antibacterial activity, carvacrol > thymol > eugenol > vanillin (Fitzgerald et al. 2004, Mastelic et al. 2008, Pei et al. 2009, Li 2011, Navarro et al. 2015). Eugenol and vanillin exhibited a poorer antibacterial activity than carvacrol and thymol, indicating that the presence of substituents in aromatic rings affected the hydroxyl group, which is essential for antibacterial activity of phenols. Eugenol has a methoxyl group in an ortho position that can form hydrogen bond intramolecularly, and as this can hinder the release of the proton, its pKa values should be raised. Nevertheless, the pKa values 10.38, 10.59 and 10.29 for carvacrol, thymol and eugenol respectively are very similar. It is hypothesized that the weak antibacterial property of eugenol cannot be attributed to difficulties in releasing its acid proton, and is probably connected to its poor capacity to permeabilize the membrane, thereby affecting the motion of K^+ ions and ATP out of the bacteria, an effect linked to the Log P value. Thus, the eugenol derivative, 4-allyl-6-(hydroxymethyl)-2-methoxyphenol with a Log P of 1.02 and pKa of 10.35, had no antibacterial effect on *E. coli* (Mastelic et al. 2008). Carvacrol and thymol destabilize the cytoplasmic membrane by acting as proton exchangers, reducing the pH gradient across the membrane and decreasing the intracellular ATP pool. This reduction of intracellular ATP is the result of the loss of inorganic phosphate by the high permeabilization of microbial membrane produced by the phenols. Also, there is an increase in the hydrolysis of ATP by ATP-ase, with the objective of recovering the electrochemical gradient by proton extrusion as a consequence of the disturbed membrane (Faleiro 2011). Vanillin produced a less severe effect on cytoplasmic membranes, as it had the lowest values of Log P (1.19), and therefore, did not result in a large disorganization of the membrane structure. It has been suggested that vanillin does not work as a trans-membrane carrier of cations and protons, although some investigators have demonstrated K^+ leakage and negative effects on pH homeostasis. The carbonyl group present on the benzene ring of vanillin is an electron with drawing group, and its effect is to reduce the pKa values, increase acidity, and possibly alter proton exchange (Ultee et al. 2002, Fitzgerald et al. 2004, Ben Arfa et al. 2006, Devi et al. 2010, Hyldgaard et al. 2012, Nazzaro et al. 2013).

Results of the antibacterial activity of carvacrol and parent compounds against *E. coli* and *S. aureus* (Veldhuizen et al. 2006) were employed for qualitative analysis of the structure-antibacterial activity. This showed that descriptors such as molar volume, the Randix index (a topological index) and, the forming ability of hydrogen bond could explain the magnitude of the MIC values. When Log

P values are between 2 and 4, other characteristics or properties of the molecules may be playing an important role in the antibacterial activity. The magnitude of microbial toxicity was not similar for all phenols or parent compounds of carvacrol, with antibacterial activity being diminished by the loss of the methyl group (3-isopropylphenol) or isopropyl group (o-cresol) (Veldhuizen et al. 2006, Alokam et al. 2014). Hence, this shows the importance of the number and position of the substituents on the aromatic ring, with the structural features being related to the Randic index and molar volume. Moreover, the absence of the phenolic group (p-cymene), or its replacement by an amino (2-amino-p-cymene) or alkyl group (3,4-dimethylcumene) revealed a loss of toxicity, suggesting an interaction by hydrogen bonds at different degrees with the polar region of the phospholipids (Nazzaro et al. 2013, Andrade-Ochoa et al. 2015).

In summary, the phenols from PhP or MT structures are mainly membrane active compounds, causing the disintegration of ion gradients, the blockage of respiration and affecting the movement of ions (K^+, Ca^{2+}, Na^+, H^+) across the membrane. Other effects include, pH homeostasis and altering the regulation of intra- or extra-cellular ATP (Fitzgerald et al. 2004, Mastelic et al. 2008, Pei et al. 2009, Li 2011, Saad et al. 2013, Navarro et al. 2015). The present chapter supports the hypothesis that a phenol with a strong antibacterial activity has Log P values between 3 and 4 and the pKa values between 10 and 11. These Log P values would imply a hydrophobic characteristic in the relationship with the bacterial membrane, and these pKa values are affected by the molecular shape and proton exchange which regulates the movement of ions, ATP and pH homeostasis.

Antibacterial monoterpene alcohols

Although MT alcohols also have a hydroxyl group, they have a lower antibacterial activity than phenols. This reduced antibacterial activity of alcohols is produced by the absence of an aromatic system, and consequently results in a strong effect on pKa and the ability of the hydroxyl group of alcohols to release its acid proton. Several authors have reported that primary aliphatic MT alcohols have a higher antibacterial activity than tertiary alcohols or secondary aliphatic or cyclic alcohols against various bacteria (Friedman et al. 2004, Soković et al. 2010, Liu and Yang 2012, Yang et al. 2013, Andrade-Ochoa et al. 2015, Lopez-Romero et al. 2015). However, the results of some other studies did not reveal a direct relationship between antibacterial activity and the position of the hydroxyl group in the structure of MT alcohols (Kotan et al. 2007, Ngan et al. 2012, Ngassapa et al. 2016). The Structure Analysis Relationship (SAR) of hydroxylated monoterpenes from data reported by Ngan et al. (2012) suggests that the antibacterial activity on *E. coli*, a Gram negative bacterium, is primarily governed by the Randic index or connectivity index. Therefore, both the number of branches and cycles present in the molecule are important characteristics to be able to cross the outer membrane. On the other hand, as a consequence of the absence of an outer membrane, the antibacterial activity on *S. aureus*, a Gram positive bacteria, was mainly due to the Log P value. Thus, the topological or lipophilic characteristics rather than the relative position of hydroxyl group in the structure of the alcohol, seem to be the most important aspects in determining the antibacterial activity. The MT alcohols revealed K^+ leakage when *S. aureus* was tested, while *E. coli* did not show any changes (Lopez-Romero et al. 2015). Finally, the MT alcohols also produced inhibition of protein isoprenylation, with the most likely mechanisms being inhibition of prenyl transferases and disruption of the membrane (Crowell et al. 1994).

Miscellaneous

The effects of EOs on the Lag phase have been interpreted as a measure of the transformation preventing a period of physiological or biochemical changes during adaptation, which takes place before exponential growth, and as a consequence of treatment with MT or PhP, the Lag phase could be continuously lengthened (Zhou et al. 2013, Petretto et al. 2014). The EOs of *Citrus limon* var.

pompia, *Citrus* x *aurantifolia*, *Mentha suaveolens* and *Melaleuca armillaris* were shown to delay the Lag phase of *Lysteria monocytogenes, S. aureus, S. xylosus* and lactic acid bacteria (Hayouni et al. 2008, Flores et al. 2014, Petretto et al. 2014, Fancello et al. 2016). On the other hand, thymol, carvacrol and 1,8-cineole were able to extend the Lag phase of *Salmonella* spp., *S. var typhimurium*, *E. coli* and *S. aureus* (Roller and Seedhar 2002, Zhou et al. 2013, Zengin and Baysal 2014).

The pH of the medium was an important aspect to consider when testing antibacterial activity. Limonene, eugenol and cinnamaldehyde revealed a higher antibacterial activity against *E. coli* and *Helicobacter pylori*, with low pH values, due to an increase of the lipophilicity of MT or PhP. As a consequence, there was an easier MT or PhP dissolution in the membrane (Espina et al. 2013). However, other authors have reported that the bioactivity of citral, menthol, (+)-carvone, (–)-carvone, eugenol and other phenols was higher with a rise in the pH of the culture media from 6 to 8 (Devi et al. 2010, Somolinos et al. 2010, Faleiro 2011). It was suggested that the pH values modified the charge on the bacterial surface, thus the polar area of charged phospholipids might have protected the bacteria from direct contact with MT or PhP, thereby reducing their antibacterial activity (Canillac and Mourey 2004, Faleiro 2011).

Conclusions

From the information obtained in the bibliography, the success of the antibacterial activity of the EOs would be based on its great molecular diversity. In chemical diversity, we find hydrocarbons, ketones, aldehydes, alcohols and aromatic structures like phenols, going to a higher level of diversification if we consider the stereochemistry of these molecules. However, the results of structure analysis relationship (SAR) of MT and PhP, where a large number of compounds with the same functional group are compared, suggests that the antibacterial activity could be primarily governed by the Randic index (connectivity index), Log P and Molar Volume. These descriptors would be showing the structural characteristics needed to reach the target, whereas functional groups (alcohols, phenols, etc.), would be showing the importance once the target was reached. In this manner, MT or PhP hydrocarbons have a appropriate Log P and Molar Volume, although they usually showed low antimicrobial activity by the absence of a functional group. However, they could open the way to other molecules with active functional groups, as the antibiotics, so that they can find in less time and concentration the target and increase their bioactivities. Thus, the EOs or their components have great potential as antimicrobial agents by themselves or because these components may lead synergistic effects, helping to prevent the development of resistance in short periods of time.

Acknowledgments

This work has been supported by research grants FONCyT (PICT 2012-2146), CONICET (PIP 11220120100661CO) and Universidad Nacional de Córdoba (SECyT).

References

Abreu, A., McBain, A. and Simões, M. 2012. Plants as sources of new antimicrobials and resistance-modifying agents. Nat. Prod. Rep. 29: 1007–1021. doi:10.1039/c2np20035j.

Adukwu, E.C., Allen, S.C.H. and Phillips, C.A. 2012. The anti-biofilm activity of lemongrass (*Cymbopogon flexuosus*) and grapefruit (*Citrus paradisi*) essential oils against five strains of *Staphylococcus aureus*. J. Appl. Microbiol. 113: 1217–1227. doi:10.1111/j.1365-2672.2012.05418.x.

Aggarwal, K.K., Khanuja, S.P.S., Ahmad, A., Santha Kumar, T.R., Gupta, V.K. and Kumar, S. 2002. Enantiomeric distribution of some monoterpenes in the essential oils of some *Salvia* species. Flavour Fragr. J. 17: 54–58. doi:10.1002/ffj.1039.

Ahmad, A. and Viljoen, A. 2015. The *in vitro* antimicrobial activity of *Cymbopogon* essential oil (lemon grass) and its interaction with silver ions. Phytomedicine 22: 657–665. doi:10.1016/j.phymed.2015.04.002.

Alipour, Z., Taheri, P. and Samadi, N. 2014. Chemical composition and antibacterial activity of the essential oils from flower, leaf and stem of *Ferula cupularis* growing wild in Iran. Pharm. Biol. 53: 483–487. doi:10.3109/13880209.2014.924149.

Alokam, R., Jeankumar, V.U., Sridevi, J.P., Matikonda, S.S., Peddi, S., Alvala, M., Yogeeswari, P. and Sriram, D. 2014. Identification and structure-activity relationship study of carvacrol derivatives as *Mycobacterium tuberculosis* chorismate mutase inhibitors. J. Enzyme Inhib. Med. Chem. 29: 547–54. doi:10.3109/14756366.2013.823958.

Altabe, G.S., Mansilla, M.C. and de Mendoza, D. 2013. Remodeling of membrane phospholipids by bacterial desaturases. pp. 209–231. *In*: Ntambi, J.M. (ed.). Stearoyl-CoA Desaturase Genes in Lipid Metabolism. Springer, New York.

Amalaradjou, M.A.R., Narayanan, A., Baskaran, S.A. and Venkitanarayanan, K. 2010. Antibiofilm effect of trans-cinnamaldehyde on uropathogenic *Escherichia coli*. J. Urol. 184: 358–363. doi:10.1016/j.juro.2010.03.006.

Amalaradjou, M.A.R. and Venkitanarayanan, K. 2011. Effect of trans-cinnamaldehyde on inhibition and inactivation of *Cronobacter sakazakii* biofilm on abiotic surfaces. J. Food Prot. 74: 200–208. doi:10.4315/0362-028X.JFP-10-296.

Anaya-López, J.L., López-Meza, J.E. and Ochoa-Zarzosa, A. 2012. Bacterial resistance to cationic antimicrobial peptides. Crit. Rev. Microbiol. 39: 1–16. doi:10.3109/1040841X.2012.699025.

Andrade-Ochoa, S., Nevárez-Moorillón, G.V., Sánchez-Torres, L.E., Villanueva-García, M., Sánchez-Ramírez, B.E., Rodríguez-Valdez, L.M. and Rivera-Chavira, B.E. 2015. Quantitative structure-activity relationship of molecules constituent of different essential oils with antimycobacterial activity against *Mycobacterium tuberculosis* and *Mycobacterium bovis*. BMC Complement Altern. Med. 15: 332. doi:10.1186/s12906-015-0858-2.

Aumeeruddy-Elalfi, Z., Gurib-Fakim, A. and Mahomoodally, F. 2015. Antimicrobial, antibiotic potentiating activity and phytochemical profile of essential oils from exotic and endemic medicinal plants of Mauritius. Ind. Crops Prod. 71: 197–204. doi:10.1016/j.indcrop.2015.03.058.

Bakkali, F., Averbeck, S., Averbeck, D. and Idaomar, M. 2008. Biological effects of essential oils—A review. Food Chem. Toxicol. 46: 446–475. doi:10.1016/j.fct.2007.09.106.

Bassolé, I., Lamien-Meda, A., Bayala, B., Obame, L.C., Ilboudo, A.J., Franz, C., Novak, J., Nebié, R.C. and Dicko, M.H. 2011. Chemical composition and antimicrobial activity of *Cymbopogon citratus* and *Cymbopogon giganteus* essential oils alone and in combination. Phytomedicine 18: 1070–1074.

Bazargani, M.M. and Rohloff, J. 2016. Antibiofilm activity of essential oils and plant extracts against *Staphylococcus aureus* and *Escherichia coli* biofilms. Food Control. 61: 156–164. doi:10.1016/j.foodcont.2015.09.036.

Bektas, E., Serdar, G., Sokmen, M. and Sokmen, A. 2016. Biological activities of extracts and essential oil of *Thymus transcaucasicus* Ronniger. J. Essent. Oil Bear Plants 19: 444–453. doi:10.1080/0972060X.2014.895208.

Ben Arfa, A., Combes, S., Preziosi-Belloy, L., Gontard, N. and Chalier, P. 2006. Antimicrobial activity of carvacrol related to its chemical structure. Lett. Appl. Microbiol. 43: 149–154. doi:10.1111/j.1472-765X.2006.01938.x.

Ben El Hadj Ali, I., Chaouachi, M., Bahri, R., Chaieb, I., Boussaïd, M. and Harzallah-Skhiri, F. 2015. Chemical composition and antioxidant, antibacterial, allelopathic and insecticidal activities of essential oil of *Thymus algeriensis* Boiss. et Reut. Ind. Crops Prod. 77: 631–639. doi:10.1016/j.indcrop.2015.09.046.

Bigos, M., Wasiela, M., Kalemba, D. and Sienkiewicz, M. 2012. Antimicrobial activity of geranium oil against clinical strains of *Staphylococcus aureus*. Molecules 17: 10276–10291. doi:10.3390/molecules170910276.

Bogavac, M., Karaman, M., Janjušević, L., Sudji, J., Radovanović, B., Novaković, Z., Simeunović, J. and Božin, B. 2015. Alternative treatment of vaginal infections—*in vitro* antimicrobial and toxic effects of *Coriandrum sativum* L. and *Thymus vulgaris* L. essential oils. J. Appl. Microbiol. 119: 697–710. doi:10.1111/jam.12883.

Brackman, G., Defoirdt, T., Miyamoto, C., Bossier, P., Van Calenbergh, S., Nelis, H. and Coenye, T. 2008. Cinnamaldehyde and cinnamaldehyde derivatives reduce virulence in *Vibrio* spp. by decreasing the DNA-binding activity of the quorum sensing response regulator LuxR. BMC Microbiol. 8: 149. doi:10.1186/1471-2180-8-149.

Branda, S.S., Vik, A., Friedman, L. and Kolter, R. 2005. Biofilms: the matrix revisited. Trends Microbiol. 13: 20–26. doi:10.1016/j.tim.2004.11.006.

Brud, W. 2010. Industrial uses of essential oils. pp. 843–854. *In*: Baser, K.H.C. and Buchbauer, G. (eds.). Handbook of Essential Oils: Science, Technology, and Applications. CRC Press, New York, USA.

Budzyńska, A., Wieckowska-Szakiel, M., Sadowska, B., Kalemba, D. and Rózalska, B. 2011. Antibiofilm activity of selected plant essential oils and their major components. Pol. J. Microbiol. 60: 35–41.

Bukvicki, D., Stojkovic, D., Sokovic, M., Vannini, L., Montanari, C., Pejin, B., Savić, A., Veljić, M., Grujić, S. and Marin, P.D. 2014. *Satureja horvatii* essential oil: *In vitro* antimicrobial and antiradical properties and *in situ* control of *Listeriamonocytogenes* in pork meat. Meat. Sci. 96: 1355–1360. doi:10.1016/j.meatsci.2013.11.024.

Burt, S.A., van der Zee, R., Koets, A.P., de Graaff, A.M., van Knapen, F., Gaastra, W., Haagsman, H.P. and Veldhuizen, E. J. A. 2007. Carvacrol induces heat shock protein 60 and inhibits synthesis of flagellin in *Escherichia coli* O157:H7. Appl Environ. Microbiol. 73: 4484–4490. doi:10.1128/AEM.00340-07.

Burt, S.A., Ojo-Fakunle, V.T.A., Woertman, J. and Veldhuizen, E.J.A. 2014. The natural antimicrobial carvacrol inhibits quorum sensing in *Chromobacterium violaceum* and reduces bacterial biofilm formation at sub-lethal concentrations. PLoS One 9: 1–6. doi:10.1371/journal.pone.0093414.

Calo, J., Crandall, P., O'Bryan, C. and Ricke, S. 2015. Essential oils as antimicrobials in food systems—A review. Food Control. 54: 111–119. doi:10.1016/j.foodcont.2014.12.040.

Canillac, N. and Mourey, A. 2004. Effects of several environmental factors on the anti-*Listeria monocytogenes* activity of an essential oil of *Picea excelsa*. Int. J. Food Microbiol. 92: 95–103. doi:10.1016/j.ijfoodmicro.2003.09.001.

Carrillo-Hormaza, L., Mora, C., Alvarez, R., Alzate, F. and Osorio, E. 2015. Chemical composition and antibacterial activity against *Enterobacter cloacae* of essential oils from Asteraceae species growing in the Paramos of Colombia. Ind. Crops Prod. 77: 108–115. doi:10.1016/j.indcrop.2015.08.047.

Cerca, N., Pier, G.B., Vilanova, M., Oliveira, R. and Azeredo, J. 2005. Quantitative analysis of adhesion and biofilm formation on hydrophilic and hydrophobic surfaces of clinical isolates of *Staphylococcus epidermidis*. Res. Microbiol. 156: 506–514. doi:10.1016/j.resmic.2005.01.007.

Chaignon, P., Sadovskaya, I., Ragunah, C., Ramasubbu, N., Kaplan, J.B. and Jabbouri, S. 2007. Susceptibility of staphylococcal biofilms to enzymatic treatments depends on their chemical composition. Appl. Microbiol. Biotechnol. 75: 125–132. doi:10.1007/s00253-006-0790-y.

Chen, Z., He, D., Deng, J., Zhu, J. and Mao, Q. 2015. Chemical composition and antibacterial activity of the essential oil from *Agathis dammara* (Lamb.) Rich fresh leaves. Nat. Prod. Res. 29: 2050–2053. doi:10.1080/14786419.2015.1022544.

Choi, O., Cho, S.K., Kim, J., Park, C.G. and Kim, J. 2016. *In vitro* antibacterial activity and major bioactive components of *Cinnamomum verum* essential oils against cariogenic bacteria, *Streptococcus mutans* and *Streptococcus sobrinus*. Asian Pac. J. Trop. Biomed. 6: 308–314. doi:10.1016/j.apjtb.2016.01.007.

Costa, R., Bisignano, C., Filocamo, A., Grasso, E., Occhiuto, F. and Spadaro, F. 2014. Antimicrobial activity and chemical composition of *Citrus aurantifolia* (Christm.) Swingle essential oil from Italian organic crops. J. Essent. Oil Res. 26: 400–408. doi:10.1080/10412905.2014.964428.

Cox, E., Michalak, A., Pagentine, S. Seaton, P. and Pokorny, A. 2014. Lysylated phospholipids stabilize models of bacterial lipid bilayers and protect against antimicrobial peptides. Biochim. Biophys. Acta - Biomembr. 1838: 2198–2204. doi:10.1016/j.bbamem.2014.04.018.

Crevelin, E., Caixeta, S., Dias, H., Groppo, M., Cunha, W.R., Martins, C.H.G. and Crotti, A.E.M. 2015. Antimicrobial activity of the essential oil of *Plectranthus neochilus* against cariogenic bacteria. Evidence-based Complement Altern. Med. 2015: 9–11. doi:10.1155/2015/102317.

Cristani, M., D'Arrigo, M., Mandalari, G., Castelli, F., Sarpietro, M.G., Micieli, D., Venuti, V., Bisignano, G., Saija, A. and Trombetta, D. 2007. Interaction of four monoterpenes cointained in essential oil with model membranes: Implications for their antibacterial activity. J. Agric. Food Chem. 55: 6300–6308. doi:10.1021/jf070094x.

Crowell, P.L., Ren, Z., Lin, S., Vedejs, E. and Gould, M.N. 1994. Structure-activity relationships among monoterpene inhibitors of protein isoprenylation and cell proliferation. Biochem. Pharmacol. 47: 1405–1415. doi:10.1016/0006-2952(94)90341-7.

Cuaron, J., Dulal, S., Song, Y., Singh, A.K., Montelongo, C.E., Yu, W., Nagarajan, V., Jayaswal, R.K., Wilkinson, B.J. and Gustafson, J.E. 2013. Tea tree oil-induced trancriptional alterations in *Staphylococcus aureus*. Phytother. Res. 27: 390–396. doi:10.1002/ptr.4738.Tea.

Dambolena, J.S., López, A.G., Meriles, J.M., Rubinstein, H.R. and Zygadlo, J.A. 2012. Inhibitory effect of 10 natural phenolic compounds on *Fusarium verticillioides*. A structure-property-activity relationship study. Food Control. 28: 163–170. doi:10.1016/j.foodcont.2012.05.008.

de Almeida, T., Rocha, J., Rodrigues, F., Campos, A.R. and da Costa, J.G. 2013. Chemical composition, antibacterial and antibiotic modulatory effect of Croton campestris essential oils. Ind. Crops Prod. 44: 630–633. doi:10.1016/j.indcrop.2012.09.010.

de Medeiros Barbosa, I., da Costa Medeiros, J.A., de Oliveira, K.Á.R., Gomes-Neto, N.J., Tavares, J.F., Magnani, M. and de Souza, E.L. 2016. Efficacy of the combined application of oregano and rosemary essential oils for the control of *Escherichia coli, Listeria monocytogenes* and *Salmonella Enteritidis* in leafy vegetables. Food Control. 59: 468–477. doi:10.1016/j.foodcont.2015.06.017.

Demirci, B., Tabanca, N. and Can Baser, H. 2002. Enantiomeric distribution of some monoterpenes in the essential oils of some *Salvia* species. Flavour Fragr. J. 17: 54–58. doi:10.1002/ffj.1039.

Denich, T.J., Beaudette, L.A., Lee, H. and Trevors, J.T. 2003. Effect of selected environmental and physico-chemical factors on bacterial cytoplasmic membranes. doi:10.1016/S0167-7012(02)00155-0.

Devi, K.P., Nisha, S.A., Sakthivel, R. and Pandian, S.K. 2010. Eugenol (an essential oil of clove) acts as an antibacterial agent against *Salmonella typhi* by disrupting the cellular membrane. J. Ethnopharmacol. 130: 107–115. doi:10.1016/j.jep.2010.04.025.

Dhar, P., Chan, P., Cohen, D.T., Khawam, F., Gibbons, S., Snyder-Leiby, T., Dickstein, E., Rai, P.K. and Watal, G. 2014. Synthesis, antimicrobial evaluation, and structure—activity relationship of α-pinene derivatives. J. Agric. Food Chem. 62: 3548–3552. doi:10.1021/jf403586t.

Di Pasqua, R., Betts, G., Hoskins, N., Edwards, M., Ercolini, D. and Mauriello, G. 2007. Membrane toxicity of antimicrobial compounds from essential oils. J. Agric. Food. Chem. 55: 4863–4870. doi:10.1021/jf0636465.

Di Pasqua, R.D., Hoskins, N., Betts, G. and Mauriello, G. 2006. Changes in membrane fatty acids composition of microbial cells induced by addiction of thymol, carvacrol, limonene, cinnamaldehyde, and eugenol in the growing media. J. Agric. Food Chem. 54: 2745–2749. doi:10.1021/jf052722l.

Diao, W.R., Hu, Q.P., Zhang, H. and Xu, J.G. 2014. Chemical composition, antibacterial activity and mechanism of action of essential oil from seeds of fennel (*Foeniculumvulgare* Mill.). Food Control. 35: 109–116. doi:10.1016/j.foodcont.2013.06.056.

Djabou, N., Lorenzi, V., Guinoiseau, E., Andreani, S., Giuliani, M.C., Desjobert, J.M., Bolla, J.M., Costa, J., Berti, L., Luciani, A. and Muselli, A. 2013. Phytochemical composition of *Corsican teucrium* essential olis and antibacterial activity against foodborne or toxi-infectius pathogens. Food Control. 30: 354–363. doi:10.1016/j.foodcont.2012.06.025.

Domadia, P., Swarup, S., Bhunia, A., Sivaraman, J. and Dasgupta, D. 2007. Inhibition of bacterial cell division protein FtsZ by cinnamaldehyde. Biochem. Pharmacol. 74: 831–840. doi:10.1016/j.bcp.2007.06.029.

Dominguez, D.C., Guragain, M. and Patrauchan, M. 2015. Calcium binding proteins and calcium signaling in prokaryotes. Cell Calcium 57: 151–165. doi:10.1016/j.ceca.2014.12.006.

Dreger, M. and Wielgus, K. 2013. Application of essential oils as natural cosmetic preservatives. Herba Pol. doi:10.2478/hepo-2013-0030.

Dubois-Brissonnet, F., Naïtali, M., Mafu, A.A. and Briandet, R. 2011. Induction of fatty acid composition modifications and tolerance to biocides in *Salmonella enterica* serovar typhimurium by plant-derived terpenes. Appl. Environ. Microbiol. 77: 906–910. doi:10.1128/AEM.01480-10.

Epand, R., Rotem, S., Mor, A., Berno, B. and Epand, R.F. 2008. Bacterial membranes as predictors of antimicrobial potency. J. Am. Chem. Soc. 130: 14346–14352. doi:10.1021/ja8062327.

Espina, L., Gelaw, T.K., de Lamo-Castellví, S., Pagán, R. and García-Gonzalo, D. 2013. Mechanism of bacterial inactivation by (+)-limonene and its potential use in food preservation combined processes. PLoS One 8: 1–10. doi:10.1371/journal.pone.0056769.

Fadli, M., Saad, A., Sayadi, S., Chevalier, J., Mezrioui, N.E., Pags, J.M. and Hassani, L. 2012. Antibacterial activity of *Thymus maroccanus* and *Thymus broussonetii* essential oils against nosocomial infection-bacteria and their synergistic potential with antibiotics. Phytomedicine 19: 464–471. doi:10.1016/j.phymed.2011.12.003.

Faleiro, M.L. 2011. The mode of antibacterial action of essential oils. Sci. Against. Microb. Pathog. Commun. Curr. Res. Technol. Adv. 3: 1143–1156.

Falsafi, T., Moradi, P., Mahboubi, M., Rahimi, E., Momtaz, H. and Hamedi, B. 2015. Chemical composition and anti-*Helicobacter pylori* effect of *Satureja bachtiarica* Bunge essential oil. Phytomedicine 22: 173–177. doi:10.1016/j.phymed.2014.11.012.

Fancello, F., Petretto, G.L., Zara, S., Sanna, M.L., Addis, R., Maldini, M., Foddai, M., Rourke, J.P., Chessa, M. and Pintore, G. 2016. Chemical characterization, antioxidant capacity and antimicrobial activity against food related microorganisms of *Citruslimon* var. pompia leaf essential oil. LWT - Food Sci. Technol. 69: 579–585. doi:10.1016/j.lwt.2016.02.018.

Fitzgerald, D.J., Stratford, M., Gasson, M.J., Ueckert, J., Bos, A. and Narbad, A. 2004. Mode of antimicrobial of vanillin against *Escherichia coli*, *Lactobacillus plantarum* and *Listeria innocua*. J. Appl. Microbiol. 97: 104–113. doi:10.1111/j.1365-2672.2004.02275.x.

Flores, R.C., Audicio, N., Sanz, M.K. and Ponzi, M. 2014. Antibacterial activity of lime (*Citrus* x *aurantifolia*) essential oil against *Listeria monocytogenes* in tyndallised apple juice. Rev. la Soc. Venez Microbiol. 34: 10–14.

Franca Orlanda, J.F. and Nascimento, A.R. 2015. Chemical composition and antibacterial activity of *Ruta graveolens* L. (Rutaceae) volatile oils, from Sao Luis, Maranhao, Brazil. South African J. Bot. 99: 103–106.

Franz, C. 2010. Essential oil research: Past, present and future. Flavour Fragr. J. 25: 112–113. doi:10.1002/ffj.1983.

Friedman, M., Henika, P.R., Levin, C.E. and Mandrell, R.E. 2004. Antibacterial activities of plant essential oils and their components against *Escherichia coli* O157:H7 and *Salmonella enterica* in apple juice. J. Agric. Food Chem. 52: 6042–6048. doi:10.1021/jf0495340.

Gani, R., Bhat, Z.A., Ahmad, M. and Zargar, M.I. 2015. GC-MS analysis, antibacterial and antifungal activity of essential oil of *Plectranthus rugosus* from Kashmir, India. Int. J. Bioassays 4: 4692–4695.

Gelband, H., Miller-Petrie, M., Pant, S., Gandra, S., Levinson, J., Barter, D., White, A. and Laxminarayan, R. 2015. State of the World's Antibiotics, 2015. Center for Disease Dynamics, Economics & Policy (CDDEP). Washington, D.C.

Gilabert, M., Marcinkevicius, K., Andujar, S., Schiavone, M., Arena, M.E. and Bardón, A. 2015. Sesqui- and triterpenoids from the liverwort *Lepidozia chordulifera* inhibitors of bacterial biofilm and elastase activity of human pathogenic bacteria. Phytomedicine 22: 77–85. doi:10.1016/j.phymed.2014.10.006.

Gill, A.O. and Holley, R.A. 2004. Mechanisms of bactericidal action of cinnamaldehyde against *Listeria monocytogenes*. Appl. Environ. Microbiol. 70: 5750–5755. doi:10.1128/AEM.70.10.5750.

Gill, A.O. and Holley, R.A. 2006a. Inhibition of membrane bound ATPases of *Escherichia coli* and *Listeria monocytogenes* by plant oil aromatics. Int. J. Food Microbiol. 111: 170–174. doi:10.1016/j.ijfoodmicro.2006.04.046.

Gill, A.O. and Holley, R.A. 2006b. Disruption of *Escherichia coli*, *Listeria monocytogenes* and *Lactobacillus sakei* cellular membranes by plant oil aromatics. Int. J. Food Microbiol. 108: 1–9. doi:10.1016/j.ijfoodmicro.2005.10.009.

Grimm, M., Stephan, R., Iversen, C., Manzardo, G.G.G., Rattei, T., Riedel, K., Ruepp, A., Frishman, D. and Lehner, A. 2008. Cellulose as an extracellular matrix component present in *Enterobacter sakazakii* biofilms. J. Food Prot. 71: 13–18.

Guenther, E. 1948. The essential oils. Vol. 1. D. van Nostrand Company Inc, New York.

Guzman-Blanco, M., Labarca, J.A., Villegas, M.V. and Gotuzzo, E. 2014. Extended spectrum a-lactamase producers among nosocomial Enterobacteriaceae in Latin America. Brazilian J. Infect. Dis. 18: 421–433. doi:10.1016/j.bjid.2013.10.005.

Hagi, A., Iwata, K., Nii, T., Nakata, H., Tsubotani, Y. and Inoue, Y. 2015. Bactericidal effects and mechanism of action of olanexidine gluconate, a new antiseptic. Antimicrob. Agents Chemother 59: 4551–4559. doi:10.1128/AAC.05048-14.

Hammer, K.A. and Heel, K.A. 2012. Use of multiparameter flow cytometry to determine the effects of monoterpenoids and phenylpropanoids on membrane polarity and permeability in staphylococci and enterococci. Int. J. Antimicrob. Agents 40: 239–245. doi:10.1016/j.ijantimicag.2012.05.015.

Harkat-Madouri, L., Asma, B., Madani, K., Bey-Ould Si Said, Z., Rigou, P., Grenier, D., Allalou, H., Remini, H., Adjaoud, A. and Boulekbache-Makhlouf, L. 2015. Chemical composition, antibacterial and antioxidant activities of essential oil of *Eucalyptus globulus* from Algeria. Ind. Crops Prod. 78: 148–153. doi:10.106/j.indcrop.2015.10.015.

Harshey, R.M. 2003. Bacterial motility on a surface: many ways to a common goal. Annu. Rev. Microbiol. 57: 249–73. doi:10.1146/annurev.micro.57.030502.091014.

Harzallah, J.H., Kouidhi, B., Flamini, G., Bakhrouf, A. and Mahjoub, T. 2011. Chemical composition, antimicrobial potential against cariogenic bacteria and cytotoxic activity of Tunisian *Nigella sativa* essential oil and thymoquinone. Food Chem. 129: 1469–1474. doi:10.1016/j.foodchem.2011.05.117.

Hassanien, M., Assiri, A., Alzohairy, A. and Oraby, H. 2015. Health-promoting value and food applications of black cumin essential oil: an overview. J. Food Sci. Technol. 52: 6136–6142. doi:10.1007/s13197-015-1785-4.
Hayouni, E.A., Bouix, M., Abedrabba, M., Leveau, J.Y. and Hamdi, M. 2008. Mechanism of action of *Melaleucaarmillaris* (Sol. Ex Gaertu) Sm. essential oil on six LAB strains as assessed by multiparametric flow cytometry and automated microtiter-based assay. Food Chem. 111: 707–718. doi:10.1016/j.foodchem.2008.04.044.
Hung, W.C., Lee, M.T., Chen, F.Y. and Huang, H.W. 2007. The condensing effect of cholesterol in lipid bilayers. Biophys. J. 92: 3960–7. doi:10.1529/biophysj.106.099234.
Husain, F.M., Ahmad, I., Asif, M. and Tahseen, Q. 2013. Influence of clove oil on certain quorum-sensing-regulated functions and biofilm of *Pseudomonas aeruginosa* and *Aeromonas hydrophila*. J. Biosci. 38: 835–844. doi:10.1007/s12038-013-9385-9.
Hussain, A., Anwar, F., Hussain Sherazi, S. and Przybylski, R. 2008. Chemical composition, antioxidant and antimicrobial activities of basil (*Ocimum basilicum*) essential oils depends on seasonal variations. Food Chem. 108: 986–995. doi:10.1016/j.foodchem.2007.12.010.
Hyldgaard, M., Mygind, T. and Meyer, R.L. 2012. Essential oils in food preservation: Mode of action, synergies, and interactions with food matrix components. Front. Microbiol. 3: 1–24. doi:10.3389/fmicb.2012.00012.
Jaramillo-Colorado, B., Olivero-Verbel, J., Stashenko, Wagner-Döbler, I. and Kunze, B. 2012. Anti-quorum sensing activity of essential oils from Colombian plants. Nat. Prod. Res. 26: 1075–1086. doi:10.1080/14786419.2011.557376.
Jordán, M., Lax, V., Rota, M., Lorán, S. and Sotomayor, J.A. 2013. Effect of the phenological stage on the chemical composition, and antimicrobial and antioxidant properties of *Rosmarinus officinalis* L. essential oil and its polyphenolic extract. Ind. Crops Prod. 48: 144–152. doi:10.1016/j.indcrop.2013.04.031.
Juven, B.J., Kanner, J., Schved, F. and Weisslowicz, H. 1994. Factors that interact with the antibacterial action of thyme essential oil and its active constituents. J. Appl. Bacteriol. 76: 626–631. doi:10.1111/j.1365-2672.1994.tb01661.x.
Kavanaugh, N. and Ribbeck, K. 2012. Selected antimicrobial essential oils eradicate *Pseudomonas* spp. and *Staphylococcus aureus* biofilms. Appl. Environ. Microbiol. 78: 4057–4061. doi:10.1128/AEM.07499-11.
Kavoosi, G., Tafsiry, A., Ebdam, A.A. and Rowshan, V. 2013. Evaluation of antioxidant and antimicrobial activities of essential oils from *Carum copticum* seed and *Ferulaassafoetida* Latex. J. Food Sci. doi:10.1111/1750-3841.12020.
Kerekes, E.B., Deák, É., Takó, M., Tserennadmid, R., Petkovits, T., Vágvölgyi, C. and Krisch, J. 2013. Anti-biofilm forming and anti-quorum sensing activity of selected essential oils and their main components on food-related micro-organisms. J. Appl. Microbiol. 115: 933–942. doi:10.1111/jam.12289.
Khan, M.S.A., Zahin, M., Hasan, S., Husain, F.M. and Ahmad, I. 2009. Inhibition of quorum sensing regulated bacterial functions by plant essential oils with special reference to clove oil. Lett. Appl. Microbiol. 49: 354–360. doi:10.1111/j.1472-765X.2009.02666.x.
Khandelia, H., Witzke, S. and Mouritsen, O.G. 2010. Interaction of salicylate and a terpenoid plant extract with model membranes: Reconciling experiments and simulations. Biophys. J. 99: 3887–3894. doi:10.1016/j.bpj.2010.11.009.
Kim, E. and Lee, J.K. 2015. Effect of changes in the composition of cellular fatty acids on membrane fluidity of *Rhodobacter sphaeroides*. 25: 162–173. doi:10.4014/jmb.1410.10067.
Knowles, J.R., Roller, S., Murray, D.B. and Naidu, A.S. 2005. Antimicrobial action of carvacrol at different stages of dual-species biofilm development by *Staphylococcus aureus* and *Salmonella enterica* Serovar Typhimurium. American Society for Microbiology 71: 797–803. doi:10.1128/AEM.71.2.797.
Korchowiec, B., Gorczyca, M., Wojszko, K., Janikowska, M., Henry, M. and Rogalska, E. 2015. Impact of two different saponins on the organization of model lipid membranes. Biochim Biophys Acta - Biomembr. 1848: 1963–1973. doi:10.1016/j.bbamem.2015.06.007.
Koroch, A., Rodolfo Juliani, H. and Zygadlo, J. 2007. Bioactivity of essential oils and their components. pp. 87–115. *In*: Berger, R.G. (ed.). Flavours and Fragrances: Chemistry, Bioprocessing and Sustainability. Springer, Berlin Heidelberg.
Kotan, R., Kordali, S. and Cakir, A. 2007. Screening of antibacterial activities of twenty-one oxygenated monoterpenes. Zeitschrift fur Naturforsch - Sect. C. J. Biosci. 62: 507–513. doi:10.1515/znc-2007-7-808.
Kovac, J., Simunovic, K., Wu, Z., Klančnik, A., Bucar, F., Zhang, Q. and Možina, S.S. 2015. Antibiotic resistance modulation and modes of action of (–)-a-pinene in *Campylobacter jejuni*. PLoS One 10: 1–14. doi:10.1371/journal.pone.0122871.
Kubeczka, K.H. 2010. History and sources of essential oil research. pp. 3–38. *In:* Baser, K.H.C. and Buchbauer, G. (eds.). Handbook of Essential Oils: Science, Technology and Applications. CRC Press. New York, USA.
Lambert, R.J.W., Skandamis, P.N., Coote, P.J. and Nychas, G.J.E. 2001. A study of the minimum inhibitory concentration and mode of action of oregano essential oil, thymol and carvacrol. J. Appl. Microbiol. 91: 453–462. doi:10.1046/j.1365-2672.2001.01428.x.
Lang, G. and Buchbauer, G. 2012. A review on recent research results (2008–2010) on essential oils as antimicrobials and antifungals. A review. Flavour Fragr. J. 27: 13–39. doi:10.1002/ffj.2082.
Langeveld, W., Veldhuizen, E. and Burt, S. 2014. Synergy between essential oil components and antibiotics: a review. Crit Rev. Microbiol. 40: 76–94. doi:10.3109/1040841X.2013.763219.
Leite, A.M., Lima, E.O., Souza, E.L., Diniz, M.F.F.M., Trajano, V.N. and de Medeiros, I.A. 2007. Inhibitory effect of b-pinene, a-pinene and eugenol on the growth of potential infectious endocarditis causing Gram-positive bacteria. Rev. Bras Ciências Farm. 43: 121–126. doi:10.1590/S1516-93322007000100015.
Li, S. 2011. Enhancement of the antimicrobial activity of eugenol and carvacrol against *Escherichia coli* O157:H7 by lecithin in microbiological media and food. 72. Master's Thesis, University of Tennessee. Available at http://trace.tennessee.edu/utk_gradthes/996.

Lin, Z. and Rye, H.S. 2006. GroEL-Mediated Protein Folding: Making the Impossible, Possible. Crit. Rev. Biochem. Mol. Biol. 41: 211–239. doi:10.1038/nmeth.2250.Digestion.
Liu, T.T. and Yang, T.S. 2012. Antimicrobial impact of the components of essential oil of *Litsea cubeba* from Taiwan and antimicrobial activity of the oil in food systems. Int. J. Food Microbiol. 156: 68–75. doi:10.1016/j.ijfoodmicro.2012.03.005.
Lopachin, R.M. and Gavin, T. 2014. Molecular mechanisms of aldehyde toxicity: A chemical perspective. Chem. Res. Toxicol. 27: 1081–1091. doi:10.1021/tx5001046.
Lopez, S., Lima, B., Aragón, L., Ariza Espinar, L., Tapia, A., Zacchino, S., Zygadlo, J., Feresin Egly, G. and Lopez, M.L. 2012. Essential oil of *Azorella cryptantha* collected in two different locations from San Juan province, Argentina: chemical variability and anti-insect and antimicrobial activities. Chem. Biodivers 9: 1452–1464.
Lopez, S., Lima, B., Agüero, M.B., Lopez, M.L., Hadad, M., Zygadlo, J.A., Caballero, D., Stariolo, R., Suero, E., Egly Feresin, G. and Tapia, A. 2014. Chemical composition, antibacterial and repellent activities of *Azorella trifurcata*, *Senecio pogonias*, and *Seneciooreophyton* essential oils. Arab. J. Chem. In press. doi:10.1016/j.arabjc.2014.11.022.
Lopez-Romero, J.C., González-Ríos, H., Borges, A. and Simões, M. 2015. Antibacterial effects and mode of action of selected essential oils components against *Escherichia coli* and *Staphylococcus aureus*. Evidence-Based Complement Altern. Med. 2015: 1–9. doi:10.1155/2015/795435.
Lorenzi, V., Muselli, A., Bernardini, A., Berti, L., Pagès, J.M., Amaral, L. and Bolla, J.M. 2009. Geraniol restores antibiotic activities against multidrug-resistant isolates from gram-negative species. Antimicrob. Agents Chemother. 53: 2209–2211. doi:10.1128/AAC.00919-08.
Luis, A., Duarte, A., Gominho, J., Domingues, F. and Duarte, A.P. 2016. Chemical composition, antioxidant, antibacterial and anti-quorum sensing activities of *Eucalyptus globolus* and *Eucalyptus radiata* essentail oils. Ind. Crops Prod. 79: 274–282. doi:10.1016/j.indcrop.2015.10.055.
Lv, F., Liang, H., Yuan, Q. and Li, C. 2011. *In vitro* antimicrobial effects and mechanism of action of selected plant essential oil combinations against four food-related microorganisms. Food Res. Int. 44: 3057–3064. doi:10.1016/j.foodres.2011.07.030.
Machado de Araújo, D.A., Freitas, C. and Cruz, J.S. 2011. Essential oils components as a new path to understand ion channel molecular pharmacology. Life Sci. 89: 540–544. doi:10.1016/j.lfs.2011.04.020.
Maffei, M., Camusso, W. and Sacco, S. 2001. Effect of *Mentha x piperita* essential oil and monoterpenes on cucumber root membrane potential. Phytochemistry 58: 703–707. doi:10.1016/S0031-9422(01)00313-2.
Magalhaes, M.A.O. and Glogauer, M. 2010. Pivotal Advance: Phospholipids determine net membrane surface charge resulting in differential localization of active Rac1 and Rac2. J. Leukoc. Biol. 87: 545–55. doi:10.1189/jlb.0609390.
Maggi, F., Bramucci, M., Cecchini, C., Coman, M.M., Tirillini, B., Sagratini, G. and Papa, F. 2009. Composition and biological activity of essential oil of *Achillea ligustica* All. (Asteraceae) naturalized in central Italy: Ideal candidate for anti-cariogenic formulations. Fitoterapia 80: 313–319. doi:10.1016/j.fitote.2009.04.004.
Mah, T.F.C. and O'Toole, G.A. 2001. Mechanisms of biofilm resistance to antimicrobial agents. Trends Microbiol. 9: 34–39. doi:10.1016/S0966-842X(00)01913-2.
Martucci, J.F., Gende, L.B., Neira, L.M. and Ruseckaite, R.A. 2015. Oregano and lavender essential oils as antioxidant and antimicrobial additives of biogenic gelatin films. Ind. Crops Prod. 71: 205–213. doi:10.1016/j.indcrop.2015.03.079.
Mastelic, J., Jerkovic, I., Blazevic, I., Poljak-Blazi, M., Borovic, S., Ivancic-Bace, I., Smrecki, V., Zarkovic, N., Brcic-Kostic, K., Vikic-Topic, D. and Muller, N. 2008. Comparative study on the antioxidant and biological activities of carvacrol, thymol, and eugenol derivatives. J. Agric. Food Chem. 56: 3989–3996. doi:10.1021/jf073272v.
Miguel, M., García-Bores, A., Meraz, S., Piedra, E., Avila, M., Serrano, R., Orozco, J., Jimenez-Estrada, M., Chavarría, J.C., Peñalosa, I., Ávila, J.G. and Hernandez, T. 2016. Antimicrobial activity of essential oil of Cordia globosa. African J. Pharm. Pharmacol. 10: 179–184. doi:10.5897/AJPP2015.4444.
Mileva, M.M., Kusovski, V.K. and Krastev, D.S. 2014. Chemical composition, *in vitro* antiradical and antimicrobial activities of *Bulgarian Rosa alba* L. essential oil against some oral pathogens. Int. J. Curr. Microbiol. App. Sci. 3: 11–20.
Mingeot-Leclercq, M.P. and Décout, J.L. 2016. Bacterial lipid membranes as promising targets to fight antimicrobial resistance, molecular foundations and illustration through the renewal of aminoglycoside antibiotics and emergence of amphiphilic aminoglycosides. Med. Chem. Commun. 7: 586–611. doi:10.1039/C5MD00503E.
Mun, S.H., Kang, O.H., Joung, D.K., Kim, S.B., Choi, J.G., Shin, D.W. and Kwon, D.Y. 2014. *In vitro* anti-MRSA activity of carvone with gentamicin. Exp. Ther. Med. 7: 891–896. doi:10.3892/etm.2014.1498.
Murínová, S. and Dercová, K. 2014. Response mechanisms of bacterial degraders to environmental contaminants on the level of cell walls and cytoplasmic membrane. Int. J. Microbiol. 2014: 873081. doi:10.1155/2014/873081.
Nandi, N. 2003. Molecular origin of the recognition of chiral odorant by chiral lipid: Interaction of dipalmitoyl phosphatidyl choline and carvone. J. Phys. Chem. A. 107: 4588–4591. doi:10.1021/jp030076s.
Navarro, M., Stanley, R., Cusack, A. and Sultanbawa, Y. 2015. Combinations of plant-derived compounds against *Campylobacter in vitro*. J. Appl. Poult. Res. 00: 1–12. doi:10.3382/japr/pfv035.
Nazzaro, F., Fratianni, F., De Martino, L., Coppola, R. and De Feo, V. 2013. Effect of essential oils on pathogenic bacteria. Pharmaceuticals 6: 1451–1474. doi:10.3390/ph6121451.
Ngan, L.T.M., Moon, J.K., Kim, J.H. Shibamoto, T. and Ahn, Y.J. 2012. Growth-inhibiting effects of *Paeonia lactiflora* root steam distillate constituents and structurally related compounds on human intestinal bacteria. World J. Microbiol. Biotechnol. 28: 1575–1583. doi:10.1007/s11274-011-0961-6.
Ngassapa, O.D., Runyoro, D.K.B., Vagionas, K., Graikou, K. and Chinou, I.B. 2016. Chemical composition and antimicrobial activity of *Geniosporumrotundifolium* Briq and *Haumaniastrumvillosum* (Bene) A J Paton (Lamiaceae) essential oils from Tanzania. Trop J. Pharm. Res. 15: 107–113. doi:10.4314/tjpr.v15i1.15.

Nikolić, M., Jovanović, K., Marković, T., Marković, D., Gligorijević, N., Radulović, S. and Soković, M. 2014. Chemical composition, antimicrobial, and cytotoxic properties of five Lamiaceae essential oils. Ind. Crops Prod. 61: 225–232. doi:10.1016/j.indcrop.2014.07.011.

Niu, C. and Gilbert, E.S. 2004. Colorimetric method for identifying plant essential oil components that affect biofilm formation and structure. Society 70: 6951–6956. doi:10.1128/AEM.70.12.6951.

Nowotarska, S., Nowotarski, K., Friedman, M. and Situ, C. 2014. Effect of structure on the interactions between five natural antimicrobial compounds and phospholipids of bacterial cell membrane on model monolayers. Molecules 19: 7497–7515. doi:10.3390/molecules19067497.

Okoh, O., Sadimenko, A. and Afolayan, A. 2010. Comparative evaluation of the antibacterial activities of the essential oils of *Rosmarinus officinalis* L. obtained by hydrodistillation and solvent free microwave extraction methods. Food Chem. 120: 308–312. doi:10.1016/j.foodchem.2009.09.084.

Oliva, M., Beltramino, E., Gallucci, M., Casero, C., Zygadlo, J. and Demo, M. 2010. Antimicrobial activity of essential oils of *Aloysia triphylla* (L`Her.) Britton from different regions of Argentina. Boletín Latinoam y del Caribe Plantas Med. y Aromáticas 9: 29–37.

Oliveira, M.M., Brugnera, D.F., Cardoso, M.G., Alves, E. and Piccoli, R.H. 2010. Disinfectant action of *Cymbopogon* sp. essential oils in different phases of biofilm formation by *Listeriamonocytogenes* on stainless steel surface. Food Control. 21: 549–553. doi:10.1016/j.foodcont.2009.08.003.

Olivero, J.T., Pájaro, C. and Stashenko, E. 2011. Antiquorum sensing activity of essential oils isolated from different species of the genus *Piper*. Vitae Revista La Fac. Quim. Farm. 18: 77–82.

Packiavathy, I.A.S., Agilandeswari, P., Musthafa, K.S., Pandian, S.K. and Ravi, V.A. 2012. Antibiofilm and quorum sensing inhibitory potential of *Cuminum cyminum* and its secondary metabolite methyl eugenol against Gram negative bacterial pathogens. Food Res. Int. 45: 85–92. doi:10.1016/j.foodres.2011.10.022.

Páez, P.L., Becerra, M.C. and Albesa, I. 2013. Impact of ciprofloxacin and chloramphenicol on the lipid bilayer of *Staphylococcus aureus*: Changes in membrane potential. Biomed. Res. Int. 2013: 276524. doi:10.1155/2013/276524.

Patrignani, F., Iucci, L., Belletti, N., Gardini, F., Guerzoni, M.E. and Lanciotti, R. 2008. Effects of sub-lethal concentrations of hexanal and 2-(E)-hexenal on membrane fatty acid composition and volatile compounds of *Listeria monocytogenes*, *Staphylococcus aureus*, *Salmonellaenteritidis* and *Escherichia coli*. Int. J. Food Microbiol. 123: 1–8. doi:10.1016/j.ijfoodmicro.2007.09.009.

Pei, R.S., Zhou, F., Ji, B.P. and Xu, J. 2009. Evaluation of combined antibacterial effects of eugenol, cinnamaldehyde, thymol, and carvacrol against *E. coli* with an improved method. J. Food Sci. 74: 379–383. doi:10.1111/j.1750-3841.2009.01287.x.

Petretto, G.L., Fancello, F., Zara, S., Foddai, M., Mangia, N.P., Sanna, M.L., Omer, E.A., Menghini, L., Chessa, M. and Pintore, G. 2014. Antimicrobial activity against beneficial microorganisms and chemical composition of essential oil of *Mentha suaveolens* ssp. insularis Grown in Sardinia. J. Food Sci. 79: 369–377. doi:10.1111/1750-3841.12343.

Pham, Q.D., Topgaard, D. and Sparr, E. 2015. Cyclic and linear monoterpenes in phospholipid membranes: phase behavior, bilayer structure, and molecular dynamics. Langmuir 31: 11067–11077. doi:10.1021/acs.langmuir.5b00856.

Picone, G., Laghi, L., Gardini, F., Lanciotti, R., Siroli, L. and Capozzi, F. 2013. Evaluation of the effect of carvacrol on the *Escherichia coli* 555 metabolome by using 1H-NMR spectroscopy. Food Chem. 141: 4367–4374. doi:10.1016/j.foodchem.2013.07.004.

Poger, D. and Mark, A.E. 2015. A ring to rule them all: The effect of cyclopropane fatty acids on the fluidity of lipid bilayers. J. Phys. Chem. B 119: 5487–5495. doi:10.1021/acs.jpcb.5b00958.

Popovic, V., Petrovic, S., Milenkovic, M., Drobac, M.M., Couladis, M.A. and Niketic, M.S. 2015. Composition and antimicrobial activity of the essential oils of *Laserpitium latifolium* L. and *L. ochridanum* Micevski (Apiaceae). Chem. Biodivers 12: 170–177. doi:10.1002/cbdv.201400127.

Prakash, B., Kedia, A., Mishra, P. and Dubey, N. 2015. Plant essential oils as food preservatives to control moulds, mycotoxin contamination and oxidative deterioration of agri-food commodities - potentials and challenges. Food Control. 47: 381–391. doi:10.1016/j.foodcont.2014.07.023.

Price, L., Koch, B. and Hungate, B. 2015. Ominous projections for global antibiotic use in food-animal production. Proc. Natl. Acad. Sci. 112: 5554–5555. doi:10.1073/pnas.1505312112.

Rather, M.A., Dar, B.A., Dar, M.Y., Wani, B.A., Shah, W.A., Bhat, B.A., Ganai, B.A., Bhat, K.A., Anand, R. and Qurishi, M.A. 2012. Chemical composition, antioxidant and antibacterial activities of the leaf essential oil of *Juglans regia* L. and its constituents. Phytomedicine 19: 1185–1190. doi:10.1016/j.phymed.2012.07.018.

Raut, J. and Karuppayil, S. 2014. A status review on the medicinal properties of essential oils. Ind. Crops Prod. 62: 250–264. doi:10.1016/j.indcrop.2014.05.055.

Raut, J.S., Shinde, R.B., Chauhan, N.M. and Karuppayil, S.M. 2014. Phenylpropanoids of plant origin as inhibitors of biofilm formation by *Candida albicans*. J. Microbiol. Biotechnol. 24: 1216–1225. doi:10.4014/jmb.1402.02056.

Rioba, N., Itulya, F., Saidi, M., Dudai, N. and Bernstein, N. 2015. Effects of nitrogen, phosphorus and irrigation frequency on essential oil content and composition of sage (*Salvia officinalis* L.). J. Appl. Res. Med. Aromat Plants 2: 21–29. doi:10.1016/j.jarmap.2015.01.003.

Rivas da Silva, A.C., Lopes Monteiro, P., Barros de Azevedo, M.M., Machado Costa, D.C., Sales Alviano, C. and Sales Alviano, D. 2012. Biological activities of a-pinene and b-pinene enantiomers. Molecules 17: 6305–6316. doi:10.3390/molecules17066305.

Rodrigues, A.G., Ping, L.Y., Marcato, P.D., Alves, O.L. Silva, M.C.P., Ruiz, R.C., Melo, I.S., Tasic, L. and De Souza, A.O. 2013. Biogenic antimicrobial silver nanoparticles produced by fungi. Appl. Microbiol. Biotechnol. 97: 775–782. doi:10.1007/s00253-012-4209-7.

Roller, S. and Seedhar, P. 2002. Carvacrol and cinnamic acid inhibit microbial growth in fresh-cut melon and kiwifruit at 4° and 8°C. Lett Appl. Microbiol. 35: 390–394. doi:10.1046/j.1472-765X.2002.01209.x.

Saad, N.Y., Muller, C.D. and Lobstein, A. 2013. Major bioactivities and mechanism of action of essential oils and their components. Flavour Fragr. J. 28: 269–279. doi:10.1002/ffj.3165.

Sadgrove, N., Jones, G. and Nair, M. 2015. A Contemporary introduction to essential oils: chemistry, bioactivity and prospects for australian agriculture. Agriculture 5: 48–102. doi:10.3390/agriculture5010048.

Saïdana, D., Mahjoub, M., Boussaada, O., Chriaa, J., Chéraif, I., Daami, M., Mighri, Z. and Helal, A.N. 2008. Chemical composition and antimicrobial activity of volatile compounds of *Tamarix boveana* (Tamaricaceae). Microbiol. Res. 163: 445–455. doi:10.1016/j.micres.2006.07.009.

Salem, M., Ali, H.M., El-Shanhorey, N. and Abdel-Megeed, A. 2013. Evaluation of extracts and essential oil from *Callistemon viminalis* leaves: antibacterial and antioxidant activities, total phenolic and flavonoid contents. Asian Pac. J. Trop. Biomed. 785–791. doi:10.1016/S1995-7645(13)60139-X.

Santos, G.K.N., Dutra, K.A., Barros, R.A., da Câmara, C.A.G., Lira, D.D., Gusmão, N.B. and Navarro, D. 2012. Essential oils from *Alpinia purpurata* (Zingiberaceae): Chemical composition, oviposition deterrence, larvicidal and antibacterial activity. Ind. Crops Prod. 40: 254–260. doi:10.1016/j.indcrop.2012.03.020.

Santos, T.G., Laemmle, J., Rebelo, R.A., Dalmarco, E.M., Cruz, A.B., Schmit, A.P., Cruz, R.C.B. and Zeni, A.L.B. 2015. Chemical composition and antimicrobial activity of *Aloysia gratissima* (Verbenaceae) leaf essential oil. J. Essent. Oil Res. 27: 125–130. doi:10.1080/10412905.2015.1006737.

Šarac, Z., Matejić, J.S., Stojanović-Radić, Z.Z., Veselinović, J.B., Džamić, A.M., Bojović, S. and Marin, P.D. 2014. Biological activity of *Pinus nigra* terpenes-evaluation of FtsZ inhibition by selected compounds as contribution to their antimicrobial activity. Comput. Biol. Med. 54: 72–78. doi:10.1016/j.compbiomed.2014.08.022.

Sarrazin, S.L.F., Da Silva, L.A., Oliveira, R.B., Raposo, J.D.A., da Silva, J.K.R., Salimena, F.R.G., Maia, J.G.S. and Mourão, R.H.V. 2015. Antibacterial action against food-borne microorganisms and antioxidant activity of carvacrol-rich oil from *Lippiaoriganoides* Kunth. Lipids Health Dis. 14: 145. doi:10.1186/s12944-015-0146-7.

Schultz, T.W. and Yarbrough, J.W. 2004. Trends in structure-toxicity relationships for carbonyl-containing a,b-unsaturated compounds. SAR QSAR Environ. Res. 15: 139–146. doi:10.1080/10629360410001665839.

Sell, C. 2010. Chemistry of essential oils. pp. 121–151. *In*: Base, K.H.C. and Buchbauer, G. (eds.). Handbook of Essential Oils: Science, Technology and Applications. CRC Press, New York, USA.

Sepahvand, R., Delfan, B., Ghanbarzadeh, S., Rashidipour, M., Veiskarami, G.H. and Ghasemian-Yadegari, J. 2014. Chemical composition, antioxidant activity and antibacterial effect of essential oil of the aerial parts of *Salvia sclareoides.* Asian Pac. J. Trop. Med. 7: S491–S496. doi:10.1016/S1995-7645(14)60280-7.

Sethupathy, S., Shanmuganathan, B., Kasi, P.D. and Karutha Pandian, S. 2015. Alpha-bisabolol from brown macroalga *Padina gymnospora* mitigates biofilm formation and quorum sensing controlled virulence factor production in *Serratia marcescens*. J. Appl. Phycol. doi:10.1007/s10811-015-0717-z.

Shah, W.A., Dar, M.Y., Zagar, M.I., Agnihotri, V.K., Qurishi, M.A. and Singh, B. 2012. Chemical composition and antimicrobial activity of the leaf essential oil of *Skimmia laureola* growing wild in Jammu and Kashmir, India. Nat. Prod. Res. 1–5. doi:10.1080/14786419.2012.696252.

Shahbazi, Y. 2015. Chemical composition and *in vitro* antibacterial activity of *Menthaspicata* essential oil against common food-borne pathogenic bacteria. J. Pathog. 2015: 916305. doi:10.1155/2015/916305.

Sienkiewicz, M., Poznanska-Kurowska, K., Kaszuba, A. and Kowalczyk, E. 2014. The antibacterial activity of geranium oil against Gram-negative bacteria isolated from difficult-to-heal wounds. Burns 40: 1046–1051. doi:10.1016/j.burns.2013.11.002.

Silva, F., Ferreira, S., Queiroz, J.A. and Domingues, F.C. 2011. Coriander (*Coriandrum sativum* L.) essential oil: Its antibacterial activity and mode of action evaluated by flow cytometry. J. Med. Microbiol. 60: 1479–1486. doi:10.1099/jmm.0.034157-0.

Siroli, L., Patrignani, F., Gardini, F. and Lanciotti, R. 2015. Effects of sub-lethal concentrations of thyme and oregano essential oils, carvacrol, thymol, citral and trans-2-hexenal on membrane fatty acid composition and volatile molecule profile of *Listeria monocytogenes*, *Escherichia coli* and *Salmonella enteritidis*. Food Chem. 182: 185–192. doi:10.1016/j.foodchem.2015.02.136.

Snoussi, M., Dehmani, A., Noumi, E, Flamini, G. and Papetti, A. 2016. Chemical composition and antibiofilm activity of *Petroselinum crispum* and *Ocimum basilicum* essential oils against *Vibrio* spp. strains. Microb. Pathog. 90: 13–21. doi:10.1016/j.micpath.2015.11.004.

Snoussi, M., Noumi, E., Trabelsi, N., Flamini, G. Papetti, A. and De Feo, V. 2015. *Mentha spicata* essential oil: Chemical composition, antioxidant and antibacterial activities against planktonic and biofilm cultures of *Vibrio* spp. strains. Molecules 20: 14402–14424. doi:10.3390/molecules200814402.

Sohlenkamp, C. and O. Geiger. 2016. Bacterial membrane lipids: Diversity in structures and pathways. FEMS Microbiol. Rev. 40: 133–159. doi:10.1093/femsre/fuv008.

Soković, M., Glamočlija, J., Marin, P.D., Brkić, D. and van Griensven, L. 2010. Antibacterial effects of the essential oils of commonly consumed medicinal herbs using an *in vitro* model. Molecules 15: 7532–46. doi:10.3390/molecules15117532.

Somolinos, M., García, D., Condón, S., MacKey, B. and R. Pagán. 2010. Inactivation of *Escherichia coli* by citral. J. Appl. Microbiol. 108: 1928–1939. doi:10.1111/j.1365-2672.2009.04597.x.

Sun, Y., Wilkinson, B.J., Standiford, T.J., Akinbi, H.T. and O'Riordan, M.X.D. 2012. Fatty acids regulate stress resistance and virulence factor production for *Listeria monocytogenes*. J. Bacteriol. 194: 5274–5284. doi:10.1128/JB.00045-12.

Suzuki, É., Soldati, P., Chaves, M.D. and Raposo, N. 2015. Essential oil from *Origanum vulgare* Linnaeus: an alternative against microorganisms responsible for bad perspiration odor. J. Young Pharm. 7: 12–20. doi:10.5530/jyp.2015.1.4.

Szabo, M., Varga, G., Hohmann, J., Schelz, Z., Szegedi, E., Amaral, L. and Molnár, J. 2010. Inhibition of Quorum-sensing signals by essential oils. Phyther. Res. 24: 782–786. doi:10.1002/ptr.3010.

Tabanelli, G., Patrignani, F., Vinderola, G., Reinheimer, J., Gardini, F. and Lanciotti, R. 2013. Effect of sub-lethal high pressure homogenization treatments on the *in vitro* functional and biological properties of lactic acid bacteria. Food Sci. Technol. 53: 580–586. doi:10.3389/fmicb.2015.01006.

Tan, X.C., Chua, K.H., Ravishankar, R.M. and Kuppusamy, U.R. 2016. Monoterpenes: novel insights into their biological effects and roles on glucose uptake and lipid metabolism in 3T3-L1 adipocytes. Food Chem. 196: 242–250. doi:10.1016/j.foodchem.2015.09.042.

Trombetta, D., Castelli, F., Sarpietro, M.G., Venuti, V., Cristani, M., Daniele, C., Saija, A., Mazzanti, G. and Bisignano, G. 2005. Mechanisms of antibacterial action of three monoterpenes. J. Antimicrob. Agents Chemother. 49: 2474–2478. doi:10.1128/AAC.49.6.2474.

Tsuchiya, H. and Mizogami, M. 2012. The membrane interaction of drugs as one of mechanisms for their enantioselective effects. Med. Hypotheses 79: 65–67. doi:10.1016/j.mehy.2012.04.001.

Turina, A.V., Nolan, M.V., Zygadlo, J.A. and Perillo, M.A. 2006. Natural terpenes: self-assembly and membrane partitioning. Biophys Chem. 122: 101–113. doi:10.1016/j.bpc.2006.02.007.

Ultee, A., Kets, E.P.W. and Smid, E.J. 1999. Mechanisms of action of carvacrol on the food-borne pathogen *Bacillus cereus*. Appl. Environ. Microbiol. 65: 4606–4610. doi:10.1128/AEM.68.4.1561-1568.2002.

Ultee, A., Bennik, M.H.J. and Moezelaar, R. 2002. The phenolic hydroxyl group of carvacrol is essential for action against the food-borne pathogen *Bacillus cereus*. Appl. Environ. Microbiol. 68: 1561–1568. doi:10.1128/AEM.68.4.1561-1568.2002.

Van Boeckel, T., Gandra, S., Ashok, A., Caudron, Q., Grenfell, B.T., Levin, S.A. and Laxminarayan, R. 2014. Global antibiotic consumption 2000 to 2010: An analysis of national pharmaceutical sales data. Lancet Infect. Dis. 14: 742–750. doi:10.1016/S1473-3099(14)70780-7.

Veldhuizen, E.J.A., Tjeerdsma-Van Bokhoven, J.L.M., Zweijtzer, C., Zweijtzer, C., Burt, S.A. and Haagsman, H.P. 2006. Structural requirements for the antimicrobial activity of carvacrol. J. Agric. Food Chem. 54: 1874–1879. doi:10.1021/jf052564y.

Verma, R.S., Padalia, R.C., Goswami, P., Verma, S.K., Chauhan, A. and Darokar, M.P. 2016. Chemical composition and antibacterial activity of the essential oil of Kauri Pine [*Agathis robusta* (C. Moore ex F. Muell.) F.M. Bailey] from India. J. Wood Chem. Technol. 36: 270–277. doi:10.1080/02773813.2015.1137946.

Villa-Ruano, N., Pacheco-Hernández, Y., Rubio-Rosas, E., Lozoya-Gloria, E., Mosso-González, C., Ramón-Canul, L.G. and Cruz-Durán, R. 2015. Essential oil composition and biological/pharmacological properties of *Salmea scandens* (L.) DC. Food Control. 57: 177–184. doi:10.1016/j.foodcont.2015.04.018.

Vizcaino, M.I., Xun, G. and Crawford, J.M. 2014. Merging chemical ecology with bacterial genome mining for secondary metabolite discovery. J. Ind. Microbiol. Biotechnol. 41: 285–299. doi:10.3851/IMP2701.Changes.

Wang, P., Zhou, D., Kinraide, T.B., Luo, X., Li, L., Li, D. and Zhang, H. 2008. Cell membrane surface potential (psi0) plays a dominant role in the phytotoxicity of copper and arsenate. Plant Physiol. 148: 2134–2143. doi:10.1104/pp.108.127464

Wenqiang, G., Shufen, L., Ruixiang, Y. Shaokun, T. and Can, Q. 2007. Comparison of essential oils of clove buds extracted with supercritical carbon dioxide and other three traditional extraction methods. Food Chem. 1558–1564. doi:10.1016/j.foodchem.2006.04.009

Wolska, K., Grzes, K. and Anna, K. 2012. Synergy between novel antimicrobials and conventional antibiotics or bacteriocins. Polish J. Microbiol 61: 95–104.

Wu, J.G., Peng, W., Yi, J., Wu, Y.B., Chen, T.Q., Wong, K.H. and Wu, J.Z. 2014. Chemical composition, antimicrobial activity against *Staphylococcus aureus* and a pro-apoptotic effect in SGC-7901 of the essential oil from *Toona sinensis* (A. Juss.) Roem. leaves. J. Ethnopharmacol. 154: 198–205. doi:10.1016/j.jep.2014.04.002.

Wydro, P., Flasinski, M. and Broniatowski, M. 2012. Molecular organization of bacterial membrane lipids in mixed systems-A comprehensive monolayer study combined with grazing incidence X-ray diffraction and brewster angle microscopy experiments. Biochim Biophys Acta 1818: 1745–1754. doi:10.1016/j.bbamem.2012.03.010.

Yang, C., Hu, D.H. and Feng, Y. 2015a. Antibacterial activity and mode of action of the *Artemisia capillaris* essential oil and its constituents against respiratory tract infection-causing pathogens. Mol. Med. Rep. 11: 2852–2860. doi:10.3892/mmr.2014.3103.

Yang, X.N., Khan, I. and Kang, S.C. 2015b. Chemical composition, mechanism of antibacterial action and antioxidant activity of leaf essential oil of *Forsythia koreana* deciduous shrub. Asian Pac. J. Trop. Biomed. 8: 694–700.

Yang, T.S., Liou, M.L., Hu, T.F., Peng, C. and Liu, T. 2013. Antimicrobial activity of the essential oil of *Litsea cubeba* on cariogenic bacteria. J. Essent. Oil. Res. 25: 120–128. doi:10.1080/10412905.2012.758602.

Zeng, W.C., Zhu, R.X., Jia, L.R., Gao, H., Zheng, Y. and Sun, Q. 2011. Chemical composition, antimicrobial and antioxidant activities of essential oil from *Gnaphlium affine*. Food Chem. Toxicol. 49: 1322–1328. doi:10.1016/j.fct.2011.03.014.

Zengin, H. and Baysal, A.H. 2014. Antibacterial and antioxidant activity of essential oil terpenes against pathogenic and spoilage-forming bacteria and cell structure-activity relationships evaluated by SEM microscopy. Molecules 19: 17773–17798. doi:10.3390/molecules191117773.

Zhang, Y., Muend, S. and Rao, R. 2012. Dysregulation of ion homeostasis by antifungal agents. Front Microbiol. 3: 1–6. doi:10.3389/fmicb.2012.00133.
Zhou, S., Sheen, S., Pang, Y.H., Liu, L. and Yam, K.L. 2013. Antimicrobial effects of vapor phase thymol, modified atmosphere, and their combination against *Salmonella* spp. on Raw Shrimp. J. Food Sci. doi:10.1111/1750-3841.12098.
Zunino, M.P., Turina, A.V., Zygadlo, J.A. and Perillo, M.A. 2011. Stereoselective effects of monoterpenes on the microviscosity and curvature of model membranes assessed by DPH steady-state fluorescence anisotropy and light scattering analysis. Chirality 23: 867–877. doi:10.1002/chir.
Zygadlo, J.A. (ed.). 2011. Aceites esenciales. Química, Ecología, Comercio, Producción y Salud. Universitas. Editorial Científica Universitaria. Córdoba, Argentina.

5

Role of Essential Oils for the Cure of Human Pathogenic Fungal Infections

Melina G. Di Liberto,[1] *Laura A. Svetaz,*[1] *María V. Castelli,*[1] *Mahendra Rai*[3] and *Marcos G. Derita*[1,2,*]

Introduction

Although it is widely recognized that fungal pathogens have a great influence on plant and animal life, making a significant impact on species extinctions, food security, and ecosystem disorders; the serious problems that fungal infections have on human health is not widely recognized, and deaths resulting from these infections are often overlooked (Fisher et al. 2012).

In this sense, fungal pathogens are probably an even greater contributor to human morbidity and mortality than current estimates since epidemiological data for fungal infections are unacceptably poor due to a lack of standards for reporting fungal disease and problems of misdiagnosis. The increase of fungal diseases is strongly related to the rising number of patients with compromised immune systems (Kathiravan et al. 2012, Darius et al. 2014). Fungi with low virulence for immunocompetent people, can be life-threatening for neonates, cancer patients receiving chemotherapy, organ transplant and burnt patients, added to those affected by the Acquired Immunodeficiency Syndrome (AIDS) (Segal et al. 2006). Other risk factors to acquire fungal infections include: corticosteroid and antibiotic treatments, diabetes, epidermal and dermal injuries, malnutrition, neutropenia and surgery (Nucci and Marr 2005). In addition, an increasing number of healthy individuals even children in third-world nations that receive deficient sanitary attention and education, suffer superficial fungal infections. All these factors diminish the quality of their lives producing diseases that are very difficult to control (Freixa et al. 1998, Ablordeppey et al. 1999, Naik et al. 2015).

At present, pathogenic fungi are responsible for diseases ranging from superficial infections affecting ~ 1.7 billion individuals worldwide, and invasive infections that, despite the availability of several antifungal drugs, kill 1.5 million humans per year (Brown et al. 2012, Spitzer et al. 2016). Fungi of the genera *Epidermophyton*, *Microsporum*, *Trichophyton*, *Candida* and *Malassezia* cause

[1] CONICET, Universidad Nacional de Rosario/Facultad de Ciencias Bioquímicas y Farmacéuticas/Cátedra de Farmacognosia, Suipacha 531, Rosario, Santa Fe, Argentina.
[2] CONICET, Universidad Nacional del Litoral/Facultad de Ciencias Agrarias/Cátedra de Cultivos Intensivos, Kreder 2805, Esperanza, Santa Fe, Argentina.
[3] Department of Biotechnology, Sant Gadge Baba Amravati University, Amravati 444602, Maharashtra, India.
* Corresponding author: mgderita@hotmail.com

superficial infections, in which the fungus is confined to the outer layers of the skin. On the other hand, systemic mycoses that affect internal organs of the body (Kavanagh 2011), are mainly produced by *Candida* or *Aspergillus* spp., *Cryptococcus neoformans* and by the emerging pathogens of the genera *Trichosporon*, *Mucor*, *Rhizopus*, *Fusarium*, *Scedosporium*, etc. (Wisplinghoff et al. 2004, Richardson and Lass-Flörl 2008, Vincent et al. 2009, Pappas et al. 2010, Brown et al. 2012).

Among yeasts, *Candida albicans* is the main causative agent of fungal infections in immunocompromised hosts, followed by other non-*albicans Candida* species such as *C. tropicalis*, *C. parapsilosis*, *C. glabrata* and *C. krusei*, which have been frequently isolated from immunodepressed patients (Cuenca-Estrella and Rodríguez-Tudela 2007). On the other hand, the main cause of meningoencephalitis in AIDS patients at present is the fungus *Cryptococcus neoformans* (Rodríguez-Tudela and Cuenca-Estrella 1999). *Malassezia* spp. are considered opportunistic yeasts of increasing clinical importance. These lipophilic yeasts are associated with various human diseases, especially pityriasis versicolor, a chronic superficial scaling dermatomycosis (Ramadán et al. 2012). *Trichosporon* spp. are opportunistic yeast-like fungi reported as the most common cause of non-candidal yeast infection in patients with hematological malignancies (Pfaller and Diekema 2004).

Among molds, *Aspergillus fumigatus* is one of the main pathogens responsible for invasive and frequently lethal mycoses in immunocompromised patients. Some other *Aspergillus* species such as *A. flavus* and *A. terreus* are the cause of severe fungal infections as well (Gómez-López et al. 2003). Infections due to less common but antifungal-resistant species such as *Mucor*, *Rhizopus*, *Fusarium* and *Scedosporium* are being reported with greater frequency and impact. These opportunistic organisms are usually marked by a poor response to antifungal therapy, resistance *in vitro* to most available antifungal agents, and an overall poor outcome with excessive mortality (Pfaller and Diekema 2004).

In addition, dermatophytosis such as tinea unguium, tinea pedis and tinea manuum are mainly produced by *Trichophyton rubrum* and *T. interdigitale* (ex-*mentagrophytes*). In turn, *T. rubrum* and *Epidermophyton floccosum* are the cause of tinea cruris while *Microsporum canis* and *T. interdigitale* produce tinea corporis, tinea capitis and tinea barbae (Monzón et al. 2003).

Antifungal Drugs Currently in Clinical Use

The most important discoveries of novel chemical scaffolds produced by bacteria and fungi that revolutionized modern medicine were done in the golden era of antibiotics that is from the 1940s to 1970s. However, the only novel type of antifungal drugs that reached the clinic in approximately 30 years resulted to be the echinocandins (Shapiro et al. 2011). The limited number of effective antifungal drugs (Vicente and Peláez 2007, Roemer and Krysan 2014) is in large part due to the close evolutionary relatedness between humans and fungi, limiting the number of unique fungal cellular targets that can be exploited for drug development (Spitzer et al. 2016). Among the different drugs for treating mycoses, polyenes, allylamines, azoles and echinocandins along with griseofulvin and 5-flurocytosine are frequently used (Table 5.1). Regarding the mechanism of action, three of them (azoles, allylamines and polyenes) attack the fungal membrane while echinocandins inhibit the fungal cell wall synthesis which is an ideal target since it is absent in mammalian cells (Table 5.1).

There is no perfect drug that meets all the desirable requirements. Besides the limited spectrum of action, some of them are fungistatic but not fungicidal, produce recurrence, resistance or develop toxicity (Francois et al. 2005, Brown and Wright 2016). The widespread and ubiquitous evolution of antifungal drug resistance, both intrinsic and acquired, are common among all previously mentioned species across multiple antifungal drug types. For example, the echinocandins are ineffective against *Cryptococcus* species leaving the treatment of choice for cryptococcal meningitis reliant on medications developed in the 1950s that are plagued with problems of host toxicity (Day et al. 2013).

Regarding amphotericin B, in addition to its nephrotoxicity, it has two main drawbacks: (1) it must be used intravenously in its conventional form and (2) it is very expensive in its new lipid forms. About azoles, their excellent oral bioavailability, stable parenteral formulation and low toxicity makes

Table 5.1. Main antifungal drugs available for the treatment of fungal infections: their mechanisms and spectra of action and main drawbacks.

Fungal target	Antifungal drug	Mechanism of action	Spectrum of activity	Drawbacks
Nucleus	**Pirimidines;** e.g., 5-Fluorocytosine	Interferes with the biosynthesis of RNA, preventing the synthesis of essential proteins. Fluorocytosine penetrates the fungal cell through the cytosine permease enzyme. Once inside the cell, it is deaminated to 5-fluorouracil. After several phosphorylations, fluorardílic acid is incorporated into the RNA chain, resulting in the production of aberrant RNA.	Active against yeasts. Inactive against filamentous fungi.	Medular toxicity and problems in the digestive tract. Fungistatic. Certain yeast strains show primary resistance. In addition, when used in monotherapy, it generates secondary resistance easily. Reduced use in clinical practice (Ruiz-Camps and Cuenca Estrella 2009).
Nucleus	**Griseofulvin:** naturally occurring compound first isolated in 1939 from *Penicillium griseofulvum* Griseofulvin	Tricyclic spiro-diketone with two chiral centers in its structure that binds to tubulin inhibiting the formation of the mitotic spindle. Altered microtubule function by this compound, interrupts the cell cycle at metaphase and then prevents proliferation (Brian et al. 1949).	Active against dermatophytes	Fungal and mammalian tubulin are very similar, with 70% of identical amino acids (Roobol et al. 1976), making this inhibitor non selective.
Cytoplasmic membrane: Drugs that binds to ergosterol	**Polyenes;** e.g., Amphotericin B (natural product isolated from *Streptomyces nodosus*) in 1955 at the Squibb Institute for Medical Research (Gold et al. 1955).	Due to its amphipathic nature, it binds to ergosterol of the fungal membrane forming pores or channels. Thus, cell permeability is increased and the intracellular components are lost generating irreversible cell damage (Selitrennikoff 1992).	Active against yeasts and filamentous fungi.	Significant toxicity. No chemical stability. Low penetration to the central nervous system. No oral formulation available. Since its deoxycholate formulation poses a dose-related nephrotoxicity that limit its wide-spread use, lipid formulations such as liposomal AmpB (LAmB), AmpB lipid complex (ABLC) and AmpB colloid dispersion (ABCD) were further developed although the renal toxicity persisted at high cumulative doses (Saravolatz et al. 2003).

Table 5.1 contd....

...Table 5.1 contd.

Fungal target	Antifungal drug	Mechanism of action	Spectrum of activity	Drawbacks
Cytoplasmic membrane: Drugs acting on ergosterol biosynthesis	**Allylamines;** e.g., terbinafine 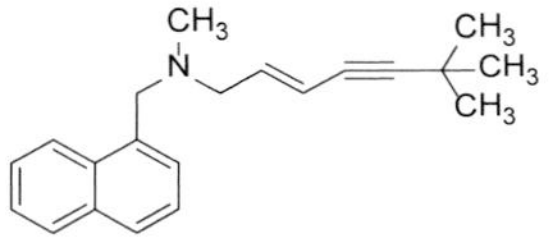	Reversible and non-competitive inhibitor of the enzyme squalene epoxidase.	*Candida* spp. (except *C. glabrata*, *C. krusei* or *C. tropicalis*), dermatophytes, *Aspergillus* spp.	Mainly used for superficial fungal infections. It has few systemic disponibility and causes hepatic toxicity. (Ruiz-Camps and Cuenca-Estrella 2009).
Cytoplasmic membrane: Drugs acting on ergosterol biosynthesis	**Azoles:** Imidazoles; e.g., ketoconazole 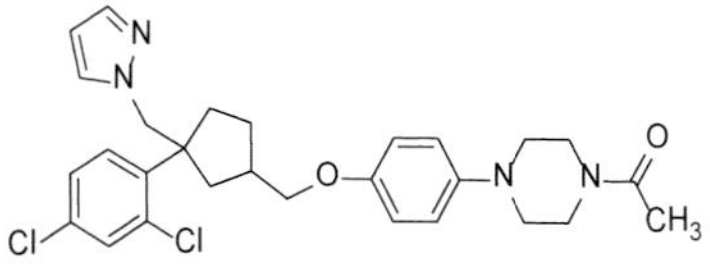	Inhibition of the conversion of lanosterol to ergosterol at the level of the cytochrome P450 dependent enzyme lanosterol 14-α-demethylase. As a result there is a decrease of ergosterol and accumulation of 14-α-methylated sterols, which are toxic to the fungal cells (Ruiz-Camps and Cuenca-Estrella 2009).	Active against yeasts, dermatophytes and filamentous fungi (variable).	Significant toxicity. Generally used for superficial fungal infections.
	Azoles: First generation Triazoles; e.g., fluconazole 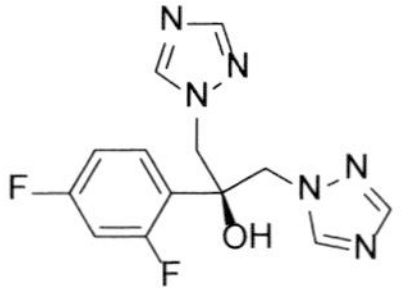		Active against yeasts.	Certain yeast strains show intrinsic resistance. It generates secondary resistance when used in prolonged treatments.
	Azoles: Second generation Triazoles; e.g., voriconazole 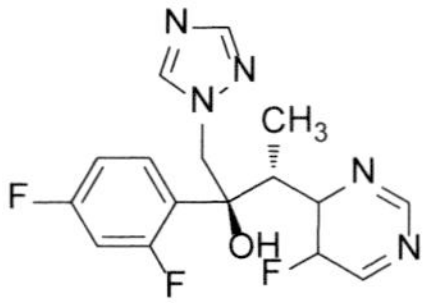		Active against yeasts, and *Aspergillus* spp. (variable).	Fungistatic against *Candida* spp. Significative interaction with several drugs.

Cell wall	**Echinocandins:** semisynthetic lipopeptides derived from natural fungal compounds (Pelaez et al. 2000); e.g., caspofungin. 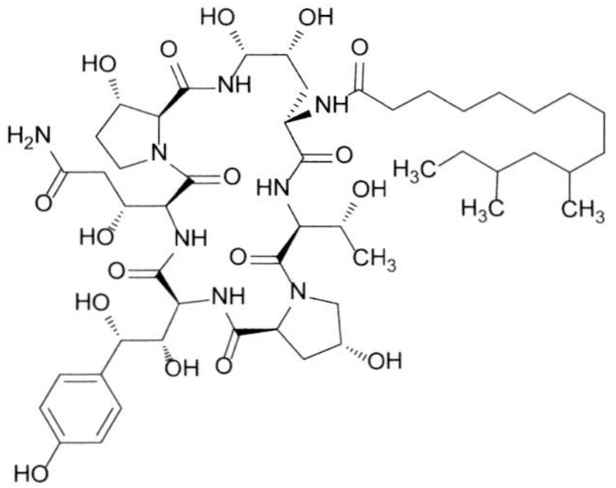	Specific and non-competitive inhibitor of the synthesis of (1,3)-ß-glucan through the inhibition of the enzyme (1,3)-ß-D-glucan synthase, enzyme complex that is essential for the biosynthesis of glucan polymers by the fungal cell wall (Ruiz-Camps and Cuenca-Estrella 2009).	*Candida* spp. (including species resistant to azoles), *Aspergillus* spp.	Limited spectrum of action. No activity against *Cryptococcus* spp. No oral formulation available. High cost (Gómez-López et al. 2003)
	Nikkomycin 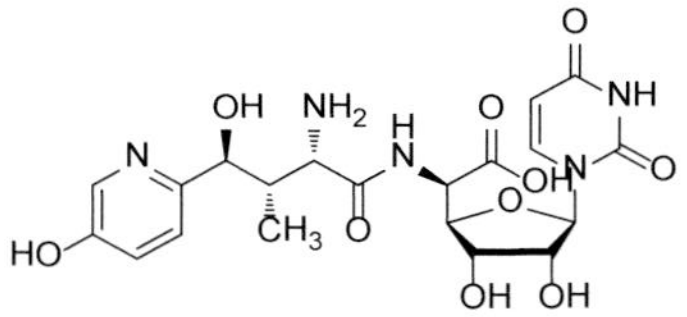	Competitive inhibitor of the fungal cell wall synthesis at the level of the enzyme chitin synthase.	Moderately active against *C. albicans* and *C. neoformans*	Low *in vivo* activity (Mathew and Nath 2009).

them the first line antifungal drugs (Graybill 1997). However, resistance to azoles has emerged in clinical isolates from immunocompromised patients, probably because they are fungistatic rather than fungicide and they need long periods of time to eradicate the fungal infections (Odds 1996). In addition, since azoles are substrates and inhibitors of the cytochrome P-450 family CY3A4 (Van Peer et al. 1989, Zonios and Bennett 2008), they cause adverse interactions with immunosuppressants, chemotherapeutics drugs, benzodiazepines, tricyclic antidepressants macrolides and selective serotonin reuptake inhibitors. Therefore, new approaches for treating these infections are greatly needed (Svetaz et al. 2016).

A promising strategy for combating the difficulty to treat fungal infections is to extend the lifespan and efficacy of our currently employed drugs by using combination therapy. Combining drugs has the potential to confer enhanced efficacy and specificity compared to individual drug treatments, lowering the toxic side effects and the development of fungal resistance. However, they can also produce lower effects than the expected (antagonism) or the effect of the sum of the actions of each drug when used alone (additivism or indifference) (Zimmermann et al. 2007, Carrillo-Muñoz et al. 2014, Hill and Cowen 2015, Svetaz et al. 2016). Although the use of drug combinations to treat fungal pathogens has garnered considerable interest over the past several years, the ideal drug to cure fungal infections has not been discovered yet (Brouwer et al. 2004). There is a real need for a next generation of safer and more potent antifungal agents (Patterson 2005, Pfaller and Diekema 2007).

Natural Products as Alternatives Therapies for Human Fungal Infections

Among the different possible sources of antifungal compounds, plants maintain a great interest because they provide unlimited opportunities for the isolation of new antifungal compounds due to their unmatched availability of chemical diversity (Cos et al. 2006, Maregesi et al. 2008). A revision of the papers published in both, the *Journal of Ethnopharmacology* and *Planta Medica* in the last three decades showed that, in the first Journal, 221 papers on antifungal screening were published in that period, increasing from 16 in 1981–1990, 66 in 1991–2000 and 139 in the 2001–2010. Regarding *Planta Medica*, 151 papers on antifungal plants were published in the last three decades, 24 in the first 10 years (1981–1990), 43 in the second (1991–2000) and 84 in the third (2001–2010) (Sortino et al. 2012) (Fig. 5.1). This analysis demonstrates the increasing interest of researchers from Academies in plants as source of antifungal compounds. In contrast, most pharmaceutical companies have abandoned programs of antifungal research based on natural products including vegetal biodiversity (Rouhi 2003).

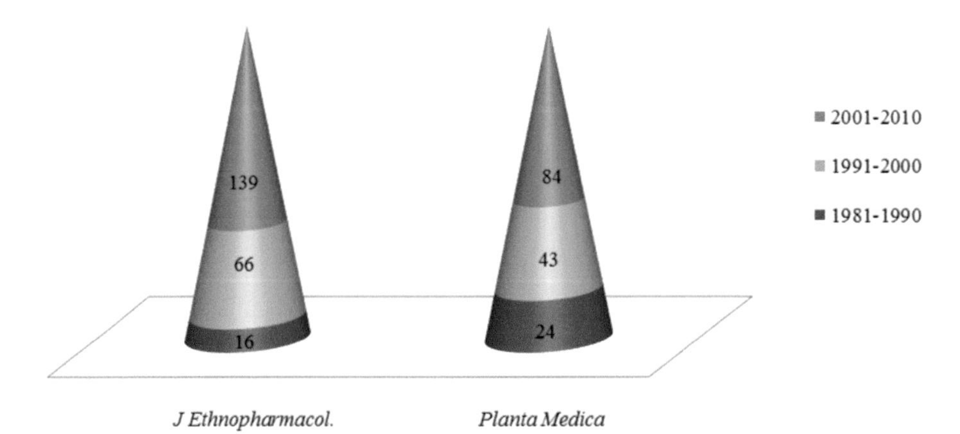

Figure 5.1. Number of articles on antifungal properties of plants published in *Journal of Ethnopharmacology* and *Planta Medica* in the last three decades (1981–2010).

Essential Oils-The Natures Gift for Cure of Diseases

Essential Oils (EOs) are natural, aromatic, volatile and oily liquids that can be obtained from several parts of plants especially the aerial ones as leaves, stems and flowers. They are synthesized from complex metabolic pathways in order to prevent infections of plants from diverse pathogenic microorganisms, repel insects that act as plague vectors or reduce the appetite of some herbivorous by conferring an unpleasant taste to the plant. On the other hand and taking into account the strong sensitivity of insects to smells, EOs generally facilitate the plant reproduction by attraction of specific insects that stimulate the dispersion of pollens and seeds (Bakkali et al. 2008). Various factors may affect the chemical composition of this natural product, since plants are composed of complex systems and different cell lines with several and extensive metabolic pathways that involve the production of these oils. Largely important is the season of the year in which the plant material is harvested for EO extraction, considering the impact of the large seasonal variation over the production of secondary metabolites. Available nutrients for plant growth have a crucial role on the definition of the biochemical reactions on plant organelles, as well as the exposition to light, temperature and water supply. Mindful of these factors, the artificial crop of plants for EOs production should consider a large amount of variables (Vigan 2010). Not all plants are able to produce EOs; it is well known that their production belongs to a small number of vegetable families such as Myrtaceae, Myristicaceae, Piperaceae, Rutaceae, Asteraceae and Lamiaceae among others. Chemical composition of EOs may also be distinctive taking into account its methods of extraction. Some of the oldest techniques for EOs production are steam and hydro distillation (Charles and Simon 1990). The first documents describing distillation are dated from the 9th century and, imported into Europe by the Arabs; the concept of EOs was limited to distilled aromatic waters obtained from plants treated with alcohol. In the 16th century the notions of fatty oils and EOs, as well as the methods for separating essences from aromatic waters, became established. In these methods, the EO released from the plant material by heating, is carried by aqueous vapor and after mutual condensation the separation between aqueous and oily drops occurs. Extraction methods mediated by organic solvents are also applied, but considering the high toxic potential of this kind of solvents, alternatives have been pursued such as inert gases and innocuous supercritical fluids (Charles and Simon 1990).

Not only the number and type of molecules presents in a given sample of EO but also their stereochemical features could be very different according to the selected method of extraction, time of the year in which the collection takes place, soil composition, plant organ, plant age and vegetative cycle stage. In this context, the quality and quantity of this natural product presents a wide range of variability (Angioni et al. 2006). Because of this fact, when an EO is commercialized, it should be chemotyped by gas chromatography and mass spectrometry analysis. Analytical monographs have been published (European Pharmacopoeia, ISO, WHO, Council of Europe) to ensure their good quality. As it was discussed above, EOs are generally complex mixtures of volatile organic compounds, produced as secondary metabolites in plants that include: hydrocarbons (terpenes and sesquiterpenes) and oxygenated compounds (alcohols, esters, ethers, aldehydes, ketones, lactones, phenols, and phenol ethers) (Guenther 1972). Generally EOs contain about 20 to 60 components up to more than 100 single substances at quite different concentrations but only a few of them are considered major components at fairly high concentrations (20 to 90%) compared to other components present in trace amounts. The main group is composed by terpenoids, phenylpropanoids and short-chain aliphatic hydrocarbon derivatives, which are all characterized by low molecular weight. Most representative structures are depicted in Fig. 5.2.

The most representative molecules constituting EOs are monoterpenes (Fig. 5.2). They are formed from the coupling of two isoprene units (C10) and the most abundant molecules constituting 90% of the EOs mixture and allow a great variety of structures. They consist of several functions including acyclic, monocyclic or bicyclic hydrocarbons; acyclic, monocyclic or bicyclic alcohols; acyclic aldehydes; acyclic, monocyclic or bicyclic ketones; acyclic, monocyclic or bicyclic esters; ethers and phenols. Sesquiterpenes are formed from the assembly of three isoprene units (C15). The

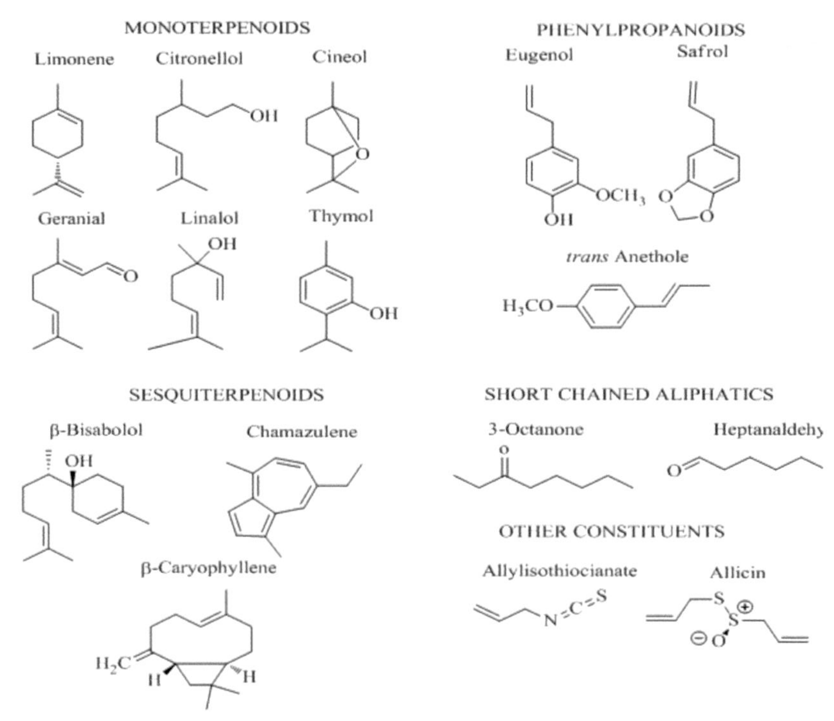

Figure 5.2. Typical representative structures of EOs chemical components.

extension of the chain increases the number of cyclisations which allows a great variety of structures (Fig. 5.2). Also sesquiterpenes include hydrocarbons, alcohols, ketones and epoxides. The second major aromatic molecules that could be present in EOs are phenylpropanoids, formed via the shikimic acid pathway (Pichersky et al. 2006) but occurring less frequently than terpenes. Aromatic compounds originated from the shikimate pathway comprise aldehydes, alcohols, phenols, methoxy derivatives and methylenedioxy compounds. Nitrogenous or sulfured components such as glucosinolates or isothiocyanate derivatives are also characteristic secondary metabolites of diverse aromatic plants or of processed, grilled or roasted products. In addition, some EOs contain photoactive molecules like coumarins and furocoumarins and short-chain aliphatic substances such as 3-octanone and heptanaldehyde (Fig. 5.2) (Pichersky et al. 2006).

Essential Oils with Antifungal Activities Against Human Pathogenic Fungi

The extensive panel of biological activities of EOs can be linked to the complexity and variability of their chemical composition. Due to their aromatic feature, they have been widely used in industries for production of soaps, fragrances and toiletries. On the other hand, their prospection and application as therapeutic products by folk medicine since the middle ages is remarkable. In observance to the popular faith on the medicinal properties of EOs, a series of ethnopharmacological investigations has been evidenced their application for treatment a wide range of pathological conditions (Hennebelle et al. 2008, Belay et al. 2011). In fact, the large bioactivity of EOs has been confirmed by several studies, including antibacterial, antiviral, anti-inflammatory, antifungal, antimutagenic, anticarcinogenic and antioxidant, as well as other miscellaneous activities (Shaaban et al. 2012).

Various human pathogenic fungi, including yeasts, were found to be susceptible to EOs. In Table 5.2 there is a summary of the most important EOs that resulted active against a panel of major human pathogenic fungi.

The efficiency of inhibition varies with the target organisms and the oil tested. For example, three members of Apiaceae family show variable anti-*Candida albicans* activity with a trend of coriander

Table 5.2. Essential Oils active againts selective human fungal pathogens.

Target fungi	Essential oil	References
Alternaria spp.	*Cedrus libani* (Cedar wood oil); *Cymbopogon martini* (Ginger grass); *C. citrates* (Lemon grass); *Tamarix boveana; Rosmarinus officinalis* (Rosemary); *Foeniculum vulgare* (Fennel)	Dikshit et al. 1983, Mimica-Dukic et al. 2004, Rota et al. 2004, Ozcan and Chalchat 2008, Rosato et al. 2007, Rasooli et al. 2008, Saidana et al. 2008, Peighami-Ashnaei et al. 2008
Aspergillus niger	*Allium sativum* (Garlic); *Artemisia judaica* (Wormwood); *A. absinthium; A. biennis; Carum nigrum* (Black caraway); *Cedrus libani* (Cedar wood oil); *Chenopodium ambrosioides; Cymbopogon martini* (Ginger grass); *C. citrates* (Lemon grass); *Syzigium aromaticum* (Clove); *Foeniculum vulgare* (Fennel); *Juniperi aetheroleum* (Juniper); *Matricaria chamomilla* (Chamomile); *Zingiber officinale* (Ginger); *Tamarix boveana*	Dikshit et al. 1983, Saikia et al. 2001, Benkeblia 2004, Mimica-Dukic et al. 2004, Kordali et al. 2005, Pepeljnjak et al. 2005, Kumar et al. 2007, Agarwal et al. 2008, Bansod and Rai 2008, Lopes-Lutz et al. 2008, Saidana et al. 2008, Singh et al. 2008, Cetin et al. 2009, Irkin and Korukluoglu 2009, Peighami-Ashnaei et al. 2008, Tolouee et al. 2010
Aspergillus flavus	*Carum nigrum* (Black caraway); *Cedrus libani* (Cedar wood oil); *Cuminum cyminum* (Cumin); *Nigella sativa* (Black cumin); *Zingiber officinale* (Ginger); *Satureja hortensis* (Summer savoury)	Dikshit et al. 1983, Singh et al. 2006, Singh et al. 2010, Razzaghi-Abyaneh et al. 2008, Khosravi et al. 2011
Aspergillus fumigatus	*Cedrus libani* (Cedar wood oil); *Chenopodium ambrosioides*; *Cuminum cyminum* (Cumin); *Syzigium aromaticum* (Clove); *Nigella sativa* (Black cumin)	Dikshit et al. 1983, Kumar et al. 2007, Bansod and Rai 2008, Khosravi et al. 2011
Candida albicans; C. glabrata; Candida spp.	*Cinnamomum* spp.; *Croton cajucara*; *Cymbopogon martini* (Ginger grass); *C. citrates* (Lemon grass); *Eucalyptus saligna* (Saligna); *Syzigium aromaticum* (Clove); *Juniperi aetherole* (Juniper); *Lavandula* spp.; *Melaleuca alternifolia*; *Melissa officinalis*; *Mentha piperita*; *M. longifolia*; *M. viridis*; *Ocimum* spp.; *Ocimum sanctum* (Holy Basil/Tulsi); *Pimpinella anisum*; *Piper nigrum* (Black Pepper); *Ziziphora clinopodioides*; *Santolina rosmarinifolia*	Mastura et al. 1999, Saikia et al. 2001, Singh et al. 2002, Dryden et al. 2004, Mimica-Dukic et al. 2004, Alviano et al. 2005, Devkatte et al. 2005, Pepeljnjak et al. 2005, Carson et al. 2006, Ioannou et al. 2007, Sartorelli et al. 2011, Zore et al. 2011, Zuzarte et al. 2011, 2012, Rabadia et al. 2012
Cryptococcus neoformans	*Lavandula* spp.; *Ziziphora clinopodioides*	Khosravi et al. 2011, Zuzarte et al. 2011, 2012
Fusarium spp.	*Allium sativum* (Garlic); *Artemisia judaica* (Wormwood); *A. absinthium*; *A. biennis*; *other Artemisia* spp.; *Chenopodium ambrosioides*; *Cymbopogon martini* (Ginger grass); *C. citrates* (Lemon grass); *Tamarix boveana*; *Rosmarinus officinalis* (Rosemary); *Zingiber officinale* (Ginger); *Salvia fruticosa*; *S. officinalis*; *S. Rosifolia*	Saikia et al. 2001, Benkeblia 2004, Rota et al. 2004, Kordali et al. 2005, Fabio et al. 2007, Kumar et al. 2007, Rosato et al. 2007, Agarwal et al. 2008, Lopes-Lutz et al. 2008, Ozcan and Chalchat 2008, Rasooli et al. 2008, Saidana et al. 2008, Singh et al. 2008, Cetin et al. 2009, Irkin and Korukluoglu 2009, Ozek et al. 2010
Fonsecaea pedrosoi	*Artemisia judaica* (Wormwood); *A. absinthium*; *A. biennis*; *other Artemisia* spp.	Kordali et al. 2005, Lopes-Lutz et al. 2008, Cetin et al. 2009, Irkin and Korukluoglu 2009

Table 5.2 contd....

...Table 5.2 contd.

Target fungi	Essential oil	References
Geotrichum candidum	*Artemisia judaica* (Wormwood); *A. absinthium*; *A. biennis*; *other Artemisia* spp.	Kordali et al. 2005, Lopes-Lutz et al. 2008, Cetin et al. 2009, Irkin and Korukluoglu 2009
Microsporum canis; *Microsporum gypseum*	*Artemisia judaica* (Wormwood); *A. absinthium*; *A. biennis*; *other Artemisia* spp.; *Cinnamomum* sp; *Croton argyrophylloides*; *C. zehntneri*; *C. cajucara*; *Syzigium aromaticum*; *Daucus carota* (Wild carrot)	Mastura et al. 1999, Dorman and Deans 2000, Alviano et al. 2005, Kordali et al. 2005, Fontenelle et al. 2008, Lopes-Lutz et al. 2008, Tavares et al. 2008, Cetin et al. 2009, Irkin and Korukluoglu 2009, Pinto et al. 2009
Trichophyton rubrum; *T. mentagrophytes*	*Artemisia judaica* (Wormwood); *A. absinthium*; *A. biennis*; *Artemisia* spp.; *Cinnamomum* spp.; *Daucus carota* (Wild carrot); *Syzigium aromaticum* (Clove)	Mastura et al. 1999, Dorman and Deans 2000, Kordali et al. 2005, Lopes-Lutz et al. 2008, Cetin et al. 2009, Irkin and Korukluoglu 2009, Pinto et al. 2009

> anise > fennel, with Minimum Inhibitory Concentrations (MICs) of 0.25, 0.5 and 1%, respectively (Hammer et al. 1999). Generally *Cymbopogon* spp. shows promising activities against pathogenic yeast (Irkin and Korukluoglu 2009).

Among different EOs, cinnamon, lemongrass, Japanese mint, ginger grass, geranium and clove oils were observed as most promising against *C. albicans*. The effective concentrations range from 0.01 to 0.15% (Devkatte et al. 2005, Hammer and Carson 2011). Growth of dermatophytes and their spore development is inhibited readily with EOs rich in phenylpropanoids like eugenol and the monocyclic sesquiterpene alcohols such as α-bisabolol (Bajpai et al. 2009, Maxia et al. 2009, Pragadheesh et al. 2013). Growth and aflatoxin production in molds like *Aspergillus flavus* is prevented by EOs (Kumar et al. 2010, Lang and Buchbauer 2012). Lemongrass (*C. citratus*) oil is one of the most effective oils against filamentous fungi with the active concentrations ranging from 0.006 to 0.03%. Orange, lemon, tangerine and grapefruit oils inhibit *A. niger* and *A. flavus*, at < 1% of concentration (Viuda Martos et al. 2008). Both drug sensitive and resistant pathogenic yeasts including the major pathogen of humans, *C. albicans*, were inhibited by terpenoid rich EOs (Devkatte et al. 2005, Zore et al. 2011). Efficacy of EOs and their components against drug resistant *C. albicans* biofilms is also significant.

The effects of EOs extracted from some spices [*Rosmarinus officinalis* (Rosemary), *Origanum vulgare* (Oregano), *Thymus vulgaris* (Thyme), *Lippia graveolens* (Mexican oregano), *Zingiber officinale* (Ginger), *Ocimum basilicum* (Basil), *Salvia officinalis* (Sage) and *Cinnamomum zeylanicum* (Cinnamon)] were examined in the inhibition of Germ Tube Formation (GTF) in three groups of yeasts: Fluconazole-Susceptible (FS) *C. albicans*, Fluconazole Resistant (FR) *C. albicans* and FS *C. dubliniensis* isolates (Pozzatti et al. 2010). *C. albicans* and *C. dubliniensis* GTF has been suggested as a potential virulence factor in their pathogenesis. GTF is a morphological characteristic that increases the ability of the fungi to adhere and penetrate into infected tissue. The authors conclude that all EOs tested in this study were able to inhibit the formation of the germ tube of *C. albicans* and *C. dubliniensis*, with highest activity observed by oregano EO, followed by the EO of Mexican oregano, thyme, cinnamon, ginger, basil and sage respectively. FS *C. albicans* were more susceptible than FR *C. albicans*, requiring lower concentrations of EO to inhibit germ tube, whereas *C. albicans* was more susceptible than *C. dubliniensis* (Pozzatti et al. 2010).

Evidences of Mechanism of Actions for the Antifungal Activities of EOs

The main mechanism of action of antimicrobial EOs is associated with its lipophilicity and consequent interaction with the microbial cell membrane. These interactions may result in changes and losses of enzymatic and structural fungal cells components, such as adenosine triphosphatase (ATPase), 1,3-β-D-glucan synthases, chitin and mannans, which are also components involved in GTF. The inhibition of chain respiration, as a result of EOs interactions with mitochondrial membranes, has also been demonstrated and can result in decreased energy production and a consequent inhibition of GTF and/or cell growth. Although no specific interaction of EOs with human membranes limits their clinical use as an antimicrobial agent in the treatment of systemic diseases, studies have shown promising clinical efficacy in the treatment of superficial mycoses, including fluconazole-refractory oral candidiasis in AIDS patients, vaginal candidiasis and pityriasis versicolor. The activities of EOs or their purified major components against fungi that demonstrate secondary resistance to antifungal agents is an area of study that has been less explored and deserves more attention (Pozzatti et al. 2010).

In addition, some authors propose that EOs activities may be mediated through inhibition of membrane ergosterol and signaling pathways involved in yeast to hyphae morphogenesis (Agarwal et al. 2008). EOs also possess cell cycle inhibitory activities against *C. albicans*. For example, citral, citronellol, geraniol and geranyl acetate which are the major constituents of eucalyptus oil, tea tree oil and geranium oil are reported to block *C. albicans* in S phase of cell cycle (Zore et al. 2011). Similarly, eugenol, thymol and carvacrol affect Ca^{2+} and H^+ homeostasis leading to loss of ions and inhibition of *Saccharomyces cerevisiae* (Rao et al. 2010). Abnormalities in membrane fluidity result in leakage of cytoplasmic contents and loss of viability of fungi. For example, membrane permeability and respiratory chain activity in *C. albicans* cells is inhibited in the presence of tea tree oil to result in cell death (Carson et al. 2006, Hammer and Carson 2011). Mitochondrial membrane permeabilization due to EOs treatment leads to apoptosis, necrosis and cell death. Also, individual constituent of EOs can interfere in the Target Of Rapamycin (TOR) signaling pathway of yeasts resulting in loss of viability (Rao et al. 2010).

Conclusions

The need of antifungal agents has grown in recent years in response to the challenge of invasive mycoses, multidrug resistant fungi and the appearance of new human diseases caused by fungal pathogens. Any antifungal drug clinically used nowadays gathers all the desirable requirements, since they have limited spectrum of action. Some of them are fungistatic but not fungicidal, produce recurrence, resistance or develop toxicity. In this context, great efforts to find out new compounds with antifungal properties have been made, but most scientists all over the world use non-targeted assays and almost all type of structures have demonstrated to be antifungal. Although many natural compounds with antifungal activities are permanently described, only very few have demonstrated to be good candidates for further research. In addition, EOs are special types of natural products obtained from certain plant families that actually offer a broad spectrum of antifungal action against human fungal infections and therefore, they should be greatly considered at the time of choosing an alternative antifungal therapy.

References

Ablordeppey, S., Fan, P., Ablordeppey, J. and Mardenborough, L. 1999. Systemic antifungal agents against AIDS-related opportunistic infections: current status and emerging drugs in development. Curr. Med. Chem. 6: 1151–1196.

Agarwal, V., Lal, P. and Pruthi, V. 2008. Prevention of *Candida albicans* biofilm by plant oils. Mycopathologia 165: 13–19.

Alviano, W., Mendonça-Filho, R., Alviano, D., Bizzo, H., Souto-Padron, T., Rodrigues, M., Bolognese, A., Alviano, C. and Souza, M. 2005. Antimicrobial activity of *Croton cajucara* Benth linaloolrich essential oil on artificial biofilms and planktonic microorganisms. Oral Microbiol. Inmunol. 20: 101–105.

Angioni, A., Barra, A., Coroneo, V., Dessi, S. and Cabras, P. 2006. Chemical composition, seasonal variability, and antifungal activity of *Lavandula stoechas* L. ssp. *stoechas* essential oils from stem/leaves and flowers. J. Agric. Food Chem. 54: 4364–4370.

Bajpai, V., Yoon, J. and Kang, S. 2009. Antifungal potential of essential oil and various organic extracts of *Nandina domestica* Thunb. against skin infectious fungal pathogens. Appl. Microbiol. Biotechnol. 83: 1127–1133.

Bakkali, F., Averbeck, S., Averbeck, D. and Idaomar, M. 2008. Biological effects of essential oils: a review. Food Chem. Toxicol. 46: 446–475.

Bansod, S. and Rai, M. 2008. Antifungal activity of essential oils from Indian medicinal plants against human pathogenic *Aspergillus fumigatus* and *A. niger*. World J. Med. Sci. 3: 81–88.

Belay, G., Tariku, Y., Kebede, T., Hymete, A. and Mekonnen, Y. 2011. Ethnopharmacological investigations of essential oils isolated from five Ethiopian medicinal plants against eleven pathogenic bacterial strains. Phytopharmacol. 1: 133–143.

Benkeblia, N. 2004. Antimicrobial activity of essential oil extracts of various onions (*Allium cepa*) and garlic (*Allium sativum*). LWT - Food Sci. Technol. 37: 263–268.

Brian, P., Curtis, P. and Hemming, H. 1949. A substance causing abnormal development of fungal hyphae produced by *Penicillium janczewskii* Zal: III. Identity of 'curling factor' with griseofulvin. Trans. Br. Mycol. Soc. 32: 30–33.

Brouwer, A., Rajanuwong, A., Chierakul, W., Griffin, G., Larsen, R., White, N. and Harrison, T. 2004. Combination antifungal therapies for HIV-associated cryptococcal meningitis: a randomised trial. Lancet. 363: 1764–1767.

Brown, E.D. and Wright, G.D. 2016. Antibacterial drug discovery in the resistance era. Nature 529: 336–343.

Brown, G.D., Denning, D.W., Gow, N.A., Levitz, S.M., Netea, M.G. and White, T.C. 2012. Hidden killers: human fungal infections. Sci. Transl. Med. 4: 165–167.

Carrillo-Muñoz, A., Finquelievich, J. and Tur-Tur, C. 2014. Combination antifungal therapy: A strategy for the management of invasive fungal infections. Rev. Esp. Quimioter. 27: 141–158.

Carson, C., Hammer, K. and Riley, T. 2006. *Melaleuca alternifolia* (Tea Tree) oil: a review of antimicrobial and other medicinal properties. Clin. Microbiol. Rev. 19: 50–62.

Cetin, B., Ozer, H., Cakir, A. and Li, D. 2009. Chemical composition of hydrodistilled essential oil of *Artemisia incana* (L.) Druce and antimicrobial activity against food borne microorganisms. Chem. Biodivers. 6: 2302–2310.

Charles, D. and Simon, J. 1990. Comparison of extraction methods for the rapid determination of essential oil content and composition. J. Am. Soc. Hortic. Sci. 115: 458–462.

Cos, P., Vlietink, A., Vanden Berghe, D. and Maes, L. 2006. Anti-infective potential of natural products: how to develop a stronger *in vitro* "proof-of-concept". J. Ethnopharmacol. 106: 290–302.

Cuenca-Estrella, M. and Rodríguez-Tudela, J.L. 2007. Elección de Especies Fúngicas para el Panel de Pruebas Antifúngicas. pp. 15–25. *In*: Zacchino, S. and Gupta, M. (eds.). Manual de Técnicas *in vitro* para la Detección de Compuestos Antifúngicos. Rosario: Edit. Corpus Argentina.

Darius, A.J., Meintjes, G. and Brown, G.D. 2014. A neglected epidemic: fungal infections in HIV/AIDS. Trends Microbiol. 22: 120–127.

Day, J.N., Chau, T.T., Wolbers, M., Mai, P.P., Dung, N.T., Mai, N.H., Phu, N.H., Nghia, H.D., Phong, N.D., Thai, C.Q. and Thai, L.H. 2013. Combination antifungal therapy for cryptococcal meningitis. N. Engl. J. Med. 368: 1291–1302.

Devkatte, A., Zore, G. and Karuppayil, S. 2005. Potential of plant oils as inhibitors of *Candida albicans* growth. FEMS Yeast Res. 5: 867–873.

Dikshit, A., Dubey, N., Tripathi, N. and Dixit, S. 1983. Cedrus oil: a promising storage fungitoxicant. J. Stored Prod. Res. 19: 159–162.

Dorman, H. and Deans, S. 2000. Antimicrobial agents from plants: antibacterial activity of plant volatile oils. J. Appl. Microbiol. 88: 308–316.

Dryden, M., Dailly, S. and Crouch, M. 2004. A randomized, controlled trial of tea tree topical preparations versus a standard topical regimen for the clearance of MRSA colonization. J. Hosp. Infect. 58: 86–87.

Fabio, A., Cermelli, C., Fabio, G., Nicoletti, P. and Quaglio, P. 2007. Screening of the antibacterial effects of a variety of essential oils on microorganisms responsible for respiratory infections. Phytother. Res. 21: 374–377.

Fisher, M.C., Henk, D.A., Briggs, C.J., Brownstein, J.S., Madoff, L.C., McCraw, S.L. and Gurr, S.J. 2012. Emerging fungal threats to animal, plant and ecosystem health. Nature 484: 186–194.

Fontenelle, R., Morais, S., Brito, E., Brilhante, R., Cordeiro, R., Nascimento, N., Rocha, M. and Sidrim, J. 2008. Antifungal activity of essential oils of *Croton* species from the Brazilian Catinga biome. J. Appl. Microbiol. 104: 1383–1390.

Francois, I., Aerts, A., Cammue, B. and Thevissen, K. 2005. Currently used antimycotics: spectrum, mode of action and resistance occurrence. Curr. Drug Tragets. 6: 895–907.

Freixa, B., Vila, R., Vargas, L., Lozano, N., Adzet, T. and Cañigueral, S. 1998. Screening for antifungal activity of nineteen Latin American plants. Phytother. Res. 12: 427–430.

Gold, W., Stout, H., Pagano, J. and Donovic, R. 1955. Amphotericins A and B, antifungal antibiotics produced by a streptomycete: I. *In vitro* studies. Antibiot. Annu. 3: 579–586.

Gómez-López, A., García-Effron, G., Mellado, E., Monzón, A., Rodríguez-Tudela, J.L. and Cuenca-Estrella, M. 2003. *In vitro* activities of three licensed antifungal agents against Spanish clinical isolates of *Aspergillus* spp. Antimicrob. Agents Chemother. 47: 3085–3088.

Graybill, J. 1997. Editorial response: can we agree on the treatment of candidiasis? Clin. Infec. Dis. 25: 60–62.

Guenther, E. 1972. The Essential Oils. Krieger Publishing Company, Malabar, Fla, USA, p. 452.

Hammer, K. and Carson, C. 2011. Antibacterial and antifungal activities of essential oils. pp. 255–306. *In*: Thormar, H. (ed.). Lipids and Essential Oils as Antimicrobial Agents. John Wiley & Sons, Ltd., UK.

Hammer, K., Carson, C. and Riley, T. 1999. Antimicrobial activity of essential oils and other plant extracts. J. Appl. Microbiol. 86: 985–990.

Hennebelle, T., Sahpaz, S., Joseph, H. and Bailleul, F. 2008. Ethnopharmacology of *Lippia alba*. J. Ethnopharmacol. 116: 211–222.

Hill, J.A. and Cowen, L.E. 2015. Using combination therapy to thwart drug resistance. Future Microbiology 10: 1719–1726.

Ioannou, E., Poiata, A., Hancianu, M. and Tzakou, O. 2007. Chemical composition and *in vitro* antimicrobial activity of the essential oils of flower heads and leaves of *Santolina rosmarinifolia* L. from Romania. Nat. Prod. Res. 21: 18–23.

Irkin, R. and Korukluoglu, M. 2009. Effectiveness of *Cymbopogon citratus* L. essential oil to inhibit the growth of some filamentous fungi and yeasts. J. Med. Food 12: 193–197.

Kathiravan, M.K., Salake, A.B., Chothe, A.S., Dudhe, P.B., Watode, R.P., Mukta, M.S. and Gadhwe, S. 2012. The biology and chemistry of antifungal agents: a review. Bioorg. Med. Chem. 20: 5678–5698.

Kavanagh, K. (ed.). 2011. Fungi: Biology and Applications. John Wiley & Sons, p. 384.

Khosravi, A., Minooeianhaghighi, M. and Shokri, H. 2011. The potential inhibitory effect of *Cuminum cyminum*, *Ziziphora clinopodioides* and *Nigella sativa* essential oils on the growth of *Aspergillus fumigatus* and *Aspergillus flavus*. Braz. J. Microbiol. 42: 216–224.

Kordali, S., Kotan, R., Mavi, A., Cakir, A., Ala, A. and Yildirim, A. 2005. Determination of the chemical composition and antioxidant activity of the essential oil of *Artemisia dracunculus* and of the antifungal and antibacterial activities of Turkish *Artemisia absinthium*, *A. dracunculus*, *Artemisia santonicum* and *Artemisia spicigera* essential oils. J. Agric. Food Chem. 53: 9452–9458.

Kumar, A., Shukla, R., Singh, P. and Dubey, N. 2010. Chemical composition, antifungal and anti-aflatoxigenic activities of *Ocimum sanctum* L. essential oil and its safety assessment as plant based antimicrobial. Food Chem. Toxicol. 48: 539–543.

Kumar, R., Mishra, A., Dubey, N. and Tripathi, Y. 2007. Evaluation of *Chenopodium ambrosioides* oil as a potential source of antifungal, antiaflatoxigenic and antioxidant activity. Int. J. Food Microbiol. 115: 159–164.

Lang, G. and Buchbauer, G. 2012. A review on recent research results (2008–2010) on essential oils as antimicrobials and antifungals. Flavour Fragr. J. 27: 13–39.

Lopes-Lutz, D., Alviano, D., Alviano, C. and Kolodziejczyk, P. 2008. Screening of chemical composition, antimicrobial and antioxidant activities of *Artemisia* essential oils. Phytochemistry 69: 1732–1738.

Maregesi, S., Pieters, L., Ngassapa, O., Apers, S., Vigerhoets, R., Cos, P., Vanden Berghe, D. and Vlietink, A. 2008. Screening of some Tanzanian medicinal plants from Bunda district for antibacterial, antifungal and antiviral activities. J. Ethnopharmacol. 119: 58–66.

Mastura, M., Azah, M., Khozirah, S., Mawardi, R. and Manaf, A. 1999. Anticandidaland antidermatophytic activity of *Cinnamomum* species essential oils. Cytobios 98: 17–23.

Mathew, B. and Nath, M. 2009. Recent approaches to antifungal therapy for invasive mycoses. ChemMedChem. 4: 310–323.

Maxia, A., Marongiu, B., Piras, A., Porcedda, S., Tuveri, E., Gonçalves, M., Cavaleiro, C. and Salgueiro, L. 2009. Chemical characterization and biological activity of essential oils from *Daucus carota* L. subsp. *carota* growing wild on the Mediterranean coast and on the Atlantic coast. Fitoterapia 80: 57–61.

Mimica-Dukic, N., Bozin, B., Sokovic, M. and Simin, N. 2004. Antimicrobial and antioxidant activities of *Melissa officinalis* L. (Lamiaceae) essential oil. J. Agric. Food Chem. 52: 2485–2489.

Monzón, A., Cuenca-Estrella, M. and Rodríguez-Tudela, J.L. 2003. Epidemiological survey of dermatophytosis in Spain (April–June 2001). Enferm. Infec. Microbiol. Clin. 21: 477–483.

Naik, V., Ahmed, F., Gupta, A., Garg, A., Sarkar, C., Sharma, B. and Mahapatra, A.K. 2015. Intracranial Fungal Granulomas: A Single Institutional Clinicopathologic Study of 66 Patients and Review of the Literature. World Neurosurg. 83(6): 1166–1172. doi: 10.1016/j.wneu.2015.01.053.

Nucci, M. and Marr, K.A. 2005. Emerging fungal diseases. Clin. Infect. Dis. 41: 521–526.

Odds, F. 1996. Resistance of clinically important yeasts to antifungal agents. Int. J. Antimicrob. Agents. 6: 145–147.

Ozcan, M. and Chalchat, J. 2008. Chemical composition and antifungal activity of rosemary (*Rosmarinus officinalis* L.) oil from Turkey. Int. J. Food Sci. Nut. 59: 691–698.

Ozek, G., Demirci, F., Ozek, T., Tabanca, N., Wedge, D., Khan, S., Baser, K., Duran, A. and Hamzaoglu, E. 2010. Gas chromatographic–mass spectrometric analysis of volatiles obtained by four different techniques from *Salvia rosifolia* Sm. and evaluation for biological activity. J. Chromatogr. A. 1217: 741–748.

Pappas, P.G., Alexander, B.D., Andes, D.R., Hadley, S., Kauffman, C.A., Freifeld, A., Anaissie, E.J., Brumble, L.M., Herwaldt, L., Ito, J. and Kontoyiannis, D.P. 2010. Invasive fungal infections among organ transplant recipients: results of the Transplant-Associated Infection Surveillance Network (TRANSNET). Clin. Infect. Dis. 50: 1101–1111.

Patterson, T.F. 2005. Advances and challenges in management of invasive mycoses. Lancet 366: 1013–1025.

Peighami-Ashnaei, S., Farzaneh, M., Sharifi-Tehrani, A. and Behboudi, K. 2008. Effect of essential oils in control of plant diseases. Comm. Agric. Appl. Biol. Sci. 74: 843–847.

Pelaez, F., Cabello, A., Platas, G., Diez, M., Gonzalez del Val, A., Martan, I., Vicente, F., Bills, G., Giacobbe, R., Schwartz, R., Onishi, J., Meinz, M., Abbruzzo, G., Flattery, A., Kong, L. and Kurtz, M. 2000. The discovery of enfumafungin, a novel antifungal compound produced by an endophytic *Hormonema* spp. biological activity and taxonomy of the producing organisms. Sys. Appl. Microbiol. 23: 333–343.

Pepeljnjak, S., Kosalec, I., Kalodera, Z. and Blazevic, N. 2005. Antimicrobial activity of juniper berry essential oil (*Juniperus communis* L., Cupressaceae). Acta Pharm. Zagreb. 55: 417.

Pfaller, M.A. and Diekema, D.J. 2004. Rare and emerging opportunistic fungal pathogens: Concern for resistance beyond *Candida albicans* and *Aspergillus fumigatus*. J. Clin. Microbiol. 42: 4419–4431.

Pfaller, M.A. and Diekema, D.J. 2007. Epidemiology of invasive candidiasis: a persistent public health problem. Clin. Microbiol. Rev. 20: 133–163.

Pichersky, E., Noel, J. and Dudareva, N. 2006. Biosynthesis of plant volatiles: nature's diversity and ingenuity. Science 3115762: 808–811.

Pinto, E., Vale-Silva, L., Cavaleiro, C. and Salgueiro, L. 2009. Antifungal activity of the clove essential oil from *Syzygium aromaticum* on *Candida*, *Aspergillus* and dermatophyte species. J. Med. Microbiol. 58: 1454–1462.

Pozzatti, P., Loreto, E., Nunes Mario, D., Rossato, L., Santurio, J. and Alves, S. 2010. Activities of essential oils in the inhibition of *Candida albicans* and *Candida dubliniensis* germ tube formation. J. Mycol. Med. 20: 185–189.

Pragadheesh, V., Saroj, A., Yadav, A., Chanotiya, C., Alam, M. and Samad, A. 2013. Chemical characterization and antifungal activity of *Cinnamomum camphora* essential oil. Ind. Crops Prod. 49: 628–633.

Rabadia, A., Kamat, S. and Kamat, D. 2012. Antifungal activity of essential oils against fluconazole resistant fungi. Int. J. Phytomed. 3: 506–510.

Ramadán, S., Sortino, M., Bulacio, L., Marozzi, M.L., López, C. and Ramos, L. 2012. Prevalence of *Malassezia* species in patients with pityriasis versicolor in Rosario, Argentina. Rev. Iberoam. Micol. 29: 14–19.

Rao, A., Zhang, Y., Muend, S. and Rao, R. 2010. Mechanism of antifungal activity of terpenoid phenols resembles calcium stress and inhibition of the TOR pathway. Antimicrob. Agents Chemother. 54: 5062–5069.

Rasooli, I., Fakoor, M., Yadegarinia, D., Gachkar, L., Allameh, A. and Rezaei, M. 2008. Antimycotoxigenic characteristics of *Rosmarinus officinalis* and *Trachyspermum copticum* L. essential oils. Int. J. Food Microbiol. 122: 135–139.

Razzaghi-Abyaneh, M., Shams-Ghahfarokhi, M., Yoshinari, T., Rezaee, M., Jaimand, K., Nagasawa, H. and Sakuda, S. 2008. Inhibitory effects of *Satureja hortensis* L. essential oil on growth and aflatoxin production by *Aspergillus parasiticus*. Int. J. Food Microbiol. 123: 228–233.

Richardson, M. and Lass-Flörl, C. 2008. Changing epidemiology of systemic fungal infections. Clin. Microbiol. Infect. 14: 5–24.

Rodríguez-Tudela, J.L. and Cuenca-Estrella, M. 1999. Originales-Estudio multicentrico sobre fungemias por levaduras en España (abril-junio de 1997). Grupo de Trabajo para el Estudio de las Fungemias. Rev. Clin. Esp. 199: 356–361.

Roemer, T. and Krysan, D.J. 2014. Antifungal drug development: challenges, unmet clinical needs, and new approaches. Cold Spring Harb. Perspect. Med. 4: 208–224.

Roobol, A., Gull, K. and Pogson, C.I. 1976. Inhibition by griseofulvin of microtubule assembly *in vitro*. FEBS Lett. 67: 248–251.

Rosato, A., Vitali, C., De Laurentis, N., Armenise, D. and Antonietta, M. 2007. Antibacterial effect of some essential oils administered alone or in combination with norfloxacin. Phytomedicine 14: 727–732.

Rota, C., Carraminana, J., Burillo, J. and Herrera, A. 2004. *In vitro* antimicrobial activity of essential oils from aromatic plants against selected food borne pathogens. J. Food Prot. 67: 1252–1256.

Rouhi, M. 2003. Rediscovering natural products. Chem. Eng. News. 81: 77–91.

Ruiz-Camps, I. and Cuenca-Estrella, M. 2009. Antifúngicos para uso sistémico. Enf. Infec. Microbiol. Clin. 27: 353–362.

Saidana, D., Mahjoub, M., Boussaada, O., Chriaa, J., Chéraif, I., Daami, M., Mighri, Z. and Helal, A. 2008. Chemical composition and antimicrobial activity of volatile compounds of *Tamarix boveanai* (Tamaricaceae). Microbiol. Res. 163: 445–455.

Saikia, D., Khanuja, S., Kahol, A., Gupta, S. and Kumar, S. 2001. Comparative antifungal activity of essential oils and constituents from three distinct genotypes of *Cymbopogon* spp. Curr. Sci. 80: 1264–1265.

Saravolatz, L., Ostrosky-Zeichner, L., Marr, K, Rex, J. and Cohen, S. 2003. Amphotericin B: time for a new "gold standard". Clin. Infect. Dis. 37: 415–425.

Sartorelli, P., Marquioreto, A., Amaral-Baroli, A., Lima, M. and Moreno, P. 2011. Chemical composition and antimicrobial activity of the essential oils from two species of *Eucalyptus*. Phytother. Res. 21: 231–233.

Segal, B.H., Kwon-Chung, J., Walsh, T.J., Klein, B.S., Battiwalla, M., Almyroudis, N.G., Holland, S.M. and Romani, L. 2006. Immunotherapy for fungal infections. Clin. Infect. Dis. 42: 507–515.

Selitrennikoff, C. 1992. Screening for antifungal drugs. pp. 189–217. *In*: Finkelstein, D. and Ball, C. (eds.). Biotechnology of Filamentous Fungi; Technology and Products. Butterworth Heinemann (Boston, USA).

Shaaban, H., El Ghorab, A. and Shibamoto, T. 2012. Bioactivity of essential oils and their volatile aroma components: Review. J. Ess. Oil Res. 24: 203–212.

Shapiro, R.S., Robbins, N. and Cowen, I.E. 2011. Regulatory circuitry governing fungal development, drug resistance, and disease. Microbiol. Mol. Biol. R. 75: 213–267.

Singh, G., Kapoor, I., Pandey, S., Singh, U. and Singh, R. 2002. Studies on essential oils: part10; antibacterial activity of volatile oils of some spices. Phytother. Res. 16: 680682.

Singh, G., Kapoor, I., Singh, P., de Heluani, C., de Lampasona, M. and Catalan, C. 2008. Chemistry, antioxidant and antimicrobial investigations on essential oil and oleoresins of *Zingiber officinale*. Food Chem. Toxicol. 46: 3295–3302.

Singh, G., Marimuthu, P., de Heluani, C. and Catalan, C. 2006. Antioxidant and biocidal activities of *Carum nigrum* (seed) essential oil, oleoresin, and their selected components. J. Agric. Food Chem. 54: 174–181.

Singh, P., Shukla, R., Kumar, A., Prakash, B., Singh, S. and Dubey, N. 2010. Effect of *Citrus reticulata* and *Cymbopogon citratus* essential oils on *Aspergillus flavus* growth and aflatoxin production on *Asparagus racemosus*. Mycopathologia 170: 195–202.

Sortino, M., Derita, M., Svetaz, L., Raimondi, M., Di Liberto, M., Petenatti, E., Gupta, M. and Zacchino, S. 2012. The role of natural products for the discovery of new anti-infective agents, with emphasis in antifungal compounds. pp. 205–229. *In*: Cechinel Filho, V. (ed.). Plants Bioactives and Drug Discovery: Principles, Practice, and Perspectives. John Wiley & Sons, New York.

Spitzer, M., Robbins, N. and Wright, G. 2016. Combinatorial strategies for combating invasive fungal infections. Virulence, just-accepted.

Svetaz, L., Postigo, A., Butassi, E., Zacchino, S. and Sortino, M. 2016. Antifungal drugs combinations: a patent review 2000–2015. Expert Opin. Ther. Pat. 26: 439–453.

Tavares, A., Goncalves, M., Cavaleiro, C., Cruz, M., Lopes, M., Canhoto, J. and Salgueiro, L. 2008. Essential oil of *Daucus carota* subsp. Halophilus: composition, antifungal activity and cytotoxicity. J. Ethnopharmacol. 119: 129–134.

Tolouee, M., Alinezhad, S., Saberi, R., Eslamifar, A., Zad, S., Jaimand, K., Taeb, J. and RazzaghiAbyaneh, M. 2010. Effect of *Matricaria chamomilla* L. flower essential oil on the growth and ultrastructure of *Aspergillus niger* van Tieghem. Int. J. Food Microbiol. 139: 127–133.

Van Peer, A., Woestenborghs, R., Heykants, J., Gasparini, R. and Gauwenbergh, G. 1989. The effects of food and dose on the oral systemic availability of itraconazole in healthy subjects. Eur. J. Clin. Pharmacol. 36: 423–426.

Vicente, M.F. and Peláez, F. 2007. Situación Actual de la Terapia Antifúngica. pp. 6–14. *In*: Zacchino, S. and Gupta, M. (eds.). Manual de Técnicas *in vitro* para la Detección de Compuestos Antifúngicos. Rosario: Edit. Corpus Argentina.

Vigan, M. 2010. Essential oils: renewal of interest and toxicity. European J. Dermatol. 20: 685–692.

Vincent, J.L., Rello, J., Marshall, J., Silva, E., Anzueto, A., Martin, C.D., Moreno, R., Lipman, J., Gomersall, C., Sakr, Y. and Reinhart, K. 2009. International study of the prevalence and outcomes of infection in intensive care units. Jama 302: 2323–2329.

Viuda Martos, M., Ruiz Navajas, Y., Fernandez Lopez, J. and Perez Alvarez, J. 2008. Antifungal activity of lemon (*Citrus lemon* L.), mandarin (*Citrus reticulata* L.), grapefruit (*Citrus paradisi* L.) and orange (*Citrus sinensis* L.) essential oils. Food Cont. 19: 1130–1138.

Wisplinghoff, H., Bischoff, T., Tallent, S.M., Seifert, H., Wenzel, R.P. and Edmond, M.B. 2004. Nosocomial bloodstream infections in US hospitals: analysis of 24,179 cases from a prospective nationwide surveillance study. Clin. Infect. Dis. 39: 309–317.

Zimmermann, G.R., Lehar, J. and Keith, C.T. 2007. Multi-target therapeutics: when the whole is greater than the sum of the parts. Drug Discov. Today 12: 34–42.

Zonios, D. and Bennett, J. 2008. Update on Azole Antifungals. Sem. Resp. Crit. Care Med. 29: 198–210.

Zore, G., Thakre, A., Rathod, V. and Karuppayil, S. 2011. Evaluation of anti-*Candida* potential of geranium oil constituents against clinical isolates of *Candida albicans* differentially sensitive to fluconazole: inhibition of growth, dimorphism and sensitization. Mycoses 54: 99–109.

Zuzarte, M., Goncalves, M. and Cruz, M. 2012. *Lavandula luisieri* essential oil as a source of antifungal drugs. Food Chem. 135: 1505–1510.

Zuzarte, M., Maria, J., Carlos, C., Canhoto, J., Vale-Silva, L., Silva, M., Pinto, E. and Salgueiro, L. 2011. Chemical composition and antifungal activity of the essential oils of *Lavandula viridis* L'Her. J. Med. Microbiol. 60: 612–618.

Essential Oils against Microbial Resistance Mechanisms, Challenges and Applications in Drug Discovery

Juan Bueno,[1,*] *Fatih Demirci*[2,3] and *K. Husnu Can Baser*[4]

> "The essence of aromaticity lies in its rarefaction; the essence of sweetness lies in its concentration and earthiness. This explain why aromatics are so much better adapted to feed the breath, while sweet substances are better fitted to nourish the body. Consequently aromaticity is more efficient for the heart then sweetness, while it is the other way round in the case of the liver."
>
> "The natural breath has a desire for aromaticity and is invigorated and refreshed by such. It is easy to see that the natural faculties will also be invigorated in consequence."
>
> —Ibn-Sīnā
>
> The Canon of Medicine, 1025 CE, pp. 548–549

Introduction

Infectious diseases caused by bacteria, viruses, parasites and fungi, and their interaction with hosts and the environment are a global public health threat. Initially, at the time when antibiotics were discovered, the infections that affected humankind could be readily eradicated. But their indiscriminate use has led to the emergence of multidrug-resistant microorganisms (MDR) which are hard to control (Hemaiswarya et al. 2008).

The infections produced by MDR cause appreciable patient mortality and morbidity. In 2005, in the United States about 95,000 people acquired methicillin-resistant *Staphylococcus aureus* (MRSA) infections and 19,000 people died of this cause (Worthington and Melander 2013). The

[1] Research Center of Bioprospecting and Biotechnology for Biodiversity Foundation (BIOLABB), Colombia.
[2] Faculty of Health Sciences, Anadolu University, Eskişehir, Turkey.
[3] Department of Pharmacognosy, Faculty of Pharmacy, Anadolu University, Eskişehir, Turkey.
[4] Department of Pharmacognosy, Faculty of Pharmacy, Near East University, Department of Pharmacognosy, Nicosia (Lefkoşa), N. Cyprus.
* Corresponding author: juanbueno@biolabb.org; juangbueno@gmail.com

microorganisms that pose a threat to human beings comprise ESKAPE pathogens (*Enterococcus faecium*, *S. aureus*, *Klebsiella pneumoniae*, *Acinetobacter baumannii*, *Pseudomonas aeruginosa* and *Enterobacter* spp.) as well as emerging and re-emerging pathogens such as carbapenem-resistant *K. pneumoniae* (CRKP), the New Delhi metallo-β-lactamase-containing Enterobacteriaceae, MDR and extensively drug-resistant (XDR) strains of *Mycobacterium tuberculosis* (MDR-TB and XDR-TB) and non-tubercular mycobacteria, *Candida albicans* and pathogenic fungi (Tegos and Hamblin 2013). Of these, the most dangerous MDR, XDR and pandrug-resistant (PDR) strains have currently been redefined as follows: MDR are the organisms that are drug-resistant at least to one medication belonging to three or more antimicrobial groups and XDR are the drug-resistant organisms to at least one medication belonging all but located in two or fewer classes of antibiotics. Also PDR are defined as drug resistant organism to all antimicrobial medications across classes. Therefore, it is assumed that a clinical isolate that is defined as XDR is a MDR and a XDR strain can be defined as PDR (Magiorakos et al. 2012). Equally, antiviral drug resistance is a rising concern particularly in immune-compromised patient populations, where prolonged exposure to drugs causes the selection of resistant mutants due to viral factors as viral mutation frequency, virus replication as well as viral replication fitness. This is compounded by the contribution of other host factors that increase the risks such as older age and no adherence to medication (Strasfeld and Chou 2010). On the other hand, the emergence of drug-resistant parasites (mainly malaria) to the commonly used drugs, has been a great cost to human health and quality of life (Lin et al. 2010). These resistance patterns are further modified by factors similar to those found in other microorganisms such as mutation rate, fitness associated with the resistance mutations, microbial load, the strength of drug selection and the treatment adherence (Petersen et al. 2011). This cause an annual cost of US$21 billion to US$ 34 billion in countries such as United States, which demonstrates that these kinds of infections are of broad spending for health system (Infectious Diseases Society of America 2011).

Currently, there is a lack of investment in research of design and development of new antimicrobial agents against MDR pathogens, due to the low amount of return compared with the medications for chronic diseases. This absence of new compounds will affect seriously many areas of medicine as surgery, premature infant care, cancer chemotherapy, care of the critically ill and transplantation medicine, which depend largely effective antimicrobial therapy. For that reason, the World Health Organization (WHO) has identified MDR microorganisms as one of the more significant threats to human health and the Infectious Disease Society of America (IDSA) has made an appeal to the biomedical community to devote great efforts to the finding of new treatments for MDR microbes (Worthington and Melander 2013). In this context, a new paradigm that involves the use a multi-drug therapy instead of mono-substance is necessary in antimicrobial chemotherapy as employed worldwide in the treatment of AIDS and TB (Worthington and Melander 2013). These multi-target therapies have the ability of blocking the mechanism of resistance to the antibiotic and prevent the emergence of MDR strains.

The multi-target therapy should involve a resistance-modifying agent (RMA) as an antibiotic's adjuvant that has the ability of targeting the mechanism of resistance and showing synergism between them when introduced into a pharmaceutical formulation. Currently, there are phytochemicals with the ability of acting as resistant modifying agents in a multi-target therapy and enhancing the antimicrobial activity of antibiotics and biocides (Abreu et al. 2012). This opens the door for the development of a new generation of antimicrobials.

Essential Oils (EOs) are complex natural mixtures of volatile secondary metabolites isolated from plant material (flowers, buds, seeds, leaves, twigs, bark, herbs, wood, fruits and roots) by hydro or steam distillation and by expression. The major components of EOs are mono and sesquiterpenes, as well as carbohydrates, alcohols, ethers, aldehydes and ketones which are responsible for the synergic interaction with antibiotics, due to that EOs antimicrobial activity can be attributed to inhibition or interaction with multiple targets in the microbial cell (Kalemba and Kunicka 2003). Specific targets have also been implicated in the activity of EOs against various microorganisms, as would be related with a mechanism of action in membrane disruption, but a new evidence is emerging which suggests the presence of other specific targets (Boire et al. 2013). The mechanisms of action of EOs and/or their components are dependent on their chemical composition. For example, thymol and carvacrol owe their antimicrobial activity to the location of functional hydroxyl groups and limonene and p-cymene depend on the alkyl group (Nazzaro et al. 2013).

This chapter will review the different antimicrobial targets described for EOs and their major components, focusing on resistance mechanisms, with the aim to use in design and development of new multi-target therapeutics useful in the control of infections caused by MDR pathogens, due to promising use of EOs as antiseptics and disinfectants.

Multi-target Combination, Translational Medicine in Essential Oils

By definition it has been suggested that Translational Medicine should be a discipline that includes the following issues that can be adapted to EOs (Mankoff et al. 2004):

1. Basic science that develop *in vitro* and *in vivo* studies, which will define the biological effects of EOs in animals and human as well as their environmental impact.
2. Investigations in animals, humans and environment which determine the biology of disease from the point of view of the interaction between pathogen, host and environment, providing the scientific basis for the development of new antimicrobial agents with higher activity and lower toxicity than the existing ones.
3. Studies conducted with the aim of bringing anti-infective medicines to clinical practice, or agricultural use or developing antimicrobial principles for application as medications that are useful in human disease and biocide formulations.
4. Clinical and environmental trials of a therapy or formulation that can be initiated based on items 1–3 for determining toxicity and/or efficacy.

In that order of ideas, it is necessary to introduce new strategies based on multi-target drugs focused in resistance mechanisms for improving antimicrobial drug discovery process. This concept applies the overlap between pathways, by study of the cellular network to make a map of potential therapeutic targets, which are interesting from the pharmacological point of view (resistance mechanisms), with the end to potentiate the current anti-infective therapy (Korcsmáros et al. 2007). This approach, that seeks to design and detect new antibiotics is also applied to different classes of promising compounds such as fungicides, anti-cancer substances as well as pesticides and biocides, for destroying and inhibiting the normal function of cellular networks (Korcsmáros et al. 2007). Since recent findings in biology and clinical outcomes indicate that the use of single-target drugs may not always produce a pharmacological effect throughout the biological system due to the fact that organisms can affect effectiveness using compensatory ways (Lu et al. 2012). For that reason, an important plan against antimicrobial drug resistance is the exploration of plant secondary metabolites and synergies for their function in biological defense acting in different targets at the same time that avoid the use of microbial adaptation ways by breaking up of the cell membrane, as well as by interaction with specific proteins or intercalating into RNA or DNA (Efferth and Koch 2011, Gertsch 2011).

Although clinical reports are insufficient, the uses of EOs in topical formulation and as carrier for penetration enhancers in antiseptics are promising (Solórzano-Santos and Miranda-Novales 2012). Likewise, in EOs, the synergy between major and minor components are responsible for antimicrobial activity (Wagner and Ulrich-Merzenich 2009, Ahmad et al. 2014). These advantages make essential oils an interesting combination for use in drug or biocide formulation to be applied in prevention of Healthcare Associated Infections (HAI) and in agriculture (Cantas et al. 2013). The aim is to look for new antimicrobial agents for use in humans, animals and crops with low impact in environmental microbiota (Bernal et al. 2013).

In the following sections, the antimicrobial therapeutic targets in which EOs have shown applications, and future prospects for antimicrobial multi-target therapy based on EOs will be explored.

Antibiotic Adjuvants

An interesting way for the development of novel antimicrobial drugs is the use of enhancers of anti-infective activity or antibiotic adjuvants (Cole et al. 2015). These bioactive compounds have the ability to form anti-infectious agents in combination with antibiotics resulting in an increase in activity (Cole et al. 2015, Bueno 2016). Adjuvants work by inhibiting microbial drug resistance mechanisms and allowing the action of antibiotics on resistant microorganisms (Bernal et al. 2013, Langeveld et al. 2014, Cole et al. 2015). In that way, there are several reports about the synergistic and additive interaction of EOs with current antibiotic treatments, showing ability to inhibit resistance mechanisms and improving antimicrobial activity (Ejim et al. 2011).

In antimicrobial drug discovery from natural products, this potentiation of the antibiotic activity has been described as in the case of clavulanic acid and recently ellagic and tannic acids against MDR strains of *Acinetobacter baumannii* (Chusri et al. 2009, Wright 2014). Equally, EOs of *Rosmarinus officinalis* L. (Lamiaceae), *Lippia alba* (Mill.) N.E.Br. ex Britton & P. Wilson (Verbenaceae), *Hyptis martiusii* Benth. (Lamiaceae) and *Lippia origanoides* Kunth (Verbenaceae) have been reported as promising adjuvants to be employed in antimicrobial therapy by their ability of restore the sensitivity of drug-resistant strains to antibiotics, and demonstrating an important field of action in pursuit of active EOs (Veras et al. 2011, de Olivera et al. 2014, Barreto et al. 2014a, 2014b).

Efflux Pumps

MDR bacteria pose an efflux mechanism that reduces the intracellular concentration of all clinically used groups of antimicrobials and other chemicals as disinfectants. This mechanism is formed by a series of carrier proteins known as efflux pumps (Nikaido and Pagès 2012). It, also, is present in other organisms such as fungi (*Candida* spp. and *Aspergillus* spp.), which takes part in expulsion of antifungal azoles (Rajendran et al. 2011), and has been shown to be one of the mechanisms involved in the resistance of biofilms (Soto 2013). For that reason, inhibition of efflux pumps is an important approach in the discovery of new antibiotic adjuvants (Van Bambeke and Lee 2006).

Some EOs such as *Helichrysum italicum* (Roth) G. Don (Compositae) reduces the multidrug resistance of *Enterobacter aerogenes*, *Escherichia coli*, *Pseudomonas aeruginosa*, and *Acinetobacter baumannii* by the inhibition of efflux, geraniol being responsible for the pharmacological effect (Fig. 6.1) (Lorenzi et al. 2009). Grapefruit oil obtained from *Citrus paradisi* Macfad (Rutaceae), contains some components (bergamottin epoxide derivative and a coumarin epoxide derivative) that modulate efflux pumps in MRSA strains (Stavri et al. 2007). Equally, coriander [*Coriandrum sativum* (Apiaceae)] oil inhibits efflux activity by the disruption of membrane cells in *Bacillus cereus* and *Enterococcus faecalis* (Silva et al. 2011). Major compounds of EOs as citral and citronellal and their monoterpene derivate amides have reportedly inhibited NorA efflux pump in *Staphylococcus aureus* (Thota et al. 2008). Although efflux pump can be an antimicrobial target of EOs, it is important to take into consideration that *Pseudomonas aeruginosa* MexAB-OprM pump can extrude monoterpenes and related alcohols, so this mechanism can be used by the microbial cells to reduce activity of the EOs (Fig. 6.1) (Papadopoulos et al. 2008).

Beta-Lactamase Inhibition

β-lactamases are enzymes that catalyze the hydrolysis of β-lactam ring of the antibiotics rendering them inactive. Up to date, more than 700 of these enzymes have been detected and are classified according to the type of target antibiotic in four groups, A–D (class A penicillins, class C cephalosporins,

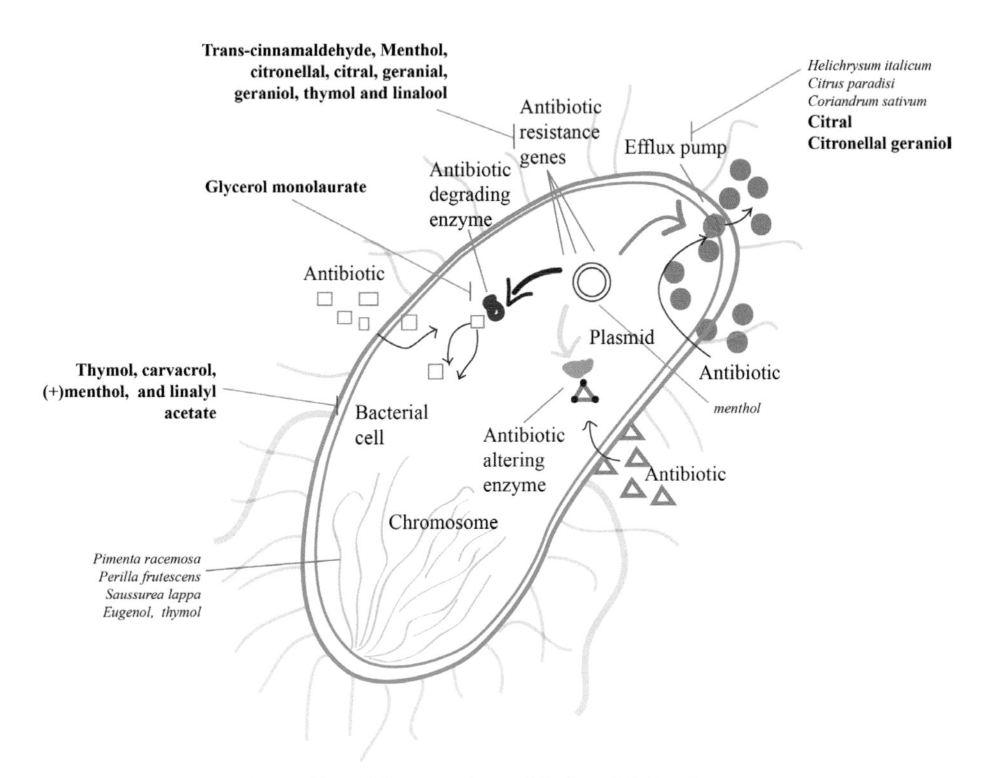

Figure 6.1. Antibacterial targets of essential oils and their major components.

class D oxacillin and class B which because they have two zinc ions, their catalytic site are called metallo-β-lactamases) (Schlesinger et al. 2011). Making β-lactamase inhibitors as an important target of action in the development of anti-infectious adjuvant chemotherapy is an essential research field task. Among the promising compounds, glycerol monolaurate, a surfactant from coconut oil, inhibits the production of β-lactamase by blocking intracellular signal in *S. aureus* (Fig. 6.1) (Projan et al. 1994), which has been proposed as a topical antimicrobial agent due to its antitoxigenic and anti-inflammatory activity (Lin et al. 2009).

Inhibition of Virulence Factors Expression

Various mechanisms of bacterial virulence have been described. Between them are adhesins for attaching and colonizing tissues, toxins that destroy cells and alter cellular signaling and specialized secretion systems for inject effectors (toxins) into the cell. For that reason inhibition of production of these factors are part of the anti-virulence strategy of screening for obtaining anti-infective drugs, that without being bactericidal or bacteriostatic may be adjuvants of conventional antibiotic therapy (Rasko and Sperandio 2010). As inhibitor of the adhesion factors from EOs, *trans*-cinnamaldehyde a major component of cinnamon EO [*Cinnamomum verum* J.Presl (Laureaceae)] have shown to decrease the expression of the major genes involved in uropathogenic *E. coli* attachment and invasion of host tissue (Amalaradjou et al. 2011). Equally, sub-inhibitory concentrations of menthol, a volatile component of peppermint [*Mentha* × *piperita* L. (Lamiaceae)] oil, alter gene expression of cpsB10 that regulates the colony mucoidity in *E. coli* for colanic acid production (Fig. 6.1) (Landau and Shapira 2012). On the other hand, the enzyme elastase produced by *Pseudomonas* spp. is employed to disrupt tight junctions in tissues allowing microbial invasion. This enzyme was inhibited by natural compounds from EOs such as citronellal, citral, geranial, geraniol, thymol and linalool (Sivamani et al. 2012).

Antibiofilm Activity

In EOs with anti-adhesion activity, a special mention of the anti-biofilm activity is required. Biofilms are microbial communities attached on a surface, which are embedded in a matrix that provides a reservoir for microbial cells and confers a protection against biocides and drugs contributing to drug resistance development (Bueno 2014). Carvacrol and thymol, two biocidal compounds, present in thyme [*Thymus vulgaris* (Lamiaceae)] oil have an important antimicrobial effect on biofilms formed by *S. aureus*, *S. epidermidis* and *Salmonella enterica* serovar *typhimurium* (Fig. 6.2) (Knowles et al. 2005, Nostro et al. 2007). Equally, carvacrol reduces biofilm biomass when used in vapor phase, but is more active in liquid contact (Nostro et al. 2009). On the other hand, EO of *Satureja thymbra* L. (Lamiaceae) whose major components are carvacrol, thymol, *p*-cymene and γ-terpinene, has potent microbicidal effect on biofilms developed on stainless steel (Fig. 6.2) (Chorianopoulos et al. 2008). Also, *Candida albicans*, *C. glabrata* and *C. parapsilosis* biofilms were treated with carvacrol, geraniol and thymol producing inhibition in biofilm formation in > 75% (Dalleau et al. 2008). Likewise, evaluating antimicrobial activity of terpenoids as linalool, nerol, isopulegol, menthol, carvone, α-thujone and farnesol in *C. albicans* biofilm model, these compounds exhibited anti-biofilm activity in the stages of planktonic growth, morphogenesis, adhesion, biofilm development and mature biofilms (Raut et al. 2013).

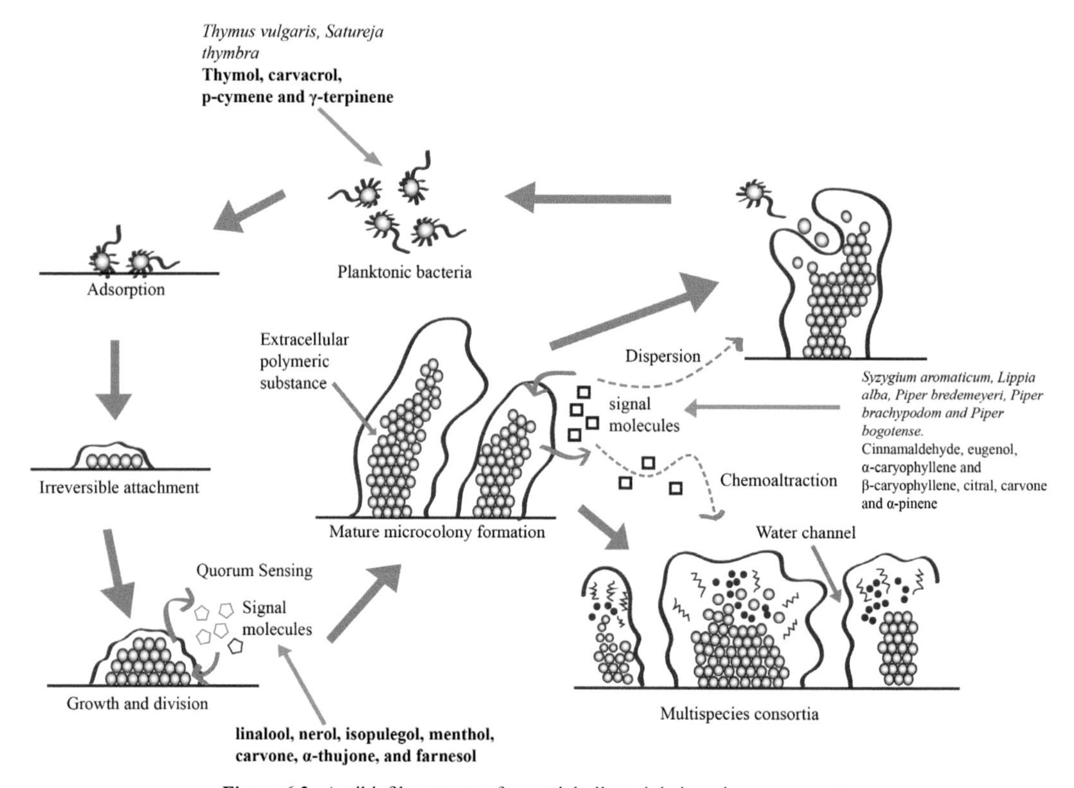

Figure 6.2. Antibiofilm targets of essential oils and their major components.

Antiquorum Sensing Activity

Bacterial communication is developed across production and detection of signaling molecules which belongs to the Quorum Sensing (QS) process that regulates the formation of virulence factors, antibiotic production and biofilm formation (Koh et al. 2013). A compound with anti-QS activity should be able to inhibit signal biosynthesis, block activity and production of Acyl Homoserine Lactones [AHL], inducing enzymatic signal degradation and inhibiting the reception of signal molecules (Khan et al., 2009). In that way, cinnamaldehyde and their derivatives have shown to interfere the QS process in *Vibrio* spp. by decreasing the DNA-binding ability of LuxR that codifies the ligand protein of AHL. In turn, *trans*-cinnamaldehyde has shown the ability to inhibit RhlI system which is necessary to synthetize N-butanoyl-L-homoserine lactone (C4-HSL) in *P. aeruginosa* (Brackman et al. 2008, Chang et al. 2014). Using biosensor strains, anti-QS activity of clove [*Syzygium aromaticum* (L.) Merr. & L.M.Perry (Myrtaceae)] oil and its major components such as eugenol α-caryophyllene and β-caryophyllene by the inhibition of violacein production in *Chromobacterium violaceum* CV12472 and CVO26 was evaluated (Khan et al. 2009). Carvacrol also reduced the production of violacein through inhibition of *cviI* gene that encodes N-acyl-L-homoserine lactone synthase in this microbial model (Fig. 6.2) (Burt 2004, Khan et al. 2009). In Colombia, various EOs have shown biological activity by the inhibition of AHL QS system in *E. coli*, particularly [*Lippia alba* (Mill.) N.E.Br. ex Britton & P. Wilson (Verbenaceae)]. Equally, moderate activity was detected in EOs components such as citral, carvone and α-pinene, citral being the most active (Jaramillo-Colorado et al. 2012). In a subsequent study with plants of the genus *Piper*, anti-QS effect was demonstrated by the inhibition of violacein production using *Chromobacterium violaceum* CV026. The most active oils were obtained from *Piper bredemeyeri* J. Jacq.; *Piper brachypodom* (Benth.) C. DC and *Piper bogotense* C. DC (Olivero-Verbel et al. 2011).

Toxin Production Inhibition

Several foodborne diseases are caused by bacterial toxins (cholera toxin, botulinum neurotoxin, Shiga toxins and *Staphylococcus* enterotoxin) that negatively affect the quality of foods, therefore the search for inhibitors is required to ensure the safety of the foods (Friedman and Rasooly 2013). EOs of bay [*Pimenta racemosa* (Mill.) J.W. Moore (Myrtaceae)], cinnamon and clove have shown a decreased production of enterotoxin A; equally the oils of clove and cinnamon also significantly decreased the production of enterotoxin B in *S. aureus* (Friedman and Rasooly 2013), but it is important to take into account that sub-inhibitory concentrations of EOs can induce the production of enterotoxins A and B and α-toxin by *S. aureus* (Fig. 6.1) (Smith-Palmer et al. 2004). Among lipids, glycerol monolaurate and lauric acid inhibit the production of toxic shock syndrome toxin-1 (TSST-1) in *Staphylococcus* spp. (Projan et al. 1994, Lin et al. 2009). Likewise, perilla [*Perilla frutescens* (L.) Britton (Lamiaceae)] oil and costus [*Saussurea costus* (Falc.) Lipsch (Asteraceae), synonym *Saussurea lappa* (Decne.) Sch.Bip.] oil reduced production of staphylococcal enterotoxins (α-toxin, enterotoxins A and B) and TSST-1 in MRSA (Ruzin and Novick 2000, Qiu et al. 2011a). Eugenol and thymol, also, inhibited staphylococcal enterotoxins, TSST-1 and α-hemolysin (toxin that cause hemolysis) (Qiu et al. 2010a, Qiu et al. 2010b, Qiu et al. 2011b). Finally, another important toxin, Listeriolysin O (LLO), a toxin produced by *Listeria monocytogenes* was inhibited by eugenol (Filgueiras and Vanetti 2006).

Disruption of Cell Membrane

It is broadly accepted that the membrane is the first target of antimicrobial activity of EOs, because of their lipophilicity as well as the fact that their components can penetrate through the double lipid layer of the membrane and can alter the permeability well as function of membrane proteins (Nazzaro et al. 2013). Thymol and carvacrol are membrane-permeabilizing compounds that disintegrate the outer membrane decreasing the intracellular ATP in *E. coli* by disruptive action on the cytoplasmic membrane (Fig. 6.1) (Quinn 2010). Equally, (+)-menthol, and linalyl acetate can perturb the lipid fraction of microbial plasma membrane, resulting in alterations of membrane permeability and leakage of intracellular materials (Helander et al. 1998). Other plant-derived terpenes such as eugenol, and citral produce changes of the fatty acid composition in a model of *Salmonella enterica* serovar *Typhimurium* causing both structural and functional damage depending on partition coefficient within the bacterial membrane lipids. For that reason, the higher amount of EO or major compound present in the membrane represents a greater physical damage on the same (Quinn 2010). For the development of biocides, it is important taking into account that the microbial cells can develop tolerance to the bactericidal activity of terpenes and produce cross-adaptation with other disinfectants (Trombetta et al. 2005).

Induction of Reactive Oxygen Species (ROS)

An interesting target for adjuvants is induction of Reactive Oxygen Species (ROS) that has been shown as an action mechanism of microbicidal antibiotics (Dubois-Brissonnet et al. 2011). EOs have the ability of induce ROS in eukaryotic cells which is caused by mitochondrial dysfunctions due to permeabilization of membranes that affect cell viability by apoptosis and necrosis (Farha and Brown 2013). The EO of dill [*Anethum graveolens* L. (Apiaceae)] showed both in a *Candida* and *Aspergillus flavus* model mitochondrial dysfunction resulting in an increase of ROS. Likewise decreased cell viability in this model was avoided by the addition of L-cysteine as an antioxidant, which indicates that induction of ROS is a target of EOs useful for their application in fungal infections (Tian et al. 2012, Chen et al. 2013, Pensel et al. 2014). Equally, the cytotoxicity and genotoxicity of oils of palmarosa, citronella, lemongrass and vetiver [*Cymbopogon martini* (Roxb.) W. Watson (Poaceae), *Cymbopogon nardus* (L.) Rendle (Poaceae), *Cymbopogon citratus* (DC.) Stapf (Poaceae), *Chrysopogon zizanioides* (L.) Roberty (Poaceae)] caused by oxidative stress mediated by ROS have been demonstrated (Sinha et al. 2014).

Antiplasmid Activity

Membrane transporters as efflux pumps can be encoded by genes localized on the chromosomes and on plasmids such as tetracycline transporters that mediate extrusion of antibiotics to extracellular space, for that reason plasmid replication can be an antimicrobial target for decreased spread of resistance (Spengler et al. 2006). Peppermint [*Mentha* × *piperita* L. (Lamiaceae)] oil and its major component menthol exhibit effect on plasmid curing activity in *in vitro* experiments on the metabolic plasmid of *E. coli* F'lac K12 LE140 (Fig. 6.1) (Schelz et al. 2006). Peppermint oil and menthol produced a plasmid elimination of 37.5 and 96%, resp., consolidating menthol as a compound capable of eliminating the plasmid-mediated resistance in bacteria (Schelz et al. 2006).

Antifungal Target Activity

The antifungal activity of EOs can be attributed to the presence and synergism of some components such as carvacrol, α-terpinyl acetate, p-cymene, thymol, pinene, linalool which have antimicrobial activity (Nuzhat and Vidyasagar 2014). *Thymus* oils and their components have presented fungicidal effect by disruption of cell membrane and suppression of germ tube formation (Fig. 6.3) (Pina-Vaz et al. 2004). Eugenol has blocked germ-tube formation in a *Candida albicans* model (Pinto et al. 2009). *Croton cajucara* Benth (Euphorbiaceae), a linalool-rich EO induces cellular damage and alterations of shape and size in yeast cells (Alviano et al. 2005). On the other hand, EOs of *C. verum* J. Presl (Lauraceae), *Cinnamomum martini*, as well as eugenol, cinnamaldehyde and geraniol have shown the ability of reduce elastase and keratinase activities in *A. fumigatus* and *Trychophyton rubrum* (Fig. 6.3) (Khan et al. 2011). An interesting antifungal target of EOs is the inhibition of mycotoxin production that has been associated with several cases of human poisoning and death. *Rosmarinus officinalis* L. (Lamiaceae) and *Trachyspermum copticum* (L.) Link (Apiaceae) oils have shown mycotoxin biosynthesis (aflatoxin) inhibition in *Aspergillus flavus* the major components involved being thymol, p-cymene, γ-terpinene, piperitone, α-pinene and limonene (Rasooli et al. 2008). Equally, thymol and limonene were found to be the most active terpenes to induce complete inhibition of Fumonisin

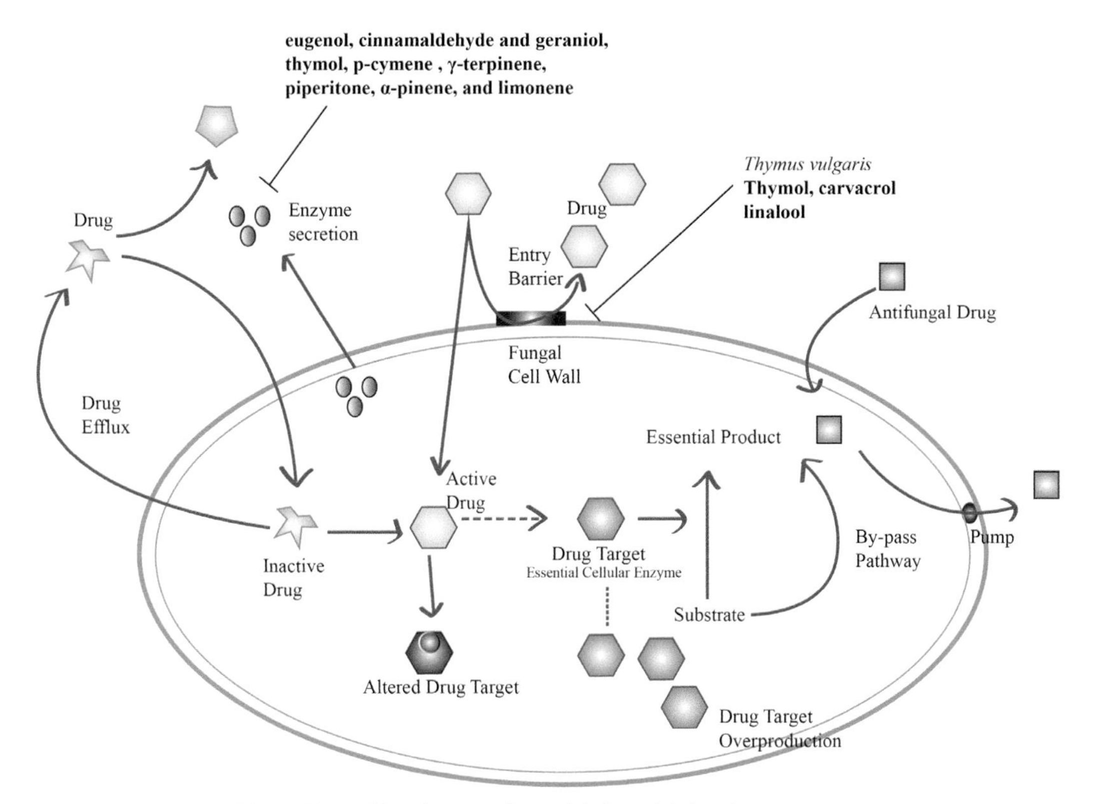

Figure 6.3. Antifungal targets of essential oils and their major components.

B1 biosynthesis a mycotoxin from *Fusarium verticillioides* in a model of maize inoculated with toxicogenic fungi during incubation. This antitoxigenic activity may be attributed to their antioxidant properties (Dambolena et al. 2008). Finally, the monoterpene carvacrol in the essential oil of *Satureja hortensis* L. (Lamiaceae) showed higher inhibitory effect toward both aflatoxins specially AFG 1 in *Aspergillus parasiticus* (Razzaghi-Abyaneh 2008).

Antiparasite Target Activity

An important target in parasites of EOs described principally in *Plasmosdium falciparum* is isoprenoid biosynthesis, which is required for the production of isopentenyl pyrophosphate and dimethylallyl pyrophosphate that participate in biosynthesis of molecules used in protein prenylation (Qidwai et al. 2014). Terpenes such as farnesol, nerolidol, limonene and linalool have antimalarial activities on intraerythrocytic stages of *P. falciparum in vitro*, inhibiting dolichol biosynthesis and biosynthesis of the isoprenic side chain in the trophozoite and plasmodium schizont stages (Fig. 6.4) (Moura et al. 2001, de Macedo et al. 2002, Goulart et al. 2004). The mechanism of action of the components of garlic [*Allium sativum* L. (Amaryllidaceae)] oil (allicin and ajoene) has been investigated in *Entamoeba histolytica* and *Trypanosoma cruzi*. Allicin and its condensation product ajoene have antimicrobial activity through interaction with thiol-containing enzymes, allicin inhibit cysteine proteinases, alcohol dehydrogenases and thioredoxin reductases in *Entamoeba histolytica*. Ajoene blocks proliferation by inhibiting of phosphatidylcholine biosynthesis in *Trypanosoma cruzi*, equally, ajoene can interfere with interaction of protein and lipid trafficking between the parasite and host cell membranes, causing damage in the parasite (Urbina et al. 1993, Ankri et al. 1997, Rabinkov et al. 1998, Miron et al. 2000). Finally, linalool-rich essential oil from the leaves of *Croton cajucara*, has anti-leishmanial activity by the destruction of leishmanial nuclear and kinetoplast chromatin with mitochondrial swelling, followed by cell lysis (Mendonça-Filho et al. 2003).

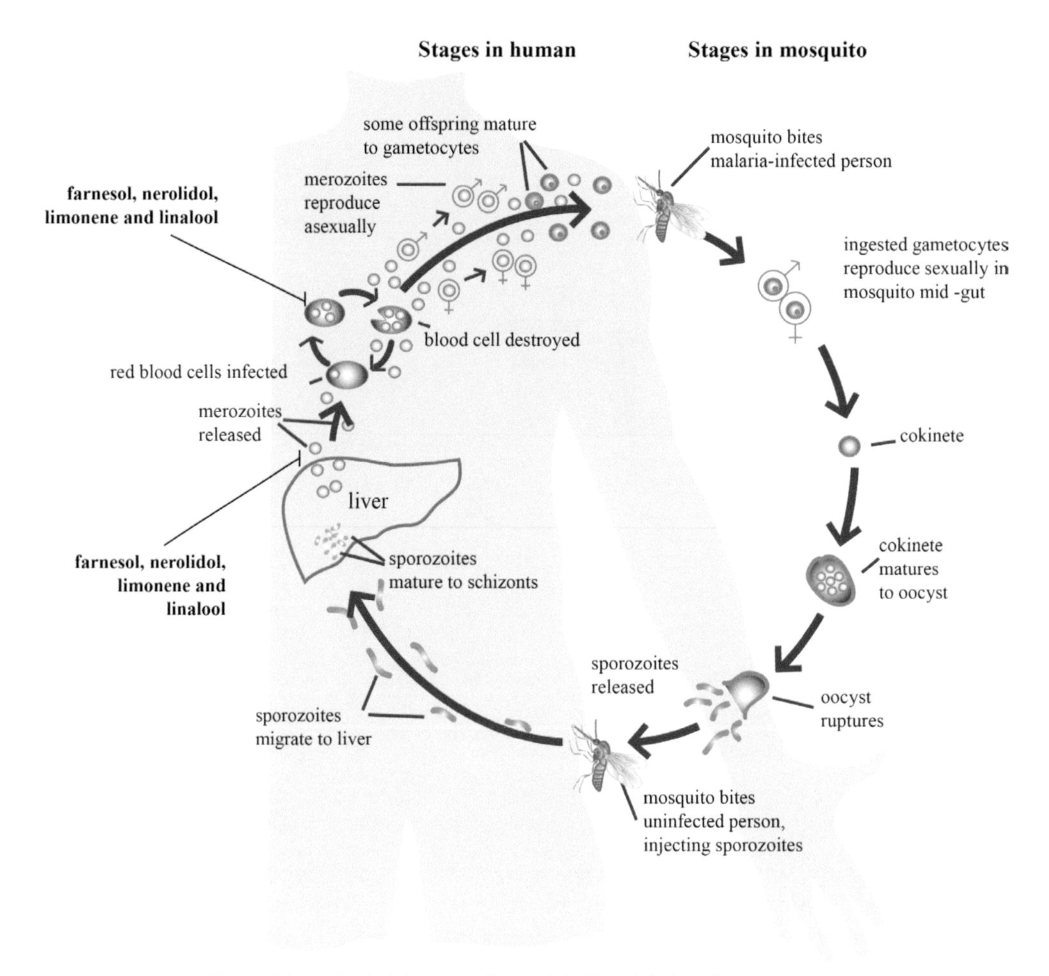

Figure 6.4. Antimalarial targets of essential oils and their major components.

Antiviral Target Activity

Volatile compounds from EOs exhibit a high level of antiviral activity. EOs affect the viral envelope that is necessary for adsorption or entry into host cells. In particular, monoterpenes have shown increased cell membrane fluidity and permeability, altering the order of membrane proteins (Schnitzler et al. 2010). EOs from ginger [*Zingiber officinale* Roscoe (Zingiberaceae)], thyme, hyssop [*Hyssopus officinalis* L. (Lamiaceae)], and sandalwood [*Santalum album* L. (Santalaceae)] have antiviral activity against HSV-1 acyclovir susceptible and acyclovir resistant. The main components involved were zingiberene, β-bisabolene, sesquiphellandrene and curcumen for ginger oil; thymol and carvacrol for thyme oil; 1-pinocamphone, isopinocamphone, pinocarvone, and α-pinene for hyssop oil; and santalol, bergamotol and santalene for sandalwood oil. The mentioned EOs were shown to interfere with virion envelope structures (Fig. 6.5) (Schnitzler et al. 2007). Equally, treatment of HSV-1 with oregano [*Origanum vulgare* L. (Lamiaceae)] oil disrupted the viral envelope necessary for entry (Schnitzler et al. 2007, Reichling et al. 2009). EO from *Melissa officinalis* L. (Lamiaceae) can inhibit the replication of HSV-2, due to the presence of citral and citronellal (Edris 2007). On the other hand, *Santolina insularis* (Gennari ex Fiori) Arrigoni (Compositae) essential oil inhibits cell-to-cell transmission of both HSV-1 and HSV-2 (De Logu et al. 2000). In conclusion, the mechanism of action of EOs against herpes virus is to destroy the virus envelope and to inhibit the cell-to-cell virus diffusion (Saddi et al. 2007). On the other hand, EOs of *Ridolfia segetum* (L.) Moris (Apiaceae)

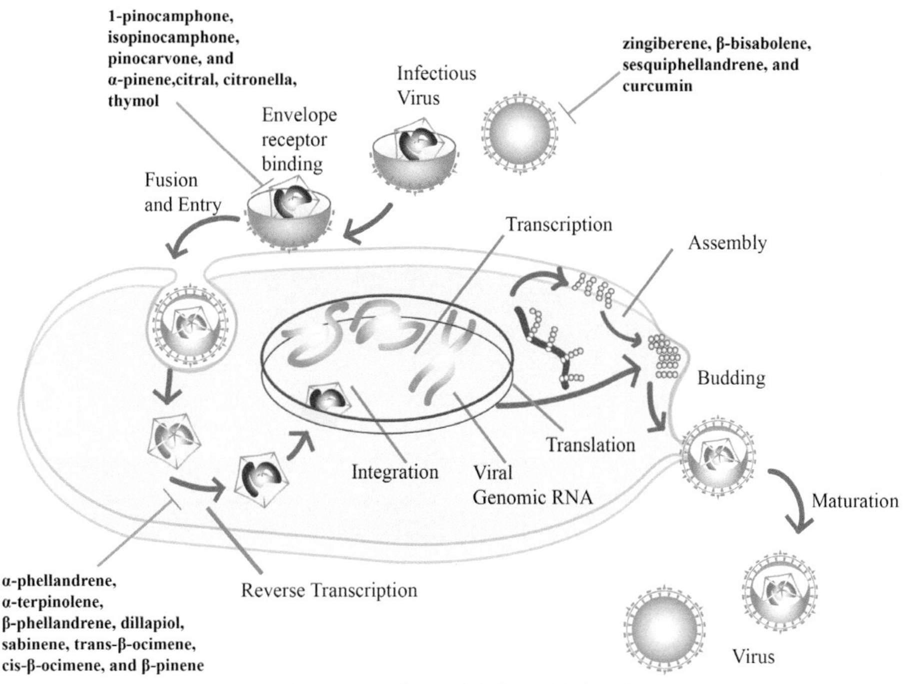

Figure 6.5. Antiviral targets of essential oils and their major components.

and *Oenanthe crocata* L. (Apiaceae) presented inhibition of the HIV-1 Reverse Transcriptase (RT) in a dose-dependent manner, the major components involved being α-phellandrene, α-terpinolene, β-phellandrene and dillapiol for *Ridolfia segetum* (L.) Moris and sabinene, trans-β-ocimene, cis-β-ocimene and β-pinene for *Oenanthe crocata* L., becoming an important source of anti-HIV compounds with potential clinical application (Bicchi et al. 2009, Krishnaveni 2012).

Conclusions and Perspectives

Synergy research of the mixture of bioactive constituents can improve the application of phytopharmaceuticals in antimicrobial drug discovery programs and should contain the following issues:

1. Studies of synergistic multi-target effects in combination
2. Design of adjuvant agents target in resistance mechanisms of bacteria for the improvement of antimicrobial chemotherapy
3. Interactions between biomedical scientists and physicians to bring EOs discoveries to clinical practice and agriculture use (Wagner and Ulrich-Merzenich 2009).

In synergism, research is important taking into account that the interactions between oils components can be determined by actions of weak and highly active molecules, and present synergistic interaction unlike combinations of highly active compounds. The presence of synergism between the strong and weaker antimicrobial constituents in EOs enhances the antimicrobial efficacy, and so, studies to determine this activity should be more exhaustive (Ahmad et al. 2014). It has been reported that antibiotics do not only kill or inhibit microbial growth, but also act as signaling molecules, affecting the behavior of bacterial communities. In this way, the recent evidence suggests that the same might be true for EOs and their constituents, for that reason, it is necessary to determine the behavior of microorganism under subinhibitory concentrations of EOs (Landau and Shapira 2012). There are various reports about the effect of sub-lethal concentrations of EOs that can lead to the development of antibiotic resistance in human pathogens (McMahon et al. 2007) causing cross-resistance with biocides and antimicrobials (Thomsen et al. 2013). Although other data suggest that EOs such as tea tree do not induce resistance to antimicrobial agents (Becerril et al. 2012), microorganisms such as *Serratia marcescens*, *Morganella morganii* and *Proteus mirabilis* have developed resistance to treatment with oregano essential oil and changed their antibiotic resistance profile (Moken et al. 1997). Equally, mutants of *E. coli* were developed in contact with disinfectant pine oil or a household product containing pine oil and showed resistance to multiple antibiotics (tetracycline, ampicillin, chloramphenicol and nalidixic acid) and overexpression the marA gene (Hammer et al. 2012).

These findings make the development of experimental models evaluating the pharmacological applicability of EO components as antimicrobial agents mandatory, evaluating antimicrobial parameters as time-kill curve, carryover, post-antimicrobial effects and mutant prevention concentrations (MPC) (Budzyńska et al. 2013, Burt et al. 2014).

It is also important to study vapor biocide action of EOs that can be used as air disinfectants in hospital rooms and surgery. Because in this phase they have the ability to adhere around microbial cells, enabling the bacterial cell wall to traverse causing increased permeability and leakage of ions and other essential molecules in bacteria (Inouye et al. 2001, Inouye et al. 2006, Al Yousef 2014).

Finally, it is important to consider that EOs are sensitive materials which can easily suffer degradation under the action of oxygen, light and moderate temperatures. Furthermore, they are insoluble in water, and for certain applications a controlled release is required. Therefore, an adequate formulation of the EO which takes into account these aspects is required for commercial applications. With the aim to protect an EO from degradation or from losses by evaporation, possible formulations include liquid forms (emulsions, micelles and liquid solutions), semi-liquid forms (gels and liposomes), and solid forms (microcapsules or microcomposites) (Cocero et al. 2009, Martín et al. 2010).

Acknowledgements

The authors acknowledges to TUBITAK program 2221, Fellowship program for Visiting Scientists and Scientists and Sabbatical Leave, and to Claudia Marcela Montes for the design of the figures contained in this paper.

References

Abreu, A.C., McBain, A.J. and Simoes, M. 2012. Plants as sources of new antimicrobials and resistance-modifying agents. Nat. Prod. Reports 29: 1007–1021.

Ahmad, A., van Vuuren, S. and Viljoen, A. 2014. Unravelling the complex antimicrobial interactions of essential oils-the case of *Thymus vulgaris* (Thyme). Molecules 19: 2896–2910.

Al Yousef, S.A. 2014. Essential oils: their antimicrobial activity and potential application against pathogens by gaseous contact–a review. Egyptian Acad. J. Biol. Sci. 6: 37–54.

Alviano, W.S., Mendonça-Filho, R.R., Alviano, D.S., Bizzo, H.R., Souto-Padrón, T., Rodrigues, M.L., Bolognese, A.M., Alviano, C.S. and Souza, M.M.G. 2005. Antimicrobial activity of *Croton cajucara* Benth linalool-rich essential oil on artificial biofilms and planktonic microorganisms. Oral Microbiol. Immunol. 20: 101–105.

Amalaradjou, M.A.R., Narayanan, A. and Venkitanarayanan, K. 2011. Trans-cinnamaldehyde decreases attachment and invasion of uropathogenic *Escherichia coli* in urinary tract epithelial cells by modulating virulence gene expression. J. Urol. 185: 1526–1531.

Ankri, S., Miron, T., Rabinkov, A., Wilchek, M. and Mirelman, D. 1997. Allicin from garlic strongly inhibits cysteine proteinases and cytopathic effects of *Entamoeba histolytica*. Antimicrob. Agents Chemother. 41: 2286–2288.

Barreto, H.M., de Lima, I.S., Coelho, K.M.R.N., Osório, L.R., de Almeida Mourão, R., dos Santos, B.H.C., Coutinho, H.D.M., de Abreu, A.P.L., de Medeiros, M.D.G.F., Citó, A.M.D.G.L. and Lopes, J.A.D. 2014. Effect of *Lippia origanoides* HBK essential oil in the resistance to aminoglycosides in methicillin resistant *Staphylococcus aureus*. Eur. J. Integrat. Med. 6: 560–564.

Barreto, H.M., Silva Filho, E.C., Lima, E.D.O., Coutinho, H.D., Morais-Braga, M.F., Tavares, C.C., Tintino, S.R., Rego, J.V., de Abreu, A.P., Lustosa, M.D.C.G. and Oliveira, R.W.G. 2014. Chemical composition and possible use as adjuvant of the antibiotic therapy of the essential oil of *Rosmarinus officinalis* L. Ind. Crops Prod. 59: 290–294.

Becerril, R., Nerín, C. and Gómez-Lus, R. 2012. Evaluation of bacterial resistance to essential oils and antibiotics after exposure to oregano and cinnamon essential oils. Foodborne Pathog. Dis. 9: 699–705.

Bernal, P., Molina-Santiago, C., Daddaoua, A. and Llamas, M.A. 2013. Antibiotic adjuvants: identification and clinical use. Microb. Biotechnol. 6: 445–449.

Bicchi, C., Rubiolo, P., Ballero, M., Sanna, C., Matteodo, M., Esposito, F., Zinzula, L. and Tramontano, E. 2009. HIV-1-inhibiting activity of the essential oil of *Ridolfia segetum* and *Oenanthe crocata*. Planta Med. 75: 1331–1335.

Boire, N., Riedel, S. and Parrish, N.M. 2013. Essential Oils and Future Antibiotics: New weapons against emerging 'Superbugs'. J. Ancient Dis. Preventive Remedies 1: 2.

Brackman, G., Defoirdt, T., Miyamoto, C., Bossier, P., Van Calenbergh, S., Nelis, H. and Coenye, T. 2008. Cinnamaldehyde and cinnamaldehyde derivatives reduce virulence in *Vibrio* spp. by decreasing the DNA-binding activity of the quorum sensing response regulator LuxR. BMC Microbiol. 8: 149.

Budzyńska, A., Sadowska, B. and Kalemba, D. 2013. Activity of selected essential oils against *Candida* spp. strains. Evaluation of new aspects of their specific pharmacological properties, with special reference to Lemon Balm. Adv. Microbiol. 3: 317.

Bueno, J. 2014. Anti-biofilm drug susceptibility testing methods: looking for new strategies against resistance mechanism. J. Microb. Biochem. Technol. S3: 1–9.

Bueno, J. 2016. Antimicrobial adjuvants drug discovery, the challenge of avoid the resistance and recover the susceptibility of multidrug-resistant strains. J. Microb. Biochem. Technol. 8: 169–176.

Burt, S. 2004. Essential oils: their antibacterial properties and potential applications in foods—a review. Int. J. Food Microbiol. 94: 223–253.

Burt, S.A., Ojo-Fakunle, V.T., Woertman, J. and Veldhuizen, E.J. 2014. The natural antimicrobial carvacrol inhibits quorum sensing in *Chromobacterium violaceum* and reduces bacterial biofilm formation at sub-lethal concentrations. PLoS One 9: e93414.

Cantas, L., Shah, S.Q., Cavaco, L.M., Manaia, C.M., Walsh, F., Popowska, M., Garelick, H., Bürgmann, H. and Sørum, H. 2013. A brief multi-disciplinary review on antimicrobial resistance in medicine and its linkage to the global environmental microbiota. Front. Microbiol. 4: 96.

Chang, C.Y., Krishnan, T., Wang, H., Chen, Y., Yin, W.F., Chong, Y.M., Tan, L.Y., Chong, T.M. and Chan, K.G. 2014. Non-antibiotic quorum sensing inhibitors acting against N-acyl homoserine lactone synthase as druggable target. Sci. Reports 4: 7245.

Chen, Y., Zeng, H., Tian, J., Ban, X., Ma, B. and Wang, Y. 2013. Antifungal mechanism of essential oil from *Anethum graveolens* seeds against *Candida albicans*. J. Medical Microbiol. 62:1175–1183.

Chorianopoulos, N.G., Giaouris, E.D., Skandamis, P.N., Haroutounian, S.A. and Nychas, G.J. 2008. Disinfectant test against monoculture and mixed-culture biofilms composed of technological, spoilage and pathogenic bacteria: bactericidal effect of essential oil and hydrosol of *Satureja thymbra* and comparison with standard acid–base sanitizers. J. Appl. Microbiol. 104: 1586–1596.

Chusri, S., Villanueva, I., Voravuthikunchai, S.P. and Davies, J. 2009. Enhancing antibiotic activity: a strategy to control *Acinetobacter* infections. J. Antimicrob. Chemother. 64: 1203–1211.

Cocero, M.J., Martín, Á., Mattea, F. and Varona, S. 2009. Encapsulation and co-precipitation processes with supercritical fluids: fundamentals and applications. J. Supercritical Fluids 47: 546–555.

Cole, M.R., Hobden, J.A. and Warner, I.M. 2015. Recycling antibiotics into GUMBOS: A new combination strategy to combat multi-drug-resistant bacteria. Molecules 20: 6466–6487.

Dalleau, S., Cateau, E., Bergès, T., Berjeaud, J.M. and Imbert, C. 2008. *In vitro* activity of terpenes against *Candida* biofilms. Int. J. Antimicrob. Agents 31: 572–576.

Dambolena, J.S., López, A.G., Cánepa, M.C., Theumer, M.G., Zygadlo, J.A. and Rubinstein, H.R. 2008. Inhibitory effect of cyclic terpenes (limonene, menthol, menthone and thymol) on *Fusarium verticillioides* MRC 826 growth and fumonisin B1 biosynthesis. Toxicon. 51: 37–44.

De Logu, A., Loy, G., Pellerano, M.L., Bonsignore, L. and Schivo, M.L. 2000. Inactivation of HSV-1 and HSV-2 and prevention of cell-to-cell virus spread by *Santolina insularis* essential oil. Antiviral Res. 48: 177–185.

de Macedo, C.S., Uhrig, M.L., Kimura, E.A. and Katzin, A.M. 2002. Characterization of the isoprenoid chain of coenzyme Q in *Plasmodium falciparum*. FEMS Microbiol. Lett. 207: 13–20.

de Oliveira, A.D.L., Rodrigue, F.F.G., Coutinho, H.D.M., da Costa, J.G.M. and de Menezes, I.R.A. 2014. Chemical composition, modulatory bacterial resistance and antimicrobial activity of essential oil the *Hyptis martiusii* benth by direct and gaseous contact. Jundishapur J. Nat. Pharma. Prod. 9: e13521.

Dubois-Brissonnet, F., Naïtali, M., Mafu, A.A. and Briandet, R. 2011. Induction of fatty acid composition modifications and tolerance to biocides in *Salmonella enterica* serovar Typhimurium by plant-derived terpenes. Appl. Environ. Microbiol. 77: 906–910.

Edris, A.E. 2007. Pharmaceutical and therapeutic potentials of essential oils and their individual volatile constituents: a review. Phytother. Res. 21: 308–323.

Efferth, T. and Koch, E. 2011. Complex interactions between phytochemicals. The multi-target therapeutic concept of phytotherapy. Curr. Drug Targets 12: 122–132.

Ejim, L., Farha, M.A., Falconer, S.B., Wildenhain, J., Coombes, B.K., Tyers, M., Brown, E.D. and Wright, G.D. 2011. Combinations of antibiotics and nonantibiotic drugs enhance antimicrobial efficacy. Nat. Chem. Boil. 7: 348–350.

Farha, M.A. and Brown, E.D. 2013. Discovery of antibiotic adjuvants. Nat. Biotechnol. 31: 120–122.

Filgueiras, C.T. and Vanetti, M.C.D. 2006. Effect of eugenol on growth and listeriolysin O production by *Listeria monocytogenes*. Braz. Arch. Biol. Technol. 49: 405–409.

Friedman, M. and Rasooly, R. 2013. Review of the inhibition of biological activities of food-related selected toxins by natural compounds. Toxins 5: 743–775.

Gertsch, J. 2011. Botanical drugs, synergy, and network pharmacology: forth and back to intelligent mixtures. Planta Med. 77: 1086–1098.

Goulart, H.R., Kimura, E.A., Peres, V.J., Couto, A.S., Duarte, F.A.A. and Katzin, A.M. 2004. Terpenes arrest parasite development and inhibit biosynthesis of isoprenoids in *Plasmodium falciparum*. Antimicrob. Agents Chemother. 48: 2502–2509.

Hammer, K.A., Carson, C.F. and Riley, T.V. 2012. Effects of *Melaleuca alternifolia* (tea tree) essential oil and the major monoterpene component terpinen-4-ol on the development of single-and multistep antibiotic resistance and antimicrobial susceptibility. Antimicrob. Agents Chemother. 56: 909–915.

Helander, I.M., Alakomi, H.L., Latva-Kala, K., Mattila-Sandholm, T., Pol, I., Smid, E.J., Gorris, L.G. and von Wright, A. 1998. Characterization of the action of selected essential oil components on Gram-negative bacteria. J. Agri. Food Chem. 46: 3590–3595.

Hemaiswarya, S., Kruthiventi, A.K. and Doble, M. 2008. Synergism between natural products and antibiotics against infectious diseases. Phytomedicine 15: 639–652.

Infectious Diseases Society of America. 2011. Combating antimicrobial resistance: policy recommendations to save lives. Clin. Infect. Dis. 52: S397–S428.

Inouye, S., Takizawa, T. and Yamaguchi, H. 2001. Antibacterial activity of essential oils and their major constituents against respiratory tract pathogens by gaseous contact. J. Antimicrob. Chemother. 47: 565–573.

Inouye, S., Uchida, K. and Abe, S. 2006. Vapor activity of 72 essential oils against a *Trichophyton mentagrophytes*. J. Infect. Chemother. 12: 210–216.

Jaramillo-Colorado, B., Olivero-Verbel, J., Stashenko, E.E., Wagner-Döbler, I. and Kunze, B. 2012. Anti-quorum sensing activity of essential oils from Colombian plants. Nat. Prod. Res. 26: 1075–1086.

Kalemba, D. and Kunicka, A. 2003. Antibacterial and antifungal properties of essential oils. Curr. Med. Chem. 10: 813–829.

Khan, M.S.A. and Ahmad, I. 2011. *In vitro* antifungal, anti-elastase and anti-keratinase activity of essential oils of *Cinnamomum-*, *Syzygium-* and *Cymbopogon*-species against *Aspergillus fumigatus* and *Trichophyton rubrum*. Phytomedicine 19: 48–55.

Khan, M.S.A., Zahin, M., Hasan, S., Husain, F.M. and Ahmad, I. 2009. Inhibition of quorum sensing regulated bacterial functions by plant essential oils with special reference to clove oil. Lett. Appl. Microbiol. 49: 354–360.

Knowles, J.R., Roller, S., Murray, D.B. and Naidu, A.S. 2005. Antimicrobial action of carvacrol at different stages of dual-species biofilm development by *Staphylococcus aureus* and *Salmonella enterica* serovar Typhimurium. Appl. Environ. Microbiol. 71: 797–803.

Koh, C.L., Sam, C.K., Yin, W.F., Tan, L.Y., Krishnan, T., Chong, Y.M. and Chan, K.G. 2013. Plant-derived natural products as sources of anti-quorum sensing compounds. Sensors 13: 6217–6228.

Korcsmáros, T., Szalay, M.S., Böde, C., Kovács, I.A. and Csermely, P. 2007. How to design multi-target drugs: target search options in cellular networks. Expert Opin. Drug Discovery 2: 99–808.

Krishnaveni, M. 2012. Medicinal Plants–A Boon for HIV/AIDS. J. Pharma. Res. 5: 5367–5379.

Landau, E. and Shapira, R. 2012. Effects of subinhibitory concentrations of menthol on adaptation, morphological, and gene expression changes in enterohemorrhagic *Escherichia coli*. Appl. Environ. Microbiol. 78: 5361–5367.

Langeveld, W.T., Veldhuizen, E.J. and Burt, S.A. 2014. Synergy between essential oil components and antibiotics: a review. Crit. Rev. Microbiol. 40: 76–94.

Lin, J.T., Juliano, J.J. and Wongsrichanalai, C. 2010. Drug-resistant malaria: the era of ACT. Curr. Infect. Dis. Reports 12: 165–173.

Lin, Y.C., Schlievert, P.M., Anderson, M.J., Fair, C.L., Schaefers, M.M., Muthyala, R. and Peterson, M.L. 2009. Glycerol monolaurate and dodecylglycerol effects on *Staphylococcus aureus* and toxic shock syndrome toxin-1 *in vitro* and *in vivo*. PloS one 4: e7499.

Lorenzi, V., Muselli, A., Bernardini, A.F., Berti, L., Pagès, J.M., Amaral, L. and Bolla, J.M. 2009. Geraniol restores antibiotic activities against multidrug-resistant isolates from gram-negative species. Antimicrob. Agents Chemother. 53: 2209–2211.

Lu, J.J., Pan, W., Hu, Y.J. and Wang, Y.T. 2012. Multi-target drugs: the trend of drug research and development. PloS one 7: e40262.

Magiorakos, A.P., Srinivasan, A., Carey, R.B., Carmeli, Y., Falagas, M.E., Giske, C.G., Harbarth, S., Hindler, J.F., Kahlmeter, G., Olsson-Liljequist, B. and Paterson, D.L. 2012. Multidrug-resistant, extensively drugresistant and pandrugresistant bacteria: aninternational expert proposal for interim standard definitions for acquired resistance. Clin. Microbiol. Infect. 18: 268–281.

Mankoff, S.P., Brander, C., Ferrone, S. and Marincola, F.M. 2004. Lost in translation: obstacles to translational medicine. J. Transl. Med. 2: 14.

Martín, Á., Varona, S., Navarrete, A. and Cocero, M.J. 2010. Encapsulation and co-precipitation processes with supercritical fluids: applications with essential oils. Open Chem. Eng. J. 4: 31–41.

McMahon, M.A.S., Blair, I.S., Moore, J.E. and McDowell, D.A. 2007. Habituation to sub-lethal concentrations of tea tree oil (*Melaleuca alternifolia*) is associated with reduced susceptibility to antibiotics in human pathogens. J. Antimicrob. Chemother. 59: 125–127.

Mendonça-Filho, R.R., Bizzo, H.R., de Almeida Rodrigues, I., Soares, R.M.A., Souto-Padrón, T., Alviano, C.S. and Lopes, A.H.C. 2003. Antileishmanial activity of a linalool-rich essential oil from *Croton cajucara*. Antimicrob. Agents Chemother. 47: 1895–1901.

Miron, T., Rabinkov, A., Mirelman, D., Wilchek, M. and Weiner, L. 2000. The mode of action of allicin: its ready permeability through phospholipid membranes may contribute to its biological activity. Biochim. Biophys. Acta (BBA)-Biomembranes 1463: 20–30.

Moken, M.C., McMurry, L.M. and Levy, S.B. 1997. Selection of multiple-antibiotic-resistant (mar) mutants of *Escherichia coli* by using the disinfectant pine oil: roles of the mar and acrAB loci. Antimicrob. Agents Chemother. 41: 2770–2772.

Moura, I.C., Wunderlich, G., Uhrig, M.L., Couto, A.S., Peres, V.J., Katzin, A.M. and Kimura, E.A. 2001. Limonene arrests parasite development and inhibits isoprenylation of proteins in *Plasmodium falciparum*. Antimicrob. Agents Chemother. 45: 2553–2558.

Nazzaro, F., Fratianni, F., De Martino, L., Coppola, R. and De Feo, V. 2013. Effect of essential oils on pathogenic bacteria. Pharmaceuticals 6: 1451–1474.

Nikaido, H. and Pagès, J.M. 2012. Broad-specificity efflux pumps and their role in multidrug resistance of Gram-negative bacteria. FEMS Microbiol. Rev. 36: 340–363.

Nostro, A., Marino, A., Blanco, A.R., Cellini, L., Di Giulio, M., Pizzimenti, F., Roccaro, A.S. and Bisignano, G. 2009. *In vitro* activity of carvacrol against staphylococcal preformed biofilm by liquid and vapour contact. J. Med. Microbiol. 58: 791–797.

Nostro, A., Roccaro, A.S., Bisignano, G., Marino, A., Cannatelli, M.A., Pizzimenti, F.C., Cioni, P.L., Procopio, F. and Blanco, A.R. 2007. Effects of oregano, carvacrol and thymol on *Staphylococcus aureus* and *Staphylococcus epidermidis* biofilms. J. Med. Microbiol. 56: 519–523.

Nuzhat, T. and Vidyasagar, G.M. 2014. Antifungal investigations on plant essential oils. A review. Int. J. Pharma. Pharma. Sci. 5: 19–28.

Olivero-Verbel, J., Pájaro, N. and Stashenko, E.E. 2011. Antiquorum sensing activity of essential oils isolated from different species of the genus *Piper.* Vitae 18: 77–82.

Papadopoulos, C.J., Carson, C.F., Chang, B.J. and Riley, T.V. 2008. Role of the MexAB-OprM efflux pump of *Pseudomonas aeruginosa* in tolerance to tea tree (*Melaleuca alternifolia*) oil and its monoterpene components terpinen-4-ol, 1, 8-cineole, and α-terpineol. Appl. Environ. Microbiol. 74: 1932–1935.

Pensel, P.E., Maggiore, M.A., Gende, L.B., Eguaras, M.J., Denegri, M.G. and Elissondo, M.C. 2014. Efficacy of essential oils of *Thymus vulgaris* and *Origanumvulgare* on *Echinococcusgranulosus*. Interdiscip. Persp. Infect. Dis. 2014: ID 693289.

Petersen, I., Eastman, R. and Lanzer, M. 2011. Drug-resistant malaria: Molecular mechanisms and implications for public health. FEBS Lett. 585: 1551–1562.

Pina-Vaz, C., Gonçalves Rodrigues, A., Pinto, E., Costa-de-Oliveira, S., Tavares, C., Salgueiro, L., Cavaleiro, C., Goncalves, M.J. and Martinez-de-Oliveira, J. 2004. Antifungal activity of *Thymus* oils and their major compounds. J. Eur. Acad. Dermatol. Venereol. 18: 73–78.

Pinto, E., Vale-Silva, L., Cavaleiro, C. and Salgueiro, L. 2009. Antifungal activity of the clove essential oil from *Syzygium aromaticum* on *Candida*, *Aspergillus* and dermatophyte species. J. Med. Microbiol. 58: 1454–1462.

Projan, S.J., Brown-Skrobot, S., Schlievert, P.M., Vandenesch, F. and Novick, R.P. 1994. Glycerol monolaurate inhibits the production of beta-lactamase, toxic shock toxin-1, and other staphylococcal exoproteins by interfering with signal transduction. J. Bacteriol. 176: 4204–4209.

Qidwai, T., Jamal, F., Khan, M.Y. and Sharma, B. 2014. Exploring drug targets in isoprenoid biosynthetic pathway for *Plasmodium falciparum*. Biochem. Res. Int. 2014: 657189.

Qiu, J., Feng, H., Lu, J., Xiang, H., Wang, D., Dong, J., Wang, J., Wang, X., Liu, J. and Deng, X. 2010a. Eugenol reduces the expression of virulence-related exoproteins in *Staphylococcus aureus*. Appl. Environ. Microbiol. 76: 5846–5851.

Qiu, J., Wang, D., Xiang, H., Feng, H., Jiang, Y., Xia, L., Dong, J., Lu, J., Yu, L. and Deng, X. 2010b. Subinhibitory concentrations of thymol reduce enterotoxins A and B and α-hemolysin production in *Staphylococcus aureus* isolates. PLoS One 5: e9736.

Qiu, J., Wang, J., Luo, H., Du, X., Li, H., Luo, M., Dong, J., Chen, Z. and Deng, X. 2011a. The effects of subinhibitory concentrations of costus oil on virulence factor production in *Staphylococcus aureus*. J. Appl. Microbiol. 110: 333–340.

Qiu, J., Zhang, X., Luo, M., Li, H., Dong, J., Wang, J., Leng, B., Wang, X., Feng, H., Ren, W. and Deng, X. 2011b. Subinhibitory concentrations of perilla oil affect the expression of secreted virulence factor genes in *Staphylococcus aureus*. PLoS One 6: e16160.

Quinn, P.J. 2010. Membranes as targets of antimicrobial lipids. pp. 1–24. *In*: Thormar, H. (ed.). Lipids and Essential Oils. John Wiley & Sons, Chichester, UK.

Rabinkov, A., Miron, T., Konstantinovski, L., Wilchek, M., Mirelman, D. and Weiner, L. 1998. The mode of action of allicin: trapping of radicals and interaction with thiol containing proteins. Biochim. Biophys. Acta (BBA)-General Subjects 1379: 233–244.

Rajendran, R., Mowat, E., McCulloch, E., Lappin, D.F., Jones, B., Lang, S., Majithiya, J.B., Warn, P., Williams, C. and Ramage, G. 2011. Azole resistance of *Aspergillus fumigatus* biofilms is partly associated with efflux pump activity. Antimicrob. Agents Chemother. 55: 2092–2097.

Rasko, D.A. and Sperandio, V. 2010. Anti-virulence strategies to combat bacteria-mediated disease. Nat. Rev. Drug Discov. 9: 117–128.

Rasooli, I., Fakoor, M.H., Yadegarinia, D., Gachkar, L., Allameh, A. and Rezaei, M.B. 2008. Antimycotoxigenic characteristics of *Rosmarinus officinalis* and *Trachyspermum copticum* L. essential oils. Int. J. Food Microbiol. 122: 135–139.

Raut, J.S., Shinde, R.B., Chauhan, N.M. and Mohan Karuppayil, S. 2013. Terpenoids of plant origin inhibit morphogenesis, adhesion, and biofilm formation by *Candida albicans*. Biofouling 29: 87–96.

Razzaghi-Abyaneh, M., Shams-Ghahfarokhi, M., Yoshinari, T., Rezaee, M.B., Jaimand, K., Nagasawa, H. and Sakuda, S. 2008. Inhibitory effects of *Satureja hortensis* L. essential oil on growth and aflatoxin production by *Aspergillus parasiticus*. Int. J. Food Microbiol. 123: 228–233.

Reichling, J., Schnitzler, P., Suschke, U. and Saller, R. 2009. Essential oils of aromatic plants with antibacterial, antifungal, antiviral, and cytotoxic properties—an overview. ForschendeKomplementärmedizin/Res. Complement. Med. 16: 79–90.

Ruzin, A. and Novick, R.P. 2000. Equivalence of lauric acid and glycerol monolaurate as inhibitors of signal transduction in *Staphylococcus aureus*. J. Bacteriol. 182: 2668–2671.

Saddi, M., Sanna, A., Cottiglia, F., Chisu, L., Casu, L., Bonsignore, L. and De Logu, A. 2007. Antiherpevirus activity of *Artemisia arborescens* essential oil and inhibition of lateral diffusion in Vero cells. Annals Clin. Microbiol. Antimicrob. 6: 10.

Schelz, Z., Molnar, J. and Hohmann, J. 2006. Antimicrobial and antiplasmid activities of essential oils. Fitoterapia 77: 279–285.

Schlesinger, S.R., Lahousse, M.J., Foster, T.O. and Kim, S.K. 2011. Metallo-β-lactamases and aptamer-based inhibition. Pharmaceuticals 4: 419–428.

Schnitzler, P., Astani, A. and Reichling, J. 2010. Antiviral effects of plant-derived essential oils and pure oil components. pp. 239–254. *In*: Thormar, H. (ed.). Lipids and Essential Oils. John Wiley & Sons, Chichester, UK.

Schnitzler, P., Koch, C. and Reichling, J. 2007. Susceptibility of drug-resistant clinical herpes simplex virus type 1 strains to essential oils of ginger, thyme, hyssop, and sandalwood. Antimicrob. Agents Chemother. 51: 1859–1862.

Silva, F., Ferreira, S., Queiroz, J.A. and Domingues, F.C. 2011. Coriander (*Coriandrum sativum* L.) essential oil: its antibacterial activity and mode of action evaluated by flow cytometry. J. Med. Microbiol. 60: 1479–1486.

Sinha, S., Jothiramajayam, M., Ghosh, M. and Mukherjee, A. 2014. Evaluation of toxicity of essential oils palmarosa, citronella, lemongrass and vetiver in human lymphocytes. Food Chem. Toxicol. 68: 71–77.

Sivamani, P., Singaravelu, G., Thiagarajan, V., Jayalakshmi, T. and Kumar, G.R. 2012. Comparative molecular docking analysis of essential oil constituents as elastase inhibitors. Bioinformation 8: 457–460.

Smith-Palmer, A., Stewart, J. and Fyfe, L. 2004. Influence of subinhibitory concentrations of plant essential oils on the production of enterotoxins A and B and α-toxin by *Staphylococcus aureus*. J. Med. Microbiol. 53: 1023–1027.

Solórzano-Santos, F. and Miranda-Novales, M.G. 2012. Essential oils from aromatic herbs as antimicrobial agents. Curr. Opnion. Biotechnol. 23: 136–141.

Soto, S.M. 2013. Role of efflux pumps in the antibiotic resistance of bacteria embedded in a biofilm. Virulence 4: 223–229.

Spengler, G., Molnár, A., Schelz, Z., Amaral, L., Sharples, D. and Molnár, J. 2006. The mechanism of plasmid curing in bacteria. Curr. Drug Targets 7: 823–841.

Stavri, M., Piddock, L.J. and Gibbons, S. 2007. Bacterial efflux pump inhibitors from natural sources. J. Antimicrob. Chemother. 59: 1247–1260.

Strasfeld, L. and Chou, S. 2010. Antiviral drug resistance: mechanisms and clinical implications. Infect. Dis. Clin. North America 24: 809–833.

Tegos, G.P. and Hamblin, M.R. 2013. Disruptive innovations: new anti-infectives in the age of resistance. Curr. Opin. Pharmacol. 13: 673–677.

Thomsen, N.A., Hammer, K.A., Riley, T.V., Van Belkum, A. and Carson, C.F. 2013. Effect of habituation to tea tree (*Melaleuca alternifolia*) oil on the subsequent susceptibility of *Staphylococcus* spp. to antimicrobials, triclosan, tea tree oil, terpinen-4-ol and carvacrol. Int. J. Antimicrob. Agents 41: 343–351.

Thota, N., Koul, S., Reddy, M.V., Sangwan, P.L., Khan, I.A., Kumar, A., Raja, A.F., Andotra, S.S. and Qazi, G.N. 2008. Citral derived amides as potent bacterial NorA efflux pump inhibitors. Bioorganic Med. Chem. 16: 6535–6543.

Tian, J., Ban, X., Zeng, H., He, J., Chen, Y. and Wang, Y. 2012. The mechanism of antifungal action of essential oil from dill (*Anethum graveolens* L.) on *Aspergillus flavus*. PloS One 7: e30147.

Trombetta, D., Castelli, F., Sarpietro, M.G., Venuti, V., Cristani, M., Daniele, C., Saija, A., Mazzanti, G. and Bisignano, G. 2005. Mechanisms of antibacterial action of three monoterpenes. Antimicrob. Agents Chemother. 49: 2474–2478.

Urbina, J.A., Marchan, E., Lazardi, K., Visbal, G., Apitz-Castro, R., Gil, F., Aguirre, T., Piras, M.M. and Piras, R. 1993. Inhibition of phosphatidylcholine biosynthesis and cell proliferation in *Trypanosoma cruzi* by ajoene, an antiplatelet compound isolated from garlic. Biochem. Pharmacol. 45: 2381–2387.

Van Bambeke, F. and Lee, V.J. 2006. Inhibitors of bacterial efflux pumps as adjuvants in antibiotic treatments and diagnostic tools for detection of resistance by efflux. Recent Patents Anti-Infect. Drug Discover. 1: 157–175.

Veras, H.N., Campos, A.R., Rodrigues, F.F., Botelho, M.A., Coutinho, H.D., Menezes, I.R. and da Costa, J.G.M. 2011. Enhancement of the antibiotic activity of erythromycin by volatile compounds of *Lippia alba* (Mill.) NE Brown against *Staphylococcus aureus*. Pharma. Mag. 7: 334.

Wagner, H. and Ulrich-Merzenich, G. 2009. Synergy research: approaching a new generation of phytopharmaceuticals. Phytomedicine 16: 97–110.

Worthington, R.J. and Melander, C. 2013. Combination approaches to combat multidrug-resistant bacteria. Trends Biotechnol. 31: 177–184.

Wright, G.D. 2014. Something old, something new: revisiting natural products in antibiotic drug discovery 1. Canadian J. Microbiol. 60: 147–154.

7

Essential Oils and Nanoemulsions

Alternative Tool to Biofilm Eradication

Z. Aumeeruddy-Elalfi and *F. Mahomoodally**

Introduction

Biofilms are surface-associated communities of microorganisms that have long been correlated to persistent infections. They are widespread, occurring in different environment systems including medical devices in healthcare facilities, where they can be life threatening (biomedical implant infection, haemodialysis complications and catheter associated infection, amongst others) (Donlan 2002).

Bacteria can be found in at least two discrete states (planktonic and the sessile cells). Planktonic cells refers to the free flowing bacteria in suspension. On the other hand, sessile cells are the consortium forming bacteria, whereby the microorganisms establish a structured community of bacterial cells, surrounded by a self-produced matrix, adherent to an inert or living surface. Biofilms are composed of these sessile adherent cells (Donlan 2002).

In terms of microbial communities that are attached to surfaces, biofilms have been identified since the 17th century by Anthony van Leewenhoek. In a report addressed to the Royal Society of London, he described biofilms as "*The numbers of these animalcules in the scurf of a man's teeth are so many that I believe they exceed the number of men in a kingdom*" (Dymock 2003).

Currently, the concept of biofilm is associated to the irreversible cell attachment that are the leading cause of recurrent infections, as well as, to new attributes such as modification in growth rate and gene transcription (Costerton et al. 1997, Poole 2009, Boto 2010). Research on the link existing between pathogenic bacteria and their host has shown that there are multiple tactics employed by the former to bypass the innate defence mechanism. These tactics involved primarily, the production of cytotoxic enzymes as well as other molecules, to allow their invasion in the host tissue, along with the formation of protective shields. Pathogenic bacteria that form biofilms are often referred as the chronic bacterial forms. Unlike the free floating planktonic bacteria that cause acute infections, those chronic bacterial forms have evolved and form communities to combat the cells of the host's immune system (Sotolongo et al. 2011, Roilides et al. 2015, Cole and Nizet 2016).

The occurrence of biofilm in nature, in specific environments and in hosts has been reported to be a mean of optimum colonization and appropriation of a nutrient-rich environment (Poole 2009, Roilides et al. 2015).

Department of Health Sciences, Faculty of Science, University of Mauritius, Réduit, Mauritius.
* Corresponding author: f.mahomoodally@uom.ac.mu

Plants are organisms adapting to biotic or abiotic stresses by producing secondary metabolites. For instance, one of the biotic stress to which these organisms have to fight is the development of biofilm on their surface. In this situation, one of their mode of defence is through the production of Essential Oils (EOs). For this reason, there has been increasing research performed on natural products, as a source of new bioactive molecules (Simões et al. 2010).

EOs are lipophilic and complex mixes with high terpenic and phenolic contents. These compounds come from different precursors of primary metabolism and are synthesized through different pathways conferring antimicrobial properties. Antibacterial activities of EOs and isolated components have been widely studied over the past few years. Due to their lipophilic character, they are able to slither through cell membranes, dislocating the different phospholipids, polysaccharides and fatty acid layers. In this process, the EO lead to a change in cells permeability, triggering a loss of integrity. EOs such as that extracted from clove, thyme, oregano and many others have been reported for their effect on bacterial biofilm, by interfering with quorum sensing and inhibiting the peptidoglycan synthesis or reducing cell adherence. For instance, Sandasi et al. (2011) and Valeriano et al. (2012) also purported the efficiency of *Mentha piperita* EOs against biofilms. Kim et al. (2015) highlighted the ability of cinnamaldehyde, a major component of *Cinnamomum cassia* EO, to down regulate quorum sensing in *P. aeruginosa* biofilm (Chorianopoulos et al. 2008, Simões et al. 2010, Faleiro 2011, Sandasi et al. 2011, Valeriano et al. 2012, Kim et al. 2015).

New strategies are constantly emerging in this fight against pathogenic biofilm. In particular, the use of biosolution containing enzymes, phages, interspecies competitions, microbially-derived antimicrobial compounds and natural plant molecules such as EOs in association with nanotechnology biofilm (Chorianopoulos et al. 2008, Simões et al. 2010, Faleiro 2011, Kim et al. 2015).

Even though EOs have been appraised as potential alternatives to chemical preservatives against foodborne biofilm forming pathogens, they present special limitations that impede their application in food products. For instance, EOs are not water soluble, they are highly volatile, and have a strong odour that make it difficult for their application. It is also a major challenge to integrate EO-based compounds in food products due to their physical and chemical instability. For these reasons, the association of EO as nanoemulsion with their specificity in promoting very small droplet diameter, high physical stability, high bioavailability and optically transparent products can be considered as a breakthrough in this field of research (Jo et al. 2015, Landry et al. 2015).

The main aim of this chapter is to highlight the action of EOs on biofilms eradication. We also endeavoured to elaborate on the application of nanotechnology as an effective method when combined to natural products, to fight opportunistic pathogens, particularly the tenacious biofilm-forming pathogens.

Essential Oils—Description, Application and Brief Historic

Essential oils (EOs) also known as ethereal oils, were characterized by Guenther in 1948 as a volatile natural liquid made of complex components, with a strong aroma. EO producing aromatic plants can be found among superior plants, angiosperms as well as gymnosperms and belong to a variety of 50 families such as Apiaceae, Lamiaceae, Myrtaceae, Pinaceae and Zingiberaceae (Aumeeruddy-Elalfi 2016).

EOs are obtained mostly from vegetable organs (e.g., flowers, leaves, barks, woods, roots, rhizomes, fruits and seeds) through different processes such as expression, fermentation, enfleurage or extraction. These oils consist mainly of terpenes and oxygenated compounds, obtained through a form of distillation from aromatic plant materials of a single botanical species and form (Bakkali et al. 2008). The constituents of EOs are produced in the plant through the mevalonic and shikimic acid pathways as secondary metabolites and stocked in the plants in different organs depending on the parts of the plant it is found: glandular trichomes, oil cells or ducts in plant tissues (Hunter 2009, Chaiyana and Okonogi 2012). EO occurs naturally in plants as a defence mechanism to protect themselves from pathogenic organisms, as an insect repellent. The rich and strong aromas of EOs protect the plants against herbivores. EOs excreted by some plants also help in attracting specific insects that allow the dispersion of pollens and seeds (Hunter 2009).

The source of the term EO is thought to exist since the 16th century, named after Paracelsus Von Hohenheim, a physician and alchemist who founded the role of chemistry in medicine (Kim et al. 2015). EOs have been used since ancient times in medicine and as part of rituals. Egyptians were using aromatic plant materials for preserving mummies while the Chinese valued aromatic plants for their vast curative purposes and also as perfume. In India, the Ayurveda philosophy was built based upon a variety of herbs among which the literature mention scented substances. Humans have been using aromatic plants since ancient times. However, the distillation of EOs for the extraction of medicinal and fragrant components was not widely practiced until the 18th century (Lawless 1995). In fact, distillation as a method for the extraction of EOs was first used in Egypt and Persia over 2000 years ago (Guenther 1948, Burt and Reinders 2004). The first authentic written proof of distillation of EO was illustrated by Villanova (ca. 1235–1311), a Catalan physician. Even though EOs were not well known in Europe until the 16th century, EOs obtained by the distillation process made their first appearance in pharmacies since the 13th century, with their pharmacological properties available as part of the pharmacopoeias (Simões et al. 2010, Valeriano et al. 2012).

Nowadays, EOs have been used for a multitude of applications such as their key role in the cosmetic industry, exploited for their rich and diverse fragrances, in the food industry as food preservatives and flavourings and in the pharmaceutical industry for their pharmacological properties (Nychas 1995, Schmidt et al. 2008). EO has shown its worth in the agro-industry, as biofertilizers. EOs have also become popular due to their eco-friendly attributes, rapid volatility properties and minimal residual activity, which are less detrimental to health (Lubbe and Verpoorte 2011).

Moreover, there is a vast amount of *in vitro* data which put forward that plants derived natural products such as EOs are valuable sources of novel treatments for multidrug resistant organisms (Kon and Rai 2010, Sienkiewicz et al. 2014, Zenati et al. 2014). Multidrug resistance related diseases are known nowadays as one of the major and leading cause of global morbidity and mortality, particularly in developing countries (WHO 2015). Currently, the use of EOs is widespread in the development of natural products, especially combined to nanoparticles for the enhancement of the biological properties, to improve the solubility, efficacy and stability. From polymeric nanoparticulate to solid-lipid nanoparticles and nanoemulsions (Al-Haj 2010, Mihai Grumezescu 2013, Anwer et al. 2014) the potential of nanotechnology, in different forms and sizes, combined with EO for the improvement of natural bioactive ingredient, is warranted. Metal based nanoparticles have also received substantial attention for their potential action against pathogenic bacteria. This includes silver, titanium dioxide and zinc oxide nanoparticles. Another advantage of the association of EO and nanotechnology resides in its use as nanovehicles for the enhancement of the solubility potential of EOs, which are natural products with poor bioavailability and immiscible in water.

Future research in nanotechnology applied to EO is gaining amplitude for developing EOs related strategies in overcoming multi-drug resistance. This target was also set for decreasing EO concentrations while achieving a particular antimicrobial and antibiofilm effect, for safety and efficacy purposes with respect to human health. Table 7.1 summarizes EOs used in food and pharmaceuticals, integrated in formulations consisting of different types of nanovehicles, as well as the microorganisms against which these different types of formulations have been reported as effective.

The broad panel of biological activities of EOs can be correlated to their complex and variable chemical compositions. The composition of EOs is variable from one EO to another and from one species to another. There are different factors affecting EO composition, among which there is the extraction type (hydrodistillation, steam distillation, supercritical fluid extraction, short-path molecular distillation, microwave assisted extraction, *in situ* microwave-generated hydrodistillation, microwave steam diffusion and ultrasound assisted extraction amongst others) (Aumeeruddy-Elalfi et al. 2016), the climatic, seasonal and geographic conditions as well as harvesting period (Hussain et al. 2008,

Table 7.1. EOs and Nanotechnology.

Essential Oil	Formulation	Microorganisms tested	Use	References
Syzygium aromaticum L. (Clove)	Nanoemulsions	*B. subtilis, S. aureus, P. vulgaris, P. aeruginosa, K. pneumoniae*	Pharmaceuticals	Anwer et al. 2014
Thymus vulgaris, Cymbopogon citratus, Salvia officinalis	Nanoemulsions	*E. coli*	Food	Acevedo-Fani et al. 2015
Ocimum basilicum	Nanoemulsions	*E. coli*	Food	Ghosh et al. 2013
Cymbopogon citratus	Nanoemulsions	*E. coli* O157:H7 *S. Typhimurium*	Food	Salvia et al. 2014, Kim et al. 2013
Cinnamaldehyde	Nanospheres	*Salmonella* spp. *Listeria* spp.	Food	Gomes et al. 2011
Eucalyptus globulus	Nanoemulsions	*S. aureus*	Pharmaceuticals	Sugumar et al. 2014
Eugenol	Nanospheres	*Salmonella* spp. *Listeria* spp.	Food	Gomes et al. 2011
Satureja hortensis	Agar-cellulose bionanocomposite	*S. aureus, L. monocytogenes,* and *B. cereus* and *E. coli*	Food	Atef et al. 2015
Melaleuca alternifolia	Nanocapsules, Nanoemulsions	*T. rubrum*	Pharmaceuticals	Flores et al. 2013
Origanum dictamnus	Lipid Nanoparticles	*L. monocytogenes*	Food	Gortzi et al. 2007

Aumeeruddy-Elalfi et al. 2016). Different parts of a plant are made of different types of system having each different metabolic pathways involved in the production of these EOs (Zenati et al. 2014).

Antibacterial Properties of EO

EOs are natural volatile complex blends of biologically active substances and are the active ingredients of a plethora of products available on the market (Bakkali et al. 2008, Bedi et al. 2010, Raut and Karuppayil 2014). According to available scientific literature, herbal mixtures are still present in 40% of drug prescriptions (Newman and Cragg 2007). EOs of *Syzygium aromaticum, Oreganum vulgare, Thymus vulgaris, Cananga odorata, Eucalyptus globulus, Kadsuralongi pedunculata, Psiadia arguta, Lavandula stoechas, Salvia rosifolia, Pimenta dioica* and *Myristica fragrans* amongst others have been traditionally used for their antimicrobial properties in different parts of the world (Burt and Reinders 2004, Hunter 2009, Aumeeruddy-Elalfi et al. 2015, Aumeeruddy-Elalfi et al. 2016).

Also, EOs such as those extracted from commonly used spices and traditional medicinal plants are rich in compounds such as terpene, limonene, terpenoid, carvacrol, thymol, menthol, citral, geraniol, eugenol, cinnamaldehyde and others, known for their antimicrobial potential. Considerable importance has been given to EOs from different plant species these last decades due to the rising interest in their antibacterial properties and to contribute to scientific data, in the quest to fight antibiotic resistance (Burt 2007, Celiktas et al. 2007, Edward-Jones 2013, Aumeeruddy-Elalfi et al. 2015, 2016).

Antibiotics have found an important place as a public health tool since the discovery of penicillin and have helped to save the lives of countless number of people around the world. Yet, today, the emergence of drug resistance in bacteria is bringing us back to the 'pre-antibiotic' times, with drug choices for the treatment of many bacterial infections were progressively limited, costly, and, in some cases, non-existent (Centers for Disease Control and Prevention 2016).

Antibiotic resistance is a phenomenon, which is on the rise. Pathogens in particular *K. pneumoniae* and *E. coli* are becoming resistant to antibiotics by acquiring new genes. The Centers for Disease Control and Prevention (CDC) postulated that drug-resistant pathogens lead to the death of around 23 thousand persons per year in the US, along with two million illnesses (Centers for Disease Control and Prevention 2016). This emergence of bacteria with antibiotic resistant gene are referred to as pan-drug resistant bacteria and is on one side causing a widespread problem in antibiotic treatment but on the other side, research work on antimicrobial natural products such as EOs, are on the rise and are gaining in significance (Centers for Disease Control and Prevention 2016). Plants known in traditional medicine for their antimicrobial properties are being screened and phytochemicals isolated in view of medical application to fight fatal opportunistic infections (Adorjan and Buchbauer 2010).

The significance of action of EOs on bacteria is dependent on the major components present in the EOs. The concentrations in the EOs of the major components having antibacterial properties as well as the affinity of these components towards the target sites within the bacterial cells are also important features. EOs and their components are detrimental to bacterial cells in several ways. The hydrophobic nature of EOs allows them to partition in the lipids of the cell membrane and mitochondria of the bacteria (Derwich et al. 2010). The partitioning of the EO in the cell membrane induces leakage of bacterial cell contents. This leakage is often followed by massive outflow of solutes, resulting in cell death. Components of EOs such as eugenol, has been reported by Beniss et al. (2004) to induce damage in the cell wall of bacteria. The proteins present in membranes of the cytoplasm of bacteria are also the target of EOs. It has been purported that cinnamaldehyde, components present in EO, has an inhibiting potential towards amino acid decarboxylase present in the membrane of *E. aerogenes* (Wendakoon and Sakaguchi 1995, Walsh et al. 2003). Cell membranes are vital barriers that define and regulate the intra- and extracellular flow of substances. Destruction of structure present in the cell membrane, such as the electrons transport systems, by EOs and components has been reported by Tassou et al. (2000) as initiative to impede bacterial growth. Phenolic compounds present in EOs have been reported to kindle the disruption of the proton motive force, electron flow and active transport as well as to initiate the coagulation of the cell contents (Denyer and Hugo 1991).

Biofilm and EO

The definition of biofilm, previously described as "*matrix-enclosed aggregates of bacteria, comparable to the planktonic cells of the species concerned, immobilized on surfaces or at interfaces in the ecosystems in which they predominate*," has changed significantly. In the late 20th century and equipped with more sophisticated laboratory tools than Anthony van Leewenhoek, biofilms were then described as "*adherent population consisting of single cells and microcolonies of sister cells, all embedded in a highly hydrated, predominantly anionic matrix of bacterial exopolymers and trapped extraneous macromolecules*" (Costerton et al. 1987). In the same time frame, the mode of existence of biofilms was investigated and bacterial biomass belonging to different environments was scrutinized. The modes of attachment of the bacteria, interaction between microorganisms of same or different species and the notion of microcolonies with efficient heterogeneity were being highlighted (Costerton et al. 1987).

With new technologies and research work targeted to understand the mechanism of action of biofilms, the most recent description of a biofilm is that of a multicellular community composed of both prokaryotic and eukaryotic cells or only prokaryotic cells implanted in a matrix composed partly of material synthesized by the sessile cells within the community (Costerton 2007). These sessile cells have been reported to be more active than the planktonic cells, as they possessed micro-colonies composed of cells implanted in the matrix material and are crossed by water channels carrying fluids, metabolic waste and nutrients into the consortium (De Beer et al. 1994). Also, this consortium has been identified as being structurally and metabolically heterogeneous with aerobic and anaerobic processes occurring at the same time in different parts of the community (Costerton et al. 1978, De Beer et al. 1994). This heterogeneity within biofilms for instance, may allow the simultaneous action of multiple resistance mechanisms within a single community, rendering infections more complex and difficult to eradicate. Biofilms are attributed to more than 50% causes of nosocomial infections (Hall-Stoodley et al. 2004).

Action of EO on Biofilm

EOs are rich in a broad assortment of secondary metabolites, having antimicrobial properties towards a large variety of microorganisms, including both the Gram negative and positive pathogens (Burt and Reinders 2004, Bakkali et al. 2008). A plethora of approaches have been used to test EOs antimicrobial potential for microbial growth inhibition, biofilm eradication and anti-pathogenic effect. The elucidation of the mechanism of action of antimicrobials has been established in current research including new analysis methods for more effective results. One of the analysis method consist in the comparison between Minimum Biofilm Eradication Concentration (MBEC) evaluated for biofilm consortium and that of the Minimum Inhibition Concentration (MIC) of the tested EO sample, against the same bacterial species but in a planktonic state. It has been purported in several publications that the minimum biofilm eradication concentration is less than that of the minimum inhibition concentration of the EO against the planktonic bacteria.

Microorganisms inside a biofilm are thought to be more susceptible to the activity of EOs and component which can penetrate the consortium and accumulate inside. It has been reported in the literature that active components of EOs are absorbed by the extracellular matrix of the biofilm and in so doing, increasing the local concentration of these components (Kavanaugh and Ribbeck 2012). Another possibility of action of EOs on biofilms suggested by Kavanaugh and Ribbeck (2012) is that of differential gene expression that exist in planktonic bacteria with respect to those forming the biofilm. Differential gene expression has been evaluated to result in a change in the morphology and physiology of the cell membrane and cell wall in biofilm cells. Also, it was emphasized by

Kavanaugh and Ribbeck (2012) that composition of cellular components is the key to fight against different microorganisms within a biofilm. Combinations of different EO components to target different microorganisms are suggested as effective against multispecies biofilm due to the specific-specific EO activity.

EO components having an antimicrobial effect on planktonic cells also induce biofilm inhibition. For instance, it has been reported by Bersan et al. (2014) that EO of *Cyperus articulatus* had an antibiofilm effect on *Candida albicans, Fusobacterium nucleatum, Porphyromonas gingivalis, Streptococcus sanguis* and *Streptococcus mitis*. The mode of action observed for *C. articulatus* on the studied biofilms have been correlated to the presence of α- and β-pinene that destroy cell integrity, and inhibit respiration and ion transport processes, which lead to cell death (Bersan et al. 2014). *In vitro* assays for the analysis of biofilm have known significant progress during this last decade. However, the *in vivo* mechanisms remain vague and needs further in-depth investigations (Kavanaugh and Ribbeck 2012, Bai and Vittal 2014, Bersan et al. 2014).

Examples of EO with Antibiofilm Properties

Pathogenic bacteria can develop and colonize an environment in the presence of appropriate substrate and optimum environment. Pathogenic biofilm forming bacteria are known to be 10 to 1000 times more tenacious than pathogenic non-biofilm forming bacteria. However, some active EOs have been reported to be effective against both types of pathogenic bacteria. Table 7.2 illustrates the most common active EOs and the biofilm forming organism that are sensitive to the components of these EOs, along with the method used to assess the effect of these EO on the biofilm forming organisms. The mode of actions of these EOs have been linked in previous studies to the components of their phytochemical profile, that have a direct effect on the respiration and ion transport process.

Table 7.2. EOs and their action on biofilm-forming microorganisms.

Essential oil	Microorganisms	Method	Reference
Melaleuca alternifolia	*S. aureus, E. coli*	Colony-forming-unit counting, TTC reduction assay	Budzyńska et al. 2011
Lavandula angustifolia	*S. aureus, E. coli*	Colony-forming-unit counting, TTC reduction assay	Budzyńska et al. 2011
Melissa officinalis	*S. aureus, E. coli*	Colony-forming-unit counting, TTC reduction assay	Budzyńska et al. 2011
Cupressus sempervirens	*S. aureus, K. pneumoniae, S. indica*	Crystal Violet assay	Selim et al. 2014
Cymbopogon flexuosus	Methicillin susceptible *S. aureus,* Methicillin resistant *S. aureus*	Colony-forming-unit counting, XTT reduction assay	Adukwu et al. 2012
Murraya koenigii	*P. psychrophila*	Crystal-violet MTP method	Bai and Vittal 2014
Mentha x piperita	*P. aeruginosa, S. aureus*	Crystal-violet MTP method	Ceylan et al. 2014

Example of Major Antimicrobial EO Components Isolated and Their Effects on Biofilm Formation

In the search for biofilm eradication, phytochemical studies of natural products in particular EOs have been elaborated. Studies have been oriented around the isolation and identification of the components of complex EO mixtures, with the aim of investigating the structure-activity correlations. Major components of EOs have thus been screened and their mechanisms of action studied and compared to planktonic microorganisms. Table 7.3 summarizes some examples of potent EO components which have been found to be more effective against biofilm-forming microorganisms, compared to antibiotics.

Table 7.3. EO components as antibiofilm agents.

Components	Microorganisms	Method	Ref.
α-pinene	*C. albicans*	Mitochondrial activity, XTT(2,3-Bis-(2-Methoxy-4-Nitro-5-Sulfophenyl)-2H-Tetrazolium-5-Carboxanilide) colorimetric method	Da Silva et al. 2012
β-pinene			
Eugenol	*S. aureus*	Crystal violet assay	Yadav et al. 2015
Terpinen-4-ol	*S. aureus* and *E. coli*	MTT(3-(4,5-Dimethylthiazol-2-yl)-2,5-Diphenyltetrazolium Bromide) reduction assay	Budzyńska et al. 2011
α-terpineol		MTT-reduction assay	Budzyńska et al. 2011
1, 8-cineole	*P. mirabilis*	Scanning Electron Microscopy	Mathur et al. 2013
α-pinene			
Thymol	MRSA	TTC (Triphenyl tetrazolium chloride) and MTT reduction assay	Kifer et al. 2016
Menthol			
1, 8-cineole			
Cinnamaldehyde	*P. aeruginosa*, *S. aureus*	Colony-Forming units counting	Kavanaugh and Ribbeck 2012

Biofilm Physiology

Biofilm physiology has been assessed through various methods. For instance, the examination of the ultra-structure using electron microscopy tool, epifluorescent microscope, confocal laser scanning microscopy as well as staining and examination, using the light microscope for better understanding of the biofilm activities (Davis et al. 2008, Alhede 2012). There are several advantages in using the different microscope as an analysis tool. The scanning electron microscope for example allows high-magnification of spatial images which allow visualizing the location of the bacteria in the biofilm and its interaction with other microorganisms within the biofilm.

The physiology of the biofilm depends highly on the extracellular polymeric substances produced by the microorganisms inside the biofilm. This niche of microorganisms is generally composed of polysaccharides, proteins, membrane vesicles, peptidoglycans, lipids, enzymes and nucleic acids. The extracellular polymeric substances have a major role in the adherence of the consortium to different surfaces and tissues of the body, along as in the containment of the microorganism within the consortium, isolating the microorganisms from any pH variation and enzyme attack (Lembre et al. 2012). Overall, a biofilm is made up by three quarters of extracellular polymeric substances and one quarter of sessile cells.

Biofilm has a heterogeneous architecture which is constantly fluctuating in composition due to external and internal processes. The biofilm matrix has been reported to be structurally made of extracellular polysaccharides. These polysaccharides allow the biofilm to be rigid or gel-like, by interacting with other polysaccharides within the biofilm or with heterologous ions and molecules. Another type of interaction is the polysaccharide-protein interaction, which confers structural and functional properties to biofilms.

Signals between cells within the biofilm have been recognized to play a role in cell attachment and detachment from biofilms. Within the biofilm, there can exist multiple types of cell to cell signalling. For instance, it has been reported by Davies et al. (1998) that in *P. aeruginosa*, there exists two types of cell to cell signalling systems which are involved in the establishment of biofilms.

According to various literatures, the site where the microorganisms are isolated has an effect on the molecules responsible for quorum sensing (cell to cell signalling). Numerous differences have been identified among the different biofilms obtained in different locations. Biofilms present on the wounds, surgical instruments and dental plaque or in the environment (rocks, soil, and water) have been found to differ in bacterial population as well as in their extracellular matrix (Lembre et al. 2012).

Biofilm physiology is significantly influenced by the chemical gradient (variation in pH and ions) occurring within the biofilm. Chemical gradient is a reflection of the diffusion of nutrients as well as metabolic products to and from the biofilm environment. Chemical gradient affect the growth rate and development of the microorganisms consisting of the biofilms. The fewer access to nutrients, that is the less dense the water channels bringing nutrients and removing waste products, the slower will be the growth rate. In the same way, a slight variation in pH results in a disturbance in the pH homeostasis thus varying their physiological condition and acid tolerance (Pamp et al. 2009).

The formation of a biofilm starts when a bacterium adheres in a reversible manner to a surface where the environment is optimum for its growth and development. Protein kinase that phosphorylates itself using adenosine triphosphate and response regulator receiving the phosphoryl groups can be identified in the biofilm at this stage. These protein kinases allow the microorganisms to adapt more easily to an environment. Once secured to the surface, in the next stage, the attached bacteria develop and aggregate into microcolonies. These microcolonies or sessile cells produce the matrix of the biofilm. The next step is the colonization of sessile cells and development of established three-dimensional biofilm architecture.

Biofilm formation depends on various factors, which vary from environmental stress, nutritional needs and chemical agents present in its immediate environment to the phenotype and physiology of the microorganisms forming the biofilm. Factors that are directly correlated to biofilm formation are well documented in the literature. These are: the adherence factors (flagella and type IV pili) and the nutritional factors (enzymes, quorum sensing molecules, essential amino acids and others) (Béchet and Blondeau 2003).

Microorganisms in Biofilm Formation

Biofilm formation is the cause of chronic and nosocomial infections in healthcare. Acute nosocomial infections have been linked to the reversible and in most serious cases, irreversible adherence of microorganisms to tissues. These infections are caused in most of the cases by Gram positive *Staphylococcus aureus*, *Staphylococcus epidermidis*, *Streptococcus pneumoniae* and *Streptococcus pyogenes* as well as the Gram negative *Pseudomonas aeruginosa* and *Escherichia coli* (Lasaro et al. 2009).

Based on recent data collected by National Institutes of Health, up to 80% of human bacterial infections are caused by microorganisms-forming biofilms. The formation of biofilm has been associated to several chronic infections such as recurrent urinary tract infection, bacterial vaginosis, catheter infection, wound infection, osteomyelitis, recurrent tonsillitis, rhinosinusitis and endocarditis amongst others. Aside from chronic infections, common human diseases such as Crohn's disease, periodontitis and dental caries are caused by biofilm-forming microorganisms (Lasaro et al. 2009).

The significance of biofilms for public health resides in the fact that biofilm-associated microorganisms depicts reduced susceptibility to antimicrobial drugs. This susceptibility may be a natural effect of growth in the biofilm but it can also be acquired by transfer of extrachromosomal components to susceptible organisms within the biofilm. The susceptibility of biofilms to antimicrobial agents cannot be elucidated by means of standard microdilution testing, because these tests are based on the response of planktonic rather than sessile cells. Susceptibility must be thus determined directly against biofilm-associated organisms.

Mastering the mechanism and process of biofilm formation may lead to a positive impact on clinical decision-making. For instance, the way blood samples and catheter-tip samples are collected and examined in health centres or by providing a clearer picture of the restrictions of conventional therapies for treating biofilm-associated infections (Béchet and Blondeau 2003).

Effects of Chemical Agents on Biofilm

Numerous molecules with antibiofilm activities have been reported. Examples of these molecules are: indole, imidazole, phenols, triazole, furanone, bromopyrrole, sulphide and lactoferrin amongst others. The most commonly reported being indole, which is synthesized during the degradation of tryptophan by the enzyme tryptophanase. Indole has been found to inhibit significantly bacterial motility and biofilm formation of several species of bacteria (Isaacs et al. 1994, Hu et al. 2010).

Indole and lactoferrin are also considered as chemical agents influencing biofilm formation and development. Indole has a positive effect on biofilm formation in bacteria that do not synthesize it, such as *Pseudomonas* spp. On the other hand, in organisms such as *Escherichia coli* which produces indole, a negative effect was observed, leading to the reduction of biofilm. For instance, the motility and biofilm formation of *E. coli* O157:H7 has been reported to be inhibited by indole at a concentration of 500 μM (Hu et al. 2010). To understand the mechanism of action of indole on biofilm, gene expression in biofilm treated with indole has been analyzed. The result showed that indole has the potential to reduce the expression of gene such as the shock regulator gene by up to four folds (Isaacs et al. 1994).

Lactoferrin is an iron-binding protein which is part of the innate immune system (Bellamy et al. 1992). It has many roles in the immune system, such as the inhibition of neutrophil priming using bacterial lipopolysaccharide. It also favours the attachment of endothelial cells and the modulation of inflammation by amplification of apoptotic signals. Lactoferrin has been recognized for other properties, including as antibacterial, antiviral and antifungal potential (Bellamy et al. 1992, 1993). As a biofilm inhibitor, lactoferrin chelates iron in the near biofilm environment, disturbing the motility of the microorganisms. As a response, the microorganisms move to a more optimum environment in a quest for nutrient. Lactoferrin thus have an impact on biofilm attachment and development (Bellamy et al. 1993).

Resistance of Biofilm to Immune System and Antimicrobial Agents

Microorganisms within a biofilm are the cause of numerous persistent infections. Sessile cells inside the biofilms once mature, acquire resistance to the immune system and antimicrobial agents. This resistance renders biofilms health threatening, particularly in healthcare where biofilms tend to colonize medical devices, forming a niche of pathogenic bacteria. The mechanisms of resistance to antibiotics by bacterial biofilms have been reported in the literature. Some theories remain hypothetical while others have been supported by findings from studies (Davies et al. 1998, Jefferson 2004, Kifer et al. 2016).

Different possibilities have been elaborated, such as the likelihood of delayed and incomplete penetration of the antibiotic in the biofilm. In such situations, the availability of the antimicrobials inside the biofilm is reduced. *In vitro* studies have demonstrated that some antibiotics can infiltrate the biofilm, which is mostly made of water. However, other antibiotics are altered in the biofilm leading to a retarded penetration. In other situations, antibiotics without being altered are delayed in their dispersion inside the biofilm. This has accounted for the density of the biofilm matrix, which retains the antibiotics. These antibiotics have a slow penetrating rate (Bellamy et al. 1992, Günther et al. 2009, Lembre et al. 2012).

Biofilm microorganisms also propagate and grow at a slow rate, which amplifies their capacity to be less susceptible to the hosts immune mechanisms and antimicrobial agents. Along with the fact that mature biofilm are more resistant to phagocytic actions than the newer biofilm as the former possess the components of the extracellular polysaccharides which in general help to prevent phagocytosis of biofilm organisms (Günther et al. 2009).

The specific environment within biofilm such as pH, partial pressure of oxygen dissolved in the biofilm, and the presence of chemical substances have shown to impact on the activities of antimicrobials. Antimicrobials potency thus varies according to the different location within the biofilm. The rate of antimicrobial resistance has also been reported to be due to the capacity of the microorganisms within the biofilm to mutate in response to stress. This results in the transfer and acquisition of genes, among the microorganisms present inside the biofilm (Davies et al. 1998).

To summarize, the overall biofilm architecture together with the chemical substances produced by the microorganisms within the biofilm is thought to contribute to the resistance acquired by the microorganisms against immune mechanisms and antimicrobials. Finally, the mechanisms of resistance may also be due to the genetic acquisition of resistant genes (Davies et al. 1998).

Techniques of Biofilm Identification

In comparison to the methods used to demonstrate antimicrobial properties in planktonic bacteria (which involve bacterial cell death), the method used to evaluate antibiofilm properties is based on the assessment of biofilm formed (initial stage of biofilm growth), and put forward the ability of the test products to prevent adhesion and absorption of microorganisms in the test surface (usually polyvinyl chloride material). For the second and irreversible stage of biofilm development, the activity of the colonizing bacteria is determined through the inhibition of the synthesis of the insoluble polysaccharide layer by a sucrose dependent enzyme (Palombo 2011).

Recent *in vitro* assessments and susceptibility testing methods used for planktonic culture include agar diffusion, disc diffusion, broth microdilution and minimal bactericidal concentration determination while viable count of colony forming unit, dry cell weight assays, DNA quantification and XTT reduction assay is used for the determination of the antibiofilm and anti-adhesion activities. Table 7.4 summarizes the direct and indirect methods for the detection and quantitation of biofilms.

Table 7.4. Methods for biofilm detection and quantitation.

Direct methods	Indirect methods
Tissue culture plate method (Hassan et al. 2011)	Dry cell weight assays staining (Besciak and Surmacz-Górska 2014)
Congo red agar method (Hassan et al. 2011)	Colony-forming-unit counting staining (Besciak and Surmacz-Górska 2014)
Crystal violet staining (Besciak and Surmacz-Górska 2014)	DNA quantification (Elliott et al. 2007)
Visual assessment by scanning electron microscopy (Neethirajan et al. 2014)	XTT reduction assay staining (Besciak and Surmacz-Górska 2014)
Visual assessment by confocal laser scanning microscopy staining (Besciak and Surmacz-Górska 2014)	
Tube method (Hassan et al. 2011)	

Alternative Approach to Biofilm Eradication

Resistance to antimicrobial agents is representing a major concern nowadays. Extensive production and misuse of antibiotics have contributed to the rise of resistant pathogens, which are considered a next-generation concern for global public health (Pamp et al. 2009). In an attempt to eradicate the biofilm, which is the key solution to decrease nosocomial infections prevalence, several research studies proposed the prevention of biofilms formation as solution, through the utilization of nanotechnology (Gortzi et al. 2007, Gomes et al. 2011, Flores et al. 2013, Ghosh et al. 2013, Kim et al. 2013, Anwer et al. 2014, Salvia et al. 2014, Sugumar et al. 2014, Acevedo-Fani et al. 2015, Atef et al. 2015).

Nanoparticles are an important asset as antimicrobials due to their intrinsic small scale, coupled to the composition of the particles, which confer an antimicrobial potential. Pathogens are acquiring new resistant genes, engendering an increase in casualties in healthcare. The newly acquired resistance which is raising concern in the scientific world is the collistin resistance in pathogenic species such as *Escherichia coli* (Liu et al. 2016, McGann et al. 2016).

Nanotechnology offers the possibility of potentiating antimicrobial activities of fallen antibiotics through a synergistic action. It is also an alternative option to antimicrobial resistance and biofilm eradication as the encapsulation of bioactive compounds such as EOs, using nanotechniques, embodies a feasible and effective approach in the control of drug release as well as in increasing the stability of the active substance, decreasing their volatility, toxicity and enhancing their bioactivity (Namasivayam et al. 2015, McGann et al. 2016).

The most common association of natural products and nanotechnology reported in the literature is that of nanoemulsion, metal nanoparticles (silver and silver-gold alloy nanoparticles) and nanospheres. Nanoemulsions are solutions of oil, surfactant and co-surfactant which are on the average transparent, isotropic and thermodynamically stable. These nanodroplets present variances in their physical properties and structure configuration compared to other nanoparticles. However, their stability is due to the colloidal system they consist of, within nanometric size (of maximum size of 100 nm). The nanometric size of nanoemulsion particle is what differentiate them notably with conventional emulsions and conferred their specific functional activity. Their average transparent nature was explained by their capacity to scatter light in the visible region. Together with this, nanoemulsion particles offer better surface area together with kinetic stability and thus offers efficient nano-dispersion properties when associated with natural bioactives (Li et al. 2012, Jo et al. 2015).

Recently, these nano-systems have been studied widely as novel nanovehicles for the enhancement of the solubility capacity, therapeutic efficacy and bioavailability of natural product components and pharmaceuticals (McGann et al. 2016). In addition, association between metal nanoparticles and antibiotics such as the β-lactams, cephalosporin, glycopeptides, aminoglycosides, flouroquinolones, and sulphonamides have been reported to exhibit a synergistic effect and lead to an enhancement in antibacterial properties.

Metal nanoparticles are a bridge between metal compounds and a molecular structure. The metal compounds are chosen depending on the properties prerequisite, such as metal with antibacterial properties, in a nano-system. Silver nanoparticles are one such example. The silver nanoparticles experience an oxidized state, leading to the liberation of silver ions, which interferes in DNA replication of microorganisms. The silver ions also have a role in the inhibition of the synthesis of ATP, necessary to the metabolism of the microorganisms (Rai et al. 2009). Another metal nanoparticles with potent antimicrobial capacity is copper nanoparticles. Copper nanoparticles have been reported by Ingle et al. (2014) to be active against a wide range of microorganisms. Numerous metal nanoparticles have been investigated for their biomedical potential in the treatment and therapies of multiple infectious diseases such as tuberculosis and others.

Silver nanoparticles have been among the most potent combination, exhibiting significant antimicrobial efficiency against a broad range of Gram positive to Gram negative bacteria and comprising also the multidrug resistant organisms. Silver nanoparticles have thus been attributed the target and role as the next-generation antimicrobials in the fight against multidrug resistant pathogens (Rai et al. 2009).

Conclusion and Future Perspectives

Emerging and re-emerging diseases are becoming a great challenge with the increasing number of multidrug resistant organisms making their breakthrough. Several studies have highlighted the potential of nanoparticles as promising antimicrobials to combat the resistance gene acquired by pathogens. The synergistic effects of essential oil with antimicrobial properties coupled to nanoparticles may open new vistas in the fight against recurrent infectious diseases caused by the establishment of biofilms. Future works should be focused on the mechanism of action of nanoparticles on biofilm, in particular on the molecular changes of the sessile cells present within the biofilms. Based on those insights, *in vitro* and *in vivo* evaluation of the positive outcome as well as the adverse effects should be addressed.

References

Acevedo-Fani, A., Salvia-Trujillo, L., Rojas-Graü, M. and Martín-Belloso, O. 2015. Edible films from essential-oil-loaded nanoemulsions: physicochemical characterization and antimicrobial properties. Food Hydrocolloid. 47: 168–177.
Adorjan, B. and Buchbauer, G. 2010. Biological properties of essential oils: an updated review. Flavour Fragr. J. 25: 407–426.
Adukwu, E., Allen, S. and Phillips, C. 2012. The anti-biofilm activity of lemongrass (*Cymbopogon flexuosus*) and grapefruit (*Citrus paradisi*) essential oils against five strains of *Staphylococcus aureus*. J. Appl. Microbiol. 113(5): 1217–1227.
AL-Haj, N.A., Shamsudin, M.N. and Alipiah, N.M. 2010. Characterization of *Nigella sativa* L. essential oil-loaded solid lipid nanoparticles. Am. J. Pharmacol. Toxicol. 5(1): 52–57.
Alhede, M., Qvortrup, K., Liebrechts, R., Høiby, N., Givskov, M. and Bjarnsholt, T. 2012. Combination of microscopic techniques reveals a comprehensive visual impression of biofilm structure and composition. FEMS Immunol. Med. Microbiol. 65(2): 335–342.
Anwer, M., Jamil, S., Ibnouf, E. and Shakeel, F. 2014. Enhanced antibacterial effects of clove essential oil by nanoemulsion. J. Oleo Sci. 63(4): 347–354.
Atef, M., Rezaei, M. and Behrooz, R. 2015. Characterization of physical, mechanical, and antibacterial properties of agar-cellulose bionanocomposite films incorporated with savory essential oil. Food Hydrocolloid. 45: 150–157.
Aumeeruddy-Elalfi, Z., Gurib-Fakim, A. and Mahomoodally, F. 2015. Antimicrobial, antibiotic potentiating activity and phytochemical profile of essential oils from exotic and endemic medicinal plants of Mauritius. Ind Crop Prod. 71: 197–204.
Aumeeruddy-Elalfi, Z. and Mahomoodally, M. 2016. Extraction techniques and pharmacological potential of essential oils from medicinal and aromatic plants of Mauritius. pp. 51–80. *In*: Peters, M. (ed.). Essential Oils: Historical Significance, Chemical Composition and Medicinal Uses and Benefits. Nova Publisher, USA.
Aumeeruddy-Elalfi, Z., Gurib-Fakim, A. and Mahomoodally, M. 2016. Chemical composition, antimicrobial and antibiotic potentiating activity of essential oils from 10 tropical medicinal plants from Mauritius. J. Herb. Med. 6: 88–95.
Bai, A.J. and Vittal, R. 2014. Quorum sensing inhibitory and anti-biofilm activity of essential oils and their *in vivo* efficacy in food systems. Food Biotechnol. 28(3): 269–292.
Bakkali, F., Averbeck, S., Averbeck, D. and Idaomar, M. 2008. Biological effects of essential oils—a review. Food Chem. Toxicol. 46: 446–475.
Béchet, M. and Blondeau, R. 2003. Factors associated with the adherence and biofilm formation by *Aeromonas caviae* on glass surfaces. J. Appl. Microbiol. 94(6): 1072–1078.
Bedi, S., Tanuja, S. and Vyas, P. 2010. A Handbook of Aromatic and Essential Oil Plants: Cultivation, Chemistry, Processing and Uses. Agrobios, New Delhi, India, pp. 29–74.
Bellamy, W., Takase, M., Wakabayashi, H., Kawase, K. and Tomita, M. 1992. Antibacterial spectrum of lactoferricin B, a potent bactericidal peptide derived from the N-terminal region of bovine lactoferrin. J. Apply Bacteriol. 73: 472–479.
Bellamy, W., Wakabayashi, H., Takase, M., Kawase, K., Shimamura, S. and Tomita, M. 1993. Killing of *Candida albicans* by lactoferricin B, a potent antimicrobial peptide derived from the N-terminal region of bovine lactoferrin. Med. Microbiol. Immunol. 182: 97–105.
Bersan, S., Galvão, L., Goes, V., Sartoratto, A., Figueira, G. and Rehder, V. 2014. Action of essential oils from Brazilian native and exotic medicinal species on oral biofilms. BMC Complement Altern. Med. 14(1): 451.
Besciak, G. and Surmacz-Górska, J. 2014. Biofilm as a basic life form of bacteria. pp. 17–19. *In*: Besciak, G. and Surmacz-Górska, J. (eds.). Proceedings of a Polish-Swedish-Ukrainian Seminar, Krakow Poland. Available at https://www.seed.abe.kth.se/polopoly_fs/1.651085!/JPSU17P13.pdf.
Boto, L. 2010. Horizontal gene transfer in evolution: facts and challenges. Proc. Biol. Sci. 277: 819–27.
Budzyńska, A., Wieckowska-Szakie, M., Sadowska, B., Kalemba, D. and Rózalska, B. 2011. Antibiofilm activity of selected plant essential oils and their major components. Pol. J. Microbiol. 60(1): 35–41.
Burt, S.A. and Reinders, R.D. 2004. Antibacterial activity of selected plant essential oils against *Escherichia coli* O157:H7. Lett Appl. Microbiol. 36: 162–167.
Burt, S.A. 2007. Antibacterial activity of essential oils: potential application in food. Ph.D. Thesis. Netherlands, Utrecht: Utrecht University, ISBN/EAN: 978-90-393-4661-7.
Celiktas, O.Y., HamesKocabas, E.E., Bedir, E., Vardar Sukan, F., Ozek, T. and Baser, K.H.C. 2007. Antimicrobial activities of methanol extracts and essential oils of *Rosmarinus officinalis*, depending on location and seasonal variations. Food Chem. 100(2): 553–559.
Centers for Disease Control and Prevention. Superbugs threaten hospital patients. CDC Newsroom Releases. March 2016. Atlanta, USA. Available at http://www.cdc.gov/media/releases/2016/p0303-superbugs.html.
Centers for Disease Control and Prevention. 2016. CDC: 1 in 3 antibiotic prescriptions unnecessary. CDC Newsroom Release. Atlanta, USA. Available at http://www.cdc.gov/media/releases/2016/p0503-unnecessary-prescriptions.html.
Ceylan, O., Ugur, A., Sarac, N. and Donmez Sahin, M. 2014. The antimicrobial and antibiofilm activities of *Mentha* x *piperita* L. essential oil. J. Biosci. Biotech. 23–27.
Chaiyana, W. and Okonogi, S. 2012. Inhibition of cholinesterase by essential oil from food plant. Phytomedicine 19: 836–839.
Chorianopoulos, N.G., Giaouris, E.D., Skandamis, P.N., Haroutounian, S.A. and Nychas, G.J. 2008. Disinfectant test against monoculture and mixed-culture biofilms composed of technological, spoilage and pathogenic bacteria: bactericidal effect of essential oil and hydrosol of *Satureja thymbra* and comparison with standard acid-base sanitizers. J. Appl. Microbiol. 104: 1586–1596.
Cole, J. and Nizet, V. 2016. Bacterial evasion of host antimicrobial peptide defenses. Microbiol. Spectr. 4(1): VMBF-0006-2015.

Costerton, J.W., Cheng, K.J.,Geesey, G.G., Ladd, T.I., Nickel, J.C., Dasgupta, M. and Marrie, T.J. 1987. Bacterial biofilms in nature and disease. Annu. Rev. Microbiol. 41: 435–64.

Costerton, J.W, Geesey, G.G and Cheng, G.K. 1978. How bacteria stick. Sci. Am. 238: 86–95.

Costerton, J.W. 2007. The biofilm primer. pp. 129–168. *In*: Costerton, J.W. (ed). Control of all Biofilm Strategies and Behaviours. Springer-Verlag, Heidelberg, Berlin.

De Beer, D.,Stoodley, P., Lewandowski, Z. 1994. Liquid flow inheterogeneous biofilms. Biotechnol. Bioeng. 44: 636–641.

Da Silva, A., Lopes, P., Azevedo, M., Costa, D., Alviano, C. and Alviano, D. 2012. Biological activities of α-Pinene and β-Pinene enantiomers. Molecules. 17(12): 6305–6316.

Davies, D.G., Parsek, M.R., Pearson, J.P., Iglewski, B.H. and Costerton, J.W. 1998. The involvement of cell-to-cell signals in the development of a bacterial biofilm. Science 280(5361): 295–298.

Davis, S.C., Ricotti, C. and Cazzaniga, A. 2008. Microscopic and physiologic evidence for biofilm-associated wound colonization *in vivo*. Wound Repair Regen. 16(1): 23–29.

Denyer, S.P. and Hugo, W.B. 1991. Mechanisms of antibacterial action—asummary. pp. 331–334. *In*: Denyer, S.P. and Hugo, W.B. (eds.). Mechanisms of Action of Chemical Biocides. Blackwell Scientific Publications,Oxford, USA.

Derwich, E., Benziane, Z. and Boukir, A. 2010. GC/MS Analysis and antibacterial activity of the essential oil of *Mentha pulegium* grown in Morocco. Res. J. Agric. & Biol. Sci. 6: 191–198.

Donlan, R.M. 2002. Biofilms: Microbial life on surfaces. Emerg. Infect. Dis. 8(9): 881–890.

Dymock, D. 2003. Detection of microorganism in dental plaque. pp. 199–220. *In*: Jass, J., Surman, S. and Walker, J. (eds.). Medical Biofilms, Detection, Prevention and Control. Wiley, UK.

Edward-Jones, V. 2013. Alternative antimicrobial approaches to fighting multidrug-resistant infections. pp. 1–8. *In*: Rai, M. and Kon, K. (eds.). Fighting Multidrug Resistance with Herbal Extracts, Essential Oils and their Components. Elsevier Inc., UK.

Elliott, D.R., Rolfe, S.A., Scholes, J.D. and Banwart, S.A. 2007. Quantification of biofilm and planktonic communities. pp. 1–7. *In*: Gilbert, P., Allison, D., Brading, M., Pratten, J. and Spratt, D. (eds.). Biofilms: Coming of Age. The Biofilm Club, 8th Meeting. Available at: http://drelliott.net/quantification-of-biofilm-and-planktonic-communities/.

Faleiro, M.L. 2011. The mode of antibacterial action of essential oils. pp. 1143–1156. *In*: Mendez-Vilas, A. (ed.). Science Against Microbial Pathogens: Communicating Current Research and Technological Advances. Formatex Research Center, Badajoz, Spain.

Flores, F.C., de Lima, J.A., Ribeiro, R.F., Alves, S.H., Rolim, C.M., Beck, R.C. and Da Silva, C.B. 2013. Antifungal activity of nanocapsule suspensions containing tea tree oil on the growth of *Trichophyton rubrum*. Mycopathologia 175(3-4): 281–286.

Ghosh, V., Mukherjee, A. and Chandrasekaran, N. 2013. Ultrasonic emulsification of food-grade nanoemulsion formulation and evaluation of its bactericidal activity. Ultrason Sonochem. 20(1): 338–344.

Gomes, C., Moreira, R.G. and Castell-Perez, E. 2011. Poly(DL-lactide-co-glycolide) (PLGA) Nanoparticles with entrapped trans-cinnamaldehyde and eugenol for antimicrobial delivery applications. J. Food Sci. 76(2): N16–N24.

Gortzi, O., Lalas, S., Chinou, I. and Tsaknis, J. 2007. Evaluation of the antimicrobial and antioxidant activities of *Origanum dictamnus* extracts before and after encapsulation in liposomes. Molecules 12: 932–945.

Guenther, E. 1948. The Essential Oils: History-origin in Plants-production - Analysis. D. Van Nostr and Company Inc., New York. 15–77.

Günther, F., Wabnitz, G.H. and Stroh, P. 2009. Host defence against *Staphylococcus aureus* biofilms infection: phagocytosis of biofilms by polymorphonuclear neutrophils (PMN). Mol. Immunol. 46(8-9): 1805–1813.

Hall-Stoodley, L., Costerton, J.W. and Stoodley, P. 2004. Bacterial biofilms: from the natural environment to infectious diseases. Nat. Rev. Microbiol. 2: 95–108.

Hassan, A., Usman, J., Kaleem, F., Omair, M., Khalid, A. and Iqbal, M. 2011. Evaluation of different detection methods of biofilm formation in the clinical isolates. Braz. J. Infect. Dis. 15(4): 305–311.

Hunter, V.M. 2009. Essential Oils: Art, Agriculture, Science, Industry and Entrepreneurship a Focus on the Asia-Pacific Region. Malaysia: University of Malaysia Perli. Nova publishers, New York. 1–773.

Hussain, A.I., Anwar, F. and Sherazi, S.T.H. 2008. Chemical composition, antioxidant and antimicrobial activities of basil (*Ocimum basilicum*) essential oils depends on seasonal variations. Food Chem. 108: 986–995.

Hu, M., Zhang, C., Mu, Y., Shen, Q. and Feng, Y. 2010. Indole affects biofilm formation in bacteria. Indian J. Microbiol. 50(4): 362–368.

Ingle, A., Duran, N. and Rai, M. 2014. Bioactivity, mechanism of action and cytotoxicity of copper-based nanoparticles: a review. Appl. Microbiol. Biotechnol. 98: 1001–1009.

Isaacs, H., Jr, Chao, D., Yanofsky, C. and Saier, M.H., Jr. 1994. Mechanism of catabolite repression of tryptophanase synthesis in *Escherichia coli*. Microbiology 140: 2125–2134.

Jefferson, K.K. 2004. What drives bacteria to produce a biofilm? FEMS Microbial Lett. 236(2): 163–73.

Jo, Y.J., Chun, J.Y., Kwon, Y.J., Min, S.G., Hong, G.P. and Choi, M.J. 2015. Physical and antimicrobial properties of trans-cinnamaldehyde nanoemulsions in water melon juice. LWT - Food Sci. Technol. 60: 444–451.

Kavanaugh, N.L. and Ribbeck, K. 2012. Selected antimicrobial essential oils eradicate *Pseudomonas* spp. and *Staphylococcus aureus* biofilms. Appl. Environ. Microbiol. 78(11): 4057–4061.

Kifer, D., Mužinić, V. and Klarić, M. 2016. Antimicrobial potency of single and combined mupirocin and monoterpenes, thymol, menthol and 1,8-cineole against *Staphylococcus aureus* planktonic and biofilm growth. J. Antibiot. doi:10.1038/ja.2016.10.

Kim, I., Lee, H. and Kim, J. 2013. Plum coatings of lemongrass oil-incorporating carnauba wax-based nanoemulsion. J. Food Sci. 78(10): 1551–1559.

Kim, Y.G., Lee, J.H., Kim, S.I., Baek, K.H. and Lee, J. 2015. Cinnamon bark oil and its components inhibit biofilm formation and toxin production. Int. J. Food Microbiol. 195: 30–39.

Kon, K.V. and Rai, M.K. 2010. Plant essential oils and their constituents in coping with multidrug-resistant bacteria. Expert Rev. Anti Infect. Ther. 10: 775–790.

Landry, K.S., Micheli, S., McClements, D.J. and McLandsborough, L. 2015. Effectiveness of a spontaneous carvacrol nanoemulsion against *Salmonella enterica Enteritidis* and *Escherichia coli* O157:H7 on contaminated broccoli and radish seeds. Food Microbiol. 51: 10–17.

Lasaro, M.A., Salinger, N. and Zhang, J. 2009. F1C fimbriae play an important role in biofilm formation and intestinal colonization by the *Escherichia coli* commensal strain Nissle 1917. Appl. Environ. Microbiol. 75(1): 246–251.

Lawless, J. 1995. The Illustrated Encyclopedia of Essential Oils: The Complete Guide to The Use of Oils in Aromatherapy and Herbalism. Harper Collins Publishers, London, UK. 21–65.

Lembre, P., Lorentz, C. and Di Martino, P. 2012. Exopolysaccharides of the biofilm matrix: a complex biophysical world. pp. 372–392. *In*: Nedra Karunaratne, D. (ed.). The Complex World of Polysaccharides. Intech. Available at: http://www.intechopen.com/books/the-complex-world-of-polysaccharides/exopolysaccharides-of-the-biofilm-matrix-a-complex-biophysical-world.

Li, Y., Zheng, J., Xiao, H. and McClements, D.J. 2012. Nanoemulsion-based drug delivery systems for poorly-water soluble bioactive compounds: influence of formulation parameters on polymethoxyflavone crystallization. Food Hydrocoll. 27: 517–528.

Liu, Y., Wang, Y., Walsh, T., Yi, L., Zhang, R. and Spencer, J. 2016. Emergence of plasmid-mediated colistin resistance mechanism MCR-1 in animals and human beings in China: a microbiological and molecular biological study. Lancet Infect. Dis. 16(2): 161–168.

Lubbe, A. and Verpoorte, R. 2011. Cultivation of medicinal and aromatic plants for specialty industrial materials—a review. Ind. Crops Prod. 34: 785–801.

Mathur, S., Udgire, M. and Khambhapati, A. 2013. Effect of essential oils on biofilm formation by *Proteus mirabilis*. Int. J. Pharm. Bio. Sci. 4(4): 1282–1289.

McGann, P., Snesrud, E., Maybank, R., Corey, B. and Ong, A.C. 2016. *Escherichia coli* Harboring mcr-1 and blaCTX-M on a Novel IncF Plasmid: First report of mcr-1 in the USA. Antimicrob Agents Chemother. doi:10.1128/AAC.01103-16.

Mihai Grumezescu, A. 2013. Essential oils and nanotechnology for combating microbial biofilms. Curr. Org. Chem. 17(2): 90–96.

Namasivayam, S.K.R., Jayakumar, D., Kumar, R. and Bharani, R.S.A. 2015. Antibacterial and anticancerous biocompatible silver nanoparticles synthesized from the cold-tolerant strain of *Spirulina platensis*. J. Coastal Life Med. 3(4): 265–272.

Neethirajan, S., Clond, M. and Vogt, A. 2014. Medical biofilms-nanotechnology approaches. J. Biomed. Nanotechnol. 10(10): 2806–2827.

Nychas, G.J.E. 1995. Natural antimicrobials from plants. pp. 58–89. *In*: Gould, G.W. (ed.). New Methods of Food Preservation. Blackie Academic & Professional, London, UK.

Palombo, E.A. 2011. Traditional Medicinal Plant Extracts and Natural Products with Activity Against Oral Bacteria: Potential Application in the Prevention and Treatment of Oral Diseases. 2011: 680354.

Pamp, S., Sternberg, C. and Tolker-Nielsen, T. 2009. Insight into the microbial multicellular lifestyle via flow-cell technology and confocal microscopy. Cytometry 75(2): 90–103.

Poole, A.M. 2009. Horizontal gene transfer and the earliest stages of the evolution of life. Res. Microbiol. 160: 473–480.

Rai, M.K., Yadav, A.P. and Gade, A.K. 2009. Silver nanoparticles as a new generation of antimicrobials. Biotechnol. Adv. 27(1): 76–82.

Raut, J.S. and Karuppayil, S.M. 2014. A status review on the medicinal properties of essential oils. Ind. Crop Prod. 62: 250–264.

Roilides, E., Simitsopoulou, M., Katragkou, A. and Walsh, T.J. 2015. How biofilms evade host defenses. Microbiol. Spectr. 3(3): 1–10.

Salvia, L., Rojas, M., Soliva, R. and Martín, O. 2014. Impact of microfluidization or ultrasound processing on the antimicrobial activity against *Escherichia coli* of lemongrass oil-loaded nanoemulsions. Food Control. 37: 292–297.

Sandasi, M., Leonard, C.M., Van Vuuren, S.F. and Viljoen, A.M. 2011. Peppermint (*Mentha piperita*) inhibits microbial biofilms *in vitro*. South African J. Bot. 77: 80–85.

Schmidt, B., Ribnicky, D.M., Poulev, A., Logendra, S., Cefalu, W.T. and Raskin, L. 2008. A natural history of botanical therapeutics. Metab. Clin. Exp. 57: 53–59.

Selim, S., Adam, M., Hassan, S. and Albalawi, A. 2014. Chemical composition, antimicrobial and antibiofilm activity of the essential oil and methanol extract of the Mediterranean cypress (*Cupressus sempervirens* L.). BMC Complement Altern. Med. 14(1): 179.

Sienkiewicz, M., Poznańska-Kurowska, K., Kaszuba, A. and Kowalczyk, E. 2014. The antibacterial activity of geranium oil against gram-negative bacteria isolated from difficult-to-heal wounds. Burns 40(5): 1046–1051.

Simões, M., Simões, L.C. and Vieira, M.J. 2010. A review of current and emergent biofilmcontrol strategies. LWT - Food Sci. Technol. 43: 573–583.

Sotolongo, J., Ruiz, J. and Fukata, M. 2011. The role of innate immunity in the host defense against intestinal bacterial pathogens. Curr. Infect. Dis. Rep. 14(1): 15–23.

Sugumar, S., Ghosh, V., Nirmala, M., Mukherjee, A. and Chandrasekaran, N. 2014. Ultrasonic emulsification of eucalyptus oil nanoemulsion: antibacterial activity against *Staphylococcus aureus* and wound healing activity in wistar rats. Ultrason. Sonochem. 21(3): 1044–1049.

Tassou, C.C., Koutsoumanis, K. and Nychas, G.J.E. 2000. Inhibition of *Salmonella enteridis* and *Staphyloccus aureus* on nutrient both by mint essential oil. Food Res. Int. 48: 273–280.

Valeriano, C., de Oliveira, T.L.C., de Carvalho, S.M., Cardoso, M.D.G. and Alves, E. 2012. The sanitizing action of essential oil-based solutions against *Salmonella enterica* serotype *Enteritidis* S64 biofilm formation on AISI 304 stainless steel. Food Control. 25: 673–677.

Walsh, S.E., Maillard, J.Y., Russell, A.D., Catrenich, C.E., Charbonneau, D.L. and Bartolo, R.G. 2003. Activity and mechanisms of action of selected biocidal agents on Gram-positive and negative bacteria. J. Appl. Microbiol. 94: 240–247.

Wendakoon, C.N. and Sakaguchi, M. 1995. Inhibition of amino acid decarboxylase activity of *Enterobacter aerogenes* by active components in spices. J. Food Prot. 58: 280–283.

World Health Organisation. 2015. Antimicrobial Resistance. Fact Sheet N194.

Yadav, M., Chae, S., Im, G., Chung, J. and Song, J. 2015. Eugenol: A Phyto-compound effective against methicillin-resistant and methicillin-sensitive *Staphylococcus aureus* clinical strain biofilms. PLOS ONE. 10(3): p.e0119564.

Zenati, F., Benbelaid, F., Khadir, A., Bellahsene, C. and Bendahou, M. 2014. Antimicrobial effects of three essential oils on multidrug resistant bacteria responsible for urinary infections. J. App. Pharm. Sci. 4(11): 15–18.

Nano-Ag Particles and Pathogenic Microorganisms

Antimicrobial Mechanism and its Application

JiEun Yun and *Dong Gun Lee**

Introduction

Human diseases due to pathogenic microorganisms have been prevalent over the past few years and these infectious diseases are the main cause of worldwide morbidity and mortality (Ouédraogo et al. 2012). Treatment of pathogenic microorganisms such as *Escherichia coli*, *Pseudomonas aeruginosa* and *Salmonella* spp. (bacteria) and *Aspergillus fumigatus* and *Candida albicans* (fungus) is needed for human health and therefore, a variety of antibiotics have been developed and used for curing human diseases consistently (Mantareva et al. 2007, Paulo et al. 2010). Unfortunately, their continued usage in clinical application raises a serious problem, the emergence of resistant pathogens to antibiotics. To overcome the resistance, the development of new antimicrobial agents has gained increasing importance in recent years (Ahmad and Beg 2001).

Nanotechnology is a rapidly growing field in science and technology for the purpose of manufacturing new materials at the nanoscale level. Because of their high surface area to volume ration and the unique chemical and physical properties, nanoscale materials are considered as novel antimicrobial agents (Retchkiman-Schabes et al. 2006, Aruguete et al. 2013). Especially, nanoparticles are most promising as they show good antimicrobial efficacy against pathogenic microorganisms. There are various types of nanoparticles like copper, zinc, titanium, gold and silver and most of all, silver nanoparticles (nano-Ag particles) have proved to be most effective against pathogenic microorganisms. Also, it is demonstrated that nano-Ag particles are non-toxic to humans as silver ions that are non-toxic to humans through centuries of use (Oberdörster et al. 2005, Gong et al. 2007, Gurunathan et al. 2009).

Antimicrobial activity of nano-Ag particles are proved through many studies and their mechanisms on pathogenic microorganisms are discovered. Nano-Ag particles possess dual mechanisms: membrane disruption and programmed cell death called apoptosis (Fig. 8.1). Membrane damage leading to increase of membrane permeability and membrane depolarization is induced by nano-Ag particles

School of Life Sciences, College of Natural Sciences, Kyungpook National University, 80 Daehakro, Bukgu, Daegu 41566, Republic of Korea.
* Corresponding author: dglee222@knu.ac.kr

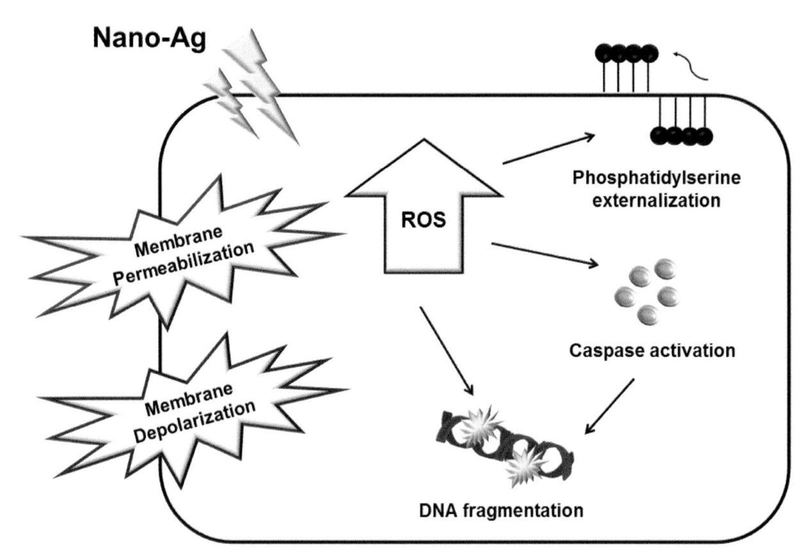

Figure 8.1. Antimicrobial mechanism of silver nanoparticles (nano-Ag particles or nano-Ag). Nano-Ag induces the cell death via dual mechanisms: membrane disruption and apoptosis in pathogenic microorganisms.

both fungal and bacterial cells (Kim et al. 2009, Li et al. 2010). Furthermore, nano-Ag particles show apoptosis hallmarks such as ROS accumulation, caspase activation, phosphatidylserine exposure and DNA fragmentation in fungal cells and especially, bacterial cells (Hwang et al. 2012, Lee et al. 2014).

Mechanism I: Membrane Disruption

Fungal cell membrane

Decrease of membrane integrity and membrane depolarization

Maintaining membrane integrity is important for cell survival because the membrane protects the cells from the external environment and regulates cellular responses (Benyagoub et al. 1996, Portet 2009). The integrity change of the fungal membrane could be detected by a fluorescence probe, 1,6-diphenyl-1,3,5-hexatriene (DPH). DPH interacts with an acyl group of the membrane lipid bilayer and do not disturb the membrane. When cells are under severe damages such as membrane disruption and pore formation, DPH becomes detached from the membrane (Vincent et al. 2004). Significant decrease of DPH fluorescence was observed in nano-Ag particles-treated cells and it indicates that nano-Ag particles interact with the fungal membrane and decrease the membrane integrity. The effect of nano-Ag particles on fungal membrane tend to increase depending on the concentration increase (Fig. 8.2A).

Fungal cells establish multiple ion gradients across the membrane for maintaining their membrane potential. The ion gradients are sustained by the restricted membrane permeability for small solutes and ions (Yu 2003). However, when membrane loses its stability, resulting from severe membrane damage, increase of membrane permeability leading to leakage of a variety of ions from the cell occurs. Then, undesired ion leakage causes the depolarization of the fungal membrane (Bolintineanu et al. 2010). The potential change of fungal membrane could be detected by a fluorescence probe, bis-(1,3-dibutylbarbituric acid)-trimethineoxonol [DiBAC$_4$(3)]. DiBAC$_4$(3) is voltage-sensitive and when it enters depolarized cells, it binds to lipid-rich intracellular components (Liao et al. 1999). The increase of DiBAC$_4$(3) fluorescence was shown after being treated with nano-Ag particles, suggesting that nano-Ag particles induce membrane depolarization via increasing of the membrane permeability (Fig. 8.2B).

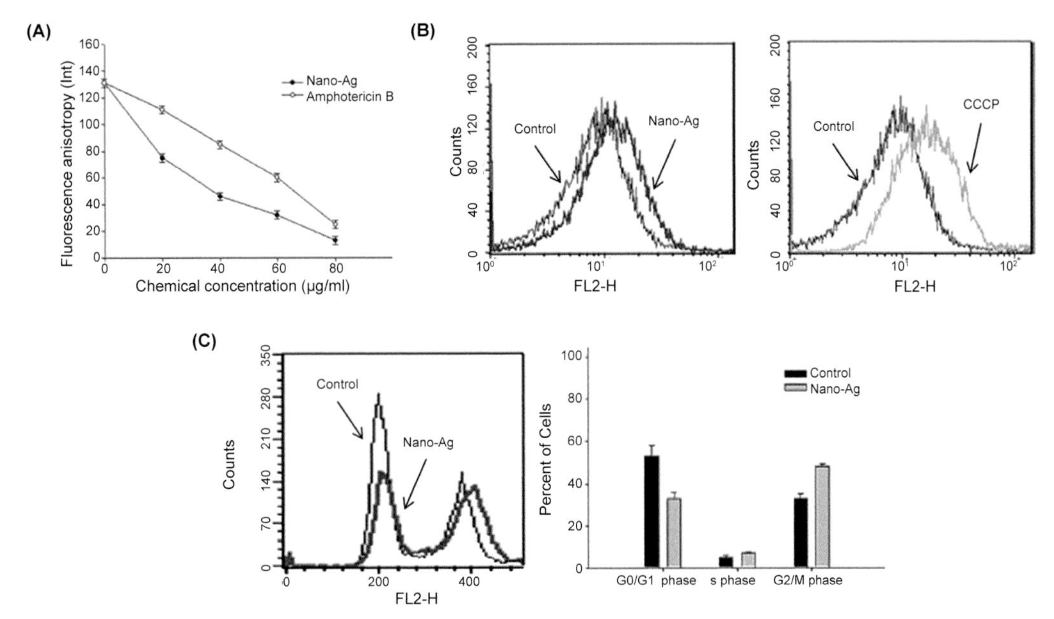

Figure 8.2. (A) Decrease of DPH fluorescence after treated with 20, 40, 60 and 80 μg/mlnano-Ag and amphotericin B shows the membrane disruption of *Candida albicans*. The error bars represent the standard deviation (SD) values of triplicate experiments. (B) Detection of membrane potential with DiBAC$_4$(3) staining. Increase of fluorescence indicates the membrane depolarization in nano-Ag-and CCCP-treated cells. (C) Analysis of cell cycle arrest by nano-Ag using PI staining. Two peaks indicate G1 (left) and G2/M phase (right), respectively. The bars represent the percent of cells in G1, S, and G2/M phase (Kim et al. 2009).

Release of intracellular glucose and trehalose and cell cycle arrest

Under stressful conditions such as desiccation, dehydration, oxidation and toxic agents, fungal cells undergo the inactivation and denaturation of proteins and membrane. To stabilize or recover these conditions, fungal cells have a trehalose, a non-reducing disaccharide of glucose that functions as a compatible solute in the stabilization of biological structures (Garg et al. 2002). When cell membrane is impaired, resulting in the loss of membrane integrity, level of intracellular trehalose rapidly increases. Therefore, the effect of nano-Ag particles on fungal membrane could be detected by measuring the glucose and trehalose released in cell suspension (Alvarez-Peral et al. 2002). After treatment of nano-Ag particles, the amount of both intracellular and released glucose and trehalose from *C. albicans* cells increased. Through these results, it is confirmed that nano-Ag particles induce cellular stress on membrane leading to release of several intracellular components such as glucose and trehalose.

Cell cycle is a collection of highly ordered processes that result in the duplication of a cell. As cells progress through the cell cycle, they undergo the check points to discriminate the disordered situations. When cellular damage is too extensive to be repaired, cells undergo cell cycle arrest to repair cellular damage or progress programmed cell death. The position of arrest within the cell cycle caries depending upon the phase in which the damage is sensed (Elledge 1996). Endo et al. (1997) have demonstrated that the growth inhibition of budding cells correlates with membrane damage and cell cycle arrest can occur by the destruction of membrane integrity. The amount of DNA contents that

Table 8.1. Amount of intracellular and released trehalose and glucose from nano-Ag-treated *C. albicans* cells.

	Intracellular glucose and trehalose (μg/mg)	Released glucose and trehalose (μg/mg)
Control	7.2	6.8
Nano-Ag	16.1	30.3
Amphotericin B	20.5	27.4

represents the G1, S and G2/M phase of fungal cell cycle could be detected with Propidium Iodide (PI). PI is a DNA-staining dye that intercalates between the bases of DNA or RNA molecules (Tas and Westerneng 1981). As shown in Fig. 8.2C, the percentage of cells in the G2/M phase increased while that in the G1 phase significantly decreased after treated with nano-Ag particles. This demonstrates that nano-Ag particles induce physiological damage leading to cell cycle arrest at the G2/M phase of *C. albicans*.

Bacterial cell membrane

Leakage of reducing sugars and proteins

Li et al. (2014) demonstrated the antibacterial mechanism of nano-Ag particles to Gram-negative bacteria. *E. coli* is selected as a model to study the effect of nano-Ag particles on membrane structure. Gram-negative bacteria have an outer membrane outside the peptidoglycan layer that is lacking in Gram-positive bacteria. The outer membrane is important for protecting bacteria from harmful agents such as degradative enzymes, toxins, drugs and detergents. The outer leaflet of the outer membrane is composed of the lipopolysaccharide (LPS) molecules that approximately cover three quarters of the surface of *E. coli* and the remaining quarter is composed of membrane proteins (Amro et al. 2000). Nano-Ag particles enhanced the membrane leakage of reducing sugars (Fig. 8.3A). Almost no reducing sugars were detected in leak from control cells, while the leakage amount of reducing sugars from nano-Ag particles treated-cells was detected. Similarly, nano-Ag particles also elevated the leakage of proteins through the membrane of *E. coli* (Fig. 8.3B). All these results demonstrate that nano-Ag particles induce the bacterial membrane damage and permeabilize the outer membrane, resulting in the leakage of cellular materials such as reducing sugars and proteins.

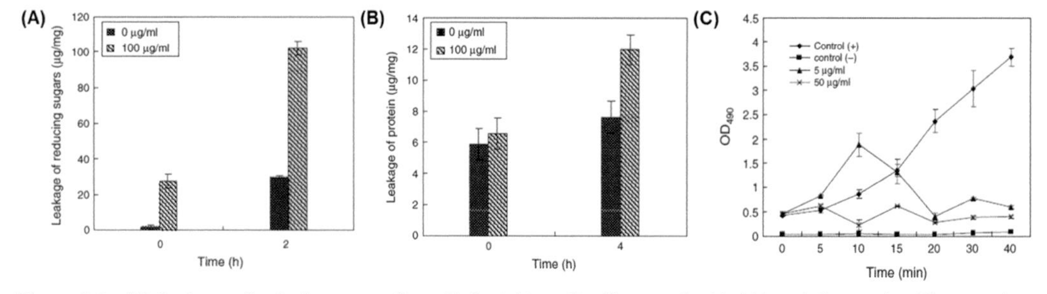

Figure 8.3. (A) Leakage of reducing sugars from *Escherichia coli* cells treated with 100 μg/ml nano-Ag. The error bars represent the SD values of duplicate experiments. (B) Leakage of proteins from nano-Ag-treated cells. The error bars represent the SD values of duplicate experiments. (C) Effect of nano-Ag on respiration chain dehydrogenases. The activity of respiration chain dehydrogenases was measured in absorbance at 490 nm. The control (–) and control (+) represent the boiled and not boiled cells, respectively (Li et al. 2010).

Effect on respiratory chain dehydrogenases

Holt et al. (2005) found that silver inhibited respiration of *E. coli* and Kim et al. (2008) found that silver interacted with thiol group of cysteine by replacing the hydrogen atom, thus inactivating the function of protein to inhibit growth of *E. coli*. As shown in Fig. 8.3C, nano-Ag particles affect the respiration chain dehydrogenases of *E. coli*. Activity of respiratory chain dehydrogenases in control (+) cells increased while that of control (–) cells that is boiled for 20 minutes showed nearly no change. Interestingly, that of nano-Ag particles-treated cells fell down rapidly and much lower. These results suggest that nano-Ag particles break the barriers of the outer membrane permeability, peptidoglycan and periplasm, and enter the inner membrane. Then, nano-Ag particles inactivate respiratory chain dehydrogenases, inhibiting respiration and growth of cells.

Action on the structures and membrane vesicle structures of bacterial cells

The structures of *E. coli* cells were observed by the electron micrographs with SEM and TEM. In micrographs by SEM, smooth surface and typical characters of rod shape were shown in control cells (Fig. 8.4A), while large leakage, others misshapen and fragmentary were shown in nano-Ag particles-treated cells (Fig. 8.4A). In micrographs by TEM, the surface of control cells was smooth and intact, and some filaments around cells were obvious and clear, while membrane of nano-Ag particles-treated cells was damaged severely; many pits and gaps appeared, and the membrane was fragmentary (Fig. 8.4A). Furthermore, membrane fragments of *E. coli* could form vesicles spontaneously and they resembled the spheroid that sizes were between 1/20 and 1/5 of the *E. coli* cells (Li et al. 2010). Membrane vesicles were dissolved and dispersed in nano-Ag particles-treated cells (Fig. 8.4B). Membrane components became disorganized and scattered from their original order and close arrangement. But, those of control cells were intact and clear-cut. These phenomena suggest that nano-Ag particles affect some proteins and phosphate lipids and induce collapse of membrane, resulting in cell decomposition and death.

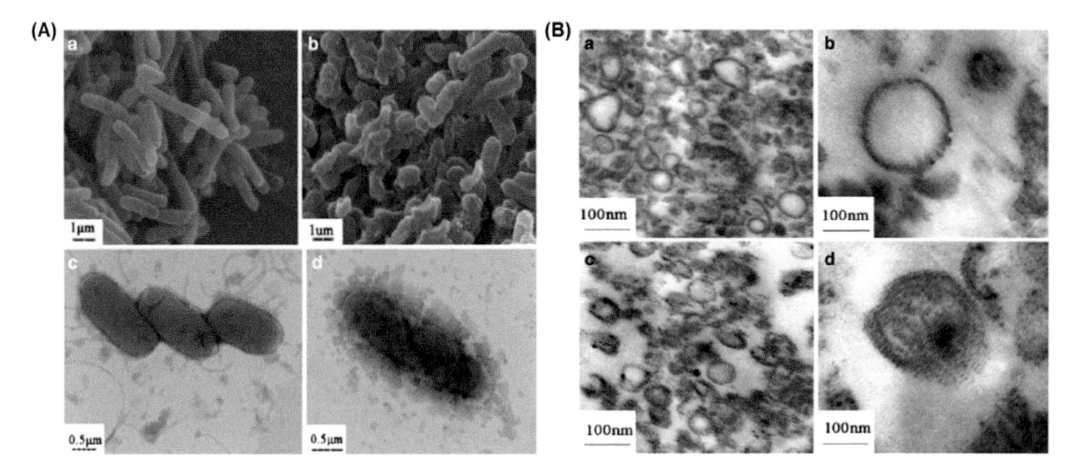

Figure 8.4. (A) Action of nano-Ag on *E. coli* cells observed by SEM (a, b) and TEM (c, d). a, c cell structure of native cells; b, d cell structure of nano-Ag-treated cells. (B) Action of nano-Ag on cell membrane vesicles observed by TEM. a, b structure of native cell membrane vesicles; c, d structure of nano-Ag-treated cell membrane vesicles (Li et al. 2010).

Mechanism II: Apoptosis

Fungal apoptosis

Reactive oxygen species (ROS) accumulation in fungal cells

Reactive Oxygen Species (ROS) including hydrogen peroxide, hydroxyl radicals and superoxide anions are continuously produced during the normal aerobic metabolism and act as second messengers in signal transduction pathways in fungal cells (Costa and Moradas-Ferreira 2001). When higher doses of ROS are produced, oxidative damage occurs because of the character of ROS, partially reduced and highly reactive (Fröhlich and Madeo 2000). According to recent research, exhibition of excessive ROS are shown in fungal cells that undergo apoptosis. ROS accumulation could be detected by the ROS-sensitive fluorescence probe, dihydrorhodamine (DHR-123). When DHR-123 enters the cells and reacts with ROS, ROS directly oxidize DHR-123 to the fluorescent derivative, rhodamine-123. The increased fluorescence intensity indicates the increase of ROS levels (Sakurada et al. 1992). After treatment of nano-Ag particles, the significant increase of DHR-123 fluorescence was shown, indicating excessive ROS levels. This result indicates that nano-Ag particles induce ROS accumulation in fungal cells and generate the oxidative stress (Fig. 8.5A).

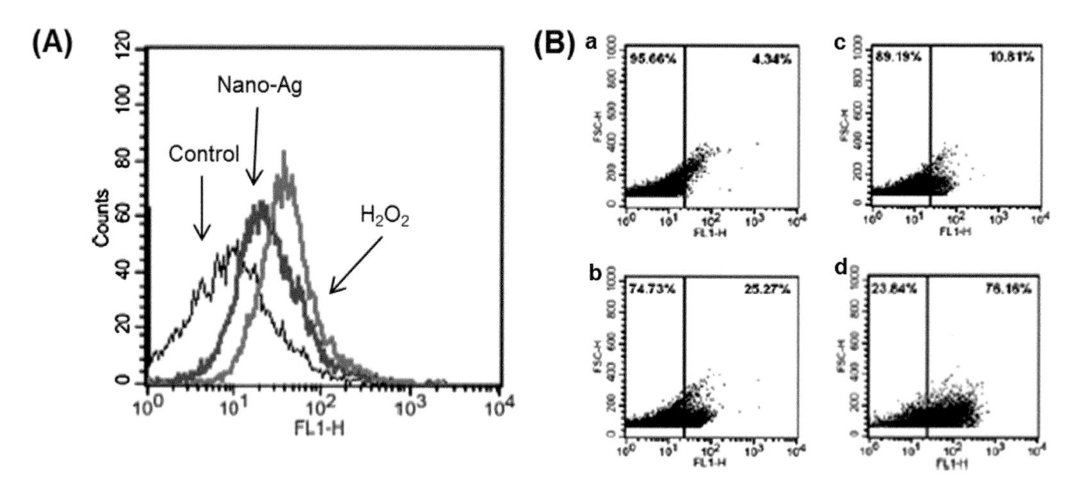

Figure 8.5. (A) Detection of ROS levels using DHR-123 staining. Increase of fluorescence indicates ROS accumulation in nano-Ag- and H_2O_2-treated cells. (B) Detection of hydroxyl radical levels using HPF staining. Increase of fluorescence indicates hydroxyl radical production in nano-Ag- and H_2O_2-treated cells. Hydroxyl radical scavenging compound, thiourea is pretreated for determining the effect of hydroxyl radical. a, control; b, nano-Ag-treated cells; c, nano-Ag- and thiourea-treated cells; d, H_2O_2-treated cells (Lee et al. 2014).

Hydroxyl radical is the neutral form of the hydroxide ion and highly reactive, promoting the cellular oxidative damage (Haruna et al. 2002). When hydrogen peroxide and metal ions exist, hydroxyl radical is produced through Fenton reaction and damages intracellular molecules such as DNA, proteins and lipids, resulting in cell death (Rollet-Labelle et al. 1998). Hydroxyl radical formation could be detected with the fluorescent probe, 3'-(p-hydroxylphenyl) fluorescein (HPF). HPF is oxidized by hydroxyl radical with high specificity and the fluorescence did not increase upon the hydroxyl radical formation (Setsukinai et al. 2003). Increase of HPF fluorescence was shown in nano-Ag particles-treated cells, suggesting that hydroxyl radical formation was induced by nano-Ag particles in *C. albicans* cells (Fig. 8.5B). To investigate the effect of hydroxyl radical, thiourea, a potent hydroxyl radical scavenger that has proved its effects of hydroxyl radical scavenging in both eukaryotes and prokaryotes was treated to nano-Ag particles-treated cells (Novogrodsky et al. 1982). Thiourea co-treatment significantly reduced hydroxyl radical formation in nano-Ag particles-treated cells. These results indicate that hydroxyl radical formation plays a key role in nano-Ag particles-induced cell death.

Caspase activation and cytochrome c release

Caspases belongs to family of cysteine proteases that use a cysteine residue as the catalytic nucleophile. It is demonstrated that action of caspases that is responsible for programmed cell death called apoptosis is conserved in fungal caspase. Caspases are activated in the early stage of apoptosis and promote the apoptotic signaling (Zivna et al. 2010). Caspase activation could be detected by the fluorescence probe, CaspACE FITC–VAD–FMK that binds to the active site of caspase (Wu et al. 2010). Nano-Ag particles-treated cells showed a significant green fluorescence of FITC–VAD–FMK and the number of activated caspase decreased in thiourea co-treated cells (Fig. 8.6A). These results suggest that nano-Ag particles lead to significant formation of strong oxidant hydroxyl radicals and then the hydroxyl radicals activated the caspase which facilitate apoptotic signals.

Many studies demonstrated that mitochondria play a key role in fungal apoptosis because of cytochrome c release. Cytochrome *c* release from mitochondria to cytosol is a crucial event in apoptotic pathway. It electrostatically binds to the outer face of the mitochondrial membrane and is the essential component of the respiratory chain in mitochondria (Dejean et al. 2006). Also, cytochrome *c* is released to the cytosol during the apoptosis and then, provokes the caspase-cascade

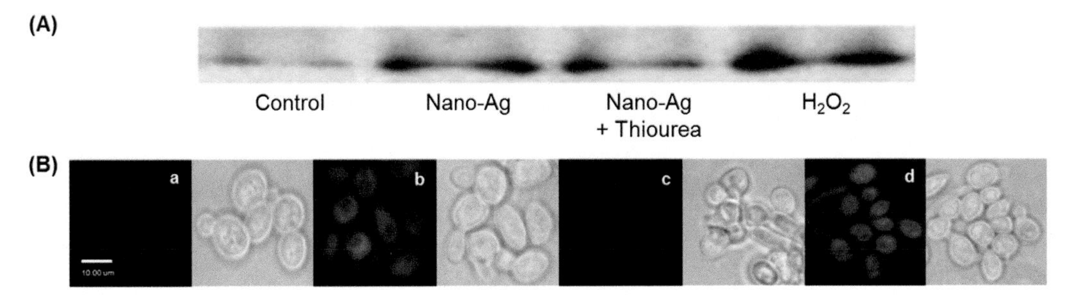

Figure 8.6. (A) Release of cytochrome *c* to cytosol after treated with nano-Ag. Cytosol was ultracentrifuged and released cytochrome *c* (the supernatant) was verified by western blotting. (B) Caspase activation in nano-Ag-treated *C. albicans* cells. Cells were stained with CaspACE FITCVAD-FMK and observed under a fluorescent microscope. a, control; b, nano-Ag-treated cells; c, nano-Ag- and thiourea-treated cells; d, H_2O_2-treated cells (Lee et al. 2014).

as a representative of the other apoptotic protease. Released cytochrome *c* from mitochondria binds to apoptotic protease-activating factors in cytosol (Pereira et al. 2007). A large amount of cytochrome *c* was detected in cytosol after treated with nano-Ag particles and a small amount of cytochrome *c* was detected in thiourea co-treated cells (Fig. 8.6B). Nano-Ag particles caused the release of cytochrome *c* from mitochondria to cytosol and hydroxyl radical formation affects cytochrome *c* release.

Phosphatidylserine exposure and DNA fragmentation

In most cells, the plasma membrane is asymmetrically consisting of phospholipids such as phosphatidylcholine, phosphatidylethanolamine, phosphatidylserine and sphingomyelin (Fadok et al. 1998). Phosphatidylcholine and sphingomyelin are concentrated on the outer leaflet of the plasma membrane, whereas phosphatidylethanolamine and phosphatidylserine are concentrated on the inner leaflet of the plasma membrane (Cerbón and Calderón 1991). In apoptotic cells, phosphatidylserine exposure is observed in early stage of apoptosis. Phosphatidylserine exposure could be detected by fluorescein isothiocyanate (FITC)–Annexin V and propidium iodide (PI) double staining. Annexin V binds to phosphatidylserine with high affinity in the presence of Ca^{2+} and therefore, is used for detecting the early apoptosis marker, phosphatidylserine exposure (Smrz et al. 2007). The cell population stained with Annexin$^+$/PI$^-$ significantly increased after treatment with nano-Ag particles, whereas no increase of Annexin$^+$/PI$^-$ staining cells in thiourea co-treated cells (Fig. 8.7A). These

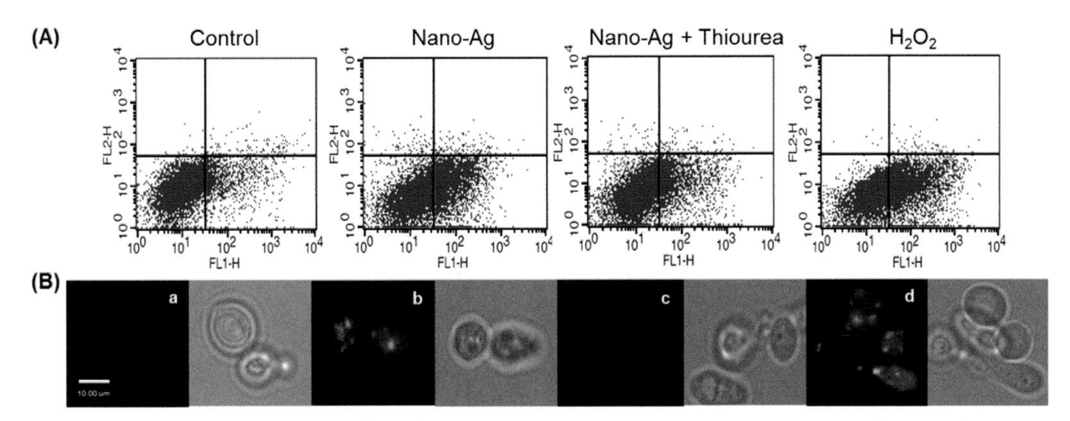

Figure 8.7. (A) Phosphatidylserine exposure in nano-Ag-treated *C. albicans* cells. Protoplasts were stained with FITC–Annexin V and PI double staining. FL1-H and FL2-H indicate Annexin V and PI fluorescence, respectively. Annexin V$^+$/PI$^-$ staining cells represent the early apoptotic cells. (B) Detection of DNA fragmentation using TUNEL staining. Cells were stained with TUNEL and observed under a fluorescent microscope. The bright green fluorescence indicates DNA fragmentation. a, control; b, nano-Ag-treated cells; c, nano-Ag- and thiourea-treated cells; d, H_2O_2-treated cells (Lee et al. 2014).

results indicate that nano-Ag particles induce the phosphatidylserine exposure of *C. albicans* cells and apoptotic cell death only in the presence of hydroxyl radical formation.

DNA is important in cell regulation and replication because it encodes genetic information. DNA can be damaged under various stressful conditions. Factors, triggering DNA damage are the oxidative stress by ROS formation and cleavage of a select group of substrates by caspases (Liu et al. 1999). Especially, DNA damage from the oxidative stress triggers the cell death via induction of apoptosis and these DNA break into short fragments by activated endonucleases during apoptosis is considered as the feature of late apoptosis (Wadskog et al. 2004). DNA fragmentation could be detected by TUNEL assay. TUNEL assay is a method for the short terminal deoxynucleotidyl transferase end labeling (Phillips et al. 2003). In nano-Ag particles-treated cells, TUNEL positive cells, showing the bright green fluorescence or intense green fluorescent spots were observed (Fig. 8.7B). But, those were not observed in thiourea co-treated cells. These results indicate that nano-Ag particles promote the oxidative condition and caspase activation of *C. albicans* cells, resulting in DNA fragmentation.

Bacterial apoptosis-like response

Time-kill kinetic assay and ROS generation

Dwyer et al. (2012) suggested that bactericidal agent can induce the apoptosis-like response in bacteria such as fungal apoptosis. Norfloxacin, targeting bacterial DNA induces apoptosis-like response in the concentration, showing bactericidal effect. In time-kill assay of *E. coli*, nano-Ag particles showed no survival within 6 hours and this result indicates that nano-Ag particles have bactericidal activity and potential for inducing apoptosis-like response (Fig. 8.8A).

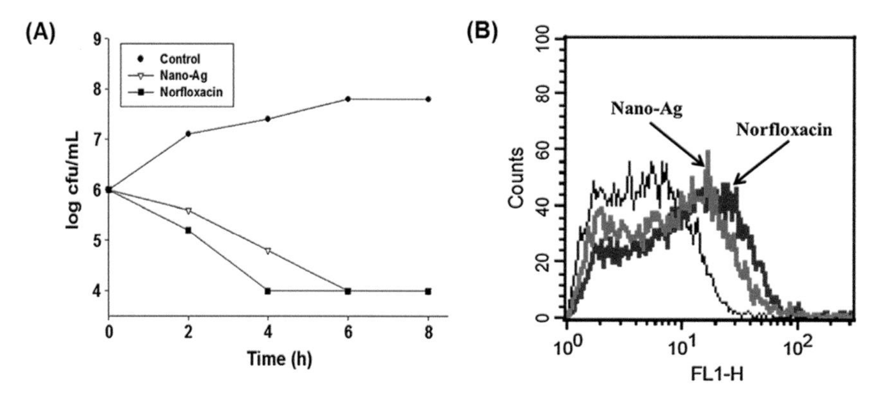

Figure 8.8. (A) Bactericidal effect of nano-Ag with time-kill kinetic analysis. After incubation for 6 hours, no survival was shown in nano-Ag-treated sample and the experiments were performed twice. (B) Detection of ROS levels using H_2DCFDA staining. Increase of fluorescence indicates ROS accumulation in nano-Ag- and norfloxacin-treated cells (Hwang et al. 2012).

ROS generation is necessary and sufficient in apoptosis of most cells (Hwang et al. 2012). Like as other cells, the induction of apoptosis-like response in *E. coli* proceeded by generation of ROS (Dwyer et al. 2012). ROS generation could be detected by a fluorescent probe, 2',7'-dichlorodihydrofluorescein diacetate (H_2DCFDA). It is a non-polar compound and becomes a non-fluorescent derivative (H_2DCF) by cellular esterase. Converted H2DCF is rapidly oxidized to highly fluorescent 2',7'-dichlorofluorescein (DCF) in the presence of intracellular ROS (LeBel et al. 1992). The increase of H_2DCFDA fluorescence was shown in nano-Ag particles-treated cells (Fig. 8.8B). This result indicates that ROS is involved in nano-Ag particles-induced bacterial apoptosis-like response.

Change in membrane potential and calcium accumulation in cytosol

During the bacterial apoptosis-like response, membrane depolarization is shown. Therefore, change of membrane potential could be measured with $DiBAC_4(3)$ staining, which is sensitive to cell membrane potential and enters into depolarized cells (Liao et al. 1999). The increase of $DiBAC_4(3)$ fluorescence was shown in nano-Ag particles-treated cells, indicating the change of membrane potential (Fig. 8.9A). This result indicates that nano-Ag particles exert its antibacterial effect via membrane depolarization.

Calcium ions are essential for the regulation of cellular processes and signal pathway, including cell death (Yoon et al. 2012). Also, calcium ions is related to some biological activities such as maintenance of cell structure, motility and transport and the increase of intracellular calcium levels was confirmed during the apoptosis-like response in *E. coli* (Lee et al. 2014). Intracellular calcium levels could be measured with the calcium indicator, Fura 2-AM. Fura 2-AM is cell permeable and selective for free cytosolic calcium ions. When Fura 2-AM crosses the membrane, it is metabolized by cellular esterase to the membrane impermeable dye, Fura 2 (Zherebitskaya et al. 2012). In nano-Ag particles-treated cells, significant increase of intracellular calcium levels was shown (Fig. 8.9B). This result indicates that calcium signaling is involved in the bacterial cell death in *E. coli* and nano-Ag particles cause the imbalance of ion homeostasis through the accumulation of calcium ions.

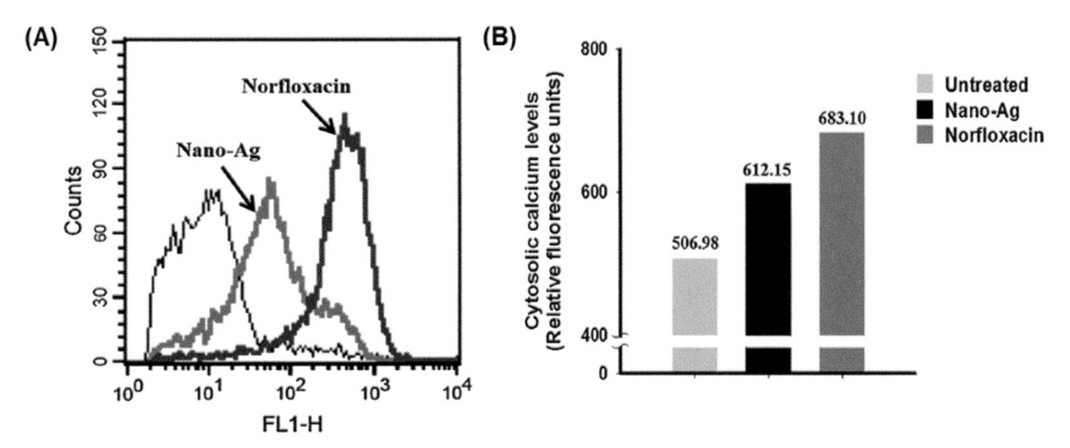

Figure 8.9. (A) Detection of membrane potential with $DiBAC_4(3)$ staining. Increase of fluorescence indicates the membrane depolarization in nano-Ag- and norfloxacin-treated cells. (B) Increase of cytosolic calcium levels in *E. coli* cells after treated with nano-Ag. Relative cytosolic calcium levels were detected with Fura 2-AM staining (Hwang et al. 2012).

Phosphatidylserine exposure and DNA fragmentation

Phosphatidylserine is located in the inner leaflet of the plasma membrane that becomes exposed to the outer leaflet during the early stages of apoptosis (van den Eijnde et al. 1998). Phosphatidylserine exposure is influenced by the calcium stimulation on enzymes that dissipates membrane arrangement. Early stage of apoptosis could be distinguished from late apoptosis and necrotic cells using Annexin V and PI double staining (Verhoven et al. 1995). The cell population stained with $Annexin^+/PI^-$ significantly increased after treated with nano-Ag particles (Fig. 8.10A). This result indicates that nano-Ag particles induce the phosphatidylserine exposure of *E. coli* cells, showing the early stage of apoptosis.

DNA fragmentation is considered as a late apoptosis hallmark and could be measured with TUNEL staining. TUNEL assay is widely used for observing DNA degradation and fragmentation. FITC-conjugated dUTP labels free 3'-OH termini with modified nucleotides and this action is catalyzed by terminal deoxynucleotidyl transferase (Labat-Moleur et al. 1998). During repair the damaged DNA, SulA stops cell division by binding to FtsZ, and this causes filamentation (Lee et al. 2014). Apoptotic

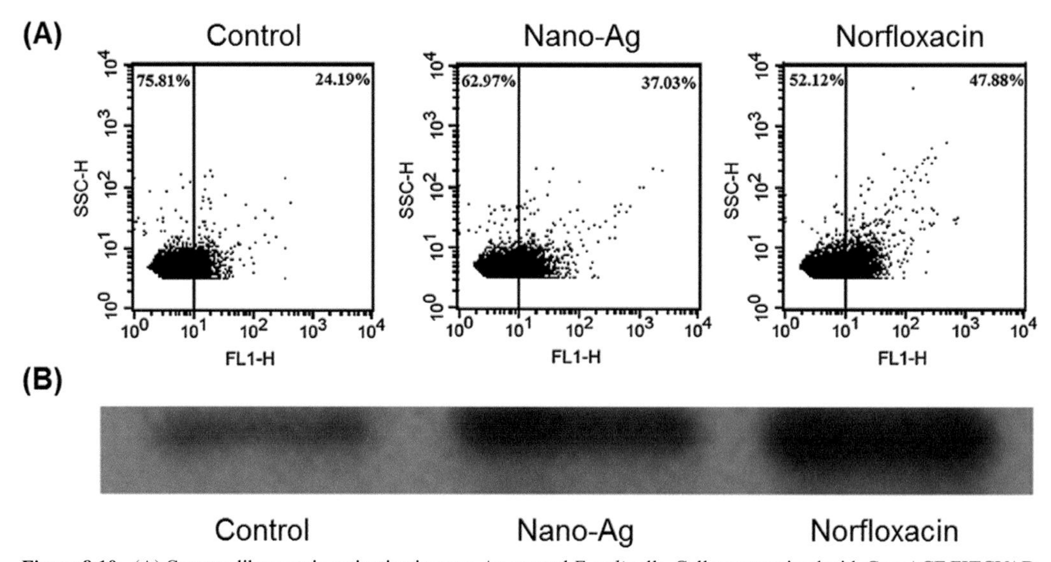

Figure 8.10. (A) Caspase-like protein activation in nano-Ag-treated *E. coli* cells. Cells were stained with CaspACE FITCVAD-FMK and increase of fluorescence indicates caspase activation in nano-Ag- and norfloxacin-treated cells. (B) Activation of SOS response in *E. coli* cells after treated with nano-Ag. RecA expression, involved in SOS response was verified by western blotting (Hwang et al. 2012).

phenomena, DNA fragmentation and filamentation were observed in nano-Ag particles-treated cells (Fig. 8.10B). Nano-Ag particles show the bright green fluorescence and prolonged cell size of *E. coli* cells. These results indicate that nano-Ag particles caused DNA fragmentation and cell filamentation due to the halt of cell division progression.

Activation of bacterial caspase-like proteins and DNA damage repair system

Caspases, cysteine-dependent aspartate-specific proteases are considered to be the major enzymes that provoke the apoptotic cascade in eukaryotes (Thornberry and Lazebnik 1998). Although bacterial sequential caspase orthologs have been not found, functional caspase orthologs of *E. coli* have been investigated (Vercammen et al. 2007). Many researchers have estimated that bacterial protein(s) activate in the procedure of bacterial apoptosis-like response and act like protease and nuclease in *E. coli* (Dwyer et al. 2012). Bacterial caspase-like protein activation could be confirmed using VAD-FMK, a FITC-conjugated peptide pan-caspase inhibitor (Madeo et al. 2002). Increase of VAD-FMK fluorescence after being treated with nano-Ag particles was shown, indicating the caspase activation in bacterial cells (Fig. 8.11A). The result of increased fluorescence suggests evidence of the presence of bacterial caspase-like proteins.

When bacterial cells suffer from severe DNA damage, SOS response is operated by Rec A protein to repair DNA damage. Rec A protein has a co-protease function in the autocatalytic cleavage of the Lex A repressor, which inhibits the expression of genes, involved in bacterial repair system (Maul et al. 2005). Dwyer et al. (2012) insisted on the relationship between Rec A and caspase and assumed that Rec A may act as a major regulator in both SOS repair system and bacterial apoptosis-like response. The activation of SOS response could be confirmed by measuring Rec A concentration via western blotting. Blots from nano-Ag particles-treated cells showed thicker Rec A band than that of control cells (Fig. 8.11B). This result indicates that nano-Ag particles-induced bacterial cell death involves the severe DNA damage and DNA repair system are activated by Rec A protein.

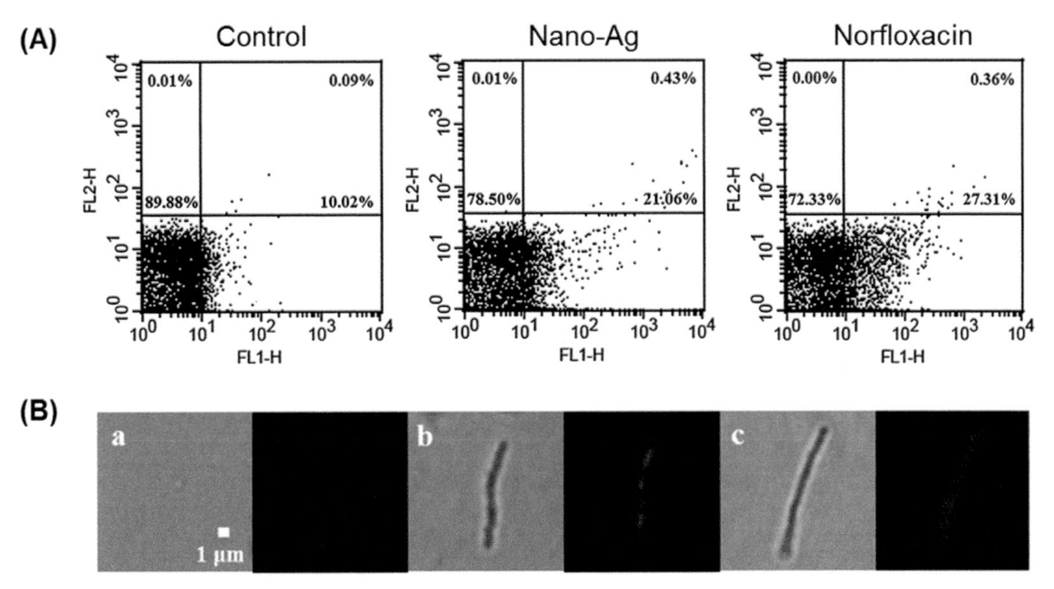

Figure 8.11. (A) Phosphatidylserine exposure in nano-Ag-treated *E. coli* cells. Cells were stained with FITC–Annexin V and PI double staining. FL1-H and FL2-H indicate Annexin V and PI fluorescence, respectively. Annexin V^+/PI^- staining cells represent the early apoptotic cells. (B) Detection of DNA fragmentation using TUNEL staining. Cells were stained with TUNEL and observed under a fluorescent microscope. The bright green fluorescence indicates DNA fragmentation. a, control; b, nano-Ag-treated cells; c, norfloxacin-treated cells (Hwang et al. 2012).

Applications

Many studies for clinical and industrial application of nano-Ag particles have been performed over the past decades (Monteiro et al. 2009, Furno et al. 2014). Application of nano-Ag particles demonstrated the possibility of a higher usage in clinical and industrial aspect (Chen and Schluesener 2007). Because dressings are major in the management of wounds, the newly designed wound dressings are required to prevent pathogenic infections. Silver nanocrystalline dressings show reduced bacterial infections in chronic wounds (Ip et al. 2006, Leaper 2006). Nano-Ag particles are also applied to coating medical devices. Silver impregnated medical devices like surgical masks and implantable devices show significant antimicrobial efficacy (Furno et al. 2004, Rai et al. 2009). In terms of industrial use, nano-Ag particles are used for water filtration and prevention and decontamination of food products (Nikawa et al. 1997, Jain and Pradeep 2005). Moreover, many researchers have great interest in the development of textile fabrics containing antimicrobial agents and nano-Ag particles-containing fibers are being prepared. The cotton fibers containing nano-Ag particles exhibited high antibacterial activity against *E. coli* (Yeo and Jeong 2003, Duran et al. 2007).

Conclusions

Because of good antimicrobial properties, silver has been used for centuries to treat various diseases, most notably infections and biological safety of them is demonstrated. Although silver is the potential antimicrobial agent, their use has a major drawback. Silver is easily inactivated due to the complexation and precipitation. However, nanotechnology resolves this problem through development of nano-Ag particles and researchers can overcome the limited usage of silver. Nano-Ag particles exhibit the unique chemical and physical properties and therefore, they can be improved as new antimicrobial agents.

In recent years, antimicrobial mechanisms of nano-Ag particles against pathogenic microorganisms have been discovered. Nano-Ag particles induced membrane disruption of fungal and bacterial cells. Furthermore, apoptosis hallmarks such as ROS accumulation, caspase activation, PS exposure and DNA fragmentation were observed in *C. albicans* cells. Interestingly, these apoptotic phenomena were also observed in *E. coli* cells and it was suggested that nano-Ag particles induced apoptosis-like response in bacterial cells. With these results, it was demonstrated that nano-Ag particles revealed dual mechanisms against pathogenic microorganisms. Considering the significant antimicrobial efficacy in clinical and industrial application, nano-Ag particles would emerge as potent antimicrobial agents.

References

Ahmad, I. and Beg, A.Z. 2001. Antimicrobial and phytochemical studies on 45 Indian medicinal plants against multi-drug resistant human pathogens. J. Ethnopharmacol. 74: 113–123.

Alvarez-Peral, F.J., Zaragoza, O., Pedreno, Y. and Argüelles, J. 2002. Protective role of trehalose during severe oxidative stress caused by hydrogen peroxide and the adaptive oxidative stress response in *Candida albicans*. Microbiology 148: 2599–2606.

Amro, N.A., Kotra, L.P., Wadu-Mesthrige, K., Bulychev, A., Mobashery, S. and Liu, G. 2000. High-resolution atomic force microscopy studies of the *Escherichia coli* outer membrane: structural basis for permeability. Langmuir 16: 2789–2796.

Aruguete, D.M., Kim, B., Hochella, M.F. Jr., Ma, Y., Cheng, Y., Hoegh, A., Liu, J. and Pruden, A. 2013. Antimicrobial nanotechnology: its potential for the effective management of microbial drug resistance and implications for research needs in microbial nanotoxicology. Environ. Sci. Process. Impacts 5: 93–102.

Benyagoub, M., Willemot, C. and Bélanger, R.R. 1996. Influence of a subinhibitory dose of antifungal fatty acids from *Sporothrix flocculosa* on cellular lipid composition in fungi. Lipids 31: 1077–1082.

Bolintineanu, D., Hazrati, E., Davis, H.T., Lehrer, R.I. and Kaznessis, Y.N. 2010. Antimicrobial mechanism of pore-forming protegrin peptides: 100 pores to kill *E. coli*. Peptides 31: 1–8.

Cerbón, J. and Calderón, V. 1991. Changes of the compositional asymmetry of phospholipids associated to the increment in the membrane surface potential. Biochim. Biophys. Acta 1067: 139–144.

Chen, X. and Schluesener, H.J. 2007. Nanosilver: a nanoproduct in medical application. Toxicol. Lett. 176: 1–12.

Costa, V. and Moradas-Ferreira, P. 2001. Oxidative stress and signal transduction in *Saccharomyces cerevisiae*: insights in to aging, apoptosis and disease. Mol. Aspects Med. 22: 217–246.

Dejean, L.M., Martinez-Caballero, S. and Kinnally, K.W. 2006. Is MAC the knife that cuts cytochrome *c* from mitochondria during apoptosis? Cell Death Differ. 13: 1387–1395.

Duran, N., Marcarto, P.D., De Souza, G.I.H., Alves, O.L. and Esposito, E. 2007. Antibacterial effect of silver nanoparticles produced by fungal process on textile fabrics and their effluent treatment. J. Biomed. Nanotechnol. 3: 203–208.

Dwyer, D.J., Camacho, D.M., Kohanski, M.A., Callura, J.M. and Collins, J.J. 2012. Antibiotic-induced bacterial cell death exhibits physiological and biochemical hallmarks of apoptosis. Mol. Cell 46: 561–572.

Elledge, S.J. 1996. Cell cycle checkpoints: preventing an identity crisis. Science 274: 1664–1672.

Endo, M., Takesako, K., Kato, I. and Yamaguchi, H. 1997. Fungicidal action of aureobasidin A, a cyclic depsipeptide antifungal antibiotic, against *Saccharomyces cerevisiae*. Antimicrob. Agents Chemother 41: 672–676.

Fadok, V.A., Bratton, D.L., Frasch, S.C., Warner, M.L. and Henson, P.M. 1998. The role of phosphatidylserine in recognition of apoptotic cells by phagocytes. Cell Death Differ. 5: 551–562.

Fröhlich, K.U. and Madeo, F. 2000. Apoptosis in yeast-a monocellular organism exhibits altruistic behavior. FEBS Lett. 473: 6–9.

Furno, F., Morley, K.S., Wong, B., Sharp, B.L., Arnold, P.L. and Howdle, S.M. 2004. Silver nanoparticles and polymeric medical devices: a new approach to prevention of infection? J. Antimicrob. Chemother. 54: 1019–1024.

Furno, F., Morley, K.S., Wong, B., Sharp, B.L., Arnold, P.L., Howdle, S.M., Bayston, R., Brown, P.D., Winship, P.D. and Reid, H.J. 2014. Silver nanoparticles and polymeric medical devices: a new approach to prevention of infection? J. Antimicrob. Chemother. 54: 1019–1024.

Garg, A.K., Kim, J.K., Owens, T.G., Ranwala, A.P., Choi, Y.D., Kochian, L.V. and Wu, R.J. 2002. Trehalose accumulation in rice plants confers high tolerance levels to different abiotic stresses. Proc. Natl. Acad. Sci. USA 99: 15898–15903.

Gong, P., Li, H., He, X., Wang, K., Hu, J. and Tan, W. 2007. Preparation and antibacterial activity of Fe3O4@Ag nanoparticles. Nanotechnology 18: 604–611.

Gurunathan, S., Lee, K.J., Kalimuthu, K., Sheikpranbabu, S., Vaidyanathan, R. and Eom, S.H. 2009. Antiangiogenic properties of silver nanoparticles. Biomaterials 30: 6341–6350.

Haruna, S., Kuroi, R., Kajiwara, K., Hashimoto, R., Matsugo, S., Tokumaru, S. and Kojo, S. 2002. Induction of apoptosis in HL-60 cells by photochemically generated hydroxyl radicals. Bioorg. Med. Chem. Lett. 12: 675–676.

Holt, K.B. and Bard, A.J. 2005. Interaction of silver (I) ions with the respiratory chain of *Escherichia coli*: an electrochemical and scanning electrochemical microscopy study of the antimicrobial mechanism of micromolar Ag^+. Biochemi. 44: 13214–13223.

Hwang, I.S., Lee, J., Hwang, J.H., Kim, K.J. and Lee, D.G. 2012. Silver nanoparticles induce apoptotic cell death in *Candida albicans* through the increase of hydroxyl radicals. FEBS J. 279: 1327–1338.

Ip, M., Lui, S.L., Poon, V.K.M., Lung, I. and Burd, A. 2006. Antimicrobial activities of silver dressings: an *in vitro* comparison. J. Med. Microbiol. 55: 59–63.

Jain, P. and Pradeep, T. 2005. Potential of silver nanoparticle-coated polyurethane foam as an antibacterial water filter. Biotechnol. Bioeng. 90: 59–63.

Kim, K.J., Sung, W.S., Moon, S.K., Choi, J.S., Kim, J.G. and Lee, D.G. 2008. Antifungal effect of silver nanoparticles on dermatophytes. J. Microbiol. Biotechnol. 18: 1482–1484.

Kim, K.J., Sung, W.S., Suh, B.K., Moon, S.K., Choi, J.S., Kim, J.G. and Lee, D.G. 2009. Antifungal activity and mode of action of silver nano-particles on *Candida albicans*. Biometals 22: 235–242.

Labat-Moleur, F., Guillermet, C., Lorimier, P., Robert, C., Lantuejoul, S., Brambilla, E. and Negoescu, A. 1998. TUNEL apoptotic cell detection in tissue sections: critical evaluation and improvement. J. Histochem. Cytochem. 46: 327–334.

Leaper, D.L. 2006. Silver dressings: their role in wound management. Int. Wound J. 3: 282–294.

LeBel, C.P., Ischiropoulos, H. and Bondy, S.C. 1992. Evaluation of the probe 2',7'-dichlorofluorescin as an indicator of reactive oxygen species formation and oxidative stress. Chem. Res. Toxicol. 5: 227–231.

Lee, W., Kim, K.J. and Lee, D.G. 2014. A novel mechanism for the antibacterial effect of silver nanoparticles on *Escherichia coli*. Biometals 27: 1191–1201.

Li, W.R., Xie, X.B., Shi, Q.S., Zeng, H.Y., Ou-Yang, Y.S. and Chen, Y.B. 2010. Antibacterial activity and mechanism of silver nanoparticles on *Escherichia coli*. Appl. Microbiol. Biotechnol. 85: 1115–1122.

Liao, R.S., Rennie, R.P. and Talbot, J.A. 1999. Assessment of the effect of amphotericin B on the vitality of *Candida albicans*. Antimicrob. Agents Chemother. 43: 1034–1041.

Liu, X., Zou, H., Widlak, P., Garrard, W. and Wang, X. 1999. Activation of the apoptotic endonuclease DFF40 (caspase-activated DNase or nuclease). Oligomerization and direct interaction with histone H1. J. Biol. Chem. 14: 13836–13840.

Madeo, F., Herker, E., Maldener, C., Wissing, S., Lächelt, S., Herlan, M., Fehr, M., Lauber, K., Sigrist, S.J., Wesselborg, S. and Fröhlich, K.U. 2002. A caspase-related protease regulates apoptosis in yeast. Mol. Cell 9: 911–917.

Mantareva, V., Kussovski, V., Angelov, I., Borisova, E., Avramov, L., Schnurpfeil, G. and Wöhrle, D. 2007. Photodynamic activity of water-soluble phthalocyanine zinc (II) complexes against pathogenic microorganisms. Bioorg. Med. Chem. 15: 4829–4835.

Maul, R.W. and Sutton, M.D. 2005. Roles of the *Escherichia coli* RecA protein and the global SOS response in effecting DNA polymerase selection *in vivo*. J. Bacteriol. 187: 7607–7618.

Monteiro, D.R., Gorup, L.F., Takamiya, A.S., Ruvollo-Filho, A.C., de Camargo, E.R. and Barbosa, D.B. 2009. The growing importance of materials that prevent microbial adhesion: antimicrobial effect of medical devices containing silver. Int. J. Antimicrob. Agents 34: 103–110.

Nikawa, H., Yamamoto Hamada, T., Rahardjo, M.B. and Murata Nakaando, S. 1997. Antifungal effect of zeolite-incorporated tissue conditioner against *Candida albicans* growth and/or acid production. J. Oral. Rehabil. 24: 350–357.

Novogrodsky, A., Ravid, A., Rubin, A.L. and Stenzel, K.H. 1982. Hydroxyl radical scavengers inhibit lymphocyte mitogenesis. Proc. Natl. Acad. Sci. USA 79: 1171–1174.

Oberdörster, G., Oberdörster, E. and Oberdörster, J. 2005. Nanotoxicology: an emerging discipline evolving from studies of ultrafine particles. Environ. Health Perspect. 113: 823–839.

Ouédraogo, M., Konaté, K., Lepengué, A.N., Souza, A., M'Batchi, B. and Sawadogo, L.L. 2012. Free radical scavenging capacity, anticandicidal effect of bioactive compounds from *Sida cordifolia* L., in combination with nystatin and clotrimazole and their effect on specific immune response in rats. Ann. Clin. Microbiol. Antimicrob. 11: 33.

Paulo, L., Ferreira, S., Gallardo, E., Queiroz, J.A. and Domingues, F. 2010. Antimicrobial activity and effects of resveratrol on human pathogenic bacteria. World J. Microbiol. Biotechnol. 26: 1533–1538.

Pereira, C., Camougrand, N., Manon, S., Sousa, M.J. and Côrte-Real, M. 2007. ADP/ATP carrier is required for mitochondrial outer membrane permeabilization and cytochrome *c* release in yeast apoptosis. Mol. Microbiol. 66: 571–582.

Phillips, A.J., Sudbery, I. and Ramsdale, M. 2003. Apoptosis induced by environmental stresses and amphotericin B in *Candida albicans*. Proc. Natl. Acad. Sci. USA 100: 14327–14332.

Portet, T., Camps iFebrer, F., Escoffre, J.M., Favard C., Rols, M.P. and Dean, D.S. 2009. Visualization of membrane loss during the shrinkage of giant vesicles under electropulsation. Biophys. J. 96: 4109–4121.

Rai, M., Yadav, A. and Gade, A. 2009. Silver nanoparticles as a new generation of antimicrobials. Biotechnol. Adv. 27: 76–83.

Retchkiman-Schabes, P.S., Canizal, G., Becerra-Herrera, R., Zorrilla, C., Liu, H.B. and Ascencio, J.A. 2006. Biosynthesis and characterization of Ti/Ni bimetallic nanoparticles. Opt. Mater 29: 95–99.

Rollet-Labelle, E., Grange, M.J., Elbim, C., Marquetty, C., Gougerot-Pocidalo, M.A. and Pasquier, C. 1998. Hydroxyl radical as a potential intracellular mediator of polymorphonuclear neutrophil apoptosis. Free Radic Biol. Med. 24: 563–572.

Sakurada, H., Koizumi, H., Ohkawara, A., Ueda, T. and Kamo, U. 1992. Use of dihydrorhodamine 123 for detecting intracellular generation of peroxides upon UV irradiation in epidermal keratinocytes. Arch Dermatol. Res. 284: 114–116.

Setsukinai, K., Urano, Y., Kakinuma, K., Majima, H.J. and Nagano, T. 2003. Development of novel fluorescence probes that can reliably detect reactive oxygen species and distinguish specific species. J. Biol. Chem. 278: 3170–3175.

Smrz, D., Dráberová, L. and Dráber, P. 2007. Non-apoptotic phosphatidylserine externalization induced by engagement of glycosylphosphatidylinositol-anchored proteins. J. Biol. Chem. 282: 10487–10497.

Tas, J. and Westerneng, G. 1981. Fundamental aspects of the interaction of propidiumdiiodide with nuclei acids studied in a model system of polyacrylamide films. J. Histochem. Cytochem. 29: 929–936.

Thornberry, N.A. and Lazebnik, Y. 1998. Caspases: enemies within. Science 281: 1312–1316.

van den Eijnde, S.M., Boshart, L., Baehrecke, E.H., De Zeeuw, C.I., Reutelingsperger, C.P. and Vermeij-Keers, C. 1998. Cell surface exposure of phosphatidylserine during apoptosis is phylogenetically conserved. Apoptosis 3: 9–16.

Vercammen, D., Declercq, W., Vandenabeele, P. and Van Breusegem, F. 2007. Are metacaspases caspases? J. Cell Biol. 179: 375–380.

Verhoven, B., Schlegel, R.A. and Williamson, P. 1995. Mechanisms of phosphatidylserine exposure, a phagocyte recognition signal, on apoptotic T lymphocytes. J. Exp. Med. 182: 1597–1601.

Vincent, M., England, L.S. and Trevors, J.T. 2004. Cytoplasmic membrane polarization in Gram-positive and Gram-negative bacteria grown in the absence and presence of tetracycline. Biochim. Biophys. Acta 1672: 131–134.

Wadskog, I., Maldener, C., Proksch, A., Madeo, F. and Adler, L. 2004. Yeast lacking the SRO7/SOP1-encoded tumor suppressor homologue show increased susceptibility to apoptosis-like cell death on exposure to NaCl stress. Mol. Biol. Cell. 15: 1436–1444.

Wu, X.Z., Chang, W.Q., Cheng, A.X., Sun, L.M. and Lou, H.X. 2010. Plagiochin E, an antifungal active macrocyclic bis(bibenzyl), induced apoptosis in *Candida albicans* through a metacaspase-dependent apoptotic pathway. Biochim. Biophys. Acta 1800: 439–447.

Yeo, S.Y. and Jeong, S.H. 2003. Preparation and characterization of polypropylene/silver nanocomposite fibres. Polymer. Int. 52: 1053–1057.

Yoon, M.J., Kim, E.H., Kwon, T.K., Park, S.A. and Choi, K.S. 2012. Simultaneous mitochondrial Ca^{2+} overload and proteasomal inhibition are responsible for the induction of paraptosis in malignant breast cancer cells. Cancer Lett. 324: 197–209.

Yu, S.P. 2003. Regulation and critical role of potassium homeostasis in apoptosis. Prog. Neurobiol. 70: 363–386.

Zherebitskaya, E., Schapansky, J., Akude, E., Smith, D.R., Van der Ploeg, R., Solovyova, N., Verkhratsky, A. and Fernyhough, P. 2012. Sensory neurons derived from diabetic rats have diminished internal Ca^{2+} stores linked to impaired reuptake by the endoplasmic reticulum. ASN Neuro. 4: e00072.

Zivna, L., Krocova, Z., Härtlova, A., Kubelkova, K., Zakova, J., Rudof, E., Hrstka, R., Macela, A. and Stulik, J. 2010. Activation of B cell apoptotic pathways in the course of *Francisella tularensis* infection. Microb. Pathog. 49: 226–236.

Section II

Nanotechnology for Treatment of Different Microbial Diseases

Nanoparticles as Therapeutic Agent for Treatment of Bacterial Infections

Mahendra Rai,[1,*] *Raksha Pandit,*[1] *Priti Paralikar,*[1] *Sudhir Shende,*[1] *Swapnil Gaikwad,*[2] *Avinash P. Ingle*[1] and *Indarchand Gupta*[1]

Introduction

Till the 20th century, antibiotics were considered as the only successful therapeutic agents to treat bacterial infections. However, the continuous and rampant use of antibiotics have led to an increasing microbial resistance to the currently used drugs thus preventing the eradication of severe, and sometimes fatal, microbial infections (Davies and Davies 2010, Lin et al. 2015). According to the US Centers for Disease Control and Prevention, millions of people are infected by Multi Drug Resistant (MDR) bacteria and about thousands of people died because of the severe bacterial infections (Ventola 2015). Another cause of severe bacterial infection are the biofilms, which leads to chronic biofilm-based infections that are mainly formed by *Staphylococcus aureus, Pseudomonas aeruginosa, Streptococcus mutans, Candida albicans*, etc. In particular, the diabetic chronic wound is a condition in which the biofilm, formed by one or more microorganisms, results in the inability to heal chronic wounds. *S. aureus*, in combination with the pathogenic yeast *C. albicans* can cause severe wound inflammations in diabetic patients (Malik et al. 2013, Cooper et al. 2014). Chronic biofilm-based infections are tremendously resistant to antibiotics and other conventionally used antimicrobial agents. Severe biofilm-based infections can invade the host immunity (Manavathu and Vazquez 2014, Gupta et al. 2016). About 80% of the infections are caused by the biofilm forming microorganisms, which causes infections on internal organs like lung, urinary tract, heart and cervix as well as oral cavities, eyes, etc. The bacteria also resides on the various medical instruments like catheters, contact lenses, etc. (Manavathu and Vazquez 2014), being *Campylobacter jejuni, Escherichia coli, Listeria monocytogenes, Salmonella* spp., *Shigella* spp., *Yersinia enterocolitica, Bordetella pertussis, Borrelia miyamotoi, Clostridium difficile, Enterococcus faecium* and *E. faecalis* among others, the responsible biofilm-forming bacteria (Sherman et al. 2010, Sirelkhatim et al. 2015).

There is a need to search for novel antimicrobial or therapeutic agents, which can fight against bacterial infections. Recent studies have focused on the role of nanoparticles as the novel tool which can help in the prevention of the bacterial infections including chronic biofilm-based infections. Nanoparticles are nanomaterials which can act as emerging therapeutics owing to their versatile

[1] Nanobiotechnology Lab., Department of Biotechnology, SGB Amravati University, Amravati-444 602, Maharashtra.
[2] Dr. D.Y. Patil Biotechnology and Bioinformatics Institute, Tathawade, Pune 411 033, Maharashtra, India.
* Corresponding author: pmkrai@hotmail.com

physical and chemical properties. Nanoparticles have high surface/volume ratio, which is the most noteworthy property and differs in this aspect, from the other therapeutic agents. At the nanometer scale, properties of nanoparticles like electrical conductivity, optical density, catalytic properties and weight, usually change. The catalytic activity of nanoparticles depends on their size as well as their structure, morphology, size distribution, chemical and physical environments (Khodashenas and Ghorbani 2015).

Metal nanoparticles such as aluminium, copper, gold, magnesium, silver, titanium and zinc are known for their antimicrobial activity. Silver is the most potent and well-studied example of nanoparticles which are responsible for inhibition of pathogenic bacteria (Tang and Feng 2014). The nanosized therapeutic agents can act as a therapeutic agent, and can be used as a vehicle for the delivery of the drug to the site of microbial infections (Gauri 2012). The efficacy of nanoparticles can also be increased by the proper functionalization of nanoparticles surface. If functionalization is properly done, the loading of the therapeutic agent is higher and can act as a promising tool against bacterial infections (Gupta et al. 2016).

Recently, scientists are working hard towards the development of nano-antibiotics and other nano-formulations (Meziane-Cherif and Courvali et al. 2014, Reardon 2014). Researchers discovered that metal and polymeric nanoparticles can act as drug carriers or vehicles for the release of different types of drugs, i.e., peptides proteins, nucleic acids and others. Nanoparticles can also be used in the sterilization of medical instruments which can be used in the prevention of the bacterial infections (Gupta et al. 2016). The aim of the present chapter is to focus on the role of nanoparticles in the treatment of the severe emerging bacterial infections mainly produced by multidrug resistant bacteria that are the main menace for curing bacterial infections. In this chapter, we discuss the use of nanoparticles as promising candidates for the cure of bacterial infections and also in drug delivery. Here, we have also highlighted the concept of antibacterial mechanisms of nanoparticles.

Nanoparticles as a Promising Tool to Combat Bacterial Infections

Bacterial infections cause millions of severe illness worldwide which leads to the mortality of people including children every year (Gupta et al. 2016). The convolution of the biofilm matrix makes biofilm-associated diseases clinically more challenging than planktonic bacteria in both diagnosis and treatment (Costerton et al. 1999). The formation and eradiation of microbial biofilms might create a number of serious problems like food safety (contamination) (VanHoudt and Michiels 2010), industrial fluid processing operations (bio-deterioration) (Mittelman 1998), as well as public health issues (infectious diseases) (Flemming and Wingender 2010).

As an outcome, nanoparticle drug delivery platforms including polymeric nanoparticles, liposomes, various inorganic nanoparticles and dendrimers have been progressively more exploited to augment the therapeutic efficiency of existing antibiotics. The areas where nanoparticle approaches a grip for significant potential to advance the treatment of bacterial infection includes combinatorial antibiotic delivery, targeted antibiotic delivery, environmentally responsive antibiotic delivery, nanoparticle-enabled antibacterial vaccination, as well as nanoparticle-based bacterial detection (Gao et al. 2014).

The size of nanoparticles is in proportion with bio-molecular and bacterial cellular systems, which provides a platform where nano-material and bacterial interactions can take place through proper surface functionalization (Daniel and Astruc 2004, Jiang et al. 2015). In addition, the high surface area/volume ratio of the nanomaterials enables high loading of drugs. Nanoparticles provide a way to deal with antibiotic resistance mechanism, which includes multi-drug efflux pumps (Li and Nikaido 2004), permeability regulation (Falagas and Kasiakou 2005, Cui et al. 2006), antibiotic degradation (Livermore 1995, Davies and Wright 1997), and target site binding affinity mutations (Courvalin 2006). Nanoparticles also provide alternative pathways to combat biofilm or Multiple Drug Resistance (MDR) infections and notably lower bacterial resistance over time (Huh and Kwon 2011, Hajipour et al. 2012, Miller et al. 2015). Nanoparticles utilize multiple mechanisms to kill bacteria, and it is difficult for them to develop resistance (Chopra 2007). Several nanoparticles-based systems have been developed to get better antimicrobial efficacy (Sondi and Salopek-Sondi 2004, Jones 2008, Simon-Deckers 2009). Since, nanoparticles showed the antimicrobial activities against both the Gram positive as well as the Gram negative bacteria, it can be used as a vehicle to combat the bacterial diseases and may lead to a promising tool in the development of nano-antibacterial agent (Gao et al. 2014, Gupta et al. 2016).

Types of Nanoparticles Involved to Cure Bacterial Infections

Metal nanoparticles like silver, gold, zinc oxide, iron oxide, nickel nanoparticles and magnetic nanoparticles have been investigated for antibacterial property (Shahverdi et al. 2007, Azam et al. 2012). The composite nanoparticles like silver-chitosan, zinc-iron oxide, polymer-antibiotics and chitosan-argenine have been the most widely used to control bacterial infections (Huang et al. 2011, Huang et al. 2011, Lellouche et al. 2012a).

Polymeric Nanoparticles

The use of biodegradable polymeric nanoparticles as an anti-infective agent can help to overcome toxicity issues, instability, side effects, etc. Polymeric nanoparticles can effectively carry drugs, proteins as well as targeted cells or organs. Another advantage of using polymeric nanoparticles is the encapsulation of bioactive molecules protecting them against enzymatic and hydrolytic degradation (Kumari et al. 2010). Polymeric nanoparticles are composed by biodegradable materials such as poly lactic acid (PLA), poly lactic co-glycolic acid (PLGA), poly ε-caprolactone (PCL), gelatin, chitosan, etc., can be used as effective antibacterial agents (Cheng et al. 2007, Gan et al. 2007, Gu et al. 2008, Misra et al. 2009, Szlek et al. 2013). Polymeric nanoparticles are used for drug delivery, solubilization and stabilization of drugs and targeting drug at a particular site (Salouti and Ahangari 2014). Liposomes are broadly used as drug delivery vehicle in antimicrobial therapy (Zhang et al. 2010). Liposomes used as an antibiotic carrier can improve pharmacokinetics and biodistribution of antibiotics, can reduce toxicity and can enhance targeted delivery at a particular site of infection, which ultimately helps to overcome bacterial drug resistance (Abeylath and Turos 2008).

Under physiological conditions, pathogenic bacteria maintain a negative surface charge so, positively charged nanoparticles can efficiently bind with bacterial cells due to electrostatic attraction. This strategy is attractive to combat pathogenic bacterial infection. Biodegradable polymeric nanoformulations have been efficiently used for antibacterial applications. Another strategy of targeting pathogenic bacteria, is the conjugation of drug or ligand on surface of nanoparticles. For example, vancomycin is conjugated on the surface of fusogenic liposome (Nicolosi et al. 2010), isoniazid and rifampicin are conjugated on the surface of chitosan nanoparticles (Garg et al. 2016), rifabutin is loaded on Solid Lipid Nanoparticles (SLN) (Gaspar et al. 2016) vancomycin is conjugated in liposomes (Onyeji et al. 1994), etc., resulting in preferable binding of biodegradable polymeric nanoparticles to pathogenic bacteria. Table 9.1 summarizes the polymeric nanoparticles involved to cure bacterial infections.

Inorganic Nanoparticles Involved in the Cure of Bacterial Infections

The inorganic nanoparticles have been categorized as metals and non-metals. Nano inorganic metal oxides have been extensively studied for their antimicrobial behavior (Loomba et al. 2013) Metal oxide nanoparticles (MONPs) which have highly potent antibacterial effects and include titanium dioxide (TiO_2), silver oxide (AgO) compounds, copper oxide (CuO), Iron oxide (Fe_3O_4), Zinc Oxide (ZnO), Magnesium Oxide (MgO), Nitric Oxide (NO), and Aluminium Oxide (Al_2O_3) (the details of the action potential of these NPs are given in Table 9.2).

Table 9.1. Polymeric nanoparticles to cure bacterial infections.

Polymeric nanoparticles	Drug	Targeted bacteria	References
Liposome	Polymyxin B	*P. aeruginosa*	Alipour et al. 2008
	Ampicillin	*Salmonella typhimurium*	Fattal et al. 1991, Schumacher et al. 1997
	Benzyl penicillin	*S. aureus*	Kim et al. 2004
	Ciprofloxacin, Fluoroquinolone	*Salmonella dublin*	Magallanes et al. 1993
	Gentamicin, Streptomycin	*Brucella* spp.	Fountain et al. 1985
	Vancomycin, Teicoplanin	Methicillin-resistant *S. aureus* (MRSA)	Onyeji et al. 1994
Fusogenic liposomes	Vancomycin	*E. coli* and *Acinetobacter baumannii*	Nicolosi et al. 2010
Glycosylated polyacrylate nanoparticle	Beta-lactam/ciprofloxacin	Methicillin-resistant *S. aureus*, *Bacillus anthracis*	Turos et al. 2007, Abeylath et al. 2008
Solid lipid nanoparticles	Rifabutin	*Mycobacterium tuberculosis*	Gaspar et al. 2016
Poly-lactide-co-glycolide (PLG) nanoparticle	Rifampicin, isoniazid, pyrazinamide, ethambutol	*M. turberculosis*	Pandey et al. 2006
Chitosan nanoparticles	Isoniazid, rifampicin	*M. turberculosis*	Garg et al. 2016
Dimeric dendrimer	Antimicrobial peptide	*Acinetobacter baumannii, Enterobacter cloacae, E. coli, Klebsiella pneumoniae, P. aeruginosa, Stenotrophomonas maltophilia, Proteus mirabilis, Serratia marcescens*	Scorciapino et al. 2012
Polycationic dendrimers	Antimicrobial peptide	Multidrug-Resistant *P. aeruginosa*	Abdel-Sayed et al. 2016
Dendrimer	Vancomycin	*S. aureus*	Choi et al. 2013
Soya ethyl morpholinium ethosulfate micelles		*S. aureus* and methicillin-resistant *S. aureus* (MRSA)	Yang et al. 2016

Most of the MONPs exhibit bactericidal properties due to Reactive Oxygen Species (ROS) generation that leads to the oxidative stress resulting in cell damage (Beyth et al. 2015). The inorganic nanoparticles often demonstrate new physical properties as their nanometer dimensions. The well-known example of inorganic nanoparticles is mesoporous silica, which possesses an exceptionally large surface area and tunable pore size that make them promising candidates as drug delivery vehicles (Kresge et al. 1992). Hetrick and his colleagues (2009) designed amine-functionalized silica nanoparticles that were able to voluntarily penetrate and exterminate pathogenic biofilms through release of nitric oxide rapidly (Hetrick et al. 2009, Gupta et al. 2016).

Titanium dioxide (TiO_2) is an inorganic metal oxide nanomaterial that has been broadly used for its antimicrobial activities (Allahverdiyev et al. 2013). It is well-known for the capability to kill Gram positive and Gram negative bacteria (Wei et al. 1994). Like gold (Au), it is photocatalytic in nature and hence, the toxicity of TiO_2 is induced by visible light, near ultra-violet or ultra-violet (Pelgrift and Friedman 2013) which encourages the ROS burst. The ROS damage the membrane of the bacterial cell and many macromolecules like DNA (Blecher et al. 2011). TiO_2 is extremely efficient against many bacteria including *Bacillus* spp. (Hamal et al. 2010).

Zinc Oxide (ZnO) is another broad spectrum bactericidal nanomaterial. ZnO nanoparticles have shown an extensive range of antimicrobial activity against a variety of microorganisms, which is significantly reliant on the specified concentration and size of nanoparticles (Palanikumar et al. 2014). Furthermore, the treatment with ZnO nanocomposite was approved by the Food and Drug Administration (FDA) and presently ZnO is offered as a food additive (Blecher et al. 2011).

Table 9.2. Inorganic nanomaterials and their action potential against microorganisms.

Materials	Method of preparation	Antimicrobial activity against	Possible reasons for antimicrobial activity	References
TiO_2	Nano-silver decorated titanium dioxide (TiO_2) nano-fibers by electro-spinning method	Gram-positive and Gram-negative bacteria, viral species and parasites	Photocatalytic activity; the toxicity, induced by visible light, near UV or UV stimulate ROS burst	Wei et al. 1994, Zan et al. 2007, Brady-Estévez et al. 2008, Srisitthiratkul et al. 2011, Brady-Allahverdiyev et al. 2013, Pelgrift and Friedman 2013
AgO	Silver nanoparticles on cotton fabric by chemical method	Gram-positive and Gram-negative bacteria *E. coli*, kills biofilm or planktonic cells	Causes 'pits' formation in the cell wall by increasing membrane permeability and inactivating the respiratory chain, Photocatalytic and can induce ROS	Sondi and Salopek-Sondi 2004, Carlson et al. 2008, Beyth et al. 2010, El-Shishtawy et al. 2011, Piao et al. 2011, Sheng and Liu 2011, Kumar et al. 2014, Ninganagouda et al. 2014
CuO	Copper oxide nanoparticles by electrochemical reduction	*E. coli* and *S. aureus*, *Bacillus subtilis* and *B. anthracis*	Membrane disruption and ROS production leads to oxidative stress	Ruparelia et al. 2008, Jadhav et al. 2011, Pelgrift and Friedman 2013, Pey et al. 2014
Fe_3O_4	Iron oxide produces magnetic composite superparamagnetic iron oxide nanoparticles by chemical method	Gram-positive and Gram-negative bacteria	Anti adherent properties reduces bacterial colonization, penetrate and destroy biofilms	Taylor and Webster 2009, Gordon et al. 2011, Chatterjee et al. 2011, Hajipour et al. 2012, Taylor et al. 2012, Durmus et al. 2013, Anghel et al. 2014
ZnO	Zinc oxide (ZnO) nanoparticles by chemical method	Methicillin-sensitive *S. aureus* (MSSA), Methicillin-resistant *S. aureus* (MRSA), and Methicillin-resistant *S. epidermidis* (MRSE) strains, an extensive range of bacteria includes pathogens such as *K. pneumoniae*, *Salmonella enteritidis*, *L. monocytogenes*, *Lactobacillus*, *S. mutans*, and *E. coli*	The white color of ZnO being able to UV-blocking, and able to prevent biofilm formation, affect bacterial cells along the two pathways, as by binding to membranes, disrupting their potential and integrity, and as by inducing ROS production	Huang et al. 2008, Jin et al. 2009, Liu et al. 2009, Dastjerdi and Montazer 2010, Gordon et al. 2011, Ansari et al. 2012, Applerot et al. 2012, Malka et al. 2013, Hakraborti et al. 2014, Kasraei et al. 2014, Pati et al. 2014, Reddy et al. 2014
MgO	Magnesium oxide (MgO) nano-wires were synthesized by microwave hydrothermal technique	Gram-positive and Gram-negative bacteria, spores and viruses, MgF_2 found to prevent bacterial biofilm formation	MgO or MgX_2 forms induces ROS, direct inhibition of essential enzymes of the bacteria	Al-Hazmi et al. 2012, Lellouche et al. 2012a, 2012b, Pelgrift and Friedman 2013, Tang and Feng 2014
NO	Nitric oxide-(-NO) releasing nanoparticles by hydrogen and glass composite technique	Methicillin resistant *S. aureus* (MRSA), Prevention of biofilm formation of multiple bacterial species	Reactive Nitrogen Species (RNS)	Barraud et al. 2006, Han et al. 2009, Hetrick et al. 2009, Martinez et al. 2009, Chouake et al. 2012
Al_2O_3	Aluminium oxide (Al_2O_3) nanoparticles	*E. coli*	Interaction of nanoparticles with cell membrane can form pits and disrupt the membrane, moreover, the accumulation of nanoparticles in the cell interact with other cellular components and affects normal cell function	Mukharjee et al. 2011, Ansari et al. 2014

Iron Oxide (Fe_3O_4) nanoparticles is a supplementary group of antimicrobial materials that are being investigated for its use in medicine and healthcare (Chatterjee et al. 2011).

Copper Oxide CuO nanoparticles have shown a significant activity against different bacterial pathogens, rather lower than that of Ag or ZnO. Hence, comparatively high concentration of CuO nanoparticles are required to get comparable results (Ren et al. 2009). Furthermore, CuO nanoparticles activities vary significantly depending on the target bacteria. However, as Cu is cheaper than other nano-sized metal materials, it can be used for efficient potentiation in their nano-composite form. Generally, Cu nanomaterials are considered less potent than Ag nanomaterials, though in some cases the vice versa is true. Thus, it seems that in exceptional cases to use the CuO nanomaterials as an alternative of others, including silver would be favorable.

Nitric Oxide (NO) serves as a promising antibacterial compound, because of the low risk of emergence of resistance, i.e., NO is concerned with manifold mechanisms of antimicrobial activity (Carpenter and Schoenfisch 2012, Schairer et al. 2012). Antimicrobial effects of metal-based nanoparticles depend upon their size and shape (Slomberg et al. 2013), being the smallest particles the most efficient. NO molecules are produced endogenously and are implicated in a variety of physiologic functions. NO is highly reactive and hence its advantages in the clinics is constrained. Nevertheless, its antimicrobial potential can be useful due to its focal delivery, encapsulation and controlled release (Kutner and Friedman 2013, Seabra et al. 2015).

Aluminum Oxide Till date, it is unclear whether aluminum oxide (Al_2O_3) nanoparticles are useful for antibacterial treatment. Since their bactericidal activity is low and thus they work merely at high concentrations (Pelgrift et al. 2013, Qiu et al. 2012) unless they are used in combination with other nanomaterials such as Ag. More disturbing is their ability to promote horizontal multi-resistance genes which has been mediated by plasmids across genera (Qiu et al. 2012).

Metal Nanoparticles

Development in nanotechnology made production of nanoparticles more cost-effective, and therefore, they have been applied as antimicrobial agents. Silver (Rai et al. 2015), gold (Boda et al. 2015), iron (Naseem et al. 2013), copper (Kruka et al. 2015), zinc (Salema et al. 2015, Tiwari et al. 2016), titanium, (Zhao et al. 2011), etc., metal nanoparticles have demonstrated antimicrobial activity. Among all the studied nanoparticles, silver nanoparticles proved to be more applied antimicrobials (Ingle et al. 2008) and one of the fastest developing category. Silver has remarkable antimicrobial property hence it has been used since ancient times as metallic silver, silver sulfadiazine, silver nitrate form for the treatment of burns, wounds and several bacterial infections (Rai et al. 2009a). In the medical area, silver nanoparticles have been introduced into various products such as wound dressings (Wu et al. 2014), contraceptive devices (Tolaymata et al. 2010), surgical instruments and bone prostheses coated with nanosilver (Sivolella et al. 2012). Silver nanoparticles have already demonstrated bactericidal activity against Gram negative and Gram-positive bacteria, *E. coli, S. aureus, B. subtilis, S. mutans* and *S. epidermidis* (Besinis et al. 2004, Kathiresan et al. 2009, Rai et al. 2009b, Bhimba et al. 2011, Li et al. 2013, Manikprabhu et al. 2013). Silver nanoparticles release silver ions that prevent the cell division and respiratory system of bacterial cell (Sondi and Salopek-Sondi et al. 2004). Feng and coworkers (2000) reported that after treating *E. coli* and *S. aureus* with silver nitrate, cytoplasmic membrane separated from the cell wall and DNA get condensed which results in fatility of the cell.

Emergence of multidrug resistant bacteria is a big challenge in the medical field, which can be overwhelmed by applying different nanoparticles. Boda et al. (2015) produced ultrasmall gold nanoparticles and tested them against multidrug resistant *S. aureus, S. epidermidis, E. coli* and *P. aeruginosa*. Finally, they concluded that gold nanoparticles can be used to combat multidrug resistant microbial infections and biofilm formation. Das and co-workers (2015) synthesized gold nanoparticles by green techniques using graft copolymer based on Hydroxy Ethyl Starch (HES) and methylacrylate (MA). Synthesized nanoparticles were tested against two Gram negative bacteria *Shigella flexneri* and *P. mirabilis* and two Gram positive bacteria *B. cereus* and *B. subtilis*. Gold nanoparticles showed remarkable antibacterial activity at lower concentrations against Gram positive bacteria.

Recently, Naseem and Farrukh (2015) fabricated iron nanoparticles by reliable experimental protocol from *Lawsonia inermis* and studied its antibacterial activity against *E. coli, Salmonella enterica, P. mirabilis* and *S. aureus* by the well-diffusion method. Distinctive properties of magnetic nanoparticles made them the center of attraction for biological applications. It was confirmed that the uptake of magnetic nanoparticles by bacteria is increased by applying a magnetic field (Sunitha et al. 2013). They synthesized iron and silver nanoparticles by the biological method, i.e., from the fungus *Fusarium oxysporum* and *Actinomycetes* sp. and compared their antimicrobial activity. Iron nanoparticles, synthesized by using *Fusarium oxysporum,* revealed maximum bactericidal activity against *Bacillus* spp., *E. coli* and *Staphylococcus* spp., whereas silver nanoparticles from *Actinomycetes* found to be more lethal against different pathogens. The authors also concluded that the antimicrobial activity of nanoparticles is due to its small size, which efficiently are attached to the surface of microorganisms and lessen oxygen supply for respiration.

Scientists focused on the possible applications of copper nanoparticles as antimicrobials because of its low cost of production. Copper has potential applications in biology and nanomedicine as antimicrobials, water purifier and antifouling agent. Copper ions also play a crucial role in burns wound healing by collagen crosslinking and preventing wounds from microbial infections (Kruka et al. 2015). Kruka and coworkers (2015) produced monodisperse copper nanoparticles at concentration of 300 ppm of average size 50 nm. Copper nanoparticles showed a great antimicrobial activity against Gram-positive bacteria, methicillin-resistant *S. aureus* and *Candida* species.

PVP composite nanofibers of silver, copper and zinc were prepared using electrospinning and tested against *E. coli* and *S. aureus*. It was demonstrated that silver loaded fibers showed strong antimicrobial activity against these microorganisms (Quirósa et al. 2015). Copper nanoparticles synthesized from peel extract of *Punica granatum* revealed remarkable antimicrobial activity against opportunistic pathogens, i.e., *P. aeruginosa* MTCC 424, *S. enterica* MTCC 1253, *Micrococcus luteus* MTCC 1809 and *Enterobactor aerogenes* MTCC 2823 (Kaur et al. 2016). Zhao and co-workers (2011) incorporated silver nanoparticles into titania nanotubes (NT-Ag). AgNPs loaded nanotubes help in the prevention of bacterial infections. These NT-Ag nanotubes have enduring antibacterial activity so, it is used in dentistry, oethopedics and other biomedical instruments.

Functionalized or Conjugated Nanoparticles

Functionalized nanoparticles or conjugated nanoparticles can act as excellent candidates for drug delivery or targeted therapy. After conjugation of the nanoparticles, the efficacy, specificity and the loading capacity of the therapeutic agent can be enhanced. Functionalization of nanoparticles with a suitable biomolecule can act as a cost-effective solution to cure bacterial infections. The interlinking or the surface interaction of nanoparticles along with the biomolecule or therapeutic agent should be proper. Biofunctionalization of nanoparticles is the most emerging field of science and technology. The functionalized nanoparticles not only enhance the antimicrobial efficacy, but also reduce the toxicity issues (Pissuwan et al. 2010, Gupta et al. 2016). The list of the conjugated nanoparticles in combating bacterial infections is given in Table 9.3.

Brown et al. (2012) reported the functionalization of metal nanoparticles like silver (AgNPs) and gold (AuNPs) along with ampicillin. It was found that functionalized metal nanoparticles showed superior antibacterial activity as compared to nanoparticles alone. Functionalized AgNPs and AuNPs can be used in the treatment of MDR bacteria such as *P. aeruginosa, E. aerogenes* and methicillin-resistant *S. aureus.* Antibiotic conjugated ZnO NPs were prepared and it was found that they act as excellent candidates which can help in the prevention of antibiotic resistant bacteria. It can also be

Table 9.3. Functionalized or conjugated nanoparticles.

Type of nanoparticles	Functionalized biomolecule	Antibacterial activity of nanoparticles	References
AgNPs	Ampicillin	*P. aeruginosa, E. aerogens, S. aureus*	Brown et al. 2012
AuNPs	Ampicillin	*P. aeruginosa, E. aerogens, S. aureus*	Brown et al. 2012
ZnO NPs	Ciprofloxacin	*E. coli*	Iram et al. 2015
ZnO NPs	Erythromycin	*E. coli*	Iram et al. 2015
ZnO NPs	Methicillin	*E. coli*	Iram et al. 2015
Carbon nanotubes	AgNPs	*E. coli*	Kazmia et al. 2014
Silica nanoparticles	Ampicillin	*P. aeruginosa, S. aureus*	Tudose et al. 2014
Silica NPs	Penicillin G	*P. aeruginosa, S. aureus*	Tudosea et al. 2014
Silica NPs	Isoniazid	*P. aeruginosa, S. aureus*	Tudoisea et al. 2014
AuNPs	Glucosamine	*E. coli*	Govindaraju et al. 2015
ZnO NPs	Salicyalichitosan	*E. coli, B. subtilis, S. aureus*	Jayandran et al. 2016
AgNPs	Glycoprotein	*V. cholera*	Gahlawat et al. 2016
Silica NPs	Aminoglycosides	*E. coli, B. subtilis, S. aureus*	Agnihotri et al. 2015
Magnetic nanoparticles	Ampicillin	*M. tuberculosis, B. subtilis, S. aureus, P. aeruginosa, Salmonella choleraesuis*	Hussein Al-Ali et al. 2014
AgNPs	Polyphosphoester	*P. aeruginosa, S. aureus, Burkholderia* spp.	Zhang et al. 2015
Silica nanoparticles	Amine	*E. coli, P. aeruginosa, S. aureus, E. cloacae*	Duncan et al. 2015
AuNPs	Gentamicin, rifampicin, methicillin, ciprofloxacin	*S. epidermis, Staphylococcus haemolyticus*	Roshmi et al. 2015
AuNPs	Gentamicin	*P. aeruginosa, S. aureus, L. monocytogenes*	Mu et al. 2016
AgNPs	Phosphotungstic acid Phosphomolybdic acid	*S. albus, E. coli*	Daima et al. 2014

used in the treatment of bacterial infections. ZnONPs when conjugated with three antibiotics such as ciprofloxacin, erythromycin and methicillin. It was found that methicillin conjugated nanoparticles were the most effective agent against the multi drug-resistant pathogens, which cause severe bacterial infections (Iram et al. 2015).

Researchers demonstrated that after functionalization of the carbon nanotubes with the metal nanoparticles like AgNPs act as a vehicle for drug delivery which can be used against bacterial infections (Kazmia et al. 2014). Tudose and co-workers (2015) reported the synthesis of AgNPs by using the citrate reduction method and after that, it was encapsulated into silica functionalized nanoparticles. The silica nanoparticles were functionalized with the antibiotics such as ampicillin, penicillin G and isoniazid. The efficiency of silica nanoparticles increased greatly because of the conjugation of antibiotics along with AgNPs. Previous research revealed that the silica functionalized nanoparticles were effective against the biofilm-based bacterial infections. The hybrid can act as a novel antimicrobial agent against bacterial infections. AuNPs were functionalized with glucosamine. Glucosamine is biocompatible, biodegradable in nature and possesses its own antimicrobial activity. Functionalized nanoparticles were activated by UV irradiation and after functionalization remarkable antibacterial activity was noted against bacterial diseases (Govindaraju et al. 2015).

Jayandran et al. (2016) reported salicyalchitosan functionalized ZnO NPs, chitosan was used for the functionalization, owing to its biocompatibility, flexibility and biodegradable nature. The activity of the functionalized ZnO NPs increased tremendously because of biofunctionalization.

Multidrug resistance increased sufficiently in *Vibrio cholerae* species so, the researchers investigated an alternative therapy which can be used as a therapeutic agent and minimize the problem of bacterial infections. Microbial glycolipoprotein coated with AgNPs was used reportedly as novel antimicrobial agents against cholera. The antibacterial activity of functionalized nanoparticles was compared with the antibiotics. It was found that the activity of the functionalized nanoparticles was better as compared to antibiotics. As the use of antibiotics resulted in antibiotic-resistance this does not occur after usage of functionalized nanoparticles (Gahlawat et al. 2016). Agnihotri and colleagues (2015) reported the aminoglycosides conjugated silica nanoparticles. Silica nanoparticles are known for their biomolecule carrying ability. The main attribute of silica nanoparticles is that they can be surface functionalized very easily, do not have any toxicity, penetrate very easily and can be used in combating bacterial diseases. Ampicillin conjugated magnetic nanoparticles were prepared. Chitosan was also used as a stabilizing agent at the time of magnetic nanoparticles synthesis. Conjugated nanoparticles were used as a nanoantimicrobial in the treatment of the biofilm producing bacterial infection (Hussein-Al-Ali et al. 2014).

Recently, polyphosphoester functionalized AgNPs were synthesized which revealed that these functionalized nanoparticles can be used as a therapeutic agent against bacterial lung infection. It can act as a safe, cost effective therapy for the treatment of lung infection (Zhang et al. 2015). Duncan et al. (2015) showed that functionalized nanoparticles based capsules act as an excellent therapeutic agent against biofilm-forming bacteria which causes chronic infections. As cationic nanoparticles exhibited good biofilm inhibition ability, amine functionalized silica nanoparticles were formulated. Peppermint oil and cinnamaldehyde oil were selected as they are known for their antibacterial activity. Both the oils and the functionalized silica nanoparticles were used in the formulation of nanocapsules. These functionalized nanocapsules can be used as novel therapeutic agents which can be used against biofilm-forming infections. Cationic nanoparticles enabled oil to help in the eradication of the biofilm-producing bacteria. Roshmi et al. (2015) reported conjugation of AuNPs with antibiotics such gentamicin, ciprofloxacin, methicillin, rifampicin, etc., and it was found that the nanoparticles antibiotic conjugates can be used against hospital bacterial infections. Phosphatidyl choline gold nanoparticles conjugated with gentamicin was found to be effective against the biofilm producing bacteria such as *P. aeruginosa, S. aureus, L. monocytogenes*. It was shown that the antibacterial activity was enhanced after conjugation with gentamicin (Mu et al. 2016). It was demonstrated that phosphotungstic acid (PTA) and phosphomolybdic acid-functionalized silver nanoparticles can be used to enhance the antimicrobial activity of nanoparticles. The activity of functionalized nanoparticles was effective against *E. coli* and *Streptomyces albus* (Daima et al. 2014).

Nanoparticles as a Vehicle for Drug Delivery Against Bacterial Infections

Nanoparticulate drug delivery approaches grasp significant potential for the treatment of bacterial infections. Nanoparticles based drug delivery systems carried out controlled drug release in organs or tissues and deliver bioactive agents. Nanoparticles drug delivery including polymeric nanoparticles, dendrimers, liposomes, various metal nanoparticles and inorganic nanoparticles have been extensively exploited to enhance effectiveness of conventional antimicrobial therapies.

Liposomes-mediated drug delivery systems are most widely used to enhance the efficacy of drugs used for the treatment of bacterial infections (Torchilin et al. 2005, Hallaj-Nezhadi and Hassan 2015). Liposomal nanoparticles have great potential in effective and selective targeting of antibiotics to bacterial cells for eradication as well as the highest safety for humans (Hallaj-Nezhadi and Hassan 2015). Liposome encapsulated antibiotics delivery systems improve antimicrobial activities against drug-resistant bacterial strains, because liposome can be easily fused with bacterial membranes and it also protects drugs or antibiotics by isolating them from degrading enzymes and promoting their diffusion across the bacterial envelope (Kohno et al. 1998, Zhang et al. 2010). Encapsulation of antibacterial drugs vancomycin and teicoplanin in liposomes resulted in a remarkable bactericidal effect against methicillin-resistant *S. aureus* infections and improved elimination of intracellular methicillin-resistant *S. aureus* infections in macrophage infection model (Onyeji et al. 1994, Pumerantz et al. 2011).

Like liposomes, polymeric nanoparticles are also extensively used for delivery of antibacterial drugs. Polymeric nanoparticles like tri-block copolymer nanoparticles composed of Poly Lactic-co-Glycolic Acid (PLGA), poly histidine, and Poly Ethylene Glycol (PEG) have been reported for acid-responsive antibiotic delivery (Radovic-Moreno et al. 2012). Nanoparticles on exposure to acidic pH, protonate the imidazole group which switches surface charge to positive that results in an enhanced bacterial binding and improved drug delivery to targeted infected cells lead to death of bacteria. Solid Lipid Nanoparticles (SLN) have also demonstrated their ability to increase the efficacy of antibiotics against infectious diseases. SLNs can act as promising carriers for sustained vancomycin release in ocular infections enhancing the bioavailability. Vancomycin incorporated in SLN is efficiently used to treat gram-positive *Staphylococcus* spp., which is a common causative agent of ocular infection (Yousry et al. 2016). Tobramycin is an antimicrobial drug for gastrointestinal tract infection caused by *P. aeruginosa*. Tobramycin-loaded nanostructured lipid carriers efficiently act against *P. aeruginosa* infection associated with cystic fibrosis (Moreno-Sastre et al. 2016).

Silica nanoparticles possess unique properties such as large surface area and pore size which make them an efficient agent for drug delivery. Nitric oxide releasing silica nanoparticles has the ability of killing bacteria's forming biofilm. On exposure of nitric oxide releasing silica nanoparticles on biofilms of *P. aeruginosa, E. coli, S. aureus, S. epidermidis* and *C. albicans,* they kill bacteria efficiently by inhibiting fibroblast proliferation (Hetrick et al. 2009). Small molecules like vancomycin have been conjugated on the surface of mesoporous silica nanoparticles, which results in efficient targeting and killing Gram-positive bacteria over macrophage-like cells (Qi et al. 2013). This strategy enhances antibacterial capability of drug for the treatment of pathogenic bacteria with minimum side effects.

Outlook on Mechanism of Antibacterial Action of NPs

Bacterial cells consist of cell wall, plasma membrane and cytoplasm. The cell wall is the outermost covering of bacteria, containing a peptidoglycan layer, giving it a specific shape (Fu et al. 2005). Gram-positive bacteria possess a cell wall with the thickness of 20–80 nm, on the contrary Gram-negative cell wall is 7–8 nm thick (Prabhu et al. 2012). Nanoparticles within this size range therefore can cross the cell wall and plasma membrane of the bacteria thereby making them vulnerable to the damage (Dutta et al. 2012).

Bacterial cell surface is the first thing with which nanoparticles interacts. Close interaction of nanoparticles with the Gram-negative cell wall results into the pits formation. Moreover, nanoparticles get adhered to bacterial cell membrane, resulting in the blocking of transport channels (Dutta et al. 2012, Prabhu et al. 2012). This mechanism is size-dependent. The smaller the nanoparticle size, the higher will be its effective interaction with bacterial cell membrane (Raghupathi et al. 2011, Ivask et al. 2014). This is due to the well-known phenomenon that smaller nanoparticles have larger surface/volume ratio and thereby increase the probability of interaction with the cell membrane. On the contrary, the bulk counterpart of the nanoparticles possess the higher absolute surface area which lets enhanced van der Waals adhesion. Therefore, it is important to note that the nanoparticle size is not the only factor that plays a role in its antibacterial mechanism. Other factors have to be considered to determine its effectiveness against bacteria. Other factors contributing to the nanoparticle activity is its composition (Panacek et al. 2006). After attachment of the nanoparticles to the cell surface, it can get internalized and may be ionized in the cytoplasm thereby, resulting into the cell death (You et al. 2012). For instance, metallic nanoparticles such as silver and copper nanoparticles form ions inside the cell which can interact with cellular proteins, altering their function, ultimately disturbing the cellular metabolism (Hoshino et al. 1999, Nath et al. 2012, Bogdanović et al. 2014).

ROS production through nanoparticles contribute a lot in increasing the effectiveness of nanoparticles. ROS are highly reactive, short-lived oxidants. Owing to their high reactivity, they can cause great damage to bacterial cell components comprising the peptidoglycan layer, membrane lipids, proteins including enzymes, DNA and RNA (Kim et al. 2007, Pelgrift et al. 2013, Sirelkhatim et al. 2015). ROS have also been reported to prevent the process of transcription, translation, enzymatic activity and electron transport chain (Raffi et al. 2008). Metallic oxide nanoparticles do not produce high amounts of ions whereas they generate ROS. The ions produced after dissolution of CuO, NiO, ZnO and Sb_2O_3 NPs have been reported to contribute negligibly in their toxicities (Baek and An 2011), signifying the higher role of ROS to show a harmful effect. On the contrary, AgNPs generate both ROS and its ion (Zhang et al. 2010, Dutta et al. 2012, Prabhu et al. 2012). For example, ZnO NPs and CuO NPs after entering into the cell, has been reported to induce ROS in the cell (Xie et al. 2011, Ingle et al. 2014).

In case of nanoparticles such as AgNPs, the metal ions form the complex with various proteins causing the upregulation of ion efflux pumps, disruption of cellular functioning thereby leading to the cell death (Nagy et al. 2011). It is well known that protein phosphorylation has a huge impact of signal transduction. Dephosphorylation of tyrosine is one of those mechanisms. Nanoparticles has been also reported to alter the signal transduction in bacteria by dephosphorylating the peptide substrate on tyrosine residues thus causing the prevention of signal transduction (Shrivastava et al. 2007). At a molecular level, it is speculated that positive charge of silver ion is responsible for its interaction with negatively charged cell membranes of microbes (Hamouda et al. 2000, Dibrov et al. 2002). The amount of released ions depends on the nanoparticles size as smaller nanoparticles release higher percentage of ions into the solution. However, ion release from metal oxide nanoparticles is dose dependent (Baek et al. 2011).

As discussed above, many studies have supported one or the other mechanisms of antibacterial activity but the exact mechanism against the bacteria is under debate. Therefore, more studies have to be undertaken to get a precise idea about the antibacterial mechanism of nanoparticles (Fig. 9.1).

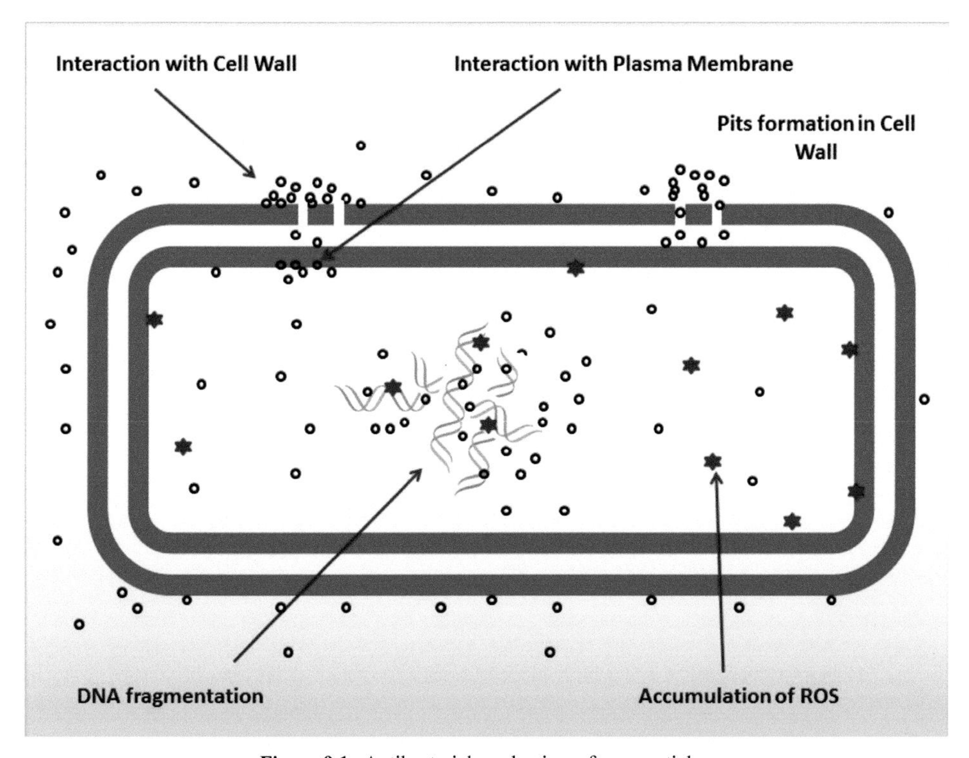

Figure 9.1. Antibacterial mechanism of nanoparticles.

Conclusions

In summary, bacterial infections are increasing at a fast rate due to the emergence of multidrug resistance bacteria. This problem has been recognized as a global threat to mankind. The MDR bacteria do not respond to the available antibiotics, which warrants the search for newer antibiotics or alternative ways to tackle the problem. More recently, there were various reports concerning the role of nanoparticles as novel tools for the prevention and control of bacterial infections including chronic biofilm-based fastidious infections. The various types of nanoparticles including polymeric nanoparticles, inorganic nanoparticles and metal nanoparticles have been proved to be effective against bacteria-borne infections. Metal nanoparticles such as silver, copper, gold, zinc, titanium and aluminium are already known for their antimicrobial potential. The nanoparticles when conjugated or functionalized with antibiotics exert enhanced and remarkable antibacterial activity. In addition, the functionalized nanoparticles are excellent candidates for drug delivery. The functionalization of nanoparticles by different drugs/antibiotics is an emerging field for combating multidrug resistance problem in bacteria. To understand the mechanism of antibacterial effect, many studies have been made, however, the exact mechanism of activity of nanoparticles against bacteria needs yet to be elucidated.

Acknowledgements

Raksha Pandit Acknowledges Department of Science and Technology, New Delhi, for providing DST-INSPIRE fellowship (IF150452) for pursuing her research work.

References

Abdel-Sayed, P., Kaeppeli, A., Siriwardena, T., Darbre, T., Perron, K., Jafari, P., Reymond, J.L., Pioletti, D.P. and Applegate, L.A. 2016. Anti-microbial dendrimers against multidrug-resistant *P. aeruginosa* enhance the angiogenic effect of biological burn-wound bandages. Sci. Rep. 6: 22020. doi:10.1038/srep22020.

Abeylath, S.C., Turos, E., Dickey, S. and Lim, D.V. 2008. Glyconanobiotics: Novel carbohydrated nanoparticle antibiotics for MRSA and *Bacillus anthracis*. Bioorg. Med. Chem. 16: 2412–2418.

Abeylath, S.C. and Turos, E. 2008. Drug delivery approaches to overcome bacterial resistance to beta-lactam antibiotics. Expert Opin. Drug Deliv. 5: 931–949.

Agnihotri, S., Pathak, R., Jha, D., Roy, I., Gautam, H.K., Sharma, A.K. and Kumar, P. 2015. Synthesis and antimicrobial activity of aminoglycoside-conjugated silica nanoparticles against clinical and resistant bacteria. New J. Chem. 39: 6746–6755.

Al-Hazmi, F., Alnowaiser, F., Al-Ghamdi, A.A., Aly, M.M., Al-Tuwirqi, R.M. and El-Tantawy, F. 2012. A new large scale synthesis of magnesium oxide nanowires: structural and antibacterial properties. Super lattices Microst. 52(2): 200–209.

Alipour, M., Halwani, M., Omri, A. and Suntres, Z.E. 2008. Antimicrobial effectiveness of liposomal polymyxin B against resistant Gram negative bacterial strains. Int. J. Pharm. 355: 293–298.

Allahverdiyev, A.M., Abamor, E.S., Bagirova, M., Baydar, S.Y, Ates, S.C., Kaya, F., Kaya, C. and Rafailovich, M. 2013. Investigation of antileishmanial activities of TiO_2@Ag nanoparticles on biological properties of *L. tropica* and *L. infantum* parasites, *in vitro*. Exp. Parasitol. 135(1): 55–63.

Anghel, A., Grumezescu, A., Chirea, M., Grumezescu, V., Socol, G., Iordache, F., Oprea, A.E., Anghel, I. and Holban, A.M. 2014. Maple fabricated Fe3O4@*Cinnamomum verum* antimicrobial surfaces for improved gastrostomy tubes. Molecules 19(7): 8981–8994.

Ansari, M.A., Khan, H.M., Khan, A.A., Cameotra, S.S., Saquib, Q. and Musarrat, J. 2014. Interaction of Al_2O_3 nanoparticles with *Escherichia coli* and their cell envelope biomolecules. J. Appl. Microbiol. 116: 772–783.

Ansari, M.A., Khan, H.M., Khan, A.A., Sultan, A. and Azam, A. 2012. Characterization of clinical strains of MSSA, MRSA and MRSE isolated from skin and soft tissue infections and the antibacterial activity of ZnO nanoparticles. World J. Microbiol. Biotechnol. 28(4):1605–1613.

Applerot, G., Lellouche, J., Perkas, N., Nitzan, Y., Gedanken, A. and Banin, E. 2012. ZnO nanoparticle-coated surfaces inhibit bacterial biofilm formation and increase antibiotic susceptibility. RSC Adv. 2(6): 2314–2321.

Azam, A., Ahmed, A.S., Oves, M., Khan, M.S., Habib, S.S. and Memic, A. 2012. Antimicrobial activity of metal oxide nanoparticles against gram-positive and gram-negative bacteria: a comparative study. Int. J. Nanomed. 7: 6003–6009.

Baek, Y.W. and An, Y.J. 2011. Microbial toxicity of metal oxide nanoparticles (CuO, NiO, ZnO, and Sb_2O_3) to *Escherichia coli, Bacillus subtilis*, and *Streptococcus aureus*. Sci. Total Environ. 409: 1603–1608.

Barraud, N., Hassett, D.J., Hwang. S.H., Rice, S.A., Kjelleberg, S. and Webb, J.S. 2006. Involvement of nitric oxide in biofilm dispersal of *Pseudomonas aeruginosa*. J. Bacteriol. 188(21): 7344–7353.

Besinis, A., De Peralta, T. and Handy, R.D. 2004. The antibacterial effects of silver, titanium dioxide and silica dioxide nanoparticles compared to the dental disinfectant chlorhexidine on *Streptococcus mutans* using a suite of bioassays. Nanotoxicology 8: 1–16.

Beyth, N., Houri-Haddad, Y., Domb, A., Khan, W. and Hazan, R. 2015. Alternative antimicrobial approach: Nano-antimicrobial materials. Evid. Based Complement Alternat Med. 2015: 1–16.

Bhimba, B.V., Nath, N. and Sinha, P. 2011. Characterization and antibacterial analysis of silver nanoparticles synthesized by the marine fungi *Hypocrea lixii* MV1 isolated from mangrove sediment soil. J. Pharm. Res. 4: 477–479.

Blecher, K., Nasir, A. and Friedmanm, A. 2011. The growing role of nanotechnology in combating infectious disease. Virulence 2(5): 395–401.

Boda, S.K., Broda, J., Schiefer, F., Heynemann, J.W., Hoss, M., Simon, U., Basu, B. and Dechent, W.J. 2015. Cytotoxicity of ultrasmall gold nanoparticles on planktonic and biofilm encapsulated gram positive *Staphylococci*. Small. 11(26): 3183–3193.

Bogdanović, U., Lazić, V., Vodnik, V., Budimir, M., Marković, Z. and Dimitrijević, S. 2014. Copper nanoparticles with high antimicrobial activity. Mater Lett. 128: 75–78.

Brady-Estévez, A.S., Kang, S. and Elimelech, M.A. 2008. Single walled-carbon-nanotube filter for removal of viral and bacterial pathogens. Small. 4(4): 481–484.

Brown, A.N., Smith, K., Samuels, T.A., Lu, J., Obare, S.O. and Scott, M.E. 2012. Nanoparticles functionalized with ampicillin destroy multiple-antibiotic-resistant isolates of *Pseudomonas aeruginosa* and *Enterobacter aerogenes* and methicillin-resistant *Staphylococcus aureus*. Appl. Environ. Microbiol. 78(8): 2768–2774.

Buckley, J.J., Gai, P.L., Lee, A.F., Olivi, L. and Wilson, K. 2008. Silver carbonate nanoparticles stabilised over alumina nano-needles exhibiting potent antibacterial properties. Chem. Commun. 34: 4013–5.

Carlson, C., Hussein, S.M., Schrand, A.M., Braydich-Stolle, L.K., Hess, K.L., Jones, R.L. and Schlager, J.J. 2008. Unique cellular interaction of silver nanoparticles: size-dependent generation of reactive oxygen species. J. Phy. Chem B. 112(43): 13608–13619.

Carpenter, A.W. and Schoenfisch, M.H. 2012. Nitric oxide release: part II. Therapeutic applications. Chem. Soc. Rev. 41(10): 3742–3752.

Chatterjee, S., Bandyopadhyay, A. and Sarkar, K. 2011. Effect of iron oxide and gold nanoparticles on bacterial growth leading towards biological application. J. Nanobiotechnol. 9: 34. doi:10.1186/1477-3155-9-34.

Cheng, J., Teply, B.A., Sherifi, I., Sung, J., Luther, G., Gu, F.X., Levy-Nissenbaum, E., Radov-ic-Moreno, A.F., Langer, R. and Farokhzad, O.C. 2007. Formulation of functionalized PLGA-PEG nanoparticles for *in vivo* targeted drug delivery. Biomaterials 28: 869–876.

Choi, S.K., Myc, A., Silpe, J.E., Sumit, M., Wong, P.T., McCarthy, K., Desai, A.M., Thomas, T.P., Kotlyar, A., Holl, M.M.B., Orr, B.G. and Baker, J.R. 2013. Dendrimer-based multivalent vancomycin nanoplatform for targeting the drug-resistant bacterial surface. ACS Nano. 7: 214–28.

Chopra, I. 2007. The increasing use of silver-based products as antimicrobial agents: a useful development or a cause for concern? J. Antimicrob. Chemother. 59(4): 587–590.

Chouake, J., Schairer, D., Kutner, A., Sanchez, D.A., Makdisi, J., Blecher-Paz, K. et al. 2012. Nitrosoglutathione generating nitric oxide nanoparticles as an improved strategy for combating *Pseudomonas aeruginosa* infected wounds. J. Drugs Dermatol. 11(12): 1471–1477.

Cooper, R.A., Bjarnsholt, T. and Alhede, M. 2014. Biofilms in wounds: a review of present knowledge. J. Wound Care. 23(11): 570–582.

Costerton, J.W., Stewart, P.S. and Greenberg, E.P. 1999. Bacterial biofilms: a common cause of persistent infections. Science 284(5418): 1318–1322.

Courvalin, P. 2006. Vancomycin resistance in gram-positive cocci. Clin. Infect. Dis. 42(Suppl 1): S25–34.

Cui, L., Iwamoto, A., Lian, J.Q., Neoh, H.M., Maruyama, T., Horikawa, Y. and Hiramatsu, K. 2006. Novel mechanism of antibiotic resistance originating in vancomycin-intermediate *Staphylococcus aureus*. Antimicrob. Agents Chemother. 50(2): 428–438.

Daima, H.K., Selvakannan, P.R., Kanljani, A.E., Shukla, R., Bhargava, S.K. and Bansal, V. 2014. Synergistic influence of polyoxymetalate surface corona towards enhancing the antibacterial performance of tyrosine capped AgNPs. Nanoscale. 6: 758–765.

Daniel, M.C. and Astruc, D. 2004. Gold nanoparticles: assembly, supramolecular chemistry, quantum-size-related properties, and applications toward biology, catalysis, and nanotechnology. Chem. Rev. 104(1): 293–346.

Das, S., Pandey, A., Pal, S., Kolya, H. and Tripathy, T. 2015. Green synthesis, characterization and antibacterial activity of gold nanoparticles using hydroxyethyl starch-g-poly(methylacrylate-co-sodium acrylate): A novel biodegradable graft copolymer. J. Molecular Liquids 2015: 259–265.

Dastjerdi, R. and Montazer, M. 2010. A review on the application of inorganic nano-structured materials in the modification of textiles: focus on antimicrobial properties. Colloids Surf B. Biointerfaces 79(1): 5–18.

Davies, J. and Davies, D. 2010. Origins and evolution of antibiotic resistance. Microbiol. Mol. Biol. Rev. 74(3): 417–443.

Davies, J. and Wright, G.D. 1997. Bacterial resistance to aminoglycoside antibiotics. Trends Microbiol. 5(6): 234–240.

Dibrov, P., Dziobam, J., Gosink, K.K. and Hase, C.C. 2002. Chemiosmotic mechanism of antimicrobial activity of Ag(+) in Vibrio cholerae. Antimicrob. Agents Chemother 46: 2668–2670.

Duncan, B., Li, X., Landis, R.F., Kim, S.T., Gupta, A., Wang, L.S., Ramanathan, R., Tang, R., Boerth, J.A. and Rotello, V.M. 2015. Nanoparticle stabilized capsules for the treatment of bacterial biofilms. ACS Nano. 9(8): 7775–7782.

Durmus, N.G., Taylor, E.N., Kummer, K.M. and Webster, T.J. 2013. Enhanced efficacy of super paramagnetic iron oxide nanoparticles against antibiotic-resistant biofilms in the presence of metabolites. Adv. Mater. 25(40): 5706–5713.

Dutta, R.K., Nenavathu, B.P., Gangishetty, M.K. and Reddy, A.V.R. 2012. Studies on antibacterial activity of ZnO nanoparticles by ROS induced lipid peroxidation. Colloids Surf B Biointerfaces 94: 143–150.

El-Shishtawy, R.M., Asiri, A.M., Abdelwahed, N.A.M. and Al-Otaibi, M.M. 2011. *In situ* production of silver nanoparticle on cotton fabric and its antimicrobial evaluation. Cellulose 18(1): 75–82.

Falagas, M.E. and Kasiakou, S.K. 2005. Colistin: the revival of polymyxins for the management of multidrug-resistant gram-negative bacterial infections. Clin. Infect Dis. 40(9): 1333–1341.

Fattal, E., Rojas, J., Youssef, M., Couvreur, P. and Andremont, A. 1991. Liposome-entrapped ampicillin in the treatment of experimental murine listeriosis and salmonellosis. Antimicrob. Agents Chemother. 35: 770–772.

Feng, Q., Wu, J., Chen, G., Cui, F., Kim, T.N. and Kim, J.O. 2000. Mechanistic study of the antibacterial effect of silver ions on *Escherichia coli* and *Staphylococcus aureus*. J. Biomed. Mater Res. 52: 662–668.

Flemming, H.C. and Wingender, J. 2010. The biofilm matrix. Nat. Rev. Microbiol. 8: 623–633.

Fountain, M.W., Weiss, S.J., Fountain, A.G., Shen, A. and Lenk, R.P. 1985. Treatment of *Brucella canis* and *Brucella abortus in vitro* and *in vivo* by stable plurilamellar vesicle-encapsulated aminoglycosides. J. Infect Dis. 1985152: 529–535.

Fu, G., Vary, P.S. and Lin, C.T. 2005. Anatase TiO_2 nanocomposites for antimicrobial coatings. J. Phys. Chem. B. 109(18): 8889–8898.

Gahlawat, G., Shikha, S., Chaddha, B.S., Chaudhuri, S.R., Mayilraj, S. and Choudhury, A.R. 2016. Microbial glycolipoprotein-capped silver nanoparticles as emerging antibacterial agents against cholera. Microb. Cell Fact. 15: 1–14.

Gan, Q. and Wang, T. 2007. Chitosan nanoparticle as protein delivery carrier–Systematic examination of fabrication conditions for efficient loading and release. Colloids Surf. B. 59(1): 24–34.

Gao, W., Thamphiwatana, S., Angsantikul, P. and Zhang, L. 2014. Nanoparticle approaches against bacterial infections. Wiley Interdiscip Rev. Nanomed. Nanobiotechnol. 6(6): 532–547.

Garg, T., Rath, G. and Goyal, AK. 2016. Inhalable chitosan nanoparticles as antitubercular drug carriers for an effective treatment of tuberculosis. Artif Cells Nanomed. Biotechnol. 44(3): 997–1001.

Gaspar, D.P., Faria,V., Gonçalves, L.M., Taboada, P., Remuñán-López, C. and Almeida, A.J. 2016. Rifabutin-loaded solid lipid nanoparticles for inhaled antitubercular therapy: Physicochemical and *in vitro* studies. Int. J. Pharm. 497(1-2): 199–209.

Gauri, B. 2012. Nanotherapeutics magic bullets—a boon or bane to human health. J. Nanomed. Nanotechol. 3: 1–10.

Gordon, T., Perlstein, B., Houbara, O., Felner, I., Banin, E. and Margel, S. 2011. Synthesis and characterization of zinc/iron oxide composite nanoparticles and their antibacterial properties. Colloid Surf. A. 374(1-3): 1–8.

Govindaraju, S., Ramasamy, M., Baskaran, R., Ahn, S.J. and Yun, K. 2015. Ultraviolet light and laser irradiation enhances the antibacterial activity of glucosamine-functionalized gold nanoparticles. Int. J. Nanomed. 10: 67–78.

Gu, F.X., Zhang, L., Teply, B.A., Mann, N., Wang, A., Radovic-Moreno, A.F., Langer, R. and Far-okhzad, O.C. 2008. Precise engineering of targeted nanoparticles by using self-assembled biointegrated block copolymers. Proc. Natl. Acad Sci. 105: 2586–2591.

Gupta, A., Landis, R.F. and Rotello, V.M. 2016. Nanoparticle based antimicrobials: surface functionality is critical. F1000 Res. 364: 1–10.

Hajipour, M.J., Fromm, K.M., Ashkarran, A.A., Aberasturi, D.J. de., Larramendi, I.R. de, Rojo, T., Serpooshan, V., Parak, W.J. and Mahmoudi, M. 2012. Antibacterial properties of nanoparticles. Trends Biotechnol. 30(10): 499–511.

Hakraborti, S., Mandal, A.K., Sarwar, S., Singh, P., Chakraborty, R. and Chakrabarti, P. 2014. Bactericidal effect of polyethyleneimine capped ZnO nanoparticles on multiple antibiotic resistant bacteria harboring genes of high-pathogenicity island. Colloids Surf B Biointerfaces. 121: 44–53.

Hallaj-Nezhadi, S. and Hassan, M. 2015. Nanoliposome-based antibacterial drug delivery. Drug Deliv. 22(5): 581–589.

Hamal, D.B., Haggstrom, J.A., Marchin, G.L., Ikenberry, M.A., Hohn, K. and Klabunde, K.J. 2010. A multifunctional biocide/sporocide and photocatalyst based on titanium dioxide (TiO_2) codoped with silver, carbon, and sulfur. Langmuir 26(4): 2805–2810.

Hamouda, T., Myc, A., Donovan, B., Shih, A., Reuter, J.D. and Baker, J.R. 2000. A novel surfactant nanoemulsion with a unique non-irritant topical antimicrobial activity against bacteria, enveloped viruses and fungi. Microbiol. Res. 156: 1–7.

Han, G., Martinez, L.R., Mihu, M.R., Friedman, A.J., Friedman, J.M. and Nosanchuk, J.D. 2009. Nitric oxide releasing nanoparticles are therapeutic for *Staphylococcus aureus* abscesses in a murine model of infection. PLoS One. 4(11): 1–7.

Hetrick, E.M., Shin, J.H., Paul, H.S. and Schoenfisch, M.H. 2009. Anti-biofilm efficacy of nitric oxide releasing silica nanoparticles. Biomaterials 30(14): 2782–2789.

Hoshino, N., Kimura, T., Yamaji, A. and Ando, T. 1999. Damage to the cytoplasmic membrane of *Escherichia coli* by catechin-copper (II) complexes. Free Radic Biol. Med. 27(11): 1245–1250.

Huang, L., Dai, T., Xuan, Y., Tegos, G.P. and Hamblin, M.R. 2011. Synergistic combination of chitosan acetate with nanoparticle silver as a topical antimicrobial: efficacy against bacterial burn infections. Antimicrob. Agents Chemother. 55: 3432–3438.

Huang, Z., Jiang, X., Guo, D. and Gu, N. 2011. Controllable synthesis and biomedical applications of silver nanomaterials. J. Nanosci. Nanotechnol. 11: 9395–9408.

Huang, Z., Zheng, X., Yan, D., Yin, G., Liao, X., Kang, Y., Yao, Y., Huang, D. and Hao, B. 2008. Toxicological effect of ZnO nanoparticles based on bacteria. Langmuir. 24(8): 4140–4144.

Huh, A.J. and Kwon, Y.J. 2011. Nanoantibiotics: a new paradigm for treating infectious diseases using nanomaterials in the antibiotics resistant era. J. Control Release 156(2): 128–145.

Hussein-Al-Ali, S., El Zowalaty, M.E., Hussein, M.Z., Geilich, B.M. and Webster, T.J. 2014. Synthesis, characterization and antimicrobial activity of an ampicillin-conjugated magnetic nanoantibiotic for medical applications. Int. J. Nanomed. 9: 3801–3814.

Ingle, A., Gade, A., Pierrat, S., Sonnichsen, C. and Rai, M. 2008. Mycosynthesis of Silver nanoparticles using the fungus *Fusarium acuminatum* and its activity against some human pathogenic bacteria. Curr. Nanosci. 4: 141–144.

Ingle, A.P., Duran, N. and Rai, M. 2014. Bioactivity, mechanism of action, and cytotoxicity of copper-based nanoparticles: a review. Appl. Microbiol. Biotechnol. 98(3): 1001–1009.

Iram, S., Nadhman, A., Akhtar, N., Zulfiqar, H.A. and Arfat yamen, M. 2015. Potentiating efficacy of antibiotic conjugates with zinc oxide nanoparticles against clinical isolates of *Staphylococcus aureus*. Digest J. Nanomat. Biostructures 10(3): 901–914.

Ivask, A., Kurvet, I., Kasemets, K., Blinova, I., Aruoja, V., Suppi, S., Vija, H., Käkinen, A., Titma, T., Heinlaan, M., Visnapuu, M., Koller, D., Kisand, V. and Kahru, A.A. 2014. Size-dependent toxicity of silver nanoparticles to bacteria, yeast, algae, crustaceans and mammalian cells *in vitro*. PLoS One. 9(7): 102–108.

Jadhav, S., Gaikwad, S., Nimse, M. and Rajbhoj, A. 2011. Copper oxide nanoparticles: synthesis, characterization and their antibacterial activity. J. Clust. Sci. 2(2): 121–129.

Jayandran, M., Muhamed, M., Haneefa, A. and Balasubramanian, V. 2016. Biosynthesis and antimicrobial activity studies of salicyclalchitosan functionalized zinc oxide nanoparticles and comparative studies with its non functionalized form. Orient. J. Chem. 32(1): 719–725.

Jiang, Z., Le, N.D., Gupta, A. and Rotello, V.M. 2015. Cell surface-based sensing with metallic nanoparticles. Chem. Soc. Rev. 44(13): 4264–4274.

Jin, T., Sun, D., Su, J.Y., Zhang, H. and Sue, H.J. 2009. Antimicrobial efficacy of zinc oxide quantum dots against *Listeria monocytogenes*, *Salmonella Enteritidis*, and *Escherichia coli* O157:H7. J. Food Sci. 74(1): M46–M52.

Jones, N., Ray, B., Ranjit, K.T. and Manna, A.C. 2008. Antibacterial activity of ZnO nanoparticle suspensions on a broad spectrum of microorganisms. FEMS Microbiol. Lett. 279(1): 71–76.

Kasraei, S., Sami, L., Hendi, S., AliKhani, M.Y., Rezaei-Soufi, L. and Khamverdi, Z. 2014. Antibacterial properties of composite resins incorporating silver and zinc oxide nanoparticles on *Streptococcus mutans* and *Lactobacillus*. Restor. Dent. Endod. 39(2): 109–14.

Kathiresan, K., Manivannan, S., Nabeel, M. and Dhivya, B. 2009. Studies on silver nanoparticles synthesized by a marine fungus, *Penicillium fellutanum* isolated from coastal mangrove sediment. Colloids Surf B. 71: 133–137.

Kaur, P., Thakur, R. and Chaudhury, A. 2016. Biogenesis of copper nanoparticles using peel extract of *Punica granatum* and their antimicrobial activity against opportunistic pathogens. Green Chem. Letter Rev. 9(1): 33–38.

Kazmia, S.J., Shehzada, M.A., Mehmooda, S., Yasara, M., Naeemb, A. and Bhattia, A.S. 2014. Effect of varied Ag nanoparticles functionalized CNTs on its Antibacterial activity against *E. coli*. Sens Actuators A: Phys. 216(1): 287–294.

Khodashenas, B. and Ghorbani, H.R. 2015. Synthesis of silver nanoparticles with different shapes. Arabian J. Chem. doi:org/10.1016/j.arabjc.2014.12.014.

Kim, H.J. and Jones, M.N. 2004. The delivery of benzyl penicillin to *Staphylococcus aureus* biofilms by use of liposomes. J. Liposome Res. 14: 123–39.

Kim, J.S., Kuk, E., Yu, K.N., Kim, J.H., Park, S.J., Lee, H.J., Kim, S.H., Park, Y.K., Park, Y.H., Hwang, C.Y., Kim, Y.K., Lee, Y.S., Jeong, D.H. and Cho, M.H. 2007. Antimicrobial effects of silver nanoparticles. Nanomed. 3(1): 95–101.

Kohno, S., Tomono, K. and Maesaki, S. 1998. Drug delivery systems for infection: liposome-incorporating antimicrobial drugs. J. Infect Chemother. 4(4): 159–173.

Kresge, C.T., Leonowicz, M.E., Roth, W.J., Vartuli, J.C. and Beck, J.S. 1992. Ordered mesoporous molecular sieves synthesized by a liquid-crystal template mechanism. Nature 359: 710–712.

Kruka, T., Szczepanowicza, K., Stefanskab, J., Sochaa, R.P. and Warszynskia, P. 2015. Synthesis and antimicrobial activity of monodisperse copper nanoparticles. Colloids Surf B. 128: 17–22.

Kumar, A.D., Palanichamy, V. and Roopan, S.M. 2014. Photocatalytic action of AgCl nanoparticles and its antibacterial activity. J. Photochem. Photobiol. B Biology. 138: 302–306.

Kumari, A., Yadav, S.K. and Yadav, S.C. 2010. Biodegradable polymeric nanoparticles based drug delivery systems. Colloids Surf B. 75(1): 1–18.

Kutner, A.J. and Friedman, A.J. 2013. Use of nitric oxide nanoparticulate platform for the treatment of skin and soft tissue infections. Wiley Interdiscipl Rev: Nanomed. Nanobiotechnol. 5(5): 502–514.

Lellouche, J., Friedman, A., Lellouche, J.P., Gedanken, A. and Banin, E. 2012a. Improved antibacterial and antibiofilm activity of magnesium fluoride nanoparticles obtained by water-based ultrasound chemistry. Nanomed. 8: 702–711.

Lellouche, J., Friedman, A., Lahmi, R., Gedanken, A. and Banin, E. 2012b. Antibiofilm surface functionalization of catheters by magnesium fluoride nanoparticles. Int. J. Nanomed. 7: 1175–1188.

Li, J., Rong, K., Zhao, H., Li, F., Lu, Z. and Chen, R. 2013. Highly selective antibacterial activities of silver nanoparticles against *Bacillus subtilis*. J. Nanosci. Nanotechnol. 13: 6806–6813.

Li, X.Z. and Nikaido, H. 2004. Efflux-mediated drug resistance in bacteria. Drugs 64(2): 159–204.

Lin, J., Nishino, K., Roberts, M.C., Tolmasky, M., Aminov, R.I. and Zhang, L. 2015. Mechanisms of antibiotic resistance. Front Microbiol. 6 (34): 1–3.

Liu, Y., He, L., Mustapha, A., Li, H., Hu, Z.Q. and Lin, M. 2009. Antibacterial activities of zinc oxide nanoparticles against *Escherichia coli* O157:H7. J. Appl. Microbiol. 107(4): 1193–1201.

Livermore, D.M. 1995. Beta-Lactamases in laboratory and clinical resistance. Clin. Microbiol. Rev. 8(4): 557–584.

Loomba, L. and Scarabelli, T. 2013. Metallic nanoparticles and their medicinal potential. Part I. gold and silver colloids. Ther. Deliv. 4(7): 859–873.

Magallanes, M., Dijkstra, J. and Fierer, J. 1993. Liposome-incorporated ciprofloxacin in treatment of murine salmonellosis. Antimicrob. Agents Chemother. 37: 2293–2297.

Malik, A., Mohammad, Z. and Ahmad, J. 2013. The diabetic foot infections: biofilms and antimicrobial resistance. Diabetes Metab Syndr. 7(2): 101–107.

Malka, E., Perelshtein, I., Lipovsky, A., Shalom, Y., Naparstek, L., Perkas, N., Patick, T., Lubart, R., Nitzan, Y., Banin, E. and Gedanken, A. 2013. Eradication of multidrug resistant bacteria by a novel Zn-doped CuO nanocomposite. Small. 9(23): 4069–4076.

Manavathu, E.K. and Vazquez, J.A. 2014. Biofilms: emerging importance in infectious diseases. J. Multidiscip. Pathol. 1(2): 1–13.

Manikprabhu, D. and Lingappa, K. 2013. Antibacterial activity of silver nanoparticles against methicillin-resistant *Staphylococcus aureus* synthesized using model *Streptomyces* sp. pigment by photo-irradiation method. J. Pharm. Res. 6: 255–260.

Martinez, L.R., Han, G., Chacko, M., Mihu, M.R., Jacobson, M., Gialanella, P., Friedman, A.J., Nosanchuk, J.D. and Friedman, J.M. 2009. Antimicrobial and healing efficacy of sustained release nitric oxide nanoparticles against *Staphylococcus aureus* skin infection. J. Invest. Dermatol. 129(10): 2463–2469.

Meziane-Cherif, D. and Courvali, P. 2014. To the rescue of old drugs. Nature 510: 477–478.

Miller, K.P., Wang, L., Benicewicz, B.C. and Decho, A.W. 2015. Inorganic nanoparticles engineered to attack bacteria. Chem. Soc. Rev. 44(21): 7787–7807.

Misra, R., Acharya, S., Dilnawaz, F. and Sahoo, S.K. 2009. Sustained antibacterial activity of doxycycline-loaded poly(d, l-lactide-co-glycolide) and poly(ε-caprolactone) nanoparticles. Nanomed. 4: 519–530.

Mittelman, M.W. 1998. Structure and functional characteristics of bacterial biofilms in fluid processing operations. J. Dairy Sci. 81: 2760–2764.

Moreno-Sastre, M., Pastor, M., Esquisabel, A., Sans, E., Viñas, M., Fleischer, A., Palomino, E., Bachiller, D. and Pedraz, J.L. 2016. Pulmonary delivery of tobramycin-loaded nanostructured lipid carriers for *Pseudomonas aeruginosa* infections associated with cystic fibrosis. Int. J. Pharm. 498(1-2): 263–273.

Mu, H., Tang, J., Sun, C., Wang, T., Duan, T. and Duan, J. 2016. Potent antibacterial nanoparticles against biofilm and intracellular bacteria. Sci. Rep. 6: 18877. doi:10.1038/srep18877.

Mukherjee, A., Sadiq, I.M., Prathna, T.C. and Chandrasekaran, N. 2011. Antimicrobial activity of aluminium oxide nanoparticles for potential clinical applications. Science against microbial pathogens: communicating current research and technological advances. A. Méndez-Vilas (Ed.) Formatex 245–251.

Nagy, A., Harrison, A., Sabbani, S., Munson, R.S., Dutta, P.K. and Waldman, W.J. 2011. Silver nanoparticles embedded in zeolite membranes: release of silver ions and mechanism of antibacterial action. Int. J. Nanomed. 6: 1833–1852.

Naseem, T. and Farrukh, M.A. 2015. Antibacterial activity of green synthesis of iron nanoparticles using *Lawsonia inermis* and *Gardenia jasminoides* leaves extract. J. Chem. doi.org/10.1155/2015/912342.

Nath, A., Das, A., Rangan, L. and Khare, A. 2012. Bacterial inhibition by Cu/Cu_2O nanocomposites prepared via laser ablation in liquids. Sci. Adv. Mat. 4(1): 106–109.

Nicolosi, D., Scalia, M., Nicolosi, V.M. and Pignatello, R. 2010. Encapsulation in fusogenic liposomes broadens the spectrum of action of vancomycin against Gram-negative bacteria. Int. J. Antimicrob. Agents. 35(6): 553–558.

Ninganagouda, S., Rathod, V., Singh, D., Hiremath, J., Singh, A.K., Mathew, J. and ul-Haq, M. 2014. Growth kinetics and mechanistic action of reactive oxygen species released by silver nanoparticles from *Aspergillus niger* on *Escherichia coli*. Bio. Med. Research Int. 2014: 1–9.

Onyeji, C.O., Nightingale, C.H. and Marangos, M.N. 1994. Enhanced killing of methicillin-resistant *Staphylococcus aureus* in human macrophages by liposome-entrapped vancomycin and teicoplanin. Infection. 22: 338–342.

Palanikumar, L., Ramasamy, S.N. and Balachandran, C. 2014. Size dependent antimicrobial response of zinc oxide nanoparticles. IET Nanobiotechnol. 8(2): 111–117.

Panacek, A., Kvítek, L., Prucek, R., Kolar, M., Vecerova, R., Pizúrova, N., Sharma, V.K., Nevecna, T. and Zboril, R. 2006. Silver colloid nanoparticles: synthesis, characterization, and their antibacterial activity. J. Phys. Chem. B. 110(33): 16248–16253.

Pandey, R. and Khuller, G.K. 2006. Oral nanoparticle-based antituberculosis drug delivery to the brain in an experimental model. J. Antimicrob. Chemother. 57:1146–1152.

Pati, R., Mehta, R.K., Mohanty, S., Padhi, A., Sengupta, M., Vaseeharan, B. et al. 2014. Topical application of zinc oxide nanoparticles reduces bacterial skin infection in mice and exhibits antibacterial activity by inducing oxidative stress response and cell membrane disintegration in macrophages. Nanomed. Nanotechnol. 10(6): 1195–1208.

Pelgrift, R.Y. and Friedman, A.J. 2013. Nanotechnology as a therapeutic tool to combat microbial resistance. Adv. Drug Deliv. Rev. 65(13-14): 1803–15.

Pey, P., Packiyaraj, M.S., Nigam, H., Agarwal, G.S., Singh, B. and Patra, M.K. 2014. Antimicrobial properties of CuO nanorods and multi-armed nanoparticles against *B. anthracis* vegetative cells and endospores. Beilstein J. Nanotechnol. 5: 789–800.

Piao, M.J., Kang, K.A., Lee, I.K., Kim, H.S., Kim, S., Choi, J. and Hyun, J.W. 2011. Silver nanoparticles induce oxidative cell damage in human liver cells through inhibition of reduced glutathione and induction of mitochondria involved apoptosis. Toxicol. Lett. 201(1): 92–100.

Pissuwan, D., Cortie, C.H., Valenzuela, S.M. and Cortie, M.B. 2010. Functionalised gold nanoparticles for controlling pathogenic bacteria. Trends Biotechnol. 28(4): 207–213.

Prabhu, S. and Poulose, E.K. 2012. Silver nanoparticles: mechanism of antimicrobial action, synthesis, medical applications, and toxicity effects. Int. Nano. Lett. 2: 32. doi:10.1186/2228-5326-2-32.
Pumerantz, A., Muppidi, K., Agnihotri, S., Guerra, C., Venketaraman, V., Wang, J. and Betageri G. 2011. Preparation of liposomal vancomycin and intracellular killing of meticillin-resistant *Staphylococcus aureus* (MRSA). Int. J. Antimicrob. Agents. 37(2): 140–144.
Qi, G., Li, L., Yu, F. and Wang, H. 2013. Vancomycin-modified mesoporous silica nanoparticles for selective recognition and killing of pathogenic gram-positive bacteria over macrophage-like cells. ACS Appl. Mater Interfaces. 5(21): 10874–10881.
Qiu, Z., Yu, Y., Chen, Z., Jin, M., Yang, D., Zuguo, Z., Wang, J., Shen, Z., Wang, X., Qian, D., Huang, A., Zhang, B. and Li, J.U. 2012. Nano alumina promotes the horizontal transfer of multi-resistance genes mediated by plasmids across genera. Proc. Natl. Acad. Sci. USA. 109(13): 4944–4949.
Quirósa, J., Borgesb, J.P., Boltesa,K., Rodea-Palomaresd, I. and Rosala, R. 2015. Antimicrobial electrospun silver-, copper- and zinc-doped polyvinylpyrrolidone nanofibers. J. Hazard Mater. 299: 298–305.
Radovic-Moreno, A.F., Lu, T.K., Puscasu, V.A., Yoon, C.J., Langer, R. and Farokhzad, O.C. 2012. Surface charge-switching polymeric nanoparticles for bacterial cell wall-targeted delivery of antibiotics. ACS Nano. 6: 4279–4287.
Raffi, M., Hussain, F., Bhatti, T.M., Akhter, J.I., Hameed, A. and Hasan, M.M. 2008. Antibacterial characterization of silver nanoparticles against *E. coli* ATCC-15224. J. Mater. Sci. Technol. 24(02): 192–196.
Raghupathi, K.R., Koodali, R.T. and Manna, A.C. 2011. Size-dependent bacterial growth inhibition and mechanism of antibacterial activity of zinc oxide nanoparticles. Langmuir. 27(7): 4020–28.
Rai, M., Yadav, A. and Gade, A. 2009a. Silver nanoparticles as a new generation of antimicrobials. Biotechnol. Adv. 27: 76–83.
Rai, M., Yadav, A., Bridge, P. and Gade, A. 2009b. Myconanotechnology: a new and emerging science. pp. 258–267. *In*: Rai, M. and Bridge, P.D. (eds.). Applied Mycology. CAB International, Oxfordshire, UK.
Rai, M., Ingle, A.P., Gade, A.K., Teixeira-Duarte, M.C. and Duran, N. 2015. Three *Phoma* spp. synthesised novel silver nanoparticles that possess excellent antimicrobial efficacy. IET Nanobiotechnol. 9(5): 280–287.
Reardon, S. 2014. Antibiotic resistance sweeping developing world. Nature. 509: 141–142.
Reddy, L.S., Nisha, M.M., Joice, M. and Shilpa, P.N. 2014. Antimicrobial activity of zinc oxide (ZnO) nanoparticle against *Klebsiella pneumoniae*. Pharma Biol. 52(11): 1388–1397.
Ren, G., Hu, D., Cheng, E.W.C., Vargas-Reus, M.A., Reip, P. and Allaker, R.P. 2009. Characterisation of copper oxide nanoparticles for antimicrobial applications. Int. J. Antimicrob. Agents. 33(6): 587–90.
Roshmi, T., Soumya, K.R., Jyothis, M. and Radhakrishnan, E.K. 2015. Effect of biofabricated gold nanoparticle-based antibiotic conjugates on minimum inhibitory concentration of bacterial isolates of clinical origin. Gold Bull. 48: 63–71.
Ruparelia, J.P., Chatterjee, A.K., Duttagupta, S.P. and Mukherji, S. 2008. Strain specificity in antimicrobial activity of silver and copper nanoparticles. Acta Biomater. 4(3): 707–716.
Salema, W., Leitnera, D.R., Zingla, F.G., Schratterc, G., Prassl, R., Goesslerd, W., Reidla, J. and Schild, S. 2015. Antibacterial activity of silver and zinc nanoparticles against *Vibrio cholerae* and enterotoxic *Escherichia coli*. Int. J. Med. Microbiol. 305: 85–95.
Salouti, M. and Ahangari, A. 2014. Nanoparticle based drug delivery systems for treatment of infectious diseases. pp. 978–953 *In*: Sezer, A.D. (ed.). Application of Nanotechnology in Drug Delivery. InTech.
Schairer, D.O., Chouake, J.S., Nosanchuk, J.D. and Friedman, A.J. 2012. The potential of nitric oxide releasing therapies as antimicrobial agents. Virulence 3(3): 271–279.
Schumacher, I. and Margalit, R. 1997. Liposome-encapsulated ampicillin: physicochemical and antibacterial properties. J. Pharm Sci. 86: 635–641.
Scorciapino, M.A., Pirri, G., Vargiu, A.V., Ruggerone, P., Giuliani, A., Casu, M., Buerck, J., Wadhwani, P., Ulrich, A.S. and Rinald, A.C. 2012. A novel dendrimeric peptide with antimicrobial properties: structure-function analysis of SB056. Biophysical J. 102: 1039–1048.
Seabra, A.B., Kitice, N.A., Pelegrino, M.T., Lancheros, C.A.C., Yamauchi, L.M., Pinge-Filho, P. and Yamada-Ogatta, S.F. 2015. Nitric oxide-releasing polymeric nanoparticles against *Trypanosoma cruzi*. J. Phys.: Conf. Ser. 617 012020. doi:10.1088/1742-6596/617/1/012020.
Shahverdi, A.R., Fakhimi, A., Shahverdi, H.R. and Minaian, S. 2007. Synthesis and effect of silver nanoparticles on the antibacterial activity of different antibiotics against *Staphylococcus aureus* and *Escherichia coli*. Nanomedicine 3(2): 168–171.
Sheng, Z. and Liu, Y. 2011. Effects of silver nanoparticles on wastewater biofilms. Water Res. 45(18): 6039–6050.
Sherman, P.M., Ossa, J.C. and Wine, E. 2010. Bacterial infections: new and emerging enteric pathogens. Curr. Opin. Gastroenterol. 26(1): 1–4.
Shrivastava, S., Bera, T., Roy, A., Singh, G., Ramachandrarao, P. and Dash, D. 2007. Characterisation of enhanced antibacterial effects of novel silver nanoparticles. Nanotechnology 18: 1–9.
Simon-Deckers, A., Loo, S., Mayne-L'hermite, M., Herlin-Boime, N., Menguy, N., Reynaud, C., Gouget, B. and Carriere, M. 2009. Size, composition- and shape-dependent toxicological impact of metal oxide nanoparticles and carbon nanotubes toward bacteria. Environ. Sci. Technol. 43(21): 8423–8429.
Sirelkhatim, A., Seeni, M.S., Mohammad Kaus, N.H., Ann, C.L., Mohd Bakhori, S.K., Hasan, H. and Mohamad, D. 2015. Review on Zinc oxide nanoparticles: antibacterial activity and toxicity mechanism. Nano-Micro. Lett. 7(3): 219–242.
Sivolella, S., Stellini, E., Brunello, G., Gardin, C., Ferroni, L., Bressan, E. and Zavan, B. 2012. Silver nanoparticles in alveolar bone surgery devices. J. Nanomater. 2012: 1–12.

Slomberg, D.L., Lu, Y., Broadnax, A.D., Hunter, R.A., Carpenter, A.W. and Schoenfisch, M.H. 2013. Role of size and shape on biofilm eradication for nitric oxide-releasing silica nanoparticles. ACS Appl. Mater Interfaces 5(19): 9322–9329.

Sondi, I. and Salopek-Sondi, B. 2004. Silver nanoparticles as antimicrobial agent: a case study on *E. coli* as a model for Gram-negative bacteria. J. Colloid Interface Sci. 275(1): 177–182.

Srisitthiratkul, C., Pongsorrarith, V. and Intasanta, N. 2011. The potential use of nanosilver-decorated titanium dioxide nanofibers for toxin decomposition with antimicrobial and self-cleaning properties. Appl. Surf. Sci. 257(21): 8850–8856.

Sunitha, A., Rimal-Isaac, R.S., Sweetly, G., Sornalekshmi, S., Arsula, R. and Praseetha, P.K. 2013. Evaluation of antimicrobial activity of biosynthesized iron and silver nanoparticles using the fungi *Fusarium Oxysporum* and *Actinomycetes* sp. On human pathogens. Nano. Biomed. Eng. 5(1): 39–45.

Szlek, J., Paclawski, A., Lau, R., Jachowicz, R. and Mendyk, A. 2013. Euristic modeling of macromo-lecule release from PLGA microspheres. Int. J. Nanomed. 8: 4601–4611.

Tang, Z.X. and Feng, Lv. B. 2014. MgO nanoparticles as antibacterial agent preparation and activity. Brazilian J. Chem. Eng. 31(03): 591–560.

Taylor, E.N., Kummer, K.M., Durmus, N.G., Leuba, K., Tarquinio, K.M. and Webster, T.J. 2012. Super paramagnetic iron oxide nanoparticles (SPION) for the treatment of antibiotic-resistant biofilms. Small 8(19): 3016–3027.

Taylor, E.N. and Webster, T.J. 2009. The use of super paramagnetic nanoparticles for prosthetic biofilm prevention. Int. J. Nanomed. 4: 145–152.

Tiwari, N., Pandit, R., Gaikwad, S., Gade, A. and Rai, M. 2016. Biosynthesis of Zinc Oxide Nanoparticles by petals extract *of Rosa indica* L., its formulation as nail paint and evaluation of antifungal activity against fungi causing onychomycosis. IET Nanobiotechnol. 2016; 1–7: doi: 10.1049/iet-nbt.2016.0003.

Tolaymata, T.M., Badawyb, E., Genaidyc, A.M., Scheckela, A., Luxtona, K.G. and Suidanb, M. 2010. An evidence-based environmental perspective of manufactured silver nanoparticle in syntheses and applications: a systematic review and critical appraisal of peer-reviewed scientific papers. Sci. Total Environ. 408: 999–1006.

Torchilin, V.P. 2005. Recent advances with liposomes as pharmaceutical carriers. Nat. Rev. Drug Discovery 4: 145–160.

Tudose, M., Culita, D.C., Munteanu, C., Pandele, J., Hristea, E., Ionita, P. Zarafu, V. and Chifiriuc, M.C. 2015. Antibacterial activity evaluation of silver nanoparticles entrapped in silica matrix functionalized with antibiotics. J. Inorg. Organomet. Polym. 25: 869–878.

Turos, E., Shim, J.Y., Wang, Y., Greenhalgh, K., Reddy, G.S., Dickey, S. and Lim, D.V. 2007. Antibiotic-conjugated polyacrylate nanoparticles: new opportunities for development of anti-MRSA agents. Bioorg. Med. Chem. Lett. 17: 53–56.

VanHoudt, R. and Michiels, C.W. 2010. Biofilm formation and the food industry, a focus on the bacterial outer surface. J. Appl. Microbiol. 109: 1117–1131.

Ventola, C.L. 2015. The antibiotic resistance crisis. PT. 40(4): 277–283.

Wei, C., Lin, W.Y., Zalnal, Z., Williams, N.E., Zhu, K., Kruzic, A.P., Smith, R.L. and Rajeshwar, K. 1994. Bactericidal activity of TiO_2 photocatalyst in aqueous media: toward a solar-assisted water disinfection system. Environ. Sci. Technol. 28(5): 934–938.

Wu, J., Zheng, Y., Wen, X., Lin, Q., Chen, X. and Wu, Z. 2014. Silver nanoparticle/bacterial cellulose gel membranes for antibacterial wound dressing: investigation *in vitro* and *in vivo.* Biomed. Mater. 9: 035005: doi:10.1088/1748-6041/9/3/035005.

Xie, Y., He, Y., Irwin, P.L., Jin, T. and Shi, X. 2011. Antibacterial activity and mechanism of action of zinc oxide nanoparticles against *Campylobacter jejuni*. Appl. Environ. Microbiol. 77(7): 2325–2331.

Yang, S.C., Aljuffali, I.A., Sung, C.T., Lin, C.F. and Fang, J.U. 2016. Antimicrobial activity of topically-applied soyaethyl morpholinium ethosulfate micelles against *Staphylococcus* species. Nanomed. 11(6): 657–671.

You, C., Han, C., Wang, X., Zheng, Y., Li, Q., Hu, X. and Sun, H. 2012. The progress of silver nanoparticles in the antibacterial mechanism, clinical application and cytotoxicity. Mol. Biol. Rep. 39(9): 9193–9201.

Yousry, C., Fahmy, R.H., Essam, T., El-laithy, H.M. and Elkheshen, S.A. 2016. Nanoparticles as tool for enhanced ophthalmic delivery of vancomycin: a multidistrict-based microbiological study, solid lipid nanoparticles formulation and evaluation. Drug Dev. Ind. Pharm. 19: 1–11.

Zan, L., Fa, W., Peng, T. and Gong, Z.K. 2007. Photocatalysis effect of nanometer TiO_2 and TiO_2-coated ceramic plate on Hepatitis B virus. J. Photochem. Photobiol. B: Biol. 86(2): 165–169.

Zhang, F., Smolen, J.A., Zhang, S., Li, R., Shah, P.N., Cho, S., Wang, H., Raymond, J.E., Carolyn, C.L., Cannon, C.L. and Wooley, K.L. 2015. Degradable polyphosphoester based Silver loaded nanoparticles as therapeutics for bacterial lung infections. Nanoscale 7: 2265–2270.

Zhang, L., Pornpattananangkul, D., Hu, M.J. and Huang, M. 2010. Development of nanoparticles for antimicrobial drug delivery. Curr. Med. Chem. 17: 585–94.

Zhao, L., Wang, H., Huo, K., Cui, L., Zhang, W., Ni, H., Zhang, Y., Wu, Z. and Chu, P.K. 2011. Antibacterial nano-structured titania coating incorporated with silver nanoparticles. Biomaterials 32: 5706–5716.

10

Anti-Adhesion Coating with Natural Products:

When the Nanotechnology Meet the Antimicrobial Prevention

Juan Bueno

Introduction

Microorganisms such as bacteria and fungi have tendency to adhere to a wide variety of surfaces and form biofilms, being a threat to public health causing 65% of nosocomial infections (Otter et al. 2015). In addition, it has been reported that over 75% of infectious diseases that affect humans are boosted by the biofilm formation (Miquel et al. 2016). Also other kinds of organisms as viruses can survive for long periods of time on surfaces despite the use of disinfectants (Kramer et al. 2006, Bagattini et al. 2015). Among the Hospital-Acquired Infections (HAI) the most common is caused by methicillin resistant *Staphylococcus aureus* (MRSA) which is present in hospital environments, with a high rate of transmission among healthcare workers with a prevalence of 4.6% (Laxminarayan et al. 2013, Dulon et al. 2014). Equally, Catheter-Associated Urinary Tract Infections (CAUTI) are the most common source of HAIs. They account for 80% of all nosocomial infections worldwide, with approximately 450000 cases in the United States alone annually (Lo et al. 2014).

Likewise, in surgical procedures sutures can lead to infections difficult to treat, because their contamination results in wounds permeable to microorganisms (Obermeier et al. 2014). For that reason the development of anti-microbial sutures using triclosan and chlorhexidine coating has been a good choice for avoiding Surgical Site Infections (SSI) (Leaper et al. 2010) but recent reports about the induction of resistance by triclosan makes the search for new options for antimicrobial coatings urgent (Weber and Rutala 2012, Obermeier et al. 2015).

Although, the spread of infectious agents in healthcare settings can be controlled through a strict skin hygiene program provided by the author and year (Allegranzi and Pittet 2009), a highly contaminated hospital environment acts as a reservoir of microbes from where they can spread through the hospital workers or patients (Chemaly et al. 2014, Mwamungule et al. 2015), therefore, it is considered as a fact that ordinary surfaces in a hospital care setting are areas susceptible to be contaminated by pathogenic microorganisms, equally hands and any other surface can be reservoirs and spread these organisms after physical contact (Page et al. 2009).

In this way, the prevention of microbial colonization of healthcare settings surfaces is fundamental for reducing the acquisition and dissemination of infections (Chen et al. 2013). To avoid and decrease the surface contamination, the design and implementation of self-disinfecting surfaces

Research Center of Bioprospecting and Biotechnology for Biodiversity Foundation (BIOLABB), Colombia.
E-mail: juanbueno@biolabb.org

have been constituted as an important field for products development (Abreu et al. 2013, Webb et al. 2015). Self-disinfecting surfaces consists in covering surfaces with heavy metals as silver or copper, biocides or with light-activated antimicrobials (photocatalysis) for obtaining an effect of microbial contact killing and provide continuous decontamination without altering the mechanical properties of materials (Weber and Rutala 2013, Bogdan et al. 2015). As an important alternative, looking for new antimicrobial strategies anti-infectious coatings are considered an interesting field of development of innovative products, with three important design approaches: antimicrobial agent liberation, microbial contact-killing and microbial anti-adhesion (Fig. 10.1) (Cloutier et al. 2015). For developing this antibiotic approach, it is necessary to perform material surface modifications as coating by impregnation and covalent grafting of bioactive molecule (Gharbi et al. 2015). It is also necessary to seek new anti-infectious compounds for functionalizing these materials, in this order, Essential Oils (EOs) due to their antimicrobial properties are an important source for study (Upadhyay et al. 2014). EOs from *Anethum graveolens, Salvia officinalis, Pogostemon cablin, Vanilla planifolia* and *Cananga odorata* have shown promising activity in self-disinfecting surfaces (Anghel et al. 2013, Bilcu et al. 2014). In the same way, terpenes as thymol and carvacrol, major components in several EOs have been used for functionalizing antimicrobial surfaces (Gharbi et al. 2012, 2015). But it is very important to note the challenge that represents the volatility of EOs for stability and release of the incorporated agent, for that reason this research field requires a multi-trans-interdisciplinary approach that integrates disciplines as material surfaces, bioactive agents' discovery and delivery strategies (Bilia et al. 2014). Recently, nanotechnology has provided efficient delivery systems for expanding the applications of EOs in clinical medicine. These include liposomes, nanoemulsions and nanoparticles (magnetic nanoparticles, iron oxide@C_{14} nanoparticles, chitosan) (Wang et al. 2011, Anghel et al. 2013, Bilcu et al. 2014, Zorzi et al. 2015), which increase therapeutic efficacy and thermostability.

In this chapter, we have explored the possibilities to use nanotechnology in combination with natural products having antimicrobial potential in order to develop functional surfaces capable of inhibiting bacterial adhesion and prevent biofilm formation. It will help to improve the translational science for design and application of antimicrobial surfaces useful for avoiding HAIs.

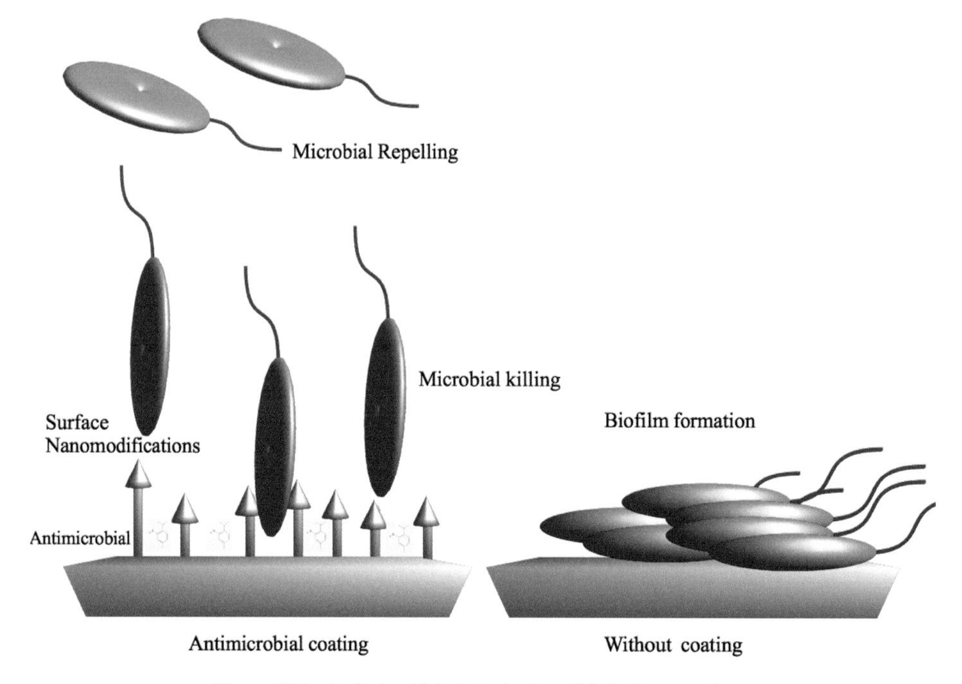

Figure 10.1. Antimicrobial strategies in anti-infectious coatings.

Environmental Surfaces and Hospital-acquired Infections (HAIs)

HAIs are infections acquired during the period of hospitalization. These diseases have been considered a major cause of patient morbidity and mortality (Zhang et al. 2016). The prevalence in developing countries is 15.5 per 100 patients, higher than Europe (7.1 per 100 patients) and USA (4.5 per 100 patients) (Allegranzi et al. 2011).

It has now become evident that the contamination of the hospital environment can hold various microorganisms, including MRSA, Vancomycin-Resistant Enterococci (VRE), *Clostridium difficile*, *Acinetobacter* spp. and norovirus (Gebel et al. 2013), with a persistence in hospital environments ranging from hours to days and even months (Weber and Rutala 2013). Also, this contamination is the major determining factor in the transfer of drug resistant organisms to healthcare workers wearing protective clothing (gloves and coats), especially in the case of *Acinetobacter baumanii* (Morgan et al. 2012, Weber et al. 2013). The best way for avoiding this health threat is constant cleaning and decontamination of all surfaces with suitable biocides, but the education and training without standard protocols have been difficult for both promotion and prevention (Siani and Maillard 2015), and has been estimated that between 15 and 30% of HAIs can be avoided by adherence with proper hygiene measures, including hand hygiene (Casey et al. 2010). But, in the present time it is very clear that the cleaning and disinfection procedures are not optimum, due to several factors as personnel issues, inappropriate dilution of disinfectants and lack of adequate monitoring of procedures (Boyce 2016).

Medical Device Related Infections

Among the HAIs, the medical device-related infection associated to biofilms that colonize medical material (intravascular catheters, urinary catheters and orthopaedic implants) may result in substantial clinical complications, including death, as well as economic consequences such as increased healthcare costs generated by prolonged hospital stays or revision surgery (Francolini and Donelli 2010). For this way the device-related infections represent in USA 60% of HAIs and produce costs between US$28 billion to $45 billion in 1 year (Salwiczek et al. 2014). Therefore, the design of medical devices with antimicrobial coating surfaces have been implemented as a strategy of prevention of HAIs, between them there are also in the market silver alloy-coated latex catheters, antimicrobial urinary catheters nitrofurazone-coated, gentamicin-loaded polymethylmethacrylate beads, antibiotic or chlorhexidine/silver sulfadiazine (CH/SSD)-coated; with activity against *Staphylococcus aureus*, *Pseudomonas aeuroginosa* and *Candida* spp. and demonstrated efficacy in HAIs prevention (Francolini and Donelli 2010). Also, an interesting strategy to avoid the emergence of antibiotic-resistant bacteria is to coat implant surfaces with non-antibiotics drugs, that block microbial targets inside the cell (Tran and Webster 2013), with the end to avoid the adhesion of the pathogens and other pathogenesis mechanisms (Knetsch and Koole 2011). This strategy appears to minimize bacterial adhesion, biofilm formation inhibition and bactericidal activity in different devices (Gallo et al. 2014). Equally, nanotechnology-based antimicrobial surfaces are an important approach for immobilization of anti-infectives keeping the physicochemical properties of the materials used (Coad et al. 2016). In this way, EOs and their major components (e.g., eugenol, carvacrol, citral and limonene) in nanoemulsion have been included in antibacterial nanofibers using electrospinning. This is an important advance in the development of surfaces with ability to permeate cell membranes of microorganisms, with applications in medical devices (Wang and Vermerris 2016).

Antimicrobial Surfaces

In design of antimicrobial surfaces for healthcare settings it is necessary to develop products with the following mechanisms of action: contact killing and anti-adhesion/microbial-repelling. In anti-adhesion coatings, it is crucial to prevent the earliest step of biofilm formation using non-cytotoxic mechanisms. In this route various compounds such as polyethylene glycol (PEG) and zwitterion have demonstrated great anti-adhesion properties *in vitro* (Cloutier et al. 2015, Lu et al. 2015). Recently, five approaches to prevent and combat the adhesion and proliferation of microorganisms on a biomaterial surface have been described. These are surface chemical modification with protein and bacteria-repellent coatings; surfaces with biocides; surfaces with antibiotics incorporated; surfaces with noble metals (silver, copper) and modification of surface checked topography (using soft lithography, microcontact printing, electron beam lithography, nanoimprint lithography, photolithography and electrode position methods) (Knetsch and Koole 2011, Graham and Cady 2014). In addition, depending on the technology used, other classifications have been proposed as passive surface, in which the microbial anti-adhesion is caused by chemical or structural changes in the final surface; active surface, in which there is incorporation of bactericidal agents; and local carriers preferably biodegradable antimicrobial agents, used at the time of surgical procedures (Romanò et al. 2015).

The polymer coatings more commonly used for antimicrobial surfaces are poly (vinyl-N-hexylpyridinium salts) (hexyl PVP), poly(1,3-bis-(p-carboxyphenoxy propane)-co-sebacic anhydride), 2-hydroxyethyl methacrylate (HEMA), poly(ethylene glycol diacrylate) (PEGDA), Acrylic Acid (AA), poly(acrylamide), poly(dimethylsiloxane) (PDMS) and PEG/cationic poly(carbonate) (Smith and Lamprou 2014). Synthetic biodegradable polymers that have been reported for various antibiotic-eluting devices include poly-(lactide-co-glycolide) copolymers, polycaprolactone, polyanhydrides, polyhydroxy-butyrate-co-hydroxyvalerate (PHBV) and polyhydroxyalkanoates. Natural polymers such as collagen and chitosan have shown major biocompatibility and increase the cell growth (Siedenbiedel and Tiller 2012).

Also, in the functionalization of the surfaces; the antimicrobial agents for incorporation are Quaternary Ammonium Compounds (QACs), triclosan, chlorohexidin, silver-sulfadiazine antibiotics (gentamycin, minocycline, norfloxacin, vancomycin and rifampin), magainin, gramicidin A, silver and copper (Knetsch and Koole 2011, Siedenbiedel and Tiller 2012). The functionalizing methods are divided in impregnation, coating methods (ultra-high vacuum deposition, electrospray deposition, spin coating) and covalent grafting (Gharbi et al. 2015).

In addition, nanotechnology has improved the potentiality of anti-microbial surfaces increasing the active contact surface. Likewise, in modification of surface topography, the nanotechnology approach have demonstrated the antimicrobial/anti-adhesive effects using nanoscale topographies which includes lotus leaves, cicada wings, shark skin, feet of gecko, wings of dragonflies and butterflies (Pogodin et al. 2013, Wang and Vermerris 2016). On the other hand, in the search for new antimicrobial agents to incorporate in the different surfaces, natural products, such plant extracts and EOs, have shown to be less toxic in comparison to inorganic antimicrobial materials. Several extracts from medicinal plants have been reported for their anti-infective properties; for example, *Ratibidal atipalearis*, *Teloxys graveolens*, *Dodonaea viscosa*, *Hyptisalbida*, *Melaleuca alternifolia* (tea tree oil) and *Sophora flavescens.* In antimicrobial surfaces, air filters coated with tea tree oil have exhibited antibacterial activity within 2–8 minutes of exposure, equally, filters coated with *S. flavescens* nanoparticles killed microorganisms in 2 minutes (Hwang et al. 2015). Another interesting advance is the design of natural products hybrid for antimicrobial surfaces, which combines biologically active fragments of two natural products, looking for synergistic activity (Wach et al. 2008). In conclusion, all the information presented demonstrate the multidisciplinary approach of antimicrobial surfaces implementation, that combines physics of materials, chemistry, nanotechnology and microbiology, looking for lack of cytotoxicity (Parvizi et al. 2015).

Microbial Anti-adhesive Surfaces

Biomaterials play a central role in management of diseases and in improvement of healthcare (Bhat and Kumar 2013). Due to surfaces of these materials having double function, represented in compatibility with the human tissues and prevention of microbial adhesion (Ribeiro et al. 2012). Microbial adhesion constitutes the first stage of biofilm formation, previous to aggregation and proliferation and is considered an important target in control of biofilm infections (Bueno 2014, Trentin et al. 2015). In surface adhesion, bacteria use extracellular organelles (flagella, curli, fimbriae and outer membrane proteins), which convert the microbial attachment into an irreversible process (Gu and Ren 2014). In a general view, an anti-adhesive agent prevents microbial adhesion by forming a highly hydrated layer on the surface of the biomaterial by taking free water molecules, producing a hydrophilic surface with negative charge that exhibit low adhesion than hydrophobic ones (Schlenoff 2014). Also, another important concept in the design of microbial anti-adhesive surfaces is the nano-roughness, because bacterial attachment and biofilm formation are influenced by surface topography (Rizzello et al. 2013, Feng et al. 2015). Nano-roughness can regulate the bacterial mechanochemical sensors that induce activation of intracellular signaling pathways (Rizzello et al. 2012, Swartjes et al. 2015).

In search of new anti-adhesives, several compounds have been studied such as carpobol 934, a hydrophilic resin; and hyaluronan, a natural anti-adhesive hydrophilic polysaccharide, both with antibiofilm activity (Hogan et al. 2015). Other anti-adhesive compounds from natural products are fimbrolides produced by red alga *Delisea pulchra* that inhibit bacterial adhesion and biofilm formation (Baveja et al. 2004). Various EOs from *Allium sativum*, *Origanum vulgare*, *Cymbopogon citratus*, *Cinnamomum verum*, *Rosmarinus officinalis*, *Ocimum basilicum* and *Cymbopogon martinii* have been incorporated on biopolymers such as alginate, chitosan, hydroxypropylmethyl cellulose, polyethylene (PE) and whey protein, finally producing antibacterial coatings (Glinel et al. 2012). Also thymol and carvacrol have been included in microcapsules and coated on polypropylene films with antimicrobial anti-adhesion properties (Guarda et al. 2011). For these reasons, anti-adhesive surfaces are resistant to microbial colonization and represents important candidates for biomaterial development, useful to prevent spread of infection (Carmona-Ribeiro et al. 2011).

Light-activated antimicrobial agents

An interesting approach for antimicrobial surfaces design is the utilization of a coating with Light-Activated Antimicrobial Agents (LAAAs) (Ismail et al. 2011). LAAAs are chemical compounds with the ability of being excited with light in a determined wavelength, this excitation transfer electrons which produce Reactive Oxygen Species (ROS) that kill microorganisms (Page et al. 2009, Gao et al. 2014). ROS act against microorganisms without interacting with a specific target, this fact is crucial to avoid the emergence of antimicrobial drug resistant strains (Sousa et al. 2011). The photoactivated approximation requires photosensitizer molecules to be attached to the surfaces. Among the more used are indocyanine green (ICG), Methylene Blue (MB), rose Bengal, toluidine blue O (TBO), and titanium dioxide (TiO_2) (Humphreys et al. 2013, Singh et al. 2015). In this order of ideas, TiO_2 have been used in numerous products as catheters to prevent urinary tract infections, coatings for bioactive surfaces, dental implants, lancets, metal pins used for skeletal traction, orthodontic wires and surgical face masks (Foster et al. 2011). Photo-destruction of pathogenic microorganisms will be a fundamental tool to develop bactericidal materials for healthcare setting (Beech et al. 2015).

Antibiofilm Activity

Microbial attachment and the resultant growth and organization on the surfaces cause the biofilm, which possess hydrophilic nano-structures in the interface near to the surface (Sugimoto et al. 2016). In the design of antimicrobial surfaces with antibiofilm activity it is necessary to achieve the following functions: avoid microbial attachment, increase microbicidal activity and remotion of dead microbial cells (Yu et al. 2015). In this way, there are two strategies based in nanotechnology for affecting the biofilm formation, first the use of nano-modifications of surface topography (roughness and nanostructure), and second the application of antimicrobial agents that inhibit primary adhesion (Desrousseaux et al. 2013, Armentano et al. 2014). An interesting approach has been evaluated using polydopamine (PDA) to covalently attach the enzyme lysostaphin (Lst) to produce antibiofilm activity (Yeroslavsky et al. 2015, 2016). Equally, the use of chitosan coatings bound to substrate produce a positive charge that alter the integrity of the cell membrane, destroying the cells and inhibiting fungal biofilm formation (Carmona-Ribeiro et al. 2013, Coad et al. 2014). Likewise, EOs are inhibitors of biofilm blocking the intracellular communications that regulates the socio-microbiology inside the same. It can also induce changes in the biofilm substrate altering the attachment process (Saviuc et al. 2015).

Antiquorum Sensing Activity

In the search for new molecules with anti-adhesive properties, researchers have tried to identify the mechanisms that regulate the attachment and colonization of pathogenic microorganisms (Anthouard et al. 2015). One actual tendency is the development of antimicrobial coatings that interfere with Quorum-Sensing (QS) mechanisms that express virulence factors (Castillo-Juárez et al. 2015). It has been reported that antiquorum-sensing activity of halogenated furanones synthesized by the red alga *Delisea pulchra* with anti-adhesive properties, because of structural similarities with AHLs (N-acyl homoserine lactones), which are the principal molecules in the quorum sensing process (Desrousseaux et al. 2013, Raphel et al. 2016). There are some evidences that support a correlation between anti-biofilm activity and interruption of QS that prevent bacterial adhesion (Husain et al. 2015, Sadekuzzaman et al. 2015). Among other active natural products is *Allium sativum* that acts as a QS inhibitor compound, due to the production of ajoene, a sulfur-rich molecule (Jakobsen et al. 2012). Also, cranberry (*Vaccinium* spp.) is an anti-QS agent, used in formulations in combination with *Solidago* spp. and *Orthosiphon* spp. (CISTIMEV PLUS™) (Cai et al. 2014, Harjai et al. 2014). Likewise, natural products such as eugenol, carvone and usnic acid, have shown anti-QS effects, which make them useful for the prevention of nosocomial infections (Francolini et al. 2004, Ion et al. 2016).

Other Applications of Antimicrobial Surfaces

Wound healing

Cutaneous wounds are classified as acute (e.g., surgical/incisional wounds) or chronic (e.g., pressure ulcers, diabetic foot ulcers and venous leg ulcers) (Percival et al. 2016). Clinical evidence shows that the biofilm formation delay the wound healing, altering epithelialization and formation of granulation tissue (Metcalf and Bowler 2015). For wound care and to avoid the infection, an important treatment strategy is the use of wound dressings that are a protective barrier used to aid and improve the healing process (Rieger et al. 2013). The dressings can be classified for their use in cleaning, granulation and epithelialization; also, the technology used to develop microbial anti-adhesive surfaces is very useful in obtaining new materials, coating processes and antimicrobial agents (Skórkowska-Telichowska et al. 2013). This combination of scientific disciplines has achieved a new approach that have developed a new generation of wound dressings, with anti-infective agents eluted and innovative materials that have improved wound curing ability (Zilberman and Elsner 2008, Elsner and Zilberman 2010). Equally, wound dressing permits to introduce nano-modifications in dressing surface, increasing anti-adhesive and anti-biofilm activities (Anghel et al. 2012). Also, the inclusion of biomaterials with EOs incorporated have shown to be effective in wound dressing development as chitosan films incorporated with EO from *Thymus vulgaris* with antimicrobial activity against *Escherichia coli*, *Klebsiella pneumoniae*, *Pseudomonas aeruginosa* and *Staphylococcus aureus*. Additionally, the antioxidant activity due to carvacrol, a major component present in the *Thymus vulgaris* EO, represent an added value for their wound epithelialization effects (Dursun et al. 2003, Altiok et al. 2010). In this way commercial dressing such as Sorbact® have been immersed in EOs solution (*Melissa citrate indica*, *Pelargonium graveolens*, *Cymbopogon nardus* and *Eugenia caryophyllata*) increasing antimicrobial activity, showing that the EOs encapsulated in electrospun nanofibers are a promising subject in wound dressing development and in combination of natural products improve the wound repair (Payzar et al. 2014, Andreu et al. 2015).

Antimicrobial sutures

Surgical Site Infection (SSI) is a large threat to health with a prevalence estimated around 5%. Between the predisposing factors to SSI are wound contamination, contaminated instruments and contamination of sutures. The prevention of the SSI includes prophylactic antibiotics and wound care (Sajid et al. 2013). An alternative in SSI prevention is the use of antimicrobial coated sutures being the more used triclosan coated sutures; it has been estimated that triclosan-coated sutures reduce costs in hospitalization per SSI in a range of US$4109–$13975 (Singh et al. 2014), but recent reports have shown that triclosan has the ability of inducing antimicrobial drug resistance in pathogenic microorganisms (Leaper et al. 2011, Li et al. 2011, Obermeier et al. 2015).

Antimicrobial Assays for Coated Surface Evaluation

To perform proper antimicrobial assays in self-disinfecting surfaces it is always necessary to assess the antimicrobial agent separately and the biomaterial both *in vitro* as *in vivo* and later evaluate the final coated material for validation.

In vitro microbial adhesion

It can be evaluated using radioactively labeled microorganisms on protein-coated hydroxyapatite (HA) beads. After incubation, the bound bacteria can be counted in a scintillation counter (Palombo 2011). Also, in a multi-well plate format it is possible to perform anti-adhesion methods with coating of collagen, human fibrinogen and bacteria with different concentrations of the product to be tested. After incubation, the adherent cells are fixed with methanol, stained with Crystal Violet (CV) and measured spectrophotometrically to 550 nm (Tomita and Ike 2004, Meireles et al. 2015).

Microbial viability methods

To evaluate viability of microbial cells, and determinate biocidal activity, vital staining for colorimetric and fluorescent methods can be used as radiolabeling, 5-cyano-2,3-ditolyl tetrazolium chloride (CTC) staining, resazurin assay, Syto 9 and fluorescein diacetate (FDA) assay (Kobayashi et al. 2012, Crawford et al. 2016).

Microbial morphological observation

Microscopy for morphological observation of alterations in adherent microbial cells in the presence of different compounds can be measured using image-analyzed epifluorescence microscopy, Scanning Electron Microscopy (SEM), Confocal Laser Scanning Microscopy (CLSM), Atomic Force Microscopy (AFM), and Fourier transform infrared spectroscopy (FTIR) (Armentano et al. 2014). These methods are very important to demonstrate microbicidal activity.

Molecular biological methods

These techniques are very useful to measure the quantity of total community of microbial cells attached to surfaces and it is possible to perform assays with polymicrobial biofilms too, for this evaluation it is possible to use Real Time quantitative polymerase chain reaction (RT-PCR) which is a very sensitive method (Crawford et al. 2016).

In vivo subcutaneous catheter model

This method is performed in C57BL/6 mice and one catheter segment inserted subcutaneously lateral to spine and inoculated with *S. epidermidis* and *S. aureus*. After sacrifice, microbial adherence to catheter can be quantified (Nowakowska et al. 2014).

Tissue cage infection model

In this model, cylindrical tissue cages inserted subcutaneously and perforated with 130 holes to introduce antibiotics, microorganisms and evaluate wound healing and antimicrobial surfaces (Blaser et al. 1995, Nowakowska et al. 2014).

Conclusions and Perspectives

HAIs, medical device-related infections and SSIs are a big public health problem that involves the hospital environment and is a source of antimicrobial drug resistance. In this way, it is necessary to seek for new alternatives to antibiotics as well as conventional health-care cleaning and disinfection protocols. Antimicrobial surfaces coated with anti-adhesive, anti-biofilm and anti-infective agents in combination with nano-modifications of materials are an important approximation in the design and development of new products as coats, gloves, masks, sutures, wound dressings, surgical equipment and hospital self-disinfecting places. But to perform this approach adequately a translational science conceptualization, to integrate a multi-trans-interdisciplinary research, that combine various disciplines as materials physics, chemistry, microbiology, clinical medicine in infectious diseases and pharmacology is very important. In addition, the search of new antimicrobial agents with low toxicity and microbicidal activity to be applied in these surfaces is necessary. Currently, EOs have been selected as a promising source of anti-infective agents that can be used as adjuvants in a multi target therapy that block the emergence of drug resistance (Bueno 2016). Finally, it is very important to remember that this approach will never replace the conventional procedures approved and standardized in the different healthcare settings, for that reason it should be considered as a part of biosafety protocols to decrease the patient and occupational exposure to biological agents.

References

Abreu, A.C., Tavares, R.R., Borges, A., Mergulhão, F. and Simões, M. 2013. Current and emergent strategies for disinfection of hospital environments. J. Antimicrob. Chemother. 68: 2718–2732.

Allegranzi, B. and Pittet, D. 2009. Role of hand hygiene in healthcare-associated infection prevention. J. Hosp. Infect. 73: 305–315.

Allegranzi, B., Nejad, S.B., Combescure, C., Graafmans, W., Attar, H., Donaldson, L. and Pittet, D. 2011. Burden of endemic health-care-associated infection in developing countries: systematic review and meta-analysis. Lancet 377: 228–241.

Altiok, D., Altiok, E. and Tihminlioglu, F. 2010. Physical, antibacterial and antioxidant properties of chitosan films incorporated with thyme oil for potential wound healing applications. J. Mater. Sci.: Mater. Med. 21: 2227–2236.

Andreu, V., Mendoza, G., Arruebo, M. and Irusta, S. 2015. Smart dressings based on nanostructured fibers containing natural origin antimicrobial, anti-inflammatory, and regenerative compounds. Materials 8: 5154–5193.

Anghel, I., Holban, A.M., Andronescu, E., Grumezescu, A.M. and Chifiriuc, M.C. 2013. Efficient surface functionalization of wound dressings by a phytoactive nanocoating refractory to *Candida albicans* biofilm development. Biointerphases 8: 12.

Anghel, I., Holban, A.M., Grumezescu, A.M., Andronescu, E., Ficai, A., Anghel, A.G., Maganu, M., Lazăr, V. and Chifiriuc, M.C. 2012. Modified wound dressing with phyto-nanostructured coating to prevent staphylococcal and pseudomonal biofilm development. Nanoscale Res. Lett. 7: 1–8.

Anthouard, R. and DiRita, V.J. 2015. Chemical biology applied to the study of bacterial pathogens. Infect Immunity 83: 456–469.

Armentano, I., Arciola, C.R., Fortunati, E., Ferrari, D., Mattioli, S., Amoroso, C.F., Rizzo, J., Kenny, J.M., Imbriani, M. and Visai, L. 2014. The interaction of bacteria with engineered nanostructured polymeric materials: a review. Sci. World J. 2014: 410423.

Bagattini, M., Buonocore, R., Giannouli, M., Mattiacci, D., Bellopede, R., Grimaldi, N., Nardone, A., Zarrilli, R. and Triassi, M. 2015. Effect of treatment with an overheated dry-saturated steam vapour disinfection system on multidrug and extensively drug-resistant nosocomial pathogens and comparison with sodium hypochlorite activity. BMC Res. Notes 8: 551.

Baveja, J., Willcox, M., Hume, E., Kumar, N., Odell, R. and Poole-Warren, L. 2004. Furanones as potential anti-bacterial coatings on biomaterials. Biomaterials 25: 5003–5012.

Beech, S.J., Noimark, S., Page, K., Noor, N., Allan, E. and Parkin, I.P. 2015. Incorporation of crystal violet, methylene blue and safranin O into a copolymer emulsion; the development of a novel antimicrobial paint. RSC Adv. 5: 26364–26375.

Bhat, S. and Kumar, A. 2013. Biomaterials and bioengineering tomorrow's healthcare. Biomatter 3: e24717.

Bilcu, M., Grumezescu, A.M., Oprea, A.E., Popescu, R.C., Mogoşanu, G.D., Hristu, R., Stanciu, G.A., Mihailescu, D.F., Lazar, V., Bezirtzoglou, E. and Chifiriuc, M.C. 2014. Efficiency of vanilla, patchouli and ylangylang essential oils stabilized by iron oxide@ C14 nanostructures against bacterial adherence and biofilms formed by *Staphylococcus aureus* and *Klebsiella pneumoniae* clinical strains. Molecules 19: 17943–17956.

Bilia, A.R., Guccione, C., Isacchi, B., Righeschi, C., Firenzuoli, F. and Bergonzi, M.C. 2014. Essential oils loaded in nanosystems: a developing strategy for a successful therapeutic approach. Evidence-Based Compl. Alternative Med. 2014: 651593.

Blaser, J., Vergères, P., Widmer, A.F. and Zimmerli, W. 1995. *In vivo* verification of *in vitro* model of antibiotic treatment of device-related infection. Antimicrob. Agents Chemother. 39: 1134–1139.

Bogdan, J., Zarzyńska, J. and Pławińska-Czarnak, J. 2015. Comparison of infectious agents susceptibility to photocatalytic effects of nanosized titanium and zinc oxides: a practical approach. Nanoscale Res. Lett. 10: 1–15.

Boyce, J.M. 2016. Modern technologies for improving cleaning and disinfection of environmental surfaces in hospitals. Antimicrob. Resist. Infect. Control. 5: 10.

Bueno, J. 2014. Anti-biofilm drug susceptibility testing methods: looking for new strategies against resistance mechanism. J. Microb. Biochem. Technol. 2014: S3–004.

Bueno, J. 2016. Antimicrobial adjuvants drug discovery, the challenge of avoid the resistance and recover the susceptibility of multidrug-resistant strains. J. Microb. Biochem. Technol. 8: 169–176.

Cai, T., Caola, I., Tessarolo, F., Piccoli, F., D'Elia, C., Caciagli, P., Nollo, G., Malossini, G., Nesi, G., Mazzoli, S. and Bartoletti, R. 2014. Solidago, orthosiphon, birch and cranberry extracts can decrease microbial colonization and biofilm development in indwelling urinary catheter: a microbiologic and ultrastructural pilot study. World J. Urol. 32: 1007–1014.

Carmona-Ribeiro, A.M. and de Melo Carrasco, L.D. 2013. Cationic antimicrobial polymers and their assemblies. Int. J. Mol. Sci. 14: 9906–9946.

Carmona-Ribeiro, A.M., de Melo, L.D. and Barbassa, L. 2011. Antimicrobial biomimetics. pp. 227–284. *In*: George, A. (ed.) Biomimetic Based Applications. InTech Open Access Publisher, Rijeka, Croatia.

Casey, A., Adams, D., Karpanen, T., Lambert, P., Cookson, B., Nightingale, P., Miruszenko, L., Shillam, R., Christian, P. and Elliott, T. 2010. Role of copper in reducing hospital environment contamination. J. Hosp. Infect. 74: 72–77.

Castillo-Juárez, I., Maeda, T., Mandujano-Tinoco, E.A., Tomás, M., Pérez-Eretza, B., García-Contreras, S.J., Wood, T.K. and García-Contreras, R. 2015. Role of quorum sensing in bacterial infections. World J. Clin. Cases 3: 575–598.

Chemaly, R.F., Simmons, S., Dale, C., Ghantoji, S.S., Rodriguez, M., Gubb, J., Stachowiak, J. and Stibich, M. 2014. The role of the healthcare environment in the spread of multidrug-resistant organisms: update on current best practices for containment. TherAdv. Infect. Dis. 2: 79–90.

Chen, M., Yu, Q. and Sun, H. 2013. Novel strategies for the prevention and treatment of biofilm related infections. Int. J. Mol. Sci. 14: 18488–18501.

Cloutier, M., Mantovani, D. and Rosei, F. 2015. Antibacterial coatings: challenges, perspectives, and opportunities. Trends Biotechnol. 33: 637–52.

Coad, B.R., Kidd, S.E., Ellis, D.H. and Griesser, H.J. 2014. Biomaterials surfaces capable of resisting fungal attachment and biofilm formation. Biotechnol. Adv. 32: 296–307.

Coad, B.R., Griesser, H.J., Peleg, A.Y. and Traven, A. 2016. Anti-infective surface coatings: design and therapeutic promise against device-assocciated infections. PLoSPathog 12: e1005598.

Crawford, E.C., Singh, A., Gibson, T.W. and Weese, J.S. 2016. Biofilm-associated gene expression in *Staphylococcus pseudintermedius* on a variety of implant materials. Veterinary Surg. 45: 499–506.

Desrousseaux, C., Sautou, V., Descamps, S. and Traoré, O. 2013. Modification of the surfaces of medical devices to prevent microbial adhesion and biofilm formation. J. Hosp. Infect. 85: 87–93.

Dulon, M., Peters, C., Schablon, A. and Nienhaus, A. 2014. MRSA carriage among healthcare workers in non-outbreak settings in Europe and the United States: a systematic review. BMC Infect. Dis. 14: 363.

Dursun, N., Liman, N., Özyazgan, I., Günes, I. and Saraymen, R. 2003. Role of *Thymus* oil in burn wound healing. J. Burn Care Res. 24: 395–399.

Elsner, J.J. and Zilberman, M. 2010. Novel antibiotic-eluting wound dressings: An *in vitro* study and engineering aspects in the dressing's design. J. Tissue Viability 19: 54–66.

Feng, G., Cheng, Y., Wang, S.Y., Borca-Tasciuc, D.A., Worobo, R.W. and Moraru, C.I. 2015. Bacterial attachment and biofilm formation on surfaces are reduced by small-diameter nanoscale pores: how small is small enough? Biofilms and Microbiomes 1: 15022.

Foster, H.A., Ditta, I.B., Varghese, S. and Steele, A. 2011. Photocatalytic disinfection using titanium dioxide: spectrum and mechanism of antimicrobial activity. Appl. Microbiol. Biotechnol. 90: 1847–1868.

Francolini, I. and Donelli, G. 2010. Prevention and control of biofilm-based medical-device-related infections. FEMS Immunol. Med. Microbiol. 59: 227–238.

Francolini, I., Norris, P., Piozzi, A., Donelli, G. and Stoodley, P. 2004. Usnic acid, a natural antimicrobial agent able to inhibit bacterial biofilm formation on polymer surfaces. Antimicrob. Agents Chemother 48: 4360–4365.

Gallo, J., Holinka, M. and Moucha, C.S. 2014. Antibacterial surface treatment for orthopaedic implants. Int. J. Mol. Sci. 15: 13849–13880.

Gao, L., Liu, R., Gao, F., Wang, Y., Jiang, X. and Gao, X. 2014. Plasmon-mediated generation of reactive oxygen species from near-infrared light excited gold nanocages for photodynamic therapy *in vitro*. ACS Nano. 8: 7260–7271.

Gebel, J., Exner, M., French, G., Chartier, Y., Christiansen, B., Gemein, S., Goroncy-Bermes, P., Hartemann, P., Heudorf, U., Kramer, A. and Maillard, J.Y. 2013. The role of surface disinfection in infection prevention. GMS Hyg. Infect. Control. 8: Doc 10.

Gharbi, A., Humblot, V., Turpin, F., Pradier, C.M., Imbert, C. and Berjeaud, J.M. 2012. Elaboration of antibiofilm surfaces functionalized with antifungal-cyclodextrin inclusion complexes. FEMS Immunol. Med. Microbiol. 65: 257–269.

Gharbi, A., Legigan, T., Humblot, V., Papot, S. and Berjeaud, J.M. 2015. Surface functionalization by covalent immobilization of an innovative carvacrol derivative to avoid fungal biofilm formation. AMB Exp. 5: 9.

Glinel, K., Thebault, P., Humblot, V., Pradier, C.-M. and Jouenne, T. 2012. Antibacterial surfaces developed from bio-inspired approaches. Acta Biomater 8: 1670–1684.

Graham, M.V. and Cady, N.C. 2014. Nano and microscale topographies for the prevention of bacterial surface fouling. Coatings 4: 37–59.

Gu, H. and Ren, D. 2014. Materials and surface engineering to control bacterial adhesion and biofilm formation: A review of recent advances. Front. Chem. Sci. Eng. 8: 20–33.

Guarda, A., Rubilar, J.F., Miltz, J. and Galotto, M.J. 2011. The antimicrobial activity of microencapsulated thymol and carvacrol. Int. J. Food Microbiol. 146: 144–150.

Harjai, K., Gupta, R.K. and Sehgal, H. 2014. Attenuation of quorum sensing controlled virulence of *Pseudomonas aeruginosa* by cranberry. Indian J. Med. Res. 139: 446–453.

Hogan, S., Stevens, N., Humphreys, H., O'gara, J. and O'neill, E. 2015. Current and future approaches to the prevention and treatment of staphylococcal medical device-related infections. Curr. Pharma Design. 21: 100–113.

Humphreys, H. 2013. Self-disinfecting and microbiocide-impregnated surfaces and fabrics: what potential in interrupting the spread of healthcare-associated infection? Clin. Infect. Dis. 58: 848–853.

Husain, F.M., Ahmad, I., Khan, M.S. and Al-Shabib, N.A. 2015. *Trigonella foenumgraceum* (seed) extract interferes with quorum sensing regulated traits and biofilm formation in the strains of *Pseudomonas aeruginosa* and *Aeromonashydrophila*. Evidence-Based Complementary Alternative Med. 2015: 879540.

Hwang, G.B., Heo, K.J., Yun, J.H., Lee, J.E., Lee, H.J., Nho, C.W., Bae, G.N. and Jung, J.H. 2015. Antimicrobial air filters using natural *Euscaphis japonica* nanoparticles. PloS ONE 10: e0126481.

Ion, A., Andronescu, E., Rădulescu, D., Rădulescu, M., Iordache, F., Vasile, B.Ş., Surdu, A.V., Albu, M.G., Maniu, H., Chifiriuc, M.C. and Grumezescu, A.M. 2016. Biocompatible 3D matrix with antimicrobial properties. Molecules 21: E115.

Ismail, S., Perni, S., Pratten, J., Parkin, I. and Wilson, M. 2011. Efficacy of a novel light-activated antimicrobial coating for disinfecting hospital surfaces. Infect. Control. Hosp. Epidemiol. 32: 1130–1132.

Jakobsen, T.H., van Gennip, M., Phipps, R.K., Shanmugham, M.S., Christensen, L.D., Alhede, M., Skindersoe, M.E., Rasmussen, T.B., Friedrich, K., Uthe, F. and Jensen, P.Ø. 2012. Ajoene, a sulfur-rich molecule from garlic, inhibits genes controlled by quorum sensing. Antimicrob. Agents Chemother 56: 2314–2325.

Knetsch, M.L. and Koole, L.H. 2011. New strategies in the development of antimicrobial coatings: the example of increasing usage of silver and silver nanoparticles. Polymers 3: 340–366.

Kobayashi, T., Mito, T., Watanabe, N., Suzuki, T., Shiraishi, A. and Ohashi, Y. 2012. Use of 5-cyano-2,3-ditolyl-tetrazolium chloride staining as an indicator of biocidal activity in a rapid assay for anti-Acanthamoeba agents. J. Clin. Microbiol. 50: 1606–1612.

Kramer, A., Schwebke, I. and Kampf, G. 2006. How long do nosocomial pathogens persist on inanimate surfaces? A systematic review. BMC Infect Dis. 6: 130.

Laxminarayan, R., Duse, A., Wattal, C., Zaidi, A.K., Wertheim, H.F., Sumpradit, N., Vlieghe, E., Hara, G.L., Gould, I.M. and Goossens, H. 2013. Antibiotic resistance-the need for global solutions. Lancet Infect. Dis. 13: 1057–1098.

Leaper, D., Assadian, O., Hubner, N.O., McBain, A., Barbolt, T., Rothenburger, S. and Wilson, P. 2011. Antimicrobial sutures and prevention of surgical site infection: assessment of the safety of the antiseptic triclosan. Int. Wound J. 8: 556–566.

Leaper, D., McBain, A.J., Kramer, A., Assadian, O., Sanchez, J.L.A., Lumio, J. and Kiernan, M. 2010. Healthcare associated infection: novel strategies and antimicrobial implants to prevent surgical site infection. Ann. Royal College Surgeons England 92: 453–458.

Li, Y., Kumar, K. N., Dabkowski, J. M., Corrigan, M., Scott, R. W., Nüsslein, K. and Tew, G.N. 2012. New bactericidal surgical suture coating. Langmuir 28: 12134–12139.

Lo, J., Lange, D. and Chew, B.H. 2014. Ureteral stents and foley catheters-associated urinary tract infections: The role of coatings and materials in infection prevention. Antibiotics 3: 87–97.

Lu, Y., Yue, Z., Wang, W. and Cao, Z. 2015. Strategies on designing multifunctional surfaces to prevent biofilm formation. Front. Chem. Sci. Eng. 9: 324–335.

Meireles, A., Gonçalves, A.L., Gomes, I.B., Simões, L.C. and Simões, M. 2015. Methods to study microbial adhesion on abiotic surfaces. AIMS Bioeng. 2: 297–309.
Metcalf, D.G. and Bowler, P.G. 2015. Biofilm delays wound healing: A review of the evidence. Burns Trauma 18: 5–12.
Miquel, S., Lagrafeuille, R., Souweine, B. and C. Forestier. 2016. Anti-biofilm activity as a health issue. Front. Microbiol. 7: 592.
Morgan, D.J., Rogawski, E., Thom, K.A., Johnson, J.K., Perencevich, E.N., Shardell, M., Leekha, S. and Harris, A.D. 2012. Transfer of multidrug-resistant bacteria to healthcare workers' gloves and gowns after patient contact increases with environmental contamination. Crit. Care Med. 40: 1045–1051.
Mwamungule, S., Chimana, H.M., Malama, S., Mainda, G., Kwenda, G. and Muma, J.B. 2015. Contamination of health care workers' coats at the University Teaching Hospital in Lusaka, Zambia: the nosocomial risk. J. Occupat. Med. Toxicol. 10: 34.
Nowakowska, J., Landmann, R. and Khanna, N. 2014. Foreign body infection models to study host-pathogen response and antimicrobial tolerance of bacterial biofilm. Antibiotics 3: 378–397.
Obermeier, A., Schneider, J., Föhr, P., Wehner, S., Kühn, K.D., Stemberger, A., Schieker, M. and Burgkart, R. 2015. *In vitro* evaluation of novel antimicrobial coatings for surgical sutures using octenidine. BMC Microbiol. 15: 186.
Obermeier, A., Schneider, J., Wehner, S., Matl, F.D., Schieker, M., von Eisenhart-Rothe, R., Stemberger, A. and Burgkart, R. 2014. Novel high efficient coatings for anti-microbial surgical sutures using chlorhexidine in fatty acid slow-release carrier systems. PloS ONE 9: e101426.
Otter, J.A., Vickery, K., Walker, J.T., deLanceyPulcini, E., Stoodley, P., Goldenberg, S.D., Salkeld, J.A.G., Chewins, J., Yezli, S. and Edgeworth, J.D. 2015. Surface-attached cells, biofilms and biocide susceptibility: Implications for hospital cleaning and disinfection. J. Hosp. Infect. 89: 16–27.
Page, K., Wilson, M. and Parkin, I.P. 2009. Antimicrobial surfaces and their potential in reducing the role of the inanimate environment in the incidence of hospital-acquired infections. J. Mater. Chem. 19: 3819–3831.
Palombo, E.A. 2011. Traditional medicinal plant extracts and natural products with activity against oral bacteria: potential application in the prevention and treatment of oral diseases. Evidence-Based Complem. Alter. Med. 2011: 680354.
Parvizi, J., Alijanipour, P., Barberi, E.F., Hickok, N.J., Phillips, K.S., Shapiro, I.M., Schwarz, E.M., Stevens, M.H., Wang, Y. and Shirtliff, M.E. 2015. Novel developments in the prevention, diagnosis, and treatment of periprosthetic joint infections. J. Am. Acad. Orthop. Surg. 23: S32–S43.
Pazyar, N., Yaghoobi, R., Rafiee, E., Mehrabian, A. and Feily, A. 2014. Skin wound healing and phytomedicine: a review. Skin Pharmacol. Physiol. 27: 303–310.
Percival, S.L., Finnegan, S., Donelli, G., Vuotto, C., Rimmer, S. and Lipsky, B.A. 2016. Antiseptics for treating infected wounds: efficacy on biofilms and effect of pH. Crit. Rev. Microbiol. 42: 293–309.
Pogodin, S., Hasan, J., Baulin, V.A., Webb, H.K., Truong, V.K., Nguyen, T.H.P., Boshkovikj, V., Fluke, C.J., Watson, G.S., Watson, J.A. and Crawford, R.J. 2013. Biophysical model of bacterial cell interactions with nano-patterned cicada wing surfaces. Biophys. J. 104: 835–840.
Raphel, J., Holodniy, M., Goodman, S.B. and Heilshorn, S.C. 2016. Multifunctional coatings to simultaneously promote osseointegration and prevent infection of orthopaedic implants. Biomaterials 84: 301–314.
Ribeiro, M., Monteiro, F.J. and Ferraz, M.P. 2012. Infection of orthopedic implants with emphasis on bacterial adhesion process and techniques used in studying bacterial-material interactions. Biomatter 2: 176–194.
Rieger, K.A., Birch, N.P. and Schiffman, J.D. 2013. Designing electrospunnanofiber mats to promote wound healing—a review. J. Mater. Chem. B. 1: 4531–4541.
Rizzello, L., Galeone, A., Vecchio, G., Brunetti, V., Sabella, S. and Pompa, P.P. 2012. Molecular response of *Escherichia coli* adhering onto nanoscale topography. Nanoscale Res. Lett. 7: 575.
Rizzello, L., Cingolani, R. and Pompa, P.P. 2013. Nanotechnology tools for antibacterial materials. Nanomedicine 8: 807–821.
Romanò, C.L., Scarponi, S., Gallazzi, E., Romanò, D. and Drago, L. 2015. Antibacterial coating of implants in orthopaedics and trauma: a classification proposal in an evolving panorama. J. Orthop. Surg. Res. 10: 157.
Sadekuzzaman, M., Yang, S., Mizan, M.F.R. and Ha, S.D. 2015. Current and recent advanced strategies for combating biofilms. Comp RevFood SciFood Safety. 14: 491–509.
Sajid, M.S., Craciunas, L., Sains, P., Singh, K.K. and Baig, M.K. 2013. Use of antibacterial sutures for skin closure in controlling surgical site infections: a systematic review of published randomized, controlled trials. Gastroenterol. Report. 2013: 1–9.
Salwiczek, M., Qu, Y., Gardiner, J., Strugnell, R.A., Lithgow, T., Mclean, K.M. and Thissen, H. 2014. Emerging rules for effective antimicrobial coatings. Trends Biotechnol. 32: 82–90.
Saviuc, C.M., Drumea, V., Olariu, L., Chifiriuc, M.C., Bezirtzoglou, E. and Lazar, V. 2015. Essential oils with microbicidal and antibiofilm activity. Curr. Pharma Biotechnol. 16: 137–151.
Schlenoff, J.B. 2014. Zwitteration: Coating surfaces with zwitter-ionic functionality to reduce nonspecific adsorption. Langmuir 30: 9625–9636.
Siani, H. and Maillard, J.Y. 2015. Best practice in healthcare environment decontamination. European J. Clin. Microbiol. Infect Dis. 34: 1–11.
Siedenbiedel, F. and Tiller, J.C. 2012. Antimicrobial polymers in solution and on surfaces: overview and functional principles. Polymers 4: 46–71.
Singh, A., Bartsch, S.M., Muder, R.R. and Lee, B.Y. 2014. An economic model: value of antimicrobial-coated sutures to society, hospitals, and third-party payers in preventing abdominal surgical site infections. Infect Control HospEpidemiol. 35: 1013–1020.

Singh, S., Nagpal, R., Manuja, N. and Tyagi, S.P. 2015. Photodynamic therapy: An adjunct to conventional root canal disinfection strategies. Australian Endodontic J. 41: 54–71.
Skórkowska-Telichowska, K., Czemplik, M., Kulma, A. and Szopa, J. 2013. The local treatment and available dressings designed for chronic wounds. J. Am. Acad. Dermatol. 68: e117–e126.
Smith, J.R. and Lamprou, D.A. 2014. Polymer coatings for biomedical applications: a review. Trans. IMF 92: 9–19.
Sousa, C., Henriques, M. and Oliveira, R. 2011. Mini-review: antimicrobial central venous catheters–recent advances and strategies. Biofouling 27: 609–620.
Sugimoto, S., Okuda, K.I., Miyakawa, R., Sato, M., Arita-Morioka, K.I., Chiba, A., Yamanaka, K., Ogura, T., Mizunoe, Y. and Sato, C. 2016. Imaging of bacterial multicellular behaviour in biofilms in liquid by atmospheric scanning electron microscopy. Sci. Reports 6: 25889.
Swartjes, J.J.T.M., Sharma, P.K., Kooten, T.G., van der Mei, H.C., Mahmoudi, M., Busscher, H.J. and Rochford, E.T.J. 2015. Current developments in antimicrobial surface coatings for biomedical applications. Curr. Med. Chem. 22: 2116–2129.
Tomita, H. and Ike, Y. 2004. Tissue-specific adherent *Enterococcus faecalis* strains that show highly efficient adhesion to human bladder carcinoma T24 cells also adhere to extracellular matrix proteins. Infect Immunity 72: 5877–5885.
Tran, P.A. and Webster, T.J. 2013. Antimicrobial selenium nanoparticle coatings on polymeric medical devices. Nanotechnology 24: 155101.
Trentin, D.S., Silva, D.B., Frasson, A.P., Rzhepishevska, O., da Silva, M.V., Pulcini, E.D.L., James, G., Soares, G.V., Tasca, T., Ramstedt, M. and Giordani, R.B. 2015. Natural Green coating inhibits adhesion of clinically important bacteria. Sci. Reports 5: 8287.
Upadhyay, A., Upadhyaya, I., Kollanoor-Johny, A. and Venkitanarayanan, K. 2014. Combating pathogenic microorganisms using plant-derived antimicrobials: a minireview of the mechanistic basis. BioMed Res. Int. 2014: 761741.
Wach, J.Y., Bonazzi, S. and Gademann, K. 2008. Antimicrobial surfaces through natural product hybrids. AngewandteChemie Int Ed. 47: 7123–7126.
Wang, J. and Vermerris, W. 2016. Antimicrobial nanomaterials derived from natural products—A review. Materials 9: 255.
Wang, L., Liu, F., Jiang, Y., Chai, Z., Li, P., Cheng, Y., Jing, H. and Leng, X. 2011. Synergistic antimicrobial activities of natural essential oils with chitosan films. J. Agri. Food Chem. 59: 12411–12419.
Webb, H.K., Crawford, R.J. and Ivanova, E.P. 2015. Introduction to antibacterial surfaces. pp. 1–8. *In*: Ivanova, E.P. and Crawford, R. (eds.). Antibacterial Surfaces. Springer International Publishing, Switzerland.
Weber, D.J. and Rutala, W.A. 2012. Self-disinfecting surfaces. InfectControl. 33: 10–13.
Weber, D.J. and Rutala, W.A. 2013. Self-disinfecting surfaces: Review of current methodologies and future prospects. Am. J. Infect. Control. 41: S31–S35.
Weber, D.J. and Rutala, W.A. 2013. Understanding and preventing transmission of healthcare-associated pathogens due to the contaminated hospital environment. InfectControl. 34: 449–452.
Weber, D.J., Anderson, D. and Rutala, W.A. 2013. The role of the surface environment in healthcare-associated infections. Curr. Opin. Infect. Dis. 26: 338–344.
Yeroslavsky, G., Girshevitz, O., Foster-Frey, J., Donovan, D.M. and Rahimipour, S. 2015. Antibacterial and antibiofilm surfaces through polydopamine-assisted immobilization of lysostaphin as an antibacterial enzyme. Langmuir 31: 1064–1073.
Yeroslavsky, G., Lavi, R., Alishaev, A. and Rahimipour, S. 2016. Sonochemically produced metal-containing polydopamine nanoparticles and their antibacterial and antibiofilm activity. Langmuir 32: 5201–5212.
Yu, Q., Wu, Z. and Chen, H. 2015. Dual-function antibacterial surfaces for biomedical applications. Acta Biomater 16: 1–13.
Zhang, Y., Zhang, J., Wei, D., Yang, Z., Wang, Y. and Yao, Z. 2016. Annual surveys for point-prevalence of healthcare-associated infection in a tertiary hospital in Beijing, China, 2012–2014. BMC Infect Dis. 16: 161.
Zilberman, M. and Elsner, J.J. 2008. Antibiotic-eluting medical devices for various applications. J. Controlled Rel. 130: 202–215.
Zorzi, G.K., Carvalho, E.L.S., von Poser, G.L. and Teixeira, H.F. 2015. On the use of nanotechnology-based strategies for association of complex matrices from plant extracts. Rev. Bras Farma 25: 426–436.

11

Nanotechnologies for the Delivery of Water-Insoluble Drugs

Omar M. Najjar# and *Rabih Talhouk**

Introduction

Although new drugs are being discovered, many suffer from various limitations in terms of their pharmacokinetics and bioavailability. A large percentage of drugs are poorly soluble in water, which limits their delivery methods and reduces their ability to be transported throughout the body (Loftsson and Brewster 2010, Kalepu and Nekkanti 2015). Other limitations of drug delivery include the difficulty of targeting specific tissues. This is particularly harmful when the drugs cause damage to healthy tissues, as is the case with some chemotherapeutic agents (Feng and Mumper 2013). Furthermore, many drugs can be metabolized or cleared by the body, reducing the half-life of the drugs and their efficacy. This has led to the development of novel drug delivery tools capable of addressing these limitations (Kalepu and Nekkanti 2015). Nanoparticle-based drug delivery systems have emerged to overcome many of the problems facing traditional drug delivery. This began when liposomes were introduced as potential drug carriers in 1978 by Bonventre and Gregoriadis.

Nanoparticles have been demonstrated to increase drug solubility (Müller et al. 2002, Lacko et al. 2015, Gupta et al. 2016), to protect against drug clearance and metabolism (Cabanes et al. 1998, Daeihamed et al. 2016), to make alternative drug delivery routes possible (Negi et al. 2013, Vaghasiya et al. 2013, Chang et al. 2015), and to target specific tissues. Targeting can reduce the side effects of drugs in healthy tissues (Cabanes et al. 1998) and can increase bioavailability in target sites. Active targeting can be achieved by conjugating targeting ligands to the surfaces of nanoparticles (Mamot et al. 2003, Zensi et al. 2009, Johnsen and Moos 2016) or by triggering drug release. This can be modulated by pH, temperature or the presence of particular proteins and enzymes, as well as by external factors such as magnetic fields, electricity, ultrasound and light (as reviewed by Zazo et al. 2016). Passive targeting, on the other hand, occurs when drug-containing nanoparticles accumulate at the target site due to their own intrinsic properties, without any ligand-mediated specificity or triggered release. Passive drug targeting is often achieved by the Enhanced Permeability and Retention (EPR) effect at sites where the vascular permeability is increased (Yuan et al. 1995). This phenomenon is illustrated in Fig. 11.1. Solid tumors and inflammation sites typically have vessels with higher permeability, as well as defective drainage by the lymphatic system. This leads to enhanced accumulation of

Department of Biology, Faculty of Arts and Sciences, American University of Beirut, Beirut, Lebanon.
Present address: The Johns Hopkins University School of Medicine, Baltimore, MD, USA, 21205.
* Corresponding author: rtalhouk@aub.edu.lb

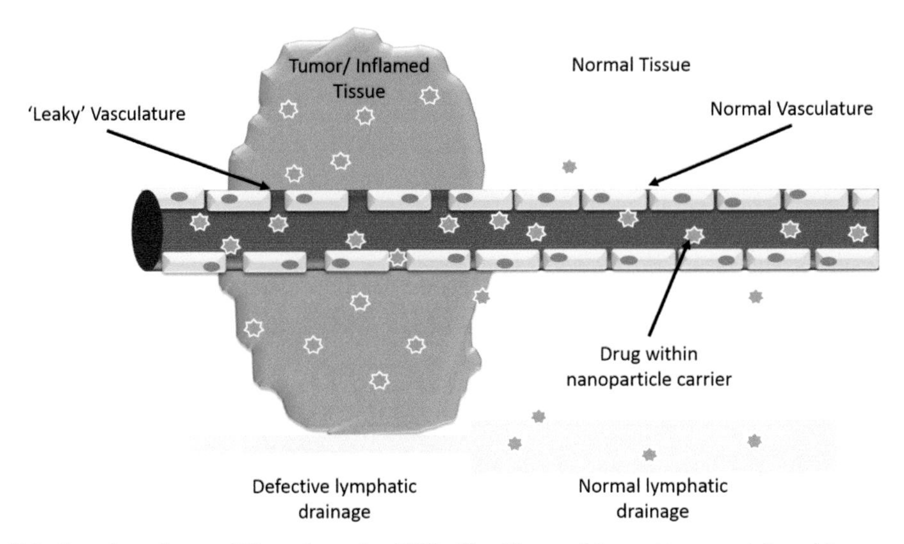

Figure 11.1. The enhanced permeability and retention (EPR) effect, illustrated. In most tumors or inflamed tissues, enhanced vascular permeability, coupled with reduced lymphatic drainage, leads to preferential accumulation of nanoparticles as compared to normal tissues. Modified from (Jhaveri and Torchilin 2014).

nanoparticles, particularly smaller nanoparticles such as micelles, in these tissues (Matsumura and Maeda 1986). This effect can be used to deliver chemotherapeutic drugs more effectively to tumor sites (as reviewed by Feng and Mumper 2013), among other applications.

Various factors must be considered when trying to develop novel drug carriers and translate them from bench to bedside. This is especially applicable to nanoparticle-based drug delivery systems. In terms of pharmacokinetics, the nanoparticles must be able to retain the drug for prolonged periods of time without excessive leakage (Lukyanov et al. 2002, Feng and Mumper 2013). Furthermore, the half-life of the drug in the body is a critical parameter, so the nanoparticles must be able to evade rapid clearance (as reviewed by Kelly et al. 2010). Other factors include the ease and cost of synthesis, the ability to produce the drug in large quantities (as reviewed by Feng and Mumper 2013, Naseri et al. 2015), the homogeneity of the particles, and the efficiency of targeting, among others. The principal factors taken into consideration when evaluating the efficacy of nanoparticles as drug carriers are illustrated in Fig. 11.2.

Many different types of nanoparticles have been studied, differing in composition, pharmacokinetic properties and overall efficacy. Some are organic, such as liposomes and micelles, while others are purely inorganic, especially metallic nanoparticles (as reviewed by Zazo et al. 2016). In some cases, both organic and inorganic components can be incorporated into the same nanoparticle (Schmidt and Ostafin 2002, Bégu et al. 2007, Yue and Dai 2014). In the case of hydrophobic drugs, organic —and particularly lipid-based—nanoparticles predominate due to their greater ability to incorporate water-insoluble drug molecules (as reviewed by Feng and Mumper 2013). Due to the emphasis of this chapter on hydrophobic drugs, organic nanoparticles will be its focus.

These nanoparticles also vary in the extent to which they have been studied. Liposomes were the first type of nanoparticles to be investigated as a drug delivery system, and they are also the most studied, while other types of nanoparticles, such as protein nanoparticles, are still in their early phases of development. With additional research efforts, the limitations of each type of nanoparticle are addressed, and newer generations of nanoparticles can be developed, each improving on its predecessors. For instance, the Mononuclear Phagocytic System (MPS), which clears most organic nanoparticles from circulation, was a major impediment in the early study of liposomes (as reviewed by Kelly et al. 2010). The conjugation of polyethylene glycol (PEG) to the surface of liposomes allowed them to evade this clearance (Senior et al. 1991). Some of these drug delivery systems have even managed to transition into the market and gain the approval of the US Food and Drug administration

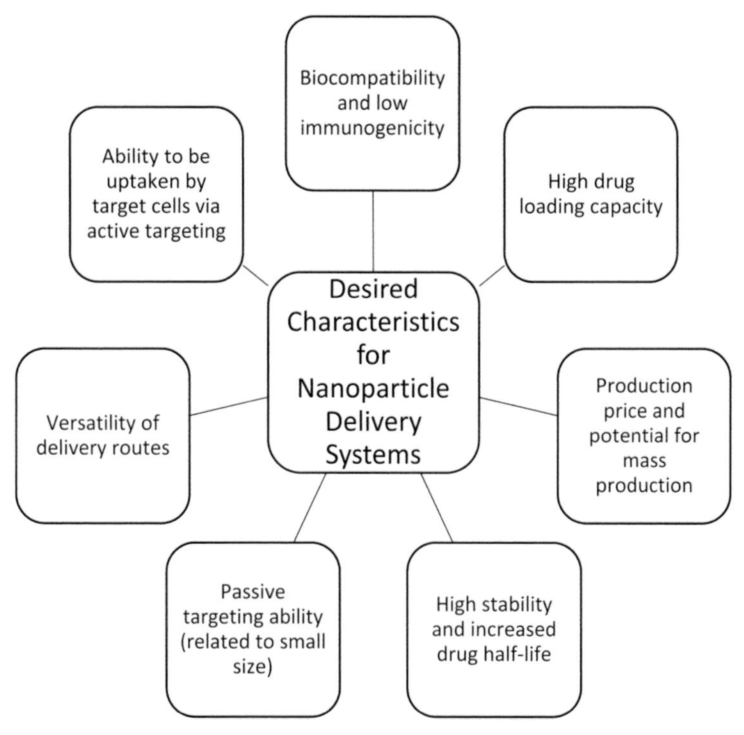

Figure 11.2. Desirable traits that offer advantages for nanoparticle-based drug delivery systems.

(FDA). These mainly include formulations based on liposomes (as reviewed by Feng and Mumper 2013) and, to a lesser extent, micelles (Werner et al. 2013).

At the same time, new types of nanoparticles are being developed, each with unique properties. While these nanoparticle types are still in early stages of development, they also exhibit great promise for the future, as is the case for protein nanoparticles and non-lamellar liquid crystalline lipid phases (as reviewed by Azmi et al. 2015, Lacko et al. 2015, Tezcaner et al. 2016).

This chapter will provide an overview of different types of organic nanoparticles, emphasizing those that have been studied as vehicles for water-insoluble drugs. Every type of nanoparticle carries its own advantages and flaws in terms of stability, targeting and pharmacokinetics. Through examples of *in vitro* experiments, animal studies and clinical trials, the efficacy of each of the main types of organic nanoparticles used to deliver hydrophobic drugs will be assessed. The potential impact of these nanoparticles on the field of antimicrobial drug delivery will be emphasized to highlight the unique advantages of nanoparticles in this domain.

Liposomes

Due to their versatility and unique properties, liposomes are the most extensively studied type of nanoparticles as vehicles for drug delivery. Liposomes are lipid bilayers, typically consisting of amphiphilic phospholipids enclosing an aqueous phase. Liposomes with one lipid bilayer are referred to as unilamellar vesicles, while those with multiple bilayers are called multilamellar vesicles and tend to have larger sizes (Emeje et al. 2012), as illustrated in Fig. 11.3. The molecules composing liposomes must be amphiphilic in order to spontaneously self-organize into bilayers. Phospholipids are the most common components in liposome synthesis, particularly phosphatidylcholine and, to a lesser extent, phosphatidylglycerol (Sharma and Straubinger 1994). The aqueous phase of liposomes may include organic solvents in which water-insoluble drugs are miscible. Thus, liposomes can be effective carriers for both water-soluble and insoluble drugs, and everything in between.

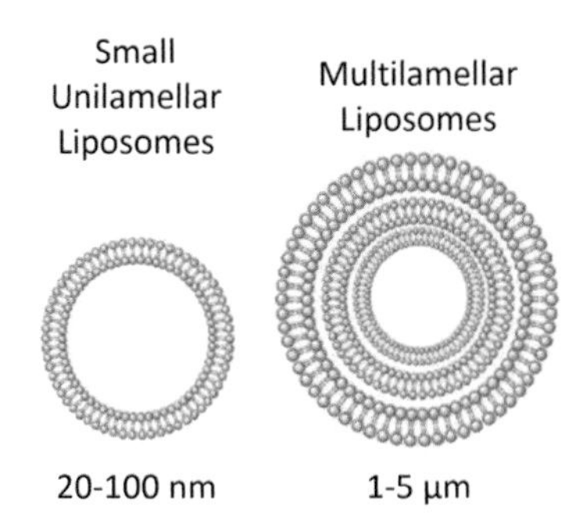

Figure 11.3. Comparison of unilamellar and multilamellar liposomes in terms of size and structure. Modified from (Pandey et al. 2016).

The preparation of drug-containing liposomes usually involves the dissolution of the drug with the lipid bilayer components in a miscible solvent. Evaporation of the solvent allows the lipids to form a film. Upon addition of an aqueous solution, liposomes form spontaneously and trap drug molecules within them. Control over liposome size is usually achieved by sonication or by passing the resulting liposomes through pores of defined sizes. The sizes of most unilamellar liposomes lie in the range of 20–150 nm (Straubinger and Balasubramanian 2005).

Many factors, including size, charge and lipid composition, need to be thoroughly considered when synthesizing liposomes. This ensures an adequately high drug-to-lipid ratio, the solubilization of the drug of interest, and the selective delivery of the drug to target cells (as reviewed by Feng and Mumper 2013). Liposomes circumvent the need to use potentially toxic organic solvents or surfactants which are commonly co-delivered with poorly water-soluble drugs. They can transport drugs of interest with relatively low toxicity or immunogenicity (Straubinger and Balasubramanian 2005). Liposomes also generally enhance the circulation times of drugs. Paclitaxel, a hydrophobic anticancer drug, exhibits much lower levels of systemic toxicity in mice when delivered in liposomes as compared to conventional paclitaxel (Cabanes et al. 1998). Furthermore, the liposomal form exhibited a much longer half-life in the blood. In fact, the first nanoparticle-based drug delivery systems to gain FDA approval were liposomes carrying anticancer drugs. The liposomal anticancer drugs with FDA approval today include Doxil® (doxorubicin), DaunoXome® (daunorubicin) and DepoCyt® (cytarabine) (as reviewed by Feng and Mumper 2013).

Another advantage of liposomes is that they can be tagged with specific ligands to actively target cells. Immunoliposomes, which are conjugated to antibodies, preferentially interact with cells expressing the antigens of their antibodies. For instance, liposomes displaying the targeting Fab fragment of a monoclonal antibody for Epidermal Growth Factor Receptor (EGFR) could be used to target cancer cells (Mamot et al. 2003). The EGFR is an important receptor tyrosine kinase overexpressed in many tumors. Normal cells, with attenuated EGFR expression, would not be significantly affected by these immunoliposomes, reducing the toxic side-effects of anticancer drugs. Another example of active targeting is using the protein transferrin. The transferrin receptor (TfR) is expressed on the surface of endothelial cells in brain capillaries, but not on endothelial cells elsewhere in the human body. Thus, transferrin-tagged liposomes can deliver drugs specifically to the brain, traversing the blood brain barrier by transcytosis across endothelial cells (as reviewed by Johnsen and Moos 2016).

Liposomes can also protect drugs from degradation in the body. They can enhance the efficacy of orally delivered drugs by improving contact with absorptive cells and by helping evade degradation in the gastrointestinal tract (as reviewed by Daeihamed et al. 2016). Liposomes have also been considered as tools for delivering drugs as aerosols to the respiratory tract (as reviewed by Resnier et al. 2016).

The major limitation of liposomes is physical and chemical instability, including the chemical modification of phospholipids, fusion of vesicles, drug leakage and other factors that could attenuate the efficacy of the liposomes (as reviewed by Yue and Dai 2014). In addition, liposomes are often cleared from the blood by the MPS; they are opsonized by serum proteins then undergo phagocytosis. This reduces circulation time significantly (as reviewed by Kelly et al. 2010). Larger liposomes are more susceptible to MPS clearance from the blood and lymph, particularly those greater than 200 nm in diameter (Harashima et al. 1994). The clearest solution would be reducing liposome size. However, limiting liposomes to smaller sizes presents a technical hurdle and reduces drug-loading capacity. Despite these difficulties, the MPS offers a potential means for targeted drug delivery to macrophages and monocytes due to their selective uptake of liposomes (as reviewed by Kelly et al. 2010).

Several modifications of the traditional liposomal structure can enhance the stability of liposomes, ensuring a longer duration of circulation and thus a higher efficacy in drug delivery. One way to enhance stability is by the polymerization of the lipids composing the liposomes (Sisson et al. 1996). Other approaches include coating the liposomes with calcium phosphate shells (Schmidt and Ostafin 2002) or with a layer of silica or ceramic material (Bégu et al. 2007, Yue and Dai 2014). This latter approach combines the biocompatibility of liposomal nanoparticles with the morphological stability of inorganic silica nanoparticles. Another example is liposomes with cationic lipids, which have enhanced interactions with the negatively charged cell membranes. The positive charges simultaneously prevent surface contact between liposomes by electrostatic repulsion (Campbell et al. 2001).

A common approach to stabilize liposomes is by coating them with polyethylene glycol (PEG) to obtain 'PEGylated' liposomes. PEGylation increases the circulation time of liposomes by steric stabilization and by limiting surface interactions between liposomes (Dos Santos et al. 2007). It also enables liposomes to avoid MPS-mediated clearance, blocking opsonization by plasma proteins (Senior et al. 1991). In fact, these liposomes were termed 'stealth' liposomes because they could evade MPS clearance (as reviewed by Immordino et al. 2006). However, a challenge that arose with PEGylation was that the PEG masked targeting ligands on the surface of the liposomes (Klibanov et al. 1991). Later approaches found that these ligands could be attached to the termini of PEG polymers rather than to the surface lipids of the liposome. This novel approach to ligand conjugation helped overcome the masking effect of PEG (Blume et al. 1993, Hansen et al. 1995, Maruyama et al. 1995).

Liposomes, as the most extensively studied type of organic nanoparticles, have undergone immense advancements since their initial use as drug delivery vehicles. Many of the techniques used to develop novel generations of liposome nanoparticles—such as PEGylation—have also been extended to other types of nanoparticles, such as micelles, protein nanoparticles and various others. Thus, liposomes have been the cornerstone for studies on organic nanoparticles used to deliver hydrophobic drugs.

Micelles

Micelles are spontaneously formed colloidal systems of amphipathic molecules in aqueous solutions. The molecules aggregate to form a hydrophilic outer shell in contact with the aqueous environment, and a hydrophobic core. Micelles can be used as drug delivery vehicles due to their ability to contain water-insoluble drug molecules in their core. One of the most important parameters of micelle-forming compounds is their Critical Micelle Concentration (CMC), above which molecules spontaneously assemble into micelles. If a colloidal solution is diluted below the CMC, the amphipathic components tend to disassemble into individual molecules (as reviewed by Feng and Mumper 2013).

The main advantages of micelles include their ease of preparation due to their spontaneous assembly and their size, which typically lies in the range of 10–100 nm. This size range makes micelles less susceptible to clearance in the kidneys and by the MPS. Furthermore, micelles hold immense potential for targeting specific tissues and controlling drug release (as reviewed by Zhang et al. 2014). Many micelle-based anticancer drugs are undergoing clinical trials to transition from bench to bedside. For instance, Genexol PM, a paclitaxel-carrying drug used to treat breast cancer, is an FDA approved micelle-based anticancer drug (Werner et al. 2013).

Micelles also face limitations as drug delivery vehicles. They tend to dissociate upon introduction to the bloodstream or tissues due to their relatively high CMC values, and they often have low loading capacities for hydrophobic drugs (as reviewed by Feng and Mumper 2013). However, these problems have been addressed through several approaches. Polymeric micelles are formed of long polymers with repeating hydrophobic and hydrophilic blocks. These exhibit a much lower CMC, around 10^{-6}–10^{-7} M, compared to more traditional lipid micelles (La et al. 1996). Micelles of PEG are among the most notable micelles that incorporate polymers into their structures. PEG forms a layer on the surface of these micelles. Varying the size of PEG chains produces micelles of variable sizes and pharmacokinetic properties (Zhang et al. 2014). PEG also reduces the unwanted aggregation of micelles, and it prolongs their half-life by helping them evade clearance by the MPS (Avgoustakis et al. 2003).

Sterically Stabilized Micelles (SSMs) are polymeric micelles composed of PEG-grafted distearoyl phosphatidylethanolamine (DSPE-PEG), and they are among the most promising categories of micelles for the delivery of water-insoluble drugs. Their long fatty acid chains create a relatively large hydrophobic interior, expanding loading size for hydrophobic drugs (Krishnadas et al. 2003). The PEG on their surface gives the micelles a higher thermodynamic stability and thus a lower CMC. Furthermore, increasing the molecular weight of the PEG moieties is associated with an increase in micelle half-life and a lower CMC (Lukyanov et al. 2002). However, elongating the PEG chains also increases the size of micelles, which may affect their ability to reach target tissues. Other modifications have also been attempted to increase the efficacy of micelle-based drug delivery. For instance, the incorporation of egg-phosphatidylcholine into SSMs allows them to form Sterically Stabilized Mixed Micelles (SSMMs), which can solubilize greater amounts of hydrophobic drugs. SSMMs were able to dissolve about 1.5 times more paclitaxel than SSMs for the same total lipid concentration (Krishnadas et al. 2003).

As in the case of liposomes, micelles can passively or actively target specific tissues for drug delivery. The small size of micelle nanoparticles, especially compared to their larger liposome counterparts, enables them to accumulate more effectively at tumor and inflammation sites via the EPR effect (see Fig. 11.1). This makes micelles an efficient drug delivery system for targeting sites with enhanced vascular permeability (Yuan et al. 1995). Active targeting is also possible using attached ligands to target specific populations of cells, facilitating the uptake of drugs. As in the case of liposomes, these ligands are often linked to the terminal ends of PEG chains (as reviewed by Zhang et al. 2014).

Micelles can also be triggered to release their drug contents in response to specific environmental stimuli. Ideally, the micelles should remain stable during circulation and in normal tissues, then release the loaded drugs upon reaching the target site. One factor that has been used to prompt drug release from micelles is differences in pH. The pH at inflammation sites and tumors is usually slightly acidic compared to other regions of the body. Some chemical groups are liable to cleavage at lower pH values, including acetal linkages (Tomlinson et al. 2002), hydrazones (Arya et al. 2009), and others. The incorporation of these pH-sensitive elements into polymeric micelles generates nanoparticles that can undergo cleavage and release their contents at tumors or sites of inflammation (Gao et al. 2010). Micelles can be synthesized to respond to other environmental factors as well, such as redox potential. Disulfide bridges incorporated into the micelles can be reduced and cleaved at target sites with high reducing potential (as reviewed by Zhang et al. 2014). Another example is ultrasound-sensitive micelles, which can be disrupted by targeted ultrasounds (Marin et al. 2002).

The next generation of drug-delivering micelles will harness the advantages of dual function micelles. In these systems, the components forming the micelles have therapeutic activity, which adds to or synergizes with the activity of the loaded drugs. Thus these micelles have two functions: solubilizing drugs to deliver them to appropriate targets, and carrying out the additional therapeutic effects of the micelle structural components (as reviewed by Zhang et al. 2014). A notable example of dual effect micelle systems is the use of Vitamin E-based micelles. Vitamin E is a hydrophobic vitamin with a wide array of biological activities and effects. D-α-tocopheryl PEG succinate (TPGS) is a PEGylated amphipathic derivative of vitamin E, capable of forming micelles. TPGS exhibits anticancer activity; it inhibits the proliferation of tumor cells (Shklar et al. 1987, Malafa and Neitzel 2000, Malafa et al. 2002), represses angiogenesis (Dong et al. 2007), and can even sensitize cells to apoptosis by suppressing the NF-κB pathway (Dalen and Neuzil 2003). TPGS micelles have been proposed as therapeutic agents capable of carrying a wide array of anticancer drugs, including paclitaxel. However, due to their high CMC, they tend to dissociate in the blood or tissues upon delivery. A more effective approach has been the synthesis of mixed micelles including TPGS along with other molecules, such as PEG-phosphatidylethanolamine (PEG-PE) (as reviewed by Duhem et al. 2014). PEG-PE enhances half-life and lowers the CMC, rendering the micelles more stable. Meanwhile, the TPGS expands the interior hydrophobic volume of the micelle allowing greater capacity for drugs (Mu et al. 2005), and it also contributes with its own antitumor activity.

Micelles as drug delivery nanoparticles have elicited much of the scientific community's attention. Research has already begun to identify and attempt to solve many of the limitations of micelles, and more micelle-based drugs are expected to soon make the transition into clinics.

Nano and Microemulsions

An emulsion generally consists of two immiscible phases and an emulsifier which dissolves droplets of one phase—the dispersed phase—in the other—the continuous phase. The emulsifier is often a surfactant which greatly reduces the surface tension at the interface between the two phases, allowing the formation of miniscule droplets. More specifically, nano and microemulsions are dispersions of liquid droplets. Despite the name, which suggests a size on the order of micrometers, microemulsion particles are similar in size distribution to the droplets in most nanoemulsions. Both lie in the range of 10–500 nm (as reviewed by Gupta et al. 2016). However, they differ in their thermodynamic and kinetic properties, as well as in their mode of preparation.

Microemulsions are formed spontaneously by self-assembly. They are thermodynamically stable and do not tend to aggregate. Meanwhile, nanoemulsions require more energetically demanding synthesis processes and are thermodynamically unstable. In other words, they do not represent the minimum energy state available to the molecules, and they tend to move towards lower energy states. Nonetheless, nanoemulsions have the advantage of high kinetic stability. They remain stable for extended periods of time before they begin to separate into aqueous and hydrophobic phases, and the shelf lives of nanoemulsion-based drugs can range from months to years (as reviewed by Gupta et al. 2016). Nanoemulsions exhibit high resistance to pH, temperature and dilution. In contrast, microemulsions are highly sensitive to dilution. Nanoemulsions are thus more suitable for parenteral delivery (as reviewed by Feng and Mumper 2013), and they will be emphasized here to a greater extent than microemulsions.

Nanoemulsions can be destabilized by different mechanisms which impede their use as drug delivery vehicles. The first is coalescence, which is when droplets fuse together. This depends on the extent to which the droplets interact favorably with one another; stronger interactions increase the likelihood of coalescence. Thus, if droplets have components with opposite charges, electrostatic attraction may favor coalescence. Meanwhile, opposite surface charges lead to repulsion, reducing the likelihood of coalescence (as reviewed by Mason et al. 2006). Another destabilization mechanism is termed Ostwald ripening, which is driven by the lower chemical potential of the dispersed phase in large droplets. This causes the molecules of the dispersed phase to move from smaller droplets to larger ones, lowering their chemical potential energy. This causes the smaller droplets to shrink further, and the larger ones to grow. This phenomenon is enhanced when the molecules are somewhat soluble in the continuous phase, which would enable them to shuttle between droplets (Taylor 2003). However, both these impediments can be largely overcome. Selecting molecules with a very low solubility in the continuous phase can reduce Ostwald ripening (Taylor 2003). Similarly, the components of nanoemulsions can be chosen so that droplets do not interact favorably with one another; this reduces the frequency of coalescence.

The preparation of these emulsions can proceed by different techniques, classified into low-energy and high-energy methods. The low-energy methods are primarily used for the spontaneous formation of microemulsions by phase inversion (as reviewed by Mason et al. 2006). This can be attained by a change in the composition of the mixture. Starting with a water-in-oil emulsion, where oil is the continuous phase and water is the dispersed phase, water can be added until an inversion point is reached. At this point, the dispersion transitions from a water-in-oil to an oil-in-water emulsion. Due to the presence of surfactant, low surface tension between the two phases allows the formation of nano-sized droplets (Forgiarini et al. 2001). Alternatively, phase inversion can by attained by a change in temperature. Rapidly cooling a water-in-oil dispersion could enable it to transition to an oil-in-water emulsion (Izquierdo et al. 2002).

On the other hand, nanoemulsions do not assemble spontaneously, and their synthesis requires the exertion of external forces, such as shear force (as reviewed by Mason et al. 2006). These high-energy preparation techniques often involve a macroemulsion of oil phase in water, containing large oil droplets on the order of micrometers. These emulsions are then subjected to procedures that reduce the sizes of the droplets, including high-pressure homogenization (Floury et al. 2004, Meleson et al. 2004) and ultrasonication (Delmas et al. 2011). There are various other preparation methods, including bubble-bursting at an oil-water interface. This produces a 'spray' of nanosized oil droplets (as reviewed by Gupta et al. 2016).

The applications of nanoemulsion in drug delivery have so far attained varying levels of success in different fields. Constantinides et al. (2000) described the preparation of the anticancer drug TOCOSOL™, a nanoemulsion composed of tocopherol (vitamin E) as an oil phase, TPGS (D-α-Tocopheryl PEG succinate) and Poloxamer 407 as surfactants, and paclitaxel as a drug loaded into the nanoparticle. TOCOSOL™, for instance, exhibited highly promising preclinical trial results (Constantinides et al. 2004) and was advanced to clinical trials. Unfortunately, during phase III clinical trials it led to higher rates of neutropenia in patients, where they exhibited a reduced neutrophil count in the blood, as compared to the traditional mode of paclitaxel delivery, Taxol™. These side effects led to the discontinuation of clinical trials of TOCOSOL™ (as reviewed by Feng and Mumper 2013).

Nanoemulsions have also witnessed studies in the domain of antimicrobial drugs, most of which have remained at the level of basic research. For instance, peppermint oil nanoemulsions displayed antimicrobial activity against strains of the Gram-positive bacteria *Listeria monocytogenes* and *Staphylococcus aureus* (Liang et al. 2012). Using long-lived, kinetically stable nanoemulsions as drug vehicles could qualify such oils to become potential candidates for treatment against bacteria. Nanosized particles are also often effective at penetrating bacterial biofilms. Teixeira et al. (2007) described the antimicrobial properties of a nanoemulsion containing soybean oil and a microemulsion of ethyl oleate. These systems both exhibit wide a range of bactericidal effects against various enteric bacterial pathogens.

Nano and microemulsions are likely to develop applications in other fields besides drug delivery. In the food industry, they can be used to prolong the shelf lives of products by preventing bacterial growth, or to improve the bioavailability of different nutrients (Silva et al. 2012). In the domain of cosmetics, they present an effective means to deliver products across the skin or to protect against UV light, and they carry the additional aesthetic advantage of being transparent. In terms of drug delivery, the major advantages of nanoemulsions stem from their resistance to dilution and changes in pH or temperature.

Protein Nanoparticles

The use of proteins as nanoparticle drug delivery systems is another domain of increasing interest. Serum proteins that circulate in the blood present powerful tools for intravenous delivery due to their biocompatibility, prolonged half-life, unique targeting abilities and high ability to penetrate cells. These include nanoparticles formed from albumin, low and high density lipoprotein (LDL and HDL respectively), fibrinogen, transferrin and others (as reviewed by Tezcaner et al. 2016).

A great deal of studies have focused on serum albumin as a drug delivery tool. Albumin has many functions, including its role as a transport mechanism for fatty acids and hydrophobic hormones in the plasma. The hydrophobic core of albumin has multiple hydrophobic binding pockets (Curry et al. 1998) where water-insoluble drugs can be bound. This enables albumin to effectively deliver hydrophobic drugs and release them slowly. Furthermore, drug molecules can also be conjugated through chemical bonds to the surface of albumin (Segura et al. 2007). Both Bovine Serum Albumin (BSA) and Human Serum Albumin (HSA) nanoparticles have been studied, and their sizes are usually in the order of 100 nm. Albumin nanoparticles can be easily prepared, and they are non-irritant and non-immunogenic. In addition, different molecules can be conjugated to their surfaces by covalent bonds with the functional groups of amino acids (as reviewed by Elzoghby et al. 2012). These molecules can be targeting ligands. For instance, HSA nanoparticles conjugated with the apolipoprotein Apo E were able to traverse the blood brain barrier in mice by transcytosis (Zensi et al. 2009). Alternatively, the drug molecules can be directly conjugated to amino acids on the surface of albumin. Wartlick et al. (2004) described antisense oligonucleotides (ASOs) conjugated to the surface of albumin nanoparticles. *In vitro* studies of several cancer cell lines demonstrated the uptake and accumulation of the ASOs in the cytoplasm.

A number of albumin-based drugs and diagnostic agents have been FDA-approved. These include Abraxane®, which contains paclitaxel for treating metastatic breast cancer, and Albunex™, which is used as a contrast agent for ultrasound imaging (as reviewed by Elsadek and Kratz 2012). Several other albumin-based drugs are undergoing clinical trials and are likely to become FDA approved in the years to come. Other serum proteins such as insulin have similarly been used as nanoparticle vehicles for drugs. For instance, Levemir® is a form of human insulin conjugated to a fatty acid chain. Victoza® is a fatty acid derivative of the glucagon-like-peptide GLP-1. They are both used for treating diabetes (as reviewed by Elsadek and Kratz 2012).

HDL and LDL have also been studied as potential vehicles for delivering hydrophobic drugs, but their applications remain limited, and further research is needed. HDL is particularly promising because it undergoes specific uptake into cells by endocytosis, then releases its contents to the cytosol before the apoprotein component is recycled. The endogenous function of HDL is the collection of cholesterol and cholesteryl esters from different body tissues and delivering them to the liver; this is referred to as reverse cholesterol transport. Highly hydrophobic molecules can thus be incorporated into the core of the HDL protein, where they can bind at the lipid binding pockets (as reviewed by Lacko et al. 2015).

The advantages of HDL as a vehicle for drugs include its stability in circulation, along with its resistance to drug leakage. It also exhibits high biocompatibility, very small particle sizes (10–50 nm) and an intrinsic targeting ability due to receptor-mediated uptake (as reviewed by Lacko et al. 2015). Using HDL and LDL as anticancer drug transporters was found to lead to increased efficiency of drug delivery and higher toxicity to carcinoma cell lines (Kader and Pater 2002). However, the major impediments to the development of HDL nanoparticles for drug delivery include the difficulty of isolating native apolipoproteins. This has initiated the search for sources of HDL apoproteins which can then be reassembled to form reconstituted HDL, or rHDL (as reviewed by Lacko et al. 2007). These efforts have been faced with the challenges of inconsistent composition and sizes of rHDL, as well as the possible danger of harboring infectious agents (as reviewed by Lacko et al. 2015). Several approaches have been used to overcome these difficulties. For instance, the insufficient incorporation of apolipoprotein Apo A-I into nanoparticles motivated the development of synthetic peptides—apo A-I mimetics—capable of mimicking the structure and function of Apo A-I, and these have been demonstrated to be an effective alternative (Zhang et al. 2010). Nonetheless, the production of these nanoparticles remains quite expensive, and the field of using LDL/HDL nanoparticles as drug delivery vehicles is still in its early phases (as reviewed by Feng and Mumper 2013).

HDL delivery systems display an intrinsic ability to target tumor cells. SR-B1 is an HDL receptor primarily expressed in liver and nonplacental steroidogenic tissues (Acton et al. 1996). However, cancer cells also generally exhibit increased levels of SR-B1 (as reviewed by Lacko et al. 2002), explained by their need for increased intake of cholesterol to allow continued proliferation (as reviewed by Cruz et al. 2013). Thus, HDL nanoparticles carrying anticancer drugs have an innate targeting ability for a wide array of cancer cells by selective SR-B1 mediated uptake. HDL also has been studied as a delivery tool for nucleic acids, which can be used for therapeutic applications by modulating gene expression. They included siRNAs, antisense RNA molecules, and many others. The negative charges of nucleic acids prevent them from traversing the cell membrane due to electrostatic repulsion. These molecules are also prone to degradation by serum nucleases (Seow and Wood 2009). However, HDL can overcome these difficulties. rHDL drug delivery systems can carry cholesterol-conjugated nucleic acids. An HDL-mimicking nanoparticle was shown to be capable of delivering siRNAs conjugated to cholesterol directly into the cytoplasm of cancer cells from different lines, bypassing endolysosomal degradation (Yang et al. 2011).

Other proteins have also been used as nanocarriers for drugs, including gelatin. The uptake of gelatin nanoparticles by the immune system can be used to target phagocytic cells. This targeting can have different aims, such as immunosuppression in the case of autoimmune disorders (as reviewed by Wang and Uludag 2008). For instance, the delivery of clodronate, an anti-inflammatory drug, was performed by conjugating bisphosphonate clodronate (CLOD) to type A gelatin nanospheres (GNS) to form CLOD-GNS. The drug successfully reduced the populations of hepatic and splenic macrophages in rats suffering from Immune Thrombocytopenic Purpura (ITP), a disorder where the immune system targets platelets leading to a high risk of hemorrhage (Li et al. 2006).

Branched amphiphilic peptide capsules are a novel peptide-based nanoparticle designed to resemble liposomes, and they consist of two branched peptides which mimic diacyl phosphoglycerides in their molecular structure. They assemble into bilayers which form nanosized capsules ranging from 20 nm to 2 μm in diameter. They can encapsulate drugs, dyes and other molecules in a similar manner to liposomes, coupled with their higher stability and robustness compared to liposomes (Sukthankar et al. 2014, Barros et al. 2016).

Overall, the field of protein-based nanoparticles holds promise for the future due to the high biocompatibility of these particles and their ability to be transported in the blood. Mimicking various types of endogenous transport proteins such as HDL, LDL and insulin has led to promising results, even though the technicalities of synthesizing these nanoparticles have yet to be optimized. Furthermore, other serum proteins such as fibrinogen and transferrin have yet to be sufficiently investigated as drug delivery nanoparticles, and they could prove to carry unforeseen advantages upon further inspection.

Solid Lipid Nanoparticles

Solid Lipid Nanoparticles (SLNs) were first prepared and studied by (Schwarz et al. 1994). Other laboratories soon followed, and the first generation of SLNs emerged (Bunjes et al. 1996, Morel et al. 1996, Cavalli et al. 1997). SLNs are colloidal systems consisting of small particles of a lipid matrix dispersed in an aqueous solution, often with the aid of surfactants. The lipids used in these particles can be widely diverse, including mono-, di-, and triglycerides, as well as phospholipids, waxes and others. A critical factor is that these particles must be in the solid phase at body temperature and, accordingly, they must be solid at lower temperatures such as room temperature. The efficacy of these nanoparticles depends on the proper choice of lipids with which the drug of interest is miscible. This can allow prolonged release time and higher levels of drug loading in the nanoparticles (as reviewed by Feng and Mumper 2013). SLNs are roughly spherical in shape and highly variable in size, often in the range of 40–1000 nm. Three models for drug incorporation into SLNs have been described, depending on the mode of preparation. These include (1) the solid solution model where the drug is dispersed throughout the lipid matrix of the nanoparticle, (2) the drug-enriched shell model where most of the drug is concentrated at the periphery of the molecule, and (3) the drug-enriched core model where most of the drug is found near the center of the particle (as reviewed by Naseri et al. 2015).

The advantages of SLNs over other nanoparticle drug delivery systems include their higher stability due to their solid nature, which reduces drug leakage and enables prolonged release. For instance, SLNs carrying paclitaxel were found to release only 10% of their drug contents after 24 hours in serum (Lee et al. 2007). Furthermore, SLNs are generally safer than emulsions and liposomes because their preparation does not require the use of potentially toxic organic solvents. Another advantage is the ease of preparation, which can be expanded to larger scale production with acceptable costs (as reviewed by Feng and Mumper 2013, Naseri et al. 2015).

The primary limitations of SLNs include low drug-loading capacity, as well as drug release during storage or due to burst releases. This is mostly due to the tendency of the lipid components of these nanoparticles to form highly ordered crystalline arrangements which leave little room to load drug molecules (as reviewed by Weber et al. 2014). To overcome this, a novel system termed the 'nanostructured lipid carrier' (NLC) was synthesized composed of spatially incompatible lipids to prevent the formation of highly ordered crystals (Müller et al. 2002). This incompatibility can be accomplished by using mono-, di-, or triglycerides with different chain lengths or by incorporating liquid lipid components into the nanoparticles. Thus, NLCs constitute a second generation SLN, with higher drug-incorporation ability and more prolonged release (as reviewed by Feng and Mumper 2013).

SLNs and NLCs are hydrophobic particles liable to clearance from the circulation by the MPS. However, similar to the case of liposomes and micelles, modification of the surface with more hydrophilic moieties can help the particles evade MPS clearance. PEGylation here reappears as a powerful means to produce 'stealth SLNs' which have prolonged circulation in the blood (as reviewed by Feng and Mumper 2013). PEG can be incorporated directly by conjugation to the lipid components of the nanoparticles, or as part of the surfactants. For instance, Chen et al. (2001) described two types of stealth SLNs carrying paclitaxel and demonstrate their anticancer efficacy in mice.

Various mechanisms can also be employed to enable drug targeting using SLNs. The first mode is the conjugation of a monoclonal antibody or the variable Fab fragment of the antibody to the surface of the SLN, enabling specific uptake of the nanoparticles at the level of target cells (as reviewed by Rostami et al. 2014). Another targeting mechanism is magnetic targeting via nanoparticles that incorporate magnetized materials such as magnetite (Fe_3O_4), an iron oxide which can be permanently magnetized. Spherical and uniform magnetic SLNs containing ibuprofen were prepared by Panga et al. (2009). When magnetized nanoparticles are used, the application of an external magnetic field near the target site enables the accumulation of nanoparticles at this site, and drug release occurs gradually (as reviewed by Rostami et al. 2014). Alternatively, pH-sensitive SLNs could enable drug release at target sites with particular pH values (as reviewed by Rostami et al. 2014). This can be effective for the treatment of tumors and inflammation due to the lower pH values at these sites. Finally, a noteworthy development in SLN technology is the emerging importance of cationic SLNs which can bind and deliver nucleic acids. Olbrich et al. (2001) reported that cationic SLNs could efficiently bind plasmid DNA through electrostatic attraction. These SLNs delivered the DNA strands into cells. This made them potential tools for transfection, where they could deliver genetic material to cells for the treatment of conditions varying from cancer to microbial infections (Rostami et al. 2014).

SLNs and NLCs can be administered via different routes for drug delivery purposes. For oral delivery, SLNs enhance intestinal lymphatic uptake of drugs. For instance, the antiretroviral drug lopinavir exhibited improved oral bioavailability when delivered with SLNs as compared to bulk lopinavir delivery (Negi et al. 2013). SLNs can also protect drugs from the acidic pH of the stomach, such as the antitubercular drug rifampicin, which exhibited reduced gastric decomposition when delivered in SLNs (Singh et al. 2013).

Parenteral modes of delivery are also possible, including intravenous, intramuscular and subcutaneous routes (as reviewed by Dolatabadi et al. 2015). In fact, SLN encapsulation not only improves circulation of drugs, but it also enhances cellular uptake. SLNs loaded with the anticancer drug doxorubicin, which exhibited improved uptake by human lung tumor cell lines (Mussi et al. 2013).

SLNs have also been used for topical drug administration, where they enable direct delivery of loaded drugs to the target site, generating higher local drug concentrations (as reviewed by Dolatabadi et al. 2015). Vaghasiya et al. (2013) described a SLN-based gel carrying the antifungal drug terbinafine hydrochloride. This gel, applied to the skin of rats, reduced the burden of *Candida albicans* infections more rapidly and efficiently than commercial products.

The pulmonary administration of drugs has advantages for both local and systemic delivery of drugs. It allows rapid absorption of drugs into the bloodstream due to the high alveolar surface area, the thin alveolar epithelium, the extensive vasculature and the prolonged release of drugs from SLNs. Owing to their small size, SLNs can deposit deep in the lungs, at the level of alveoli. For instance, SLNs incorporating rifampicin, isoniazid and pyrazinamide—all antitubercular drugs—were delivered by nebulization to guinea pigs (Pandey and Khuller 2005). Following a single nebulization, prolonged absorption of the SLN-based drugs at the level of the lungs maintained their levels in the plasma for a much longer time compared to free drugs. Furthermore, less frequent administrations of the SLN-based drugs over a shorter period were sufficient to eliminate tubercle bacilli in the lungs, while orally administered drugs required a longer delivery period and higher doses. Thus, pulmonary delivery of SLN-based drugs could be particularly useful as a highly efficient means for both local and systemic drug delivery. SLNs exhibit a high degree of diversity and flexibility which makes them versatile drug delivery vehicles in terms of targeting, modes of delivery and the drugs they carry.

Non-lamellar Liquid Crystalline Lipid Phases

Lipids tend to associate with one another to minimize contact between their hydrophobic portions and the surrounding aqueous phase. This aggregation is also governed by repulsion between the polar or charged head-groups of many lipids, and by the packing constraints of hydrophobic tails such as acyl chains (as reviewed by Chang et al. 2015). Usually, lipids spontaneously aggregate into lipid bilayers —also called lamellar phases—and micelles. However, many lipids assemble into other structures termed non-lamellar phases. These include arrangements that have been termed hexagonal, cubic or sponge phases, based on the arrangement of molecules within them. The hexagonal arrangement is best described as tubular, while cubic and other phases tend to exhibit more complicated architectures (as reviewed by Chang et al. 2015). Cardiolipin (Powell and Marsh 1985), phosphatidyl inositol (Furse et al. 2012), and phosphatidyl ethanolamine (Cullis and De Kruijff 1978), among various other lipids, form non-lamellar phases under specific conditions. In general, the lipids remain in a liquid crystalline state, combining the fluidity of the liquid state with the organized molecular architecture of the solid state. In other words, the lipid molecules move freely within their defined geometrical arrangements.

Another important distinction is between lyotropic and thermotropic non-lamellar phases. Lyotropic phases assemble upon addition of an aqueous solvent, while the thermotropic ones assemble in response to temperature changes (as reviewed by Azmi et al. 2015). Glycerol mononucleate is commonly used as a component of non-lamellar lipid phases due to its ability to form a wide array of lytropic structures.

Although the study of non-lamellar liquid crystalline lipid phases as drug delivery vehicles is very recent, these lipid arrangements carry various advantages. They incorporate aqueous channels that can form networks within the nanoparticles, creating a large surface area of lipid-water interface to allow high levels of drug loading (Zeng et al. 2012a).

Liquid crystalline phases also have very diverse applications in drug delivery. They can be utilized for topical application to the skin, oral delivery and pulmonary delivery, among others (as reviewed by Chang et al. 2015). However, the high viscosity of these phases has limited their use in intravenous delivery; other vehicles have so far proven more effective (as reviewed by Yaghmur et al. 2013).

Lamellar and micellar lipid particles can be disrupted by changes in the water potential of the surrounding medium. Meanwhile, non-lamellar, non-micellar phases exhibit greater resistance to increased water potential (as reviewed by Azmi et al. 2015). However, their bioavailability is limited due to clearance by the MPS. Stealth nanoparticles—as in the case of liposomes and micelles—can be synthesized by conjugating polar moieties such as PEG to the surfaces of the nanoparticles to evade MPS clearance Zeng et al. (2012b) described the use of PEG-grafted distearoyl-phosphatidylethanolamine (DSPE-PEG) non-lamellar liquid crystalline nanoparticles. These nanoparticles exhibited sustained release of a sample hydrophobic drug (the anticancer drug paclitaxel) with improved bioavailability compared to the commercial form, Taxol. Another study showed that PEGylation of paclitaxel-bearing non-lamellar liquid crystalline nanoparticles composed of glyceryl monooleate (GMO) helped reduce cytotoxicity of the vehicle and maintain prolonged release of paclitaxel. The PEGylated nanoparticles were more effective at inhibiting tumor growth than free drug and GMO nanoparticles without PEGylation (Jain et al. 2012).

Although the field is still a recent one, it could have diverse applications including antimicrobial drug delivery, food preservation and cosmetics. For instance, Archana et al. (2014) described a hydrogel which was loaded with cubosome nanoparticles carrying the antibacterial agent curcumin. Topical application to the skin of rats showed that incorporation of curcumin into cubosomes improved its antibacterial activity. Another study by Souza et al. (2014) investigated a non-lamellar lipid nanoparticle system capable of adhering to the buccal mucosa. When loaded with poly (hexamethylene biguanide) hydrochloride (PHMB), slow release of the drug was observed, effectively reducing microbial growth. Another example is that of cubosomes carrying silver sulfadiazine (SSD), a drug used for burn treatment through topical application as a gel. Results indicated that the gel containing cubosomes was more effective at healing second-degree burns in rats when compared to the commercial product (Morsi et al. 2014).

Non-lamellar liquid crystalline nanoparticles such as cubosomes and hexosomes can also be tagged with surface ligands or antibody fragments to allow active targeting of cells and tissues. For example, PEGylated liquid crystalline nanoparticles were capable of targeting Epidermal Growth Factor Receptor (EGFR) using a conjugated anti-EGFR Fab antibody fragment (Zhai et al. 2015).

The field of studying non-lamellar liquid crystalline lipid phases as nanoparticles remains a nascent one. The information gathered from other types of nanoparticles can be implemented in advancing this field, including information on tagging with targeting ligands, preventing clearance from the body, and ensuring adequate drug incorporation and release kinetics. While it remains unclear how successful this novel type of nanoparticles will be in the drug market, research efforts are expected to expand greatly in the years to come. A summary of the most salient advantages, limitations and function-enhancing developments for each of the discussed organic nanoparticle types can be found in Table 11.1.

Applications of Organic Nanoparticles in Antimicrobial Drug Delivery

Although most nanoparticle applications in drug delivery have been mostly focused on oncology, the domain of treating microbial diseases comes in second. Traditional means of antimicrobial drug delivery face various problems, including the difficulty of targeting intracellular pathogens, microbial drug resistance and the inability to penetrate bacterial biofilms. Furthermore, the high doses and frequent administrations required may cause toxicity and increase drug resistance (as reviewed by Zazo et al. 2016).

As discussed earlier, various modes of cellular targeting can be employed when using nanoparticles as drug delivery vehicles. In addition to the ability to target particular tissues, drugs can be transported into the cytoplasmic compartments of cells via uptake by phagocytic cells and receptor-mediated endocytosis (as reviewed by Hillaireau and Couvreur 2009). This is particularly important to target intracellular pathogens.

Passive targeting is based on the physical properties of the nanoparticle, particularly the nanoparticle size and zeta potential, a property related to the charge and electrostatic interactions of the nanoparticles with their surroundings. Small size is required for movement through capillaries, for diffusion through mucosal pores and to prevent elimination by the MPS (Gunaseelan et al. 2010, Abdulkarim et al. 2015). Zeta potential also affects where the nanoparticles will be retained and which

Table 11.1. Summary of main advantages, limitations, and enhancements associated with each type of nanoparticle discussed in the chapter.

Nanoparticle Type	Benefits	Limitations	Enhancements
Liposomes	Can carry a wide range of drugs (hydrophobic and hydrophilic). Many drugs are currently FDA-approved. Most extensively studied, cornerstone for studies on other nanoparticle types.	MPS clearance was a major limitation until the advent of PEGylation. Stability remains an issue (fusion of vesicles, drug leakage).	PEGylation increased stability and allowed evasion of MPS clearance. Cationic lipids help block fusion and enhance interactions with cells. Immunoliposomes have antibodies to enable active targeting.
Micelles	Ease of preparation. Smaller size (10–100 nm) improves ability to harness EPR effect.	Dissociation due to high CMC.	Polymeric micelles have enhanced stability. PEGylation serves similar function as in liposomes. Dual function micelles: structural components have added therapeutic function.
Nano/Micro Emulsions	Nanoemulsions are resistant to pH and temperature changes, as well as dilution. Diverse applications outside drug delivery, especially in the food and cosmetics industries.	Have witnessed mixed success in most experimental trials. More optimization is needed.	Fatty acid nanoemulsions with high efficacy at penetrating bacterial biofilms.
Protein Nanoparticles	High biocompatibility, prolonged half-life, and enhanced cellular uptake. Albumin: can carry hydrophobic drugs in its core or water soluble drugs on its surface. Several drugs are FDA-approved. HDL and LDL: low drug leakage, intrinsic targeting for most tumor cells, can bypass endolysosomal degradation. Gelatin: Targets phagocytic cells.	HDL/LDL drugs are still expensive to produce and face difficulties in reconstitution.	Branched amphiphilic peptide capsules are a more stable, robust variation of liposomes.
SLNs	Prolonged release time. Ease of preparation, easily expanded to large-scale. Preparation does not require the use of toxic organic solvents. Wide array of delivery routes: parenteral, pulmonary, topical.	Low drug-loading capacity. Drug release during storage or burst releases.	NLCs: disordered structure increases drug loading capacity. PEGylation to avoid MPS clearance. Cationic SLNs can bind and deliver nucleic acids
Non-Lamellar Liquid Crystalline Lipid Nanoparticles	Contain aqueous channels that increase area of lipid-water interface, enhancing drug loading. Resistance to increased water potential. Could have diverse applications: drug delivery, food preservation and cosmetics.	High viscosity has limited intravenous use. Nascent field, much more research required.	

cells they will target. Normally, cells have negatively charged membranes, so cytoplasmic uptake is favored for positively charged or neutral nanoparticles. Meanwhile, HIV-infected cells usually have a more positive charge, so they can be targeted using negatively charged NPs (Gunaseelan et al. 2010).

On the other hand, active targeting involves attaching ligands either directly to the surface of the nanoparticle or using a spacer molecule such as PEG to link the two. To target specific cell types, nanoparticles bearing ligands specifically found on those cells can be used. For instance, mannosylated nanoparticles have been used to deliver the antibiotic isoniazid to alveolar macrophages (Saraogi et al. 2011) and the anti-HIV drug zidovudine to leukocytes in lymph nodes (Gajbhiye et al. 2013). Furthermore, infected cells can be targeted using ligands that specifically bind to a component of the microbial or viral surface, which are typically expressed on the membranes of infected cells (as reviewed by Zazo et al. 2016). Passive and active targeting can thus promote the delivery of the drugs to specific populations of cells, reducing the potentially toxic effects of the drugs on other cells and tissues.

Targeting can also improve the bioavailability of antimicrobial drugs in specific tissues that they would normally not be able to reach, most notably in the brain due to the obstacle of the blood brain barrier. This is usually done by tagging nanoparticles with apolipoproteins (Kreuter et al. 2002), or with ligands that bind to insulin receptors (Kuo and Ko 2013) or transferrin receptors (Mishra et al. 2006). Receptors for these ligands at the endothelial cells of the brain enable the nanoparticles to traverse the blood brain barrier by transcytosis (as reviewed by Kreuter 2014). Thus, these targeting mechanisms not only impart specificity, but also allow drugs to penetrate tissues that were previously more difficult to access.

Nanoparticles can also overcome some microbial resistance mechanisms. Some nanoparticles protect drugs from beta-lactamases, which typically hydrolyze beta-lactam antibiotics (Brown et al. 2012, Zhao and Jiang 2013). Polymeric micelles delivering antiretroviral drugs also help overcome drug efflux mechanisms in infected cells (Shaik et al. 2008). Furthermore, multiple drugs can be packaged together in the same nanoparticle, increasing potency against drug-resistant strains (as reviewed by Pelgrift and Friedman 2013). For instance, Ramteke et al. (2009) described nanoparticles used to target *H. pylori* infections by co-delivering omeprazole, clarithromycin and amoxicillin.

A biofilm is a layer of bacterial cells growing on a surface, where the cells are clustered tightly together in a polymeric matrix that is difficult to penetrate (as reviewed by Donlan 2002). This makes bacterial biofilms resistant to most antibiotics. Nanoparticles are a potential solution for this problem. Baelo et al. (2015) assessed DNAse-coated nanoparticles that could degrade extracellular DNA in the biofilm matrix of *P. aeruginosa* biofilms. These nanoparticles could disassemble the biofilms and deliver their antibiotic contents to the bacterial cells. Other nanoparticles have also been demonstrated to penetrate biofilms, including cationic liposomes against *S. aureus* (Kim and Jones 2004) and solid lipid nanoparticles eliminating *C. albicans* (Sun et al. 2012).

In addition, nanoparticles reduce the need for high doses and frequent administrations of various drugs. This was demonstrated for antifilarials used to treat Lymphatic Filariasis (LF) (Ali et al. 2013) and for liposome-encapsulated chloroquine used to treat malaria in mice (Owais et al. 1995). The antitubercular drugs rifampicin, isoniazid and pyrazinamide require less frequent doses in mice when delivered orally through liposomes (Pandey et al. 2005). Moreover, alternative administration routes other than parenteral ones can be made possible when using nanoparticle drug delivery systems to target microbial infections, including ocular (Chhonker et al. 2015), oral (Pandey et al. 2005, Dodiya et al. 2011, Patel and Patravale 2011), topical (Vaghasiya et al. 2013) and pulmonary (Pandey and Khuller 2005, Weber et al. 2014) routes, among others. Lower and less frequent doses coupled with more patient-friendly administration routes can greatly improve patient compliance for antimicrobial drug regimens. This is particularly important in bacterial infections to prevent the emergence of resistant microbial strains.

Overall, nanoparticles present potential solutions for many of the problems faced by antimicrobial drug delivery mechanisms. In a time where the incidence of microbial resistance seems to be outpacing the rate of development of new drugs, nanoparticles could be a critical asset in the war against infectious diseases.

Conclusion

Nanoparticles have become a promising mean to overcome the limitations of more traditional modes of drug delivery, in particular for drugs with low solubility in water. In addition to their role in enhancing drug solubility, nanoparticles also present various other benefits. They make targeting tissues and the controlled release of drugs possible. They can lower the effective dosages needed in different therapeutic regimens, as well as improve the biocompatibility and pharmacokinetics of drugs. Nanoparticles can even allow drugs to reach sites that would not be accessible otherwise, through means such as transcytosis.

Nonetheless, other limitations arise with different nanoparticle types, often related to the difficulty of efficient production, limited drug-loading capacity, low stability upon delivery, nanoparticle clearance from the body and various other factors. For nanoparticle types that have accumulated a large body of research, newer generations of the nanoparticles have been developed, each of which addresses the limitations of its predecessors. The most notable example of this is the PEGylation of liposomes, micelles and other nanoparticle types. At the same time, novel types of nanoparticles generally display great promise, but they require more research before they can begin the transition from bench to bedside, as is the case with non-lamellar liquid crystalline lipid nanoparticles. Overall, most nanoparticle types used to deliver hydrophobic drugs are still in their early phases of study, and only a few have been implemented in the production of marketable drugs.

Many hurdles must be overcome before the potential of nanoparticles is adequately harnessed by medicine. Technologically, new forms of nanoparticles must be developed, and synthesis procedures must be optimized to enable control over different factors, such as size, homogeneity, drug loading capacity, purity, lack of structural defects and others. Furthermore, nanoparticle-based drug delivery systems can only transition to clinics if large scale production becomes financially feasible.

However, with the rapid pace at which research in this domain is expanding, it seems certain that nanoparticles will have a sizeable impact on medicine. Already, many nanoparticle-based drugs have been FDA approved, such as liposomal daunorubicin used to treat leukemia and solid tumors (as reviewed by Chang and Yeh 2012), micellar Genexol-PM for the treatment of breast cancer (as reviewed by Oerlemans et al. 2010), and liposomal amphotericin B (AMB) for the treatment of fungal infections such as *Candida*, *Aspergillus* and *Fusarium* species, among many others (as reviewed by Zhang et al. 2010). Furthermore, many other hydrophobic drug formulations using organic nanoparticle delivery vehicles are undergoing clinical studies. The growth of the field of nanotechnology in drug delivery depends strongly on interdisciplinary collaboration. Optimizing the synthesis of nanoparticles, developing novel generations of nanoparticles with improved efficacy, testing them *in vitro* and *in vivo*, and eventually translating them into clinical studies, are all necessary steps to benefit from the advantages of nanoparticle drug delivery systems.

Acknowledgements

The authors acknowledge Dr. Isabelle Fakhoury and Dr. Farah Yassine for their critical reading of the manuscript. The authors also acknowledge the Lebanese National Council for Scientific Research (CNRS-L) for scholarship awarded to OMN.

References

Abdulkarim, M., Aguló, N., Cattoz, B., Griffiths, P., Bernkop-Schnürch, A., Borros, S.G. and Gumbleton, M. 2015. Nanoparticle diffusion within intestinal mucus: Three-dimensional response analysis dissecting the impact of particle surface charge, size and heterogeneity across polyelectrolyte, pegylated and viral particles. Eur. J. of Pharm. and Bio. 97, Part A: 230–238.

Acton, S., Rigotti, A., Landschulz, K.T., Xu, S., Hobbs, H.H. and Krieger, M. 1996. Identification of scavenger receptor SR-BI as a high density lipoprotein receptor. Science 271(5248): 518–520.

Ali, M., Afzal, M., Bhattacharya, S.M., Ahmad, F.J. and Dinda, A.K. 2013. Nanopharmaceuticals to target antifilarials: a comprehensive review. Expert Opin. Drug Deliv. 10(5): 665–678.

Archana, A., Vijayasri, K., Madhurim, M. and Kumar, C. 2014. Curcumin loaded nano cubosomal hydrogel: preparation, *in vitro* characterization and antibacterial activity. Chem. Sci. Trans. 4(1): 75–80.

Aryal, S., Hu, C.M.J. and Zhang, L. 2009. Polymer – cisplatin conjugate nanoparticles for acid-responsive drug delivery. Acs. Nano. 4(1): 251–258.

Avgoustakis, K., Beletsi, A., Panagi, Z., Klepetsanis, P., Livaniou, E., Evangelatos, G. and Ithakissios, D.S. 2003. Effect of copolymer composition on the physicochemical characteristics, *in vitro* stability, and biodistribution of PLGA-mPEG nanoparticles. Int. J. Pharm. 259(1-2): 115–127.

Azmi, I.D., Moghimi, S.M. and Yaghmur, A. 2015. Cubosomes and hexosomes as versatile platforms for drug delivery. Drug Deliv. 6(12): 1347–1364.

Baelo, A., Levato, R., Julián, E., Crespo, A., Astola, J., Gavaldà, J., Engel, E., Mateos-Timoneda, M.A. and Torrents, E. 2015. Disassembling bacterial extracellular matrix with DNase-coated nanoparticles to enhance antibiotic delivery in biofilm infections. J. Control. Release 209: 150–158.

Barros, S.M., Whitaker, S.K., Sukthankar, P., Avila, L.A., Gudlur, S., Warner, M., Beltrão, E.I. and Tomich, J.M. 2016. A review of solute encapsulating nanoparticles used as delivery systems with emphasis on branched amphipathic peptide capsules. Arch. of Biochem. and Biophys. 596: 22–42.

Bégu, S., Pouëssel, A.A., Lerner, D.A., Tourné-Péteilh, C. and Devoisselle, J.M. 2007. Liposil, a promising composite material for drug storage and release. J. Control. Release 118(1): 1–6.

Blume, G., Cevc, G., Crommelin, M.D.J.A., Bakker-Woudenberg, I.A.J.M., Kluft, C. and Storm, G. 1993. Specific targeting with poly(ethylene glycol)-modified liposomes: coupling of homing devices to the ends of the polymeric chains combines effective target binding with long circulation times. Biochimica et Biophysica Acta (BBA) - Biomembranes 1149(1): 180–184.

Bonventre, P.F. and Gregoriadis, G. 1978. Killing of intraphagocytic *Staphylococcus aureus* by dihydrostreptomycin entrapped within liposomes. Antimicrob. Agents Chemother. 13(6): 1049.

Brown, A.N., Smith, K., Samuels, T.A., Lu, J., Obare, S.O. and Scott, M.E. 2012. Nanoparticles functionalized with ampicillin destroy multiple-antibiotic-resistant isolates of *Pseudomonas aeruginosa* and *Enterobacter aerogenes* and methicillin-resistant *Staphylococcus aureus*. Appl. Environ. Microbiol. 78(8): 2768–2774.

Bunjes, H., Westesen, K. and Koch, M.H.J. 1996. Crystallization tendency and polymorphic transitions in triglyceride nanoparticles. Int. J. Pharm. 129(1-2): 159–173.

Cabanes, A., Briggs, K.E., Gokhale, P.C., Treat, J. and Rahman, A. 1998. Comparative *in vivo* studies with paclitaxel and liposome-encapsulated paclitaxel. Int. J. Oncol. 12(5): 1035–1075.

Campbell, R.B., Balasubramanian, S.V. and Straubinger, R.M. 2001. Influence of cationic lipids on the stability and membrane properties of paclitaxel-containing liposomes. J. Pharm. Sci. 90(8): 1091–1105.

Cavalli, R., Caputo, O., Carlotti, M.E., Trotta, M., Scarnecchia, C. and Gasco, M.R. 1997. Sterilization and freeze-drying of drug-free and drug-loaded solid lipid nanoparticles. Int. J. Pharm. 148(1): 47–54.

Chang, D.P., Barauskas, J., Dabkowska, A.P., Wadsäter, M., Tiberg, F. and Nylander, T. 2015. Non-lamellar lipid liquid crystalline structures at interfaces. Adv. Colloid Interface Sci. 222: 135–147.

Chang, H.I. and Yeh, M.K. 2012. Clinical development of liposome-based drugs: formulation, characterization, and therapeutic efficacy. Int. J. Nanomedicine 7: 49–60.

Chen, D.-B., Yang, T.-z., Lu, W.-L. and Zhang, Q. 2001. *In vitro* and *in vivo* study of two types of long-circulating solid lipid nanoparticles containing paclitaxel. Chem. Pharm. Bull. 49(11): 1444–1447.

Chhonker, Y.S., Prasad, Y.D., Chandasana, H., Vishvkarma, A., Mitra, K., Shukla, P. K. and Bhatta, R.S. 2015. Amphotericin-B entrapped lecithin/chitosan nanoparticles for prolonged ocular application. Int. J. Biol. Macromolec. 72: 1451–1458.

Constantinides, P.P., Lambert, K.J., Tustian, A.K., Schneider, B., Lalji, S., Ma, W., Wentzel, B., Kessler, D., Worah, D. and Quay, S.C. 2000. Formulation development and antitumor activity of a filter-sterilizable emulsion of paclitaxel. Pharm. Res. 17(2): 175–182.

Constantinides, P.P., Tustian, A. and Kessler, D.R. 2004. Tocol emulsions for drug solubilization and parenteral delivery. Adv. Drug Deliv. Rev. 56(9): 1243–1255.

Cruz, P.M., Mo, H., McConathy, W.J., Sabnis, N. and Lacko, A. 2013. The role of cholesterol metabolism and cholesterol transport in carcinogenesis: a review of scientific findings, relevant to future cancer therapeutics. Front Pharmacol. 4: 119.

Cullis, P.R. and De Kruijff, B. 1978. The polymorphic phase behaviour of phosphatidylethanolamines of natural and synthetic origin. A 31P NMR study. Biochim. Biophys. Acta (BBA) - Biomembranes 513(1): 31–42.

Curry, S., Mandelkow, H., Brick, P. and Franks, N. 1998. Crystal structure of human serum albumin complexed with fatty acid reveals an asymmetric distribution of binding sites. Nat. Struct. Mol. Biol. 5(9): 827–835.
Daeihamed, M., Dadashzadeh, S., Haeri, A. and Akhlaghi, M. 2016. Potential of liposomes for enhancement of oral drug absorption. Curr. Drug Deliv. 14(2): 289–303.
Dalen, H. and Neuzil, J. 2003. Alpha-tocopheryl succinate sensitises a T lymphoma cell line to TRAIL-induced apoptosis by suppressing NF-kappa B activation. Br J. Cancer 88(1): 153–158.
Delmas, T., Piraux, H., Couffin, A.-C., Texier, I., Vinet, F., Poulin, P., Cates, M.E. and Bibette, J. 2011. How to prepare and stabilize very small nanoemulsions. Langmuir 27(5): 1683–1692.
Dodiya, S.S., Chavhan, S.S., Sawant, K.K. and Korde, A.G. 2011. Solid lipid nanoparticles and nanosuspension formulation of Saquinavir: preparation, characterization, pharmacokinetics and biodistribution studies. J. Microencapsul. 28(6): 515–527.
Dolatabadi, J.E.N., Valizadeh, H. and Hamishehkar, H. 2015. Solid lipid nanoparticles as efficient drug and gene delivery systems: Recent breakthroughs. Adv. Pharm. Bull. 5(2): 151.
Dong, L.-F., Swettenham, E., Eliasson, J., Wang, X.-F., Gold, M., Medunic, Y., Stantic, M., Low, P., Prochazka, L. and Witting, P.K. 2007. Vitamin E analogues inhibit angiogenesis by selective induction of apoptosis in proliferating endothelial cells: the role of oxidative stress. Cancer Res. 67(24): 11906–11913.
Donlan, R.M. 2002. Biofilms: microbial life on surfaces. Emerg. Infect. Diseases 8(9): 881–890.
Dos Santos, N., Allen, C., Doppen, A.-M., Anantha, M., Cox, K.A., Gallagher, R.C., Karlsson, G., Edwards, K., Kenner, G. and Samuels, L. 2007. Influence of poly(ethylene glycol) grafting density and polymer length on liposomes: relating plasma circulation lifetimes to protein binding. Biochim. Biophys. Acta (BBA) - Biomembranes 1768(6): 1367–1377.
Duhem, N., Danhier, F. and Préat, V. 2014. Vitamin E-based nanomedicines for anti-cancer drug delivery. J. Control. Release 182: 33–44.
Elsadek, B. and Kratz, F. 2012. Impact of albumin on drug delivery—new applications on the horizon. J. Control. Release 157(1): 4–28.
Elzoghby, A.O., Samy, W.M. and Elgindy, N.A. 2012. Albumin-based nanoparticles as potential controlled release drug delivery systems. J. Control. Release 157(2): 168–182.
Emeje, M.O., Akpabio, E.I., Obidike, I.C. and Ofoefule, S.I. 2012. Nanotechnology in drug delivery, INTECH Open Access Publisher.
Feng, L. and Mumper, R.J. 2013. A critical review of lipid-based nanoparticles for taxane delivery. Cancer Lett. 334(2): 157–175.
Floury, J., Bellettre, J., Legrand, J. and Desrumaux, A. 2004. Analysis of a new type of high pressure homogeniser. A study of the flow pattern. Chem. Eng. Sci. 59(4): 843–853.
Forgiarini, A., Esquena, J., Gonzalez, C. and Solans, C. 2001. Formation of nano-emulsions by low-energy emulsification methods at constant temperature. Langmuir 17(7): 2076–2083.
Furse, S., Brooks, N.J., Seddon, A.M., Woscholski, R., Templer, R.H., Tate, E.W., Gaffney, P.R. and Ces, O. 2012. Lipid membrane curvature induced by distearoyl phosphatidylinositol 4-phosphate. Soft Matter 8(11): 3090–3093.
Gajbhiye, V., Ganesh, N., Barve, J. and Jain, N.K. 2013. Synthesis, characterization and targeting potential of zidovudine loaded sialic acid conjugated-mannosylated poly(propyleneimine) dendrimers. Eur. J. Pharm. Sci. 48(4-5): 668–679.
Gao, W., Chan, J.M. and Farokhzad, O.C. 2010. pH-responsive nanoparticles for drug delivery. Mol. Pharm. 7(6): 1913–1920.
Gunaseelan, S., Gunaseelan, K., Deshmukh, M., Zhang, X. and Sinko, P.J. 2010. Surface modifications of nanocarriers for effective intracellular delivery of anti-HIV drugs. Adv. Drug Deliv. Rev. 62(4-5): 518–531.
Gupta, A., Eral, H.B., Hatton, T.A. and Doyle, P.S. 2016. Nanoemulsions: formation, properties and applications. Soft Matter 12(11): 2826–2841.
Hansen, C.B., Kao, G.Y., Moase, E.H., Zalipsky, S. and Allen, T.M. 1995. Attachment of antibodies to sterically stabilized liposomes: evaluation, comparison and optimization of coupling procedures. Biochim. Biophys. Acta (BBA) - Biomembranes 1239(2): 133–144.
Harashima, H., Sakata, K., Funato, K. and Kiwada, H. 1994. Enhanced hepatic uptake of liposomes through complement activation depending on the size of liposomes. Pharm Res. 11(3): 402–406.
Hillaireau, H. and Couvreur, P. 2009. Nanocarriers' entry into the cell: relevance to drug delivery. Cell. Mol. Life Sci. 66(17): 2873–2896.
Immordino, M.L., Dosio, F. and Cattel, L. 2006. Stealth liposomes: review of the basic science, rationale, and clinical applications, existing and potential. Int. J. Nanomedicine 1(3): 297.
Izquierdo, P., Esquena, J., Tadros, T.F., Dederen, C., Garcia, M., Azemar, N. and Solans, C. 2002. Formation and stability of nano-emulsions prepared using the phase inversion temperature method. Langmuir 18(1): 26–30.
Jain, V., Swarnakar, N.K., Mishra, P.R., Verma, A., Kaul, A., Mishra, A.K. and Jain, N.K. 2012. Paclitaxel loaded PEGylated glyceryl monooleate based nanoparticulate carriers in chemotherapy. Biomaterials 33(29): 7206–7220.
Jhaveri, A.M. and Torchilin, V.P. 2014. Multifunctional polymeric micelles for delivery of drugs and siRNA. Front. Pharmacol. 5: 77.
Johnsen, K.B. and Moos, T. 2016. Revisiting nanoparticle technology for blood–brain barrier transport: Unfolding at the endothelial gate improves the fate of transferrin receptor-targeted liposomes. J. Control. Release 222: 32–46.
Kader, A. and Pater, A. 2002. Loading anticancer drugs into HDL as well as LDL has little affect on properties of complexes and enhances cytotoxicity to human carcinoma cells. J. Control. Release 80(1-3): 29–44.

Kalepu, S. and Nekkanti, V. 2015. Insoluble drug delivery strategies: review of recent advances and business prospects. Acta Pharm. Sin. B. 5(5): 442–453.

Kelly, C., Jefferies, C. and Cryan, S.-A. 2010. Targeted liposomal drug delivery to monocytes and macrophages. J. Drug Deliv. 2011.

Kim, H.J. and Jones, M.N. 2004. The delivery of benzyl penicillin to *Staphylococcus aureus* biofilms by use of liposomes. J. Liposome Res. 14(3-4): 123–139.

Klibanov, A.L., Maruyama, K., Beckerleg, A.M., Torchilin, V.P. and Huang, L. 1991. Activity of amphipathic poly(ethylene glycol) 5000 to prolong the circulation time of liposomes depends on the liposome size and is unfavorable for immunoliposome binding to target. Biochim. Biophys. Acta (BBA) - Biomembranes 1062(2): 142–148.

Kreuter, J. 2014. Drug delivery to the central nervous system by polymeric nanoparticles: What do we know? Adv. Drug Deliv. Rev. 71: 2–14.

Kreuter, J., Shamenkov, D., Petrov, V., Ramge, P., Cychutek, K., Koch-Brandt, C. and Alyautdin, R. 2002. Apolipoprotein-mediated transport of nanoparticle-bound drugs across the blood-brain barrier. J. Drug Target. 10(4): 317–325.

Krishnadas, A., Rubinstein, I. and Önyüksel, H. 2003. Sterically stabilized phospholipid mixed micelles: *in vitro* evaluation as a novel carrier for water-insoluble drugs. Pharm. Res. 20(2): 297–302.

Kuo, Y.-C. and Ko, H.-F. 2013. Targeting delivery of saquinavir to the brain using 83-14 monoclonal antibody-grafted solid lipid nanoparticles. Biomaterials 34(20): 4818–4830.

La, S.B., Okano, T. and Kataoka, K. 1996. Preparation and characterization of the micelle-forming polymeric drug indomethacin-incorporated poly(ethylene oxide)–poly(β-benzyl L-aspartate) block copolymer micelles. J. Pharm. Sci. 85(1): 85–90.

Lacko, A.G., Nair, M., Paranjape, S., Johnson, S. and McConathy, W.J. 2002. High density lipoprotein complexes as delivery vehicles for anticancer drugs. Anticancer Res. 22(4): 2045–2050.

Lacko, A.G., Nair, M., Prokai, L. and McConathy, W.J. 2007. Prospects and challenges of the development of lipoprotein-based formulations for anti-cancer drugs. Expert Opin. Drug Deliv. 4(6): 665–675.

Lacko, A.G., Sabnis, N.A., Nagarajan, B. and McConathy, W.J. 2015. HDL as a drug and nucleic acid delivery vehicle. Front. Pharmacol. 6.

Lee, M.-K., Lim, S.-J. and Kim, C.-K. 2007. Preparation, characterization and *in vitro* cytotoxicity of paclitaxel-loaded sterically stabilized solid lipid nanoparticles. Biomaterials 28(12): 2137–2146.

Li, P., Tan, Z., Zhu, Y., Chen, S., Ding, S. and Zhuang, H. 2006. Targeting study of gelatin adsorbed clodronate in reticuloendothelial system and its potential application in immune thrombocytopenic purpura of rat model. J. Control Release 114(2): 202–208.

Liang, R., Xu, S., Shoemaker, C.F., Li, Y., Zhong, F. and Huang, Q. 2012. Physical and antimicrobial properties of peppermint oil nanoemulsions. J. Agric. Food Chem. 60(30): 7548–7555.

Loftsson, T. and Brewster, M.E. 2010. Pharmaceutical applications of cyclodextrins: basic science and product development. J. Pharm. Pharmacol. 62(11): 1607–1621.

Lukyanov, A.N., Gao, Z., Mazzola, L. and Torchilin, V.P. 2002. Polyethylene glycol-diacyllipid micelles demonstrate increased acculumation in subcutaneous tumors in mice. Pharm Res. 19(10): 1424–1429.

Malafa, M.P., Fokum, F.D., Smith, L. and Louis, A. 2002. Inhibition of angiogenesis and promotion of melanoma dormancy by vitamin E succinate. Ann. Surg. Oncol. 9(10): 1023–1032.

Malafa, M.P. and Neitzel, L.T. 2000. Vitamin E succinate promotes breast cancer tumor dormancy. J. Surg. Res. 93(1): 163–170.

Mamot, C., Drummond, D.C., Greiser, U., Hong, K., Kirpotin, D.B., Marks, J.D. and Park, J.W. 2003. Epidermal growth factor receptor (EGFR)-targeted immunoliposomes mediate specific and efficient drug delivery to EGFR-and EGFRvIII-overexpressing tumor cells. Cancer Res. 63(12): 3154–3161.

Marin, A., Sun, H., Husseini, G.A., Pitt, W.G., Christensen, D.A. and Rapoport, N.Y. 2002. Drug delivery in pluronic micelles: effect of high-frequency ultrasound on drug release from micelles and intracellular uptake. J. Control. Release 84(1): 39–47.

Maruyama, K., Takizawa, T., Yuda, T., Kennel, S.J., Huang, L. and Iwatsuru, M. 1995. Targetability of novel immunoliposomes modified with amphipathic poly(ethylene glycol)s conjugated at their distal terminals to monoclonal antibodies. Biochim. Biophys. Acta (BBA) - Biomembranes 1234(1): 74–80.

Mason, T.G., Wilking, J., Meleson, K., Chang, C. and Graves, S. 2006. Nanoemulsions: formation, structure, and physical properties. J. Phys: Condensed Matter 18(41): R635.

Matsumura, Y. and Maeda, H. 1986. A new concept for macromolecular therapeutics in cancer chemotherapy: mechanism of tumoritropic accumulation of proteins and the antitumor agent smancs. Cancer Res. 46(12 Pt 1): 6387–6392.

Meleson, K., Graves, S. and Mason, T.G. 2004. Formation of concentrated nanoemulsions by extreme shear. Soft Mater. 2(2-3): 109–123.

Mishra, V., Mahor, S., Rawat, A., Gupta, P.N., Dubey, P., Khatri, K. and Vyas, S.P. 2006. Targeted brain delivery of AZT via transferrin anchored pegylated albumin nanoparticles. J. Drug Target. 14(1): 45–53.

Morel, S., Ugazio, E., Cavalli, R. and Gasco, M.R. 1996. Thymopentin in solid lipid nanoparticles. Int. J. Pharm. 132(1): 259–261.

Morsi, N.M., Abdelbary, G.A. and Ahmed, M.A. 2014. Silver sulfadiazine based cubosome hydrogels for topical treatment of burns: Development and *in vitro*/*in vivo* characterization. Eur. J. Pharm. Biopharm. 86(2): 178–189.

Mu, L., Elbayoumi, T.A. and Torchilin, V.P. 2005. Mixed micelles made of poly(ethylene glycol)–phosphatidylethanolamine conjugate and d-α-tocopheryl polyethylene glycol 1000 succinate as pharmaceutical nanocarriers for camptothecin. Int. J. Pharm. 306(1-2): 142–149.
Müller, R., Radtke, M. and Wissing, S. 2002. Nanostructured lipid matrices for improved microencapsulation of drugs. Int. J. Pharm. 242(1): 121–128.
Mussi, S.V., Silva, R.C., Oliveira, M.C.d., Lucci, C.M., Azevedo, R.B.d. and Ferreira, L.A.M. 2013. New approach to improve encapsulation and antitumor activity of doxorubicin loaded in solid lipid nanoparticles. Eur. J. Pharm. Sci. 48(1-2): 282–290.
Naseri, N., Valizadeh, H. and Zakeri-Milani, P. 2015. Solid lipid nanoparticles and nanostructured lipid carriers: structure, preparation and application. Adv. Pharm. Bull. 5(3): 305.
Negi, J.S., Chattopadhyay, P., Sharma, A.K. and Ram, V. 2013. Development of solid lipid nanoparticles (SLNs) of lopinavir using hot self nano-emulsification (SNE) technique. Eur. J. Pharm. Sci. 48(1-2): 231–239.
Oerlemans, C., Bult, W., Bos, M., Storm, G., Nijsen, J.F.W. and Hennink, W.E. 2010. Polymeric micelles in anticancer therapy: targeting, imaging and triggered release. Pharm. Res. 27(12): 2569–2589.
Olbrich, C., Bakowsky, U., Lehr, C.-M., Müller, R.H. and Kneuer, C. 2001. Cationic solid-lipid nanoparticles can efficiently bind and transfect plasmid DNA. J. Control. Release 77(3): 345–355.
Owais, M., Varshney, G.C., Choudhury, A., Chandra, S. and Gupta, C.M. 1995. Chloroquine encapsulated in malaria-infected erythrocyte-specific antibody-bearing liposomes effectively controls chloroquine-resistant Plasmodium berghei infections in mice. Antimicrob. Agents Chemother. 39(1): 180–184.
Pandey, H., Rani, R. and Agarwal, V. 2016. Liposome and their applications in cancer therapy. Braz. Arch. Biol. Technol. 59.
Pandey, R. and Khuller, G. 2005. Solid lipid particle-based inhalable sustained drug delivery system against experimental tuberculosis. Tuberculosis 85(4): 227–234.
Pandey, R., Sharma, S. and Khuller, G.K. 2005. Oral solid lipid nanoparticle-based antitubercular chemotherapy. Tuberculosis 85(5-6): 415–420.
Panga, X., Cui, F., Tian, J., Chen, J., Zhou, J. and Zhou, W. 2009. Preparation and characterization of magnetic solid lipid nanoparticles loaded with ibuprofen. Asian J. Pharm. Sci. 4(2): 132–137.
Patel, P.A. and Patravale, V.B. 2011. AmbiOnp: solid lipid nanoparticles of amphotericin B for oral administration. J. Biomed. Nanotechnol. 7(5): 632–639.
Pelgrift, R.Y. and Friedman, A.J. 2013. Nanotechnology as a therapeutic tool to combat microbial resistance. Adv. Drug Deliv. Rev. 65(13-14): 1803–1815.
Powell, G.L. and Marsh, D. 1985. Polymorphic phase behavior of cardiolipin derivatives studied by phosphorus-31 NMR and X-ray diffraction. Biochemistry 24(12): 2902–2908.
Ramteke, S., Ganesh, N., Bhattacharya, S. and Jain, N.K. 2009. Amoxicillin, clarithromycin, and omeprazole based targeted nanoparticles for the treatment of *H. pylori*. J. Drug Target. 17(3): 225–234.
Resnier, P., Mottais, A., Sibiril, Y., Le Gall, T. and Montier, T. 2016. Challenges and successes using nanomedicines for aerosol delivery to the airways. Curr. Gene Ther. 16(1): 34–46.
Rostami, E., Kashanian, S., Azandaryani, A.H., Faramarzi, H., Dolatabadi, J.E.N. and Omidfar, K. 2014. Drug targeting using solid lipid nanoparticles. Chem. Phys. Lipids 181: 56–61.
Saraogi, G.K., Sharma, B., Joshi, B., Gupta, P., Gupta, U.D., Jain, N.K. and Agrawal, G.P. 2011. Mannosylated gelatin nanoparticles bearing isoniazid for effective management of tuberculosis. J. Drug Target. 19(3): 219–227.
Schmidt, H.T. and Ostafin, A.E. 2002. Liposome directed growth of calcium phosphate nanoshells. Adv. Mater. 14(7): 532–535.
Schwarz, C., Mehnert, W., Lucks, J.S. and Müller, R.H. 1994. Solid lipid nanoparticles (SLN) for controlled drug delivery. I. Production, characterization and sterilization. J. Control. Release 30(1): 83–96.
Segura, S., Gamazo, C., Irache, J.M. and Espuelas, S. 2007. Gamma interferon loaded onto albumin nanoparticles: *in vitro* and *in vivo* activities against Brucella abortus. Antimicrob. Agents Chemother. 51(4): 1310–1314.
Senior, J., Delgado, C., Fisher, D., Tilcock, C. and Gregoriadis, G. 1991. Influence of surface hydrophilicity of liposomes on their interaction with plasma protein and clearance from the circulation: studies with poly(ethylene glycol)-coated vesicles. Biochim. Biophys. Acta (BBA) - Biomembranes 1062(1): 77–82.
Seow, Y. and Wood, M.J. 2009. Biological gene delivery vehicles: beyond viral vectors. Mol. Ther. 17(5): 767–777.
Shaik, N., Pan, G. and Elmquist, W.F. 2008. Interactions of pluronic block copolymers on P-gp efflux activity: experience with HIV-1 protease inhibitors. J. Pharm. Sci. 97(12): 5421–5433.
Sharma, A. and Straubinger, R.M. 1994. Novel taxol formulations: preparation and characterization of taxol-containing liposomes. Pharm. Res. 11(6): 889–896.
Shklar, G., Schwartz, J., Trickler, D.P. and Niukian, K. 1987. Regression by vitamin E of experimental oral cancer. J. Natl. Cancer Inst. 78(5): 987–992.
Silva, H.D., Cerqueira, M.Â. and Vicente, A.A. 2012. Nanoemulsions for food applications: development and characterization. Food Bioprocess Tech. 5(3): 854–867.
Singh, H., Bhandari, R. and Kaur, I.P. 2013. Encapsulation of Rifampicin in a solid lipid nanoparticulate system to limit its degradation and interaction with Isoniazid at acidic pH. Int. J. Pharm. 446(1-2): 106–111.
Sisson, T.M., Lamparski, H.G., Kölchens, S., Elayadi, A. and O'Brien, D.F. 1996. Cross-linking polymerizations in two-dimensional assemblies. Macromolecules 29(26): 8321–8329.

Souza, C., Watanabe, E., Borgheti-Cardoso, L.N., De Abreu Fantini, M.C. and Lara, M.G. 2014. Mucoadhesive system formed by liquid crystals for buccal administration of poly(hexamethylene biguanide) hydrochloride. J. Pharm. Sci. 103(12): 3914–3923.
Straubinger, R.M. and Balasubramanian, S.V. 2005. Preparation and characterization of taxane-containing liposomes. Methods Enzymol. 391: 97–117.
Sukthankar, P., Avila, L.A., Whitaker, S.K., Iwamoto, T., Morgenstern, A., Apostolidis, C., Liu, K., Hanzlik, R.P., Dadachova, E. and Tomich, J.M. 2014. Branched amphiphilic peptide capsules: cellular uptake and retention of encapsulated solutes. Biochim. Biophys. Acta (BBA) - Biomembranes 1838(9): 2296–2305.
Sun, L.-m., Zhang, C.-l. and Li, P. 2012. Characterization, antibiofilm, and mechanism of action of novel PEG-stabilized lipid nanoparticles loaded with terpinen-4-ol. J. Agric. Food Chem. 60(24): 6150–6156.
Taylor, P. 2003. Ostwald ripening in emulsions: estimation of solution thermodynamics of the disperse phase. Adv. Colloid Interface Sci. 106(1): 261–285.
Teixeira, P.C., Leite, G.M., Domingues, R.J., Silva, J., Gibbs, P.A. and Ferreira, J.P. 2007. Antimicrobial effects of a microemulsion and a nanoemulsion on enteric and other pathogens and biofilms. Int. J. Food Microbiol. 118(1): 15–19.
Tezcaner, A., Baran, E.T. and Keskin, D. 2016. Nanoparticles based on plasma proteins for drug delivery applications. Curr. Pharm Des. 22: 3445–54.
Tomlinson, R., Klee, M., Garrett, S., Heller, J., Duncan, R. and Brocchini, S. 2002. Pendent chain functionalized polyacetals that display pH-dependent degradation: A platform for the development of novel polymer therapeutics. Macromolecules 35(2): 473–480.
Vaghasiya, H., Kumar, A. and Sawant, K. 2013. Development of solid lipid nanoparticles based controlled release system for topical delivery of terbinafine hydrochloride. Eur. J. Pharm. Sci. 49(2): 311–322.
Wang, G. and Uludag, H. 2008. Recent developments in nanoparticle-based drug delivery and targeting systems with emphasis on protein-based nanoparticles. Expert Opin. Drug Deliv. 5(5): 499–515.
Wartlick, H., Spänkuch-Schmitt, B., Strebhardt, K., Kreuter, J. and Langer, K. 2004. Tumour cell delivery of antisense oligonuclceotides by human serum albumin nanoparticles. J. Control. Release 96(3): 483–495.
Weber, S., Zimmer, A. and Pardeike, J. 2014. Solid lipid nanoparticles (SLN) and nanostructured lipid carriers (NLC) for pulmonary application: a review of the state of the art. Eur. J. Pharm Biopharm. 86(1): 7–22.
Werner, M.E., Cummings, N.D., Sethi, M., Wang, E.C., Sukumar, R., Moore, D.T. and Wang, A.Z. 2013. Preclinical evaluation of Genexol-PM, a nanoparticle formulation of paclitaxel, as a novel radiosensitizer for the treatment of non-small cell lung cancer. Int. J. Radiat Oncol. Biol. Phys. 86(3): 463–468.
Yaghmur, A., Rappolt, M. and Larsen, S.W. 2013. *In situ* forming drug delivery systems based on lyotropic liquid crystalline phases: structural characterization and release properties. J. Drug Deliv. Sci. Technol. 23(4): 325–332.
Yang, M., Jin, H., Chen, J., Ding, L., Ng, K.K., Lin, Q., Lovell, J.F., Zhang, Z. and Zheng, G. 2011. Efficient cytosolic delivery of siRNA using HDL-mimicking nanoparticles. Small 7(5): 568–573.
Yuan, F., Dellian, M., Fukumura, D., Leunig, M., Berk, D.A., Torchilin, V.P. and Jain, R.K. 1995. Vascular permeability in a human tumor xenograft: molecular size dependence and cutoff size. Cancer Res. 55(17): 3752–3756.
Yue, X. and Dai, Z. 2014. Recent advances in liposomal nanohybrid cerasomes as promising drug nanocarriers. Adv. Colloid Interface Sci. 207: 32–42.
Zazo, H., Colino, C.I. and Lanao, J.M. 2016. Current applications of nanoparticles in infectious diseases. J. Control. Release 224: 86–102.
Zeng, N., Gao, X., Hu, Q., Song, Q., Xia, H., Liu, Z., Gu, G., Jiang, M., Pang, Z., Chen, H., Chen, J. and Fang, L. 2012a. Lipid-based liquid crystalline nanoparticles as oral drug delivery vehicles for poorly water-soluble drugs: cellular interaction and *in vivo* absorption. Int. J. Nanomedicine 7: 3703–3718.
Zeng, N., Hu, Q., Liu, Z., Gao, X., Hu, R., Song, Q., Gu, G., Xia, H., Yao, L. and Pang, Z. 2012b. Preparation and characterization of paclitaxel-loaded DSPE-PEG-liquid crystalline nanoparticles (LCNPs) for improved bioavailability. Int. J. Pharm. 424(1): 58–66.
Zensi, A., Begley, D., Pontikis, C., Legros, C., Mihoreanu, L., Wagner, S., Buchel, C., von Briesen, H. and Kreuter, J. 2009. Albumin nanoparticles targeted with Apo E enter the CNS by transcytosis and are delivered to neurones. J. Control Release 137(1): 78–86.
Zhai, J., Scoble, J.A., Li, N., Lovrecz, G., Waddington, L.J., Tran, N., Muir, B.W., Coia, G., Kirby, N. and Drummond, C.J. 2015. Epidermal growth factor receptor-targeted lipid nanoparticles retain self-assembled nanostructures and provide high specificity. Nanoscale 7(7): 2905–2913.
Zhang, L., Pornpattananangkul, D., Hu, C.-M. and Huang, C.-M. 2010. Development of nanoparticles for antimicrobial drug delivery. Curr. Med. Chem. 17(6): 585–594.
Zhang, X., Huang, Y. and Li, S. 2014. Nanomicellar carriers for targeted delivery of anticancer agents. Ther. Deliv. 5(1): 53–68.
Zhang, Z., Chen, J., Ding, L., Jin, H., Lovell, J.F., Corbin, I.R., Cao, W., Lo, P.C., Yang, M. and Tsao, M.S. 2010. HDL-mimicking peptide–lipid nanoparticles with improved tumor targeting. Small. 6(3): 430–437.
Zhao, Y. and Jiang, X. 2013. Multiple strategies to activate gold nanoparticles as antibiotics. Nanoscale 5(18): 8340–8350.

12

Potential of Oils in Development of Nanostructured Lipid Carriers

Anisha A. D'Souza[1] and *Ranjita Shegokar*[2,*]

Introduction

Lipid based nano drug delivery systems have gained momentum over the decades due to their role in improving bioavailability, controlling and targeting actives, enhancing stability and also as vaccine adjuvants. Their popularity over polymeric nanoparticles has been due to their excellent biocompatibility and ease during formulation (Muchow et al. 2008). The downside of polymeric nanoparticles has been mainly both the lack of biocompatibility especially for the non-biodegradable polymers and the use of Class 2 and 3 toxic organic solvents in their preparation. However, most of the lipids frequently used in lipid delivery systems are physiological lipids. These lipids are being accepted by regulatory authorities due to their safety profile.

Liposomes have been successfully commercialized as Ambisome®, Dauno Xome® and Doxil® among others. These systems were developed to reduce side effects of potent drugs and to improve efficacy. However, low physical stability, high drug leakage and non-specific clearance by the macrophages of human host are some of their limitations. SLNs (solid lipid nanoparticles) combine the advantage of both, the emulsions and the liposomes. Technology transfer of lipid based drug delivery systems is also readily amenable and already available in the pharmaceutical industry. SLNs are prepared from lipids that are solid at room temperature. Solid lipids could be explored for controlled release delivery systems. Drugs or actives with poor water solubility can have improved bioavailability after their incorporation in lipids. SLNs can have significantly enhanced plasma stability compared to that of bilayer liposomes. Active targeting to various tissues like brain or cancer tissue is also possible. Unlike polymeric nanoparticles, SLNs exhibit excellent biocompatibility or biodegradability since they are mostly physiological lipid materials and easy to be formulated without the need of organic solvents. They exhibit good physical stability, can protect drug from degradation and control the release of drugs, depending upon the model of incorporation. SLNs, however, exhibit a low drug loading capacity (generally between 25 and 50%) which is a major

[1] Indian Institute of Technology-Bombay (IIT-B), Department of Bioscience and Biomedical Engineering, Powai, Mumbai, India.

[2] Freie Universität Berlin, Institute of Pharmacy, Department Pharmaceutics, Biopharmaceutics, and NutriCosmetics, 12169 Berlin, Germany.

* Corresponding author: ranjita@arcslive.com

drawback. Drug expulsion from SLN during storage is possible when the lipid crystallizes to their stable β-form. SLNs are composed of 0.1 to 30% w/w of solid lipid dispersed in surfactant solution typically in concentrations between 0.5 and 5% w/w.

Expulsion of drugs with restricted solubility in lipids from lipid matrix, highly limits the exploration of SLNs. These limitations are overcomed by a second generation of SLNs namely Nanostructured Lipid Carrier (NLCs), which has a controlled nanostructure. The main difference being the less ordered arrangement of lipid blend (solid lipid and oil) with amorphous drug solubilized in liquid lipids or oil inter-dispersed among solid lipid crystals compared to the perfect crystal shaped brick model of lipids. Hence, liquid lipids or oils were incorporated into the core of a solid lipid (Fig. 12.1). This facilitates the loading of a drug maintaining the physical stability thus modulating drug release.

The current chapter reviews the present state of the art regarding production of NLCs and role of oils (as a matrix component and as an active). The techniques of active incorporation are also discussed. At the end of chapter various marketed NLC based product examples from pharmaceutical and cosmetic industry have been discussed.

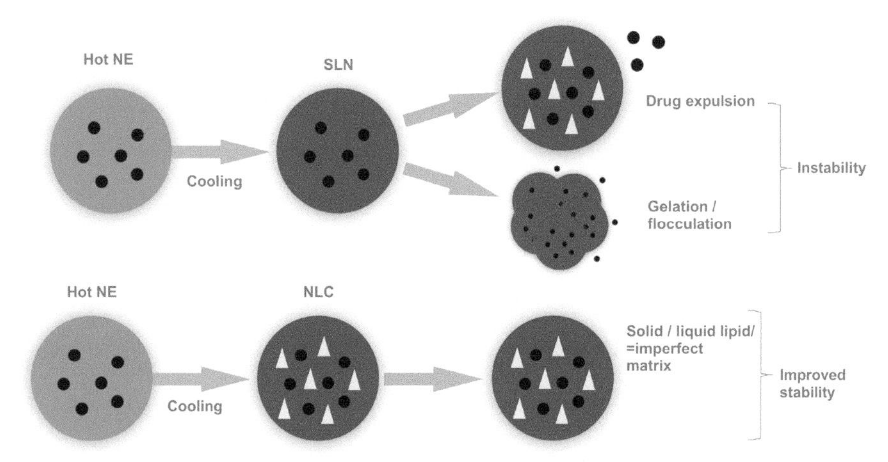

Figure 12.1. Schematic representation of drug loading in SLNs and NLCs.

Nanostructured Lipid Carriers

NLCs are mixtures of solid lipids with spatially incompatible liquid lipids. Besides the benefits of improved bioavailability, another advantage of these nanoparticles is the encapsulation of oils which forms physically modified solid NLCs with altered properties providing an ease of handling. NLCs are the new generation and advanced version of SLNs. These NLCs thus open new opportunities in the therapeutic area wherein the active ingredient is oil. Technology transfer of NLCs is readily amenable. NLCs were introduced with the intent of increasing drug loading thereby minimizing the drug expulsion (Müller and Radtke 2002). The loading of oils in nanocarriers is a challenge especially if they are essential oils, which can undergo rapid oxidation decreasing efficacy or forming of allergenic products. Also, these oils can be easily expulsed out from the solid matrix.

Types of NLC Matrices

Lipids with different structures do not fit together to form a perfect crystal thus producing a highly imperfect matrix. NLC creates a less ordered solid lipid matrix and prevents a proper packaging of the crystal matrix (Müller et al. 2002). The presence of oil also influences the crystallization, and its melting point can be probably attributed to the decrease of hydrophobic attractions (Jenning et al. 2000b). It can be assumed that the oil layers are arranged between the solid lipids and the surfactants

(Jores et al. 2003, Jores et al. 2004, Jores et al. 2005). Drug/oil can be accommodated in these imperfect spaces in their amorphous form. Thus, NLCs can increase the lipid loading. Depending upon the composition of both, the lipids combinations and methods of production, NLCs can be obtained indifferent types namely amorphous, imperfect and of multiple components (Fig. 12.2).

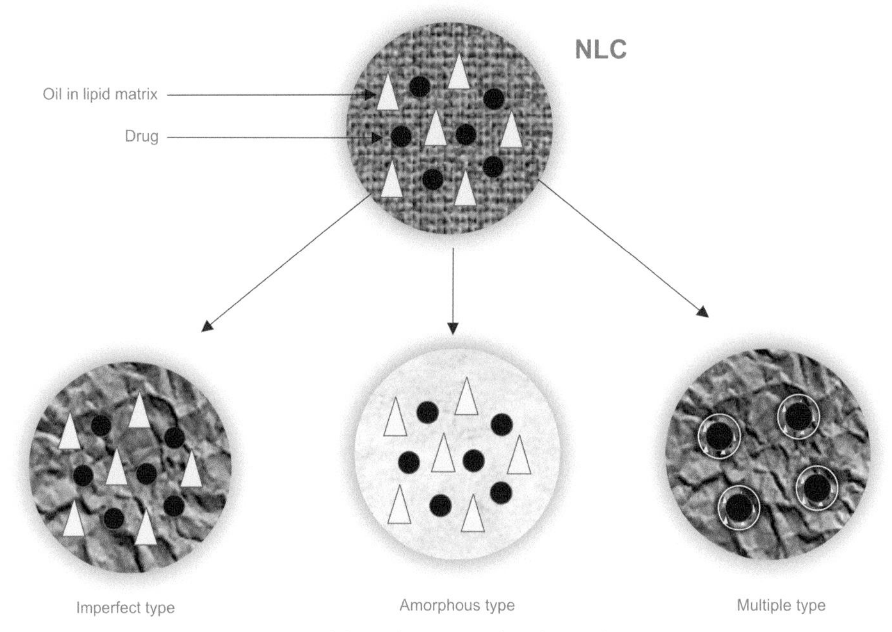

Figure 12.2. Various types of NLCs matrices.

Imperfect NLC (Imperfectly Structured Solid Matrix)

Spatially different glycerides with different fatty acid chains increase the distance between the crystals increasing the imperfections. This imperfection creates spaces in the matrix that are suitable for amorphous clusters of drugs. Thus, high incompatibility between the lipids increases the chances for drug loading.

Amorphous Type

This order deficient solid matrix can be obtained by admixture of solid lipids like medium chain triglycerides (i.e., Miglyol, isopropylmyristate or hydroxyl-octacosanylhydroxystearate). Change in crystallization to their β-forms is thus prevented due to the lipid matrix. The lipid matrix is thus retained in their amorphous form (Radtke and Müller 2001).

Multiple Type

This is a multiple 'oil-fat-water' system. The solubility of a drug decreases when the temperature of the lipid melt falls during congealing process. This expels out the drug. Drugs have higher solubility in liquid lipid and hence would be preferable than solid lipids. The oil droplets thus form nano-compartments preventing expulsion of the drug from the system (Jenning et al. 2000a). If the liquid lipid content exceeds its solubility in solid lipid, the phase separation of liquid lipid can form nano-compartments. Thus, a judicious decision is needed during the selection of type and amount of liquid and solid lipid. This type of NLC is formed when the amount of oil exceeds the solubility of crystallized lipids.

Methods for Preparation of NLCs

The most commonly used methods for preparation of NLCs are discussed in following section.

Solvent Displacement or Solvent Emulsification/Evaporation

This method is analogous to that used for the preparation of polymeric nanoparticles and essentially employs an organic solvent. Lipids and drug are dissolved in an organic solvent either miscible (in the solvent displacement method) or water-immiscible (in the case of solvent emulsification method). The organic solvent is added to aqueous medium containing stabilizers, surfactants or emulsifiers and it is later evaporated. A decrease in the size of particles can be further obtained with either sonication or by passing particulate dispersion through a High-Pressure Homogenizer (HPH). A shortcoming of this method is the use of organic solvents.

Microemulsion-based NLCs

Microemulsions (o/w) consist of low melting lipid, emulsifiers and water such that their mixture forms an optically transparent mixture at 65–70°C. This hot microemulsion is then added to cold water in range of 1:25 w/w to 1:50 w/w depending upon the composition of microemulsion under stirring. The droplet of the lipid congeals to form NLCs.

Phase Inversion Temperature Method

Oil-in-water or water-in-oil microemulsions can be phase-inverted by changing the temperature. Phase inversion can also be obtained with surfactants which are polyoxyethylated. At room temperature, the hydrophilic units are hydrated. With an increase in the temperature, a change in lipophilicity is observed. The temperature at which the affinity of surface active compounds to aqueous and a lipid phase is same, it is known as phase-inversion temperature. With an increase in the temperature beyond phase-inversion temperature, affinity of surface active compounds towards lipid phase increases, thus they can easily stabilize w/o emulsions.

Melt Dispersion Method

Drug and lipids are melted together in organic solvent and added to aqueous phase, which is kept at the same temperature as that of organic solvent. This results in formation of an emulsion, which is kept for emulsification for few minutes. This emulsion is then cooled to room temperature and nanoparticles are obtained (Reithmeier et al. 2001).

High Pressure Homogenization

HPH is the preferred method as it avoids use of organic solvents commonly used during preparation in any of the above described methods. In addition, HPH renders the process scalable from a research batch size of 20–50 mL to an industrial batch scale > 50 liters with lipid concentrations of 5–10% and in some experiments up to 40%. In HPH, the molten lipid is passed through a micron sized narrow gap by using high pressure, which accelerates the flow of the fluid through a short distance. Forces caused by high shear stress and cavitations from homogenizer breaks down particles to submicron range. Two commonly used approaches for homogenization are cold homogenization or hot homogenization, while the basic step of incorporating the drug in bulk lipid melt remains the same.

In the heat assisted homogenization technique, a hot solution containing surfactant and/or stabilizers is added to the mixture of the molten lipids (Souto and Müller 2006) resulting in development of microemulsions. Usually the temperature is selected 10°C above the melting point of lipids to obtain a clear molten mass of lipids. This pre-emulsion can also be prepared with the aid of high shear mixing device like Ultra-Turrax (IKA, Werke Staufen, Germany). The emulsion is then submitted to high pressure and homogenized at hot temperatures which result in nanoemulsion. High temperature results in emulsion of smaller particle size due to decreased viscosity of the lipid phase. However, this method is avoided in case of heat labile drugs as they undergo degradation at elevated temperature. The homogenization cycle can be repeated several times with varied pressures to get the product in the desired size range. Once cooled, the nanoemulsion lead to formation of NLCs. Excessive homogenization cycles though initially decreases the particle size but prolonged use may increase the particle size due to coalescence of particles with a high kinetic energy.

Instead, cold homogenization prevents the temperature-induced degradation of both the drug and the lipid. The initial step of dispersing or dissolving a drug in molten lipid is same as that of hot homogenization technique. The melt is rapidly cooled with dry ice or liquid nitrogen. This helps in maintaining the homogenous distribution of drug within the molten lipid matrix. This solidified melt is either ball milled or mortar milled to form microparticles in the range of 50–100 μm. Low temperature facilitates faster particle comminution. The microparticles are dispersed in a cool emulsifier solution and then subjected to HPH below or at room temperature. Compared to hot homogenization, broad distribution of polydisperse particles is usually obtained.

Spray Drying

Spray drying is a cost effective technique than the lyophilization. The method is mainly used for lipids with their melting point higher than 70°C. Alternatively, lipids can be dissolved in ethanol-water mixtures which lead to formation of heterogeneous and small sized crystals due to lower inlet temperatures. Addition of carbohydrates preserve the colloidal size during lyophilization. Spray drying is usually carried out at high temperatures; hence essential oils may undergo rancidity and evaporation.

Among the different methods described above, HPH seems to be the ideal industrial feasible method for preparation of NLCs.

Composition of NLCs

As discussed earlier, In NLCs, a lipid matrix essentially comprises a mixture of a solid lipid and a liquid lipid (Kalepu et al. 2013). They are generally prepared in the aqueous phase containing dissolved surfactant(s)/stabilizer(s). A list of GRAS (Generally Recognized as Safe) approved solid and liquid lipids used in NLC preparation is given in Table 12.1. A schematic representation of the NLC structure is depicted in Fig. 12.1. The blend of lipids used could vary from 99.9:0.1 w/w for solid lipids to liquid lipids in ratio of 70:30 to form unstructured lipid matrix.

Solid Lipid

Like in SLN, biocompatible and biodegradable lipids, which are solid at room temperature can also be used in preparation of NLCs. Solid lipids could be fatty acids, fatty alcohols, mono/di/tri-glycerides or fats. A list of various solid lipids used in NLCs is given in Table 12.1. They should exhibit solubility in molten lipids. Lipids are known to have polymorphic forms. The physical and chemical nature

Table 12.1. Various types of lipids used in preparation of NLCs.

Lipids	Examples	Trade name	Physical form	Melting point °C
Triglycerides	Tripalmitin	Dynasan 116	Solid	61–65
	Tristearin	Dynasan 118	Solid	70–73
	Trilaurin	-	Solid	-
	Trimyristin	Dynasan 114	Solid	55–58
	Triglyceride derivatives of saturated even number of fatty acids	Softisan	Solid	44–58
Long chain (14–18C)	Corn oil	-	Liquid	-
	Soybean oil	-	Liquid	-
	Safflower oil	-	Liquid	-
	Olive oil	-	Liquid	-
	Hydrogenated palm oil	Softisan®154	Solid	55–60
Medium chain triglycerides (6–12C)	Glyceryltricaprylate	Miglyol 812	Liquid	-
	Glycerylcaprate	Miglyol 810	Liquid	-
		Captex 300	Liquid	-
	Medium-chain triglycerides	Labrafac	Liquid	-
		Neobee M-5		
Mono/di glycerides	Glyceryl monocaprylocaprate	Capmul MCM	Solid	
	Glyceryl dibehenate/Glyceryl tribehenate	Compritol 888 ATO/ Dynasan® 122	Solid	70–75/ 81–85
	Cetylpalmitate, Hexadecyl decanoate	Crodamol CP, Kollicream® CP 15	Solid	43–57
	Glyceryl monostearate	Geleol, Cutins GMS, Tegin, Imwitor 900P	Solid	58–60
	Glyceryl tristearate	Dynasan® 118	Solid	70–73
	Glyceryl tripalmitate	Dynasan® 116	Solid	61–65
	Glyceryltrilaurate	Dynasan® 112	Solid	43–46
	Glyceryltrimirystate	Dynasan® 114	Solid	55–58
	Glyceryl palmitostearate/ Glyceryl distearate	Precitol ATO 5	Solid	52–55
	Glyceryl monooleate	Peceol	Liquid	-
	LauroylPolyoxylglycerides	Gelucire 44/14	Solid	>37
	Monoacylglycerols	Myverol 18–99 K		
	Glycerylmonolinoleate	Maisine 35-1	Liquid	-
	oleoyl macrogol-6 glycerides	Labrafil	Liquid	-
Propylene glycol esters	Propylene glycol monocaprylate	Capmul Capryol™ 90		-
	Propylene glycol monolaurate	Capmul PG-12, Lauroglycol™ 90		-
	Polyglyceryl-3 dioleate	PlurolOleique® CC 497		

Table 12.1 contd....

...Table 12.1 contd.

Lipids	Examples	Trade name	Physical form	Melting point °C
Fatty acids	Stearic acid	Kolliwax® S	Solid	56–72
	Oleic acid	-	Liquid	-
	Palmitic acid	-	-	-
	Linoleic acid	-	-	-
Fatty alcohol	Stearyl alcohol	Crodacol S95 EP, Ginol® 18/95, Kolliwax® SA	-	57–60
	Cetyl alcohol	Crodacol C90 EP, Ginol® 16/95, Kolliwax® CA	Solid	46–52
	Cetearyl Alcohol, Cetostearyl alcohol	Crodacol™ CS90 EP, Kolliwax™ CSA 50, Kolliwax™ CSA 70	Solid	49–56
	C14-22 alcohol	Montanov		
Phospholipids	Phosphatidylcholine Soy Lecithin Egg lecithin	Epikuron™200 Lipoid® E 80	Solid	 42–46 42–46
	Phosphatidylglycerol	-	Solid	-
	Vitamin E and derivatives	TOS, TPGS	Solid	-
Others	diethylene glycol monoethyl ether	Transcutol® HP	Liquid	-
	Polyoxyl 40 hydrogenated castor oil	Cremophor® RH40	Liquid	-
Wax	Beeswax, Carnauba wax	Witepsol Series	Solid	42–66
	Cetylpalmitate	-	-	-
	Theobroma butter	-	-	-

of lipids play an important role in drug entrapment. Triglycerides typically exhibit three crystalline forms—the α-hexagonal (almost spherical) form, β' orthorhombic form and β triclinic form (Sato and Garti 1988). The α-hexagonal form has the lowest melting point with the least stability, and hence it has low density and high mobility (Mehnert and Mader 2001). Therefore, NLCs offer a higher capacity of loading than other polymorphic forms. The β-form is the most stable with higher melting point, offering a lower loading. The β'-form exhibit properties that are between those of α and β-forms.

Solid lipid crystallizes in either α- or β'-form with higher energy. During storage they get transformed to the β-form, which possess low energy and is more ordered. The particle size also increases on transformation from spherical form to needle or platelet form and during this modification, the drug gets expelled out. Higher crystallinity leads to a poor drug entrapment and a higher drug expulsion. The rate of the drug incorporation can be briefly explained in the following sequence: super cooled melt < α-form < β'-modification < β-modification (Muller et al. 2000). Moreover, lower crystallinity increases the degradation of the drug. If the preparation is for dermal and topical use, the crystallinity of the lipid determines the occlusiveness (Wissing and Muller 2002, Muller et al. 2009). Long-chain lipids like tri-stearin and tri-palmitin slow-down the polymorphic transformation (Bunjes and Koch 2005, Rosenblatt and Bunjes 2009). Saturated phospholipids and co-surfactants also slow down the process. High melting surfactants increase α polymorph. Faster cooling rates retard polymorphic transformation (Awad et al. 2008). The presence of oil also depresses the solid lipid's melting point but, however they are solid at room temperature (Müller et al. 2000). Tables 12.2 and 12.3 enlists GRAS recorded lipids and solid lipids used in dermal delivery of active.

Table 12.2. List of GRAS recognized solid lipids.

Lipids/Brand name	Physicochemical properties	Official status
Glyceryldibehenate Compritol® 888 ATO (GatteFosse), Kolliwax® GDB (BASF, Germany) former known as Speziol® GDB PHARMA	• Mixture of tri-, di- and mono-esters of glycerols of behenic acid (C22) and small amounts of stearic or palmitic acids • Mixture shows an HLB of 2 • Release modifier • Viscosity enhancer • Nanoparticles preparation	Ph. Eur
Glyceryldistearate orglycerylpalmito-stearate Precirol® ATO 5	• Mixture of di- and triglycerides of stearic and palmitic acid (C16-C18) • Contains 40–60% as diglycerides melting point lower than glyceryldibehenate • Precirol® ATO 5 Closely fulfils the pharmacopoeia definition for type I glyceryldistearate	Ph. Eur
Glyceryl monostearate 40–55 Type I: Geleol™ mono and diglycerides NF (Gattefosse, France), Imwitor® 900P (Cremer Oleo, former Sasol, Germany) Type II: Kolliwax® GMS II (BASF, Germany earlier Cutina® GMS V PH), Imwitor®900K (Cremer Oleo, former Sasol, Germany) Type III: Imwitor® 491 (Cremer Oleo, former Sasol, Germany)	• Contains 40–55% of monoglyceride and 30–45% of diglycerides • Three different types depending on stearic acid content • Type I: Stearic acid 40–60%, and 90% palmitic acid • Type II : Stearic acid substitutes 60–80% glycerol • Type III: 90% of glyceryl monostearate	Ph. Eur

Table 12.3. Topically employed solid lipids.

Lipids/Brand name	Physicochemical properties	Applications
Gelucire® 44/14 (Gattefossé)	• Lauryl macrogol-32 glycerides • T_m > 37°C, partially solid matrix	Lipid nanoparticles skin delivery
Gelucire® 50/13 (Most popular)	• stearoyl macrogol-32 glycerides • T_m > 37°C, processed colloidal sized lipids decreases T_m to form partially crystallized lipid matrix	Useful for drugs/peptides In preparation of pediatric/geriatric rectal and vaginal suppositories Mucoadhesive lipid nanoparticles
Witepsol™ (Cremer Oleo)	• melting point 32–37°C	
Witepsol® H15	• melting range 33–35°C	Nanoparticles of a sensitive polyphenol compound
Witepsol® E 85	• T_m of 42–44°C	Nanoparticles of drugs
Witepsol® S series	• melting range up to 70°C	
Beeswax and carnauba	-	Lipid nanoparticle formulation

Liquid Lipid

Liquid lipids like Miglyol, capric and caprylic triglycerides when used at high concentrations delayed the recrystallization rate and exhibited good stability (Jenning et al. 2000b). NLC prepared with tripalmitin (solid lipid) and fish oil as liquid oil behave in the similar way (Awad et al. 2009). A similar study showed a lack of change in lipid structure due to Miglyol in the presence of Compritol ATO 888 (Jores et al. 2003, Jores et al. 2004, Jores et al. 2005). Non-volatile fixed oils and fats as well as volatile essential oils are commonly used as liquid lipids.

Fixed Oils

Fixed oils are obtained from plants such as castor oil, arachis oil, sesame oil, safflower oil, linseed oil, olive oil, coconut oil, etc., while those obtained from animals include shark liver oil. Unsaturated fatty acids such as linoleic acid, linolenic acid and arachidonic acids are called essential fatty acids. Fixed oils are also reported to be carriers of fat-soluble vitamins like A, D, E and K (Fasina et al. 2006, Dauqan et al. 2011). In the cosmetic industry, they are popular for their demulcent and emollient effects. The fixed oils and fats contain esters of glycerin with palmitic, oleic and stearic aliphatic long chain fatty acids. Of them, the fats contain higher proportion esters of glycerin with fatty acids of high degree of saturation (e.g., glycerin stearate) and are solids at room temperature. Meanwhile, fixed oils contain a high proportion of liquid unsaturated glycerides such as glycerin oleate and have lower melting points.

Essential Oils

Regarding essential oils, they are plant-derived natural aromatic, clear and volatile oily liquids. Essential oils also known as essences or volatile oils are of lower density than water, formed by several volatile compounds (Sangwan et al. 2001, Baser and Demirci 2007). Unlike fixed oils they cannot be saponified, since they do not contain glycerin esters as fixed oils and fats. Instead, they possess high refractive index and are optically active. From a chemical point of view, they contain mixtures of low molecular weight terpenes and phenyl propanoids as well as other volatile components. The components of an essential oil typically have a molecular weight less than 300 that exhibit good solubility in organic solvents. Essential oils have demonstrated antimicrobial, antioxidant, anti-mutagenic and anti-carcinogenic properties. Other medicinal properties include sedative, local anesthesia, analgesic and anti-inflammatory (São Pedro et al. 2013). According to the International Standard Organization on Essential Oils (ISO 9235: 2013) and the European Pharmacopoeia (Council of Europe 2004) an essential oil is defined as the product obtained from plant raw material by hydrodistillation, steam distillation or dry distillation or by a suitable mechanical process (Zuzarte and Salgueiro 2015).

The hydrophobic nature of essential oils aids their antimicrobial activity and can be used as an active in lotions, creams, ointments, emulsions for dermal applications. They are also used as antioxidants and preservatives in foods and foodstuff packaging material. The agriculture sector has taken advantage of essential oils as crop protectants. The issues of volatility and oxidation of essential oils can be counter acted by mixing essential oils with fixed lipids that act as a vehicle for the oil. The flavor and functional properties of the essential oil can be preserved with enhanced stability and improved shelf life by nano-encapsulation. If not protected, oxidized oils not only alter the organoleptic properties but also causes skin-sensitization and contact dermatitis (Brared Christensson et al. 2009). Cells easily interact with lipid or phospholipids of lipid particulates. Antimicrobial activity was thus eventually increase with NLCs containing essential oil (Allahverdiyev et al. 2011, Huh and Kwon 2011). Hydrophobic and relatively lower molecular weight of essential oils can easily permeate through skin and drain easily into systemic circulation (Adorjan and Buchbauer 2010). High permeation of essential oils are of concern when applied on broken skin or large surface areas due to increased systemic side effects, e.g., convulsions (Bilia et al. 2014). Following dermal pathway, respiratory tract provides a rapid entry into systemic circulation (Moss et al. 2003). Encapsulation of essential oils would increase the physical stability, protect from the environment, decrease volatility and enhance bioactivity. Drug release can be modulated with improved patient acceptability (Ravi Kumar 2000). According to United States Food and Drug Administration (USFDA), essential oils are recognized as GRAS (Generally Recognized As Safe). Poor water solubility, dispersibility and sensory properties limit their use in food products. NLCs are thus the right carrier for entrapment of essential oils. They increase the oil loading especially and minimize expulsion. Till date, essential oils are being studied either as active or as encapsulants.

Semi-synthetic Oils

Miglyol 812, tocopherol and long chain triacylglycerols/fractionated coconut oil are most commonly studied oils for encapsulation of synthetic drugs (Pardeike et al. 2009). Capmul MCM has also been reported for preparation of NLC with glyceryl dilaurate selected as solid lipid for encapsulation of artemether. The NLCs were prepared by microemulsion method. Nevertheless, artemether NLCs exhibited significant improvement in anti-malarial treatment compared to current injectable IM formulation (Larither®). The duration of action was also improved. The anti-malarial activity of placebo vehicle still is to be elucidated (Joshi et al. 2008). Table 12.4 lists various oils with potential antimicrobial, antifungal and anti-parasitic actions. Some of these oils already encapsulated in NLCs are indicated.

Allergenicity and rancidity however, holds same for solid lipids and fixed oils where in the lipids contain mixture of several compounds, which differ from supplier to supplier as well as among the batches supplied from same supplier (Mehnert and Mader 2001). An advantage of using these lipids is that all these lipids are food-grade ingredients and can be easily used in the food and beverage

Table 12.4. List of liquid lipids with antimicrobial activities, antifungal activity, anti-parasitic activity, and other activities. Italicize all botanical names.

Liquid Lipid	Biological Source	Activity
	Fixed oils	
Arachis/Peanut oil	*Arachis hypogaea*	Nutritive in injectable
Castor oil*	*Ricinus communis*	Laxative, cosmetics
Chaulmoogra oil	*Hydnocarpus wightiana*	Antibacterial – Antileprotic (Non-edible)
Cod liver oil*	*Gadusm orrhua*	Vitamin A and D deficiency
Jojoba oil*	*Simmondsia chinensis*	Cosmetics
Mustard oil	*Brassica nigra*	Rubefacient, counter irritant
Neem*/Margosa oil	*Azadirachta indica*	Cosmetics, spermicidal (non-edible oil)
Olive oil*	*Olea europoea*	Cosmetics, vehicle in injectables
Rice bran oil	*Oryza sativa*	Antioxidant, Cosmetics, food
Sesame oil*	*Sesamum indicum*	Vehicle in intramuscular injectables
Shark liver oil	*Hypoprionbrevirostris*	Vitamin A deficiency
Sunflower oil*	*Helianthus annus*	Dietary supplement, cosmetics
	Essential oils	
Cinnamon oil*	*Cinnamomum zeylanicum*	Cosmetics, dentifrices
Citronella oil*	*Cymbopogon nardus*	Perfumery, mosquito repellent
Clove oil*	*Eugenia caryophyllus*	Flavoring agent, dentistry, antiseptic
Eucalyptus oil*	*Eucalyptus globules*	Counter-irritant, expectorant, inhalations
Garlic oil	*Allium sativum*	Disinfectant in pulmonary and amoebic dysentery
Geranium oil	*Pelargonium graveolens*	Cosmetics, beverages and dairy products
Lavender oil	*Lavandula officinalis*	Flavoring agent, cosmetics
Lemon grass oil	*Cymbopogon flexuousus*	Flavoring agent
Peppermint oil*	*Mentha piperita*	Flavoring, carminative, calcium channel blocking activity, USFDA approved nasal decongestant
Thyme oil	*Thymus vulgaris*	Antifungal, antibacterial, anti-oxidant, anthelmintic

*oils used in NLCs

industries too. The third generation of lipids includes smart lipids, which is a blend of 10 different liquid or solid lipids. Such a mixture does not permit the formation of dense packed structure and facilitates the incorporation of drugs (Müller and Keck 2015).

Surfactants

Several surfactants have been studied by different research groups. Some of them are included in Table 12.5. The main role of surfactant in NLC is to prevent aggregation of the colloids and thus stabilize the dispersion by both steric and electric mechanism. This can be indicated by their zeta potential. Electrical stabilization occurs at 40 mV while steric stabilization occurs at 30 mV (Radomska-Soukharev 2007). Lecithin and lipids provide electrostatic stabilization while poloxamer provides steric stabilization. Surfactants with HLB < 9 are lipophilic in nature and those above this value are hydrophilic. Particle size of colloids can be also decreased with high concentrations of surfactant. It influences the polymorphic transition and crystallization temperature of lipids. For instance, sodium glycolate, a ionic surfactant retains α-form of lipid while cremophor, a nonionic surfactant retains the β-form (Gramdorf et al. 2010). Poloxamer and tween 80 is not affected by enzymatic degradation while cholates are affected by enzymes (Olbrich and Muller 1999).

All surfactants cannot be used for stabilization. Some surfactants generate allergic reaction or aggregates whole blood or red blood corpuscles. For instance Span 85 causes red blood corpuscle's aggregation (Olbrich et al. 1999) while cremophor EL was found to be allergic when used in paclitaxel emulsion for parenteral use (Alkan-Onyuksel et al. 1994). GRAS and parenteral use approved grade of surfactants include poloxamer 188, lecithin, sodium glycolate and tween 80 (Wissing et al. 2004). For enteral use, non-ionic surfactants are commonly used.

Liver uptake of nanoparticles can be prevented by coating them with PEG in order to prevent binding of plasma proteins and opsonins to particulates. Polysorbates like tween 80 coated upon particulates can help in targeting drug to CNS as they can easily cross the blood brain barrier. Likewise,

Table 12.5. List of surfactants used in production of lipid nanoparticles.

Hydrophilic emulsifiers	Poloxamer - Poloxamer 188, Poloxamer 407
	Polysorbates - Tween 20, Tween 40, Tween 80
	Cholate derivatives - sodium deoxycholate, sodium glycocholate
	Polyvinyl alcohol
	Other derivatives - sodium oleate, polyglycerol methyl glucose distearate
	Alkyl polyglucosides - Plantacare® 810 UP, Glucopon® 425 N/HH, Glucopon® 600 CS UP, Plantacare® 1200 UP
	Polyhydroxy surfactants - Plurol® Stearique WL1009, Plantacare® 810 PL
	Sucrose esters - Surfhope SE Pharma D1216, Surfhope SE Cosme C1416, C1616, C1816
	Quillajasaponins - Quillajasaponaria wood extract, composed mainly of predominantly of glycosides of quillaic acid
Lipophilic emulsifiers	Myverol® 18-04 K, Span 20, Span 40, Span 60
Amphiphilic emulsifiers	Phosphatidylcholines - Egg lecithin, soya lecithin
	Phosphatidylethanolamines
	Gelucire® 50/13
Surface modified surfactant	PEGylated lipids - DSPE-PEG, Stearic acid-PEG2000, Solutol® HS15
	Ligand conjugated lipid - Folate-PEG-Cholesterol, Folate – stearic acid, Folate-PEG-DSPE, transferrin-PEG-Phosphotidylethanolamine, Mannose-6-cholesterol, Mannan-PEG-Phosphatidylethanolamine, N-hexadecyllactobionamide

cationic albumin can also cross blood brain barrier. While poloxamer, cremophor and vitamin E fail to cross the barrier (Aboutaleb and Dinarvand 2012), stabilizers like Gelucire and Peceol increases the oral absorption as it is known to inhibit P-gp efflux pump (Sachs-Barrable et al. 2007). Pluronic P85, Cremophor EL and Solutol HS 15 can modulate the efflux pump activity (Batrakova et al. 1999). The exact mechanism is however not known. It is suggested that these surfactant/stabilizer interfere with ATP hydrolysis and the ATP hydrolysis cycle is made futile. This inhibits the P-gp efflux (Rege et al. 2002). They are also explored for dermal application (Datte and Nagarsenkar 2009).

Scale-up Strategies for NLCs

An important aspect of nanoparticles for commercialization is the large-scale production. Qualification, validation and acceptance by regulatory agencies are important for any cost-effective method. Large-scale production of nanoparticles is a herculean task. However, the use of HPH has been proved to be successful for lipid nanoparticles and with regulatory acceptance. Reports for scaling-up of solid nanoparticles up to a batch size of 10 kg has been possible using HPH (Shegokar et al. 2011). Scale-up of NLC is similar to that of SLN.

For cosmetics containing NLCs, the process of preparation for cosmetic base need not be altered; concentrated NLC suspensions are prepared separately and mixed to already established products or cosmetic base. The established cosmetic base are prepared as the routine process but with reduced quantity of water. The pending amount of water is added by the NLC concentrate. For creams, the addition of NLC concentrate is done when the cream is slightly cooled but above room temperature (about 30°C). This becomes a more feasible technique for already established product (Muller et al. 2007).

Commercialization of NLCs

The first NLC concentrate to be launched to market was NanoLipid Restore CLR™ (Dr. Kurt Richter's Chemisches Laboratorium, Berlin, Germany) in 2006. It contains 45% concentrate (particle mass) and is stabilized with Plantacare® 200. The NLC is made of carnauba wax blended with blackcurrant oil used for dermal applications (IOPE® from Amore Pacific, Seoul, South Korea). The liquid lipid is rich in linoleic acid and linolenic acid with beneficial effects on skin. The nanostructures exhibit good adhesion to skin due to its prolonged effect. Instead, NanoLipid Q 10 CLR contained coenzyme Q in addition to blackcurrant oil and carnauba wax.

Some other worldwide finished products include NanoRepair Q 10 cream and Serum, developed by Dr. Rimpler GmbH (Wedemark, Germany). The coenzyme Q 10 concentrations were 0.5 and 0.1% in cream and serum respectively. Nanovital contains Q10 as 0.1% concentration along with UV blocker—nanosized titanium dioxide and sunflower seed extract rich in polyphenol. Many more NLCs have been introduced to the global market for cosmetic purposes (Muller et al. 2007). Lipidex® by Capsugel is platform technology of solid, liquid and semi-solid lipid fill into hard shell or solid lipid pellets. This is mainly used for drugs with poor solubility.

Currently, essential oils are recommended for external application either in topical applications, as gargles, mouthwashes or inhalation. Rarely they have been explored for oral use though they are GRAS approved. They are usually diluted with other oils like olive or milk, before ingestion. Inhalation of concentrated oils may cause eye irritation and hence, they are frequently not recommended. Likewise, in aromatherapy, direct application of concentrated oil can cause skin irritation. Bergamot oil causes photosensitization of skin (Bilia et al. 2014). Essential oils undergo renal excretion. (–)-Menthol on oral administration is eliminated as menthol glucuronide from renal route (Bronaugh et al. 1990, Kohlert et al. 2000). Likewise, other oils like eugenol, limonene, thymol and carvacrol are metabolized to glucuronide and sulfate phase II derivatives in plasma and urine, respectively (Guenette et al. 2007,

Michiels et al. 2008). Fast metabolism followed by excretion minimizes the risk of oil accumulation through the body (Kohlert et al. 2002).

New Excipients in the Market

Lipid and surfactant manufacturers have been researched for innovative excipients to expand the applications of these excipients with multifaceted benefits in different dosage forms.

Solid Matrix

The current trend is to select a mixture of triglyceride lipids for NLC preparation depending upon the route of NLC administration. For instance, lipids used for dermal preparations would be different from the oral preparations based upon the fats used in the lipids. Oral preparations usually have lipids with well-defined structures while in dermal and cosmetic industry; lipids with a portfolio of different fats with varying characteristics are used. Many of the lipids have been officially recognized. Tables 12.2 and 12.3 enlist some of the officially accepted lipids used for topical preparations.

Oils as Matrices

Refined white oils based on mineral hydrocarbons containing saturated aliphatic hydrocarbons of Ph. Eur. and USP/NF grades are already available. Another two white oils Primol™ and Marcol™ with excellent emolliency and hydration has been launched for cosmetic applications by Exxonmobil Fuels & Lubricants (F&L) (Houston, Texas, USA). A number of oils like coconut oil, eucalyptus oil, cod liver oil, and others are commercially available.

Poor stability of drugs in liquid lipids due to oxidative impurities, free fatty acids, coloring matters have been an inhibition for use of liquid lipids in NLC preparation. Super Refined™ Oils have been especially developed for this limitation and processed by flash chromatographic processes. Croda chemicals (East Yorkshire, UK) have developed refined oils of olive, sesame, soybean, cottonseed, castor and sunflower oils among others. Besides these oils, Super Refined™ PEG and polysorbates are also available. Flash chromatographic process decreases the impurities, moisture, peroxides and aldehyde contents, thus improving oil stability. Tea tree oil obtained by extraction from *Melaleuca alternifolia* is commercially available from Reitzer Pharmaceuticals (Kempton Park, Gauteng, South Africa). It has wide applications in pharmaceutical and cosmetic industry due to the antifungal, anti-inflammatory and anti-bacterial effects in skin related disorders like acne, psoriasis and eczema.

Surfactants

Avanti Polar Lipids (Alabaster, Alabama) has been a supplier of various lipid types like fatty acid modified lipids, natural and PEGylated lipids, phosphatidylserine, phosphatidylinositols, phytosphingosine, diglycerols, glycolipids, sphingolipids, diglyceride and their pyrophosphate derivatives, lipids with biotin, fluorescent and deuterium labeled lipids for live imaging with fluorescent microscopy of radiolabeling, antigenic lipids, lanthanide chelating lipids and many more functionalized PEG-lipids. Corden Pharma (Sermoneta, Italy) offers a range of phospholipids including lyso-lipids, ether lipids and many more.

TegoBetain P 50 C (Evonik Industries AG Personal Care, North Phine-Westphalia, Germany) is an amphoteric surfactant chemically containing cocamidopropyl betaine and 38% active betaine based on palm kernel oil. It is synthesized with high purity by a patented process and contains low level of amidoamine, chloro acetic acids and RSPO-certified preservative-free. They have been widely used in lipid nanoparticles and exhibit good viscosity in hair and skin care preparations.

Pureact SLI (Innospec Speciality Chemicals, USA) is a surfactant based on sodium lauryl isethionate and is sulfate-free. It exhibits good flash foam comparable to sodium lauryl sulfate stable over a wide range of pH. Hence, it can be used either as primary or secondary surfactant.

Steposol Met-10U [collaboration efforts of two surfactant manufacturers–Stepan (New Jersey, USA) and Elevance Renewable Sciences Inc. (Bolingbrook, Illinois)] surfactants obtained from natural oils. This surfactant, commonly used to displace solvents, have been used in household applications and in adhesives/paint removals.

Sapnov™ (Naturex, Avignon, France) is a non-ionic, water-soluble foaming surfactant naturally extracted from Quillajasaponaria wood extract and sodium benzoate. Similar other surfactants include Colonial's natural coconut based surfactant, Colafax PME (monoalkyl phosphate) substitute for lauryl sulfates. Cognis's series of Plantapon, Plantaren and Plantacare can be explored for skin care preparations as SLNs.

SymSol® PF-3 (Symrise, Holzminden, Germany) is a PEG-free solubilizer for solubilization of water insoluble ingredients. Its solubilizer properties supersede the solubilization ability of PEG-40 Hydrogenated Castor Oil (Soluplus, BASF, New Jersey, USA). It is also an emulsifier for o/w preparations. Tego® Solve 61 (Evonik Industries AG Personal Care, North Phine-Westphalia, Germany) is another PEG-free solubilizer consisting of polyglyceryl derivatives of caprylic acid, capric acid and ricinoleic acid. It is a good solubilizer of fatty and natural oils.

Pemulen® (Lubrizol Advanced Materials Inc., Ohio, USA) is a polymeric emulsifier containing polyacrylic acid polymers with high molecular weights. Different grades of o/w stabilizer Pemulen® like Pemulen® TR-1, TR-2 are available based on their oil emulsifying properties. Gemini surfactants comprising of two chemically linked surfactant monomers are popular these days. The different spacer structure and morphology of micelles especially at their critical micellar concentration improve the surface active property and wetting property. These surfactants find their use mainly in dermal preparations and genetic delivery (Sekhon 2013). A non-ionic heptylglucoside, Sepiclear G7 (Seppic, Paris, France) derived from castor-seeds and sugar was a prize winner at sustainable green chemistry. It is a good solubilizer of vitamin E, perfumes and essential oils.

Though most of the excipients have been explored for oral and topical dermal applications, readers are suggested to check for their GRAS status especially for the lipids and determine its intravenous use accordingly.

Conclusions

Lipid nanoparticles offer superior drug delivery over polymeric nanoparticles. The poor stability of liposomes and SLNs alone has been overcome by NLCs—the second generation of lipid nanoparticles. NLCs not only offer improved stability but also higher drug loading capacity due to the formation of imperfect matrix due to the presence of oil. Essential oil can be studied as a composition of lipid matrix as well as for encapsulation. Oils from different categories offer improved lipid structure to incorporate drug. NLCs is industrially scalable drug delivery system and have high potential in loading lipophilic drugs and oils as active moiety. This delivery system is widely used in cosmetic as well as nutraceuticals. Pharmaceutical ingredient industries are actively launching to the market lipids with improved performance. Overall, data from the literature confirm the potential of NLCs in delivery of various active compounds with improved performance.

References

Aboutaleb, E. and Dinarvand, R. 2012. Vincristine-dextran complex loaded solid lipid nanoparticles for drug delivery to the brain. World Acad Sci. Eng. Technol. 67: 611–615.

Adorjan, B. and Buchbauer, G. 2010. Biological properties of essential oils: an updated review. Flavour Frag. J. 25: 407–426.

Alkan-Onyuksel, H., Ramakrishnan, S., Chai, H.B. and Pezzuto, J.M. 1994. A mixed micellar formulation suitable for the parenteral administration of taxol. Pharm. Res. 11: 206–212.

Allahverdiyev, A.M., Kon, K.V., Abamor, E.S., Bagirova, M. and Rafailovich, M. 2011. Coping with antibiotic resistance: combining nanoparticles with antibiotics and other antimicrobial agents. Expert Rev. Anti. Infect. Ther. 9: 1035–1052.

Awad, T., Helgason, T., Kristbergsson, K., Decker, E., Weiss, J. and McClements, D.J. 2008. Effect of cooling and heating rates on polymorphic transformations and gelation of Tripalmitin solid lipid nanoparticle (SLN) suspensions. Food Biophys. 3: 155–162.

Awad, T.S., Helgason, T., Weiss, J., Decker, E.A. and McClements, D.J. 2009. Effect of omega-3 fatty acids on crystallization, polymorphic transformation and stability of Tripalmitin solid lipid nanoparticle suspensions. Cryst. Growth Des. 9: 3405–3411.

Batrakova, E.V., Li, S., Miller, D.W. and Kabanov, A.V. 1999. Pluronic P85 increases permeability of a broad spectrum of drugs in polarized BBMEC and Caco-2 cell monolayers. Pharm. Res. 16: 1366–1372.

Bilia, A.R., Guccione, C., Isacchi, B., Righeschi, C., Firenzuoli, F. and Bergonzi, M.C. 2014. Essential oils loaded in nanosystems: A developing strategy for a successful therapeutic approach. J. Evid. Based Complementary Altern. Med. 2014: 14.

Brared Christensson, J., Forsstrom, P., Wennberg, A.M., Karlberg, A.T. and Matura, M. 2009. Air oxidation increases skin irritation from fragrance terpenes. Contact Derm. 60: 32–40.

Bronaugh, R.L., Wester, R.C., Bucks, D., Maibach, H.I. and Sarason, R. 1990. *In vivo* percutaneous absorption of fragrance ingredients in rhesus monkeys and humans. Food Chem. Toxicol. 28: 369–373.

Bunjes, H. and Koch, M.H.J. 2005. Saturated phospholipids promote crystallization but slow down polymorphic transitions in triglyceride nanoparticles. J. Control Release 107: 229–243.

Date, A. and Nagarsenkar, M. 2009. Development of Gelucire based lipid nanoparticles (GeluPearls) and their application in drug delivery AAPS Annual Meeting and Exposition.

Dauqan, E.M., Sani, H.A., Abdullah, A. and Kasim, Z.M. 2011. Fatty acids composition of four different vegetable oils (red palm olein, palm olein, corn oil and coconut oil) by gas chromatography 2nd International Conference on Chemistry and Chemical Engineering, Chengdu, China, pp. 31–34.

Fasina, O.O., Hallman, H., Craig-Schmidt, M. and Clements, C. 2006. Predicting temperature-dependence viscosity of vegetable oils from fatty acid composition. J. Am. Oil Chem. Soc. 83: 899–903.

Gramdorf, S., Kumpugdee-Vollrath, M., Bilek, H. and Perlich, J. 2010. Influence of emulsifiers on the crystallization of solid lipid nanoparticles (SLN), HASYLAB.

Guenette, S.A., Ross, A., Marier, J.F., Beaudry, F. and Vachon, P. 2007. Pharmacokinetics of eugenol and its effects on thermal hypersensitivity in rats. Eur. J. Pharmacol. 562: 60–67.

Huh, A.J. and Kwon, Y.J. 2011. Nanoantibiotics: a new paradigm for treating infectious diseases using nanomaterials in the antibiotics resistant era. J. Control. Release 156: 128–145.

Jenning, V., Gysler, A., Schafer-Korting, M. and Gohla, S.H. 2000. Vitamin A loaded solid lipid nanoparticles for topical use: occlusive properties and drug targeting to the upper skin. Eur. J. Pharm. Biopharm. 49: 211–218.

Jenning, V., Schäfer-Korting, M. and Gohla, S. 2000. Vitamin A-loaded solid lipid nanoparticles for topical use: drug release properties. J. Control Release 66: 3405–3411.

Jores, K., Haberland, A., Wartewig, S., Mader, K. and Mehnert, W. 2005. Solid lipid nanoparticles (SLN) and oil-loaded SLN studied by spectrofluorometry and Raman spectroscopy. Pharm. Res. 22: 1887–1897.

Jores, K., Mehnert, W., Drechsler, M., Bunjes, H., Johann, C. and Mader, K. 2004. Investigations on the structure of solid lipid nanoparticles (SLN) and oil-loaded solid lipid nanoparticles by photon correlation spectroscopy, field-flow fractionation and transmission electron microscopy. J. Control Release 95: 217–227.

Jores, K., Mehnert, W. and Mader, K. 2003. Physicochemical investigations on solid lipid nanoparticles and on oil-loaded solid lipid nanoparticles: a nuclear magnetic resonance and electron spin resonance study. Pharm. Res. 20: 1274–1283.

Joshi, M., Pathak, S., Sharma, S. and Patravale, V. 2008. Design and *in vivo* pharmacodynamic evaluation of nanostructured lipid carriers for parenteral delivery of artemether: Nanoject. Int. J. Pharm. 364: 119–126.

Kalepu, S., Manthina, M. and Padavala, V. 2013. Oral lipid-based drug delivery systems—an overview. Acta Pharm. Sin. B. 3: 361–372.

Kohlert, C., Schindler, G., Marz, R.W., Abel, G., Brinkhaus, B., Derendorf, H., Grafe, E.U. and Veit, M. 2002. Systemic availability and pharmacokinetics of thymol in humans. J. Clin. Pharmacol. 42: 731–737.

Kohlert, C., van Rensen, I., Marz, R., Schindler, G., Graefe, E.U. and Veit, M. 2000. Bioavailability and pharmacokinetics of natural volatile terpenes in animals and humans. Planta Med. 66: 495–505.

Mehnert, W. and Mader, K. 2001. Solid lipid nanoparticles: production, characterization and applications. Adv. Drug Deliv. Rev. 47: 165–196.

Michiels, J., Missotten, J., Dierick, N., Fremaut, D., Maene, P. and De Smet, S. 2008. *In vitro* degradation and *in vivo* passage kinetics of carvacrol, thymol, eugenol and trans-cinnamaldehyde along the gastrointestinal tract of piglets. J. Sci. Food Agr. 88: 2371–2381.

Moss, M., Cook, J., Wesnes, K. and Duckett, P. 2003. Aromas of rosemary and lavender essential oils differentially affect cognition and mood in healthy adults. Int. J. Neurosci. 113: 15–38.

Muchow, M., Maincent, P. and Muller, R.H. 2008. Lipid nanoparticles with a solid matrix (SLN, NLC, LDC) for oral drug delivery. Drug Dev. Ind. Pharm. 34: 289–300.

Müller, R.H. and Keck, C.M. 2015. Next generation after SLN® and NLC®—the "chaotic" smartLipids® Wissenschaftliche Posterausstellung 2015. p Poster 9, Berlin.

Muller, R.H., Mader, K. and Gohla, S. 2009. Lipid nanoparticles (SLN, NLC) in cosmetic and pharmaceutical dermal products. Int. J. Pharm. 366: 170–184.
Müller, R.H., Mäder, K., Lippacher, A. and Jenning, V. 2000. Solid-liquid (semi-solid) liquid particles and method of producing highly concentrated lipid particle dispersions. *In*: PCT/EP00/04565 (ed.).
Muller, R.H., Maeder, K. and Gohla, S. 2000. Solid lipid nanoparticles (SLN) for controlled drug delivery—a review of the state of the art. Eur. J. Pharm. Biopharm. 50: 161–177.
Muller, R.H., Petersen, R.D., Hommoss, A. and Pardeike, J. 2007. Nanostructured lipid carriers (NLC) in cosmetic dermal products. Adv. Drug Deliv. Rev. 59: 522–530.
Müller, R.H. and Radtke, M. 2002. Nanostructured lipid matrices for improved microencapsulation of drugs. Int. J. Pharm. 242: 121–128.
Müller, R.H., Radtke, M. and Wissing, S.A. 2002. Solid lipid nanoparticles (SLN) and nanostructured lipid carriers (NLC) in cosmetic and dermatological preparations. Adv. Drug Deliv. Rev. 54: S131–S155.
Olbrich, C., Kayser, O., Kiderlen, A.F. and Muller, R.H. 1999. Flow cytometry as a possible tool for assessing cellular binding capacities of solid lipid particles (SLN). pp. 519–520. *In*: Proc 26th International Symposium Controlled Relevant Bioactive Materials, Boston Marriott Copley Place, Boston, MA.
Olbrich, C. and Muller, R.H. 1999. Enzymatic degradation of SLN—effect of surfactant and surfactant mixtures. Int. J. Pharm. 39: 31–39.
Pardeike, J., Hommoss, A. and Muller, R.H. 2009. Lipid nanoparticles (SLN, NLC) in cosmetic and pharmaceutical dermal products. Int. J. Pharm. 366: 170–184.
Radomska-Soukharev, A. 2007. Stability of lipid excipients in solid lipid nanoparticles. Adv. Drug Deliv. Rev. 59: 8.
Radtke, M. and Müller, R.H. 2001. Nanostructured lipid carriers: the new generation of lipid drug carriers. New Drugs 2: 4.
Ravi Kumar, M.N. 2000. Nano and microparticles as controlled drug delivery devices. J. Pharm. Pharm. Sci. 3: 234–258.
Rege, B.D., Kao, J.P. and Polli, J.E. 2002. Effects of nonionic surfactants on membrane transporters in Caco-2 cell monolayers. Eur. J. Pharm. Sci. 16: 237–246.
Reithmeier, H., Herrmann, J. and Gopferich, A. 2001. Lipid microparticles as a parenteral controlled release device for peptides. J. Control Release 73: 339–350.
Rosenblatt, K.M. and Bunjes, H. 2009. Poly(vinyl alcohol) as emulsifier stabilizes solid triglyceride drug carrier nanoparticles in the alpha-modification. Mol. Pharm. 6: 105–120.
Sachs-Barrable, K., Thamboo, A., Lee, S.D. and Wasan, K.M. 2007. Lipid Excipients Peceol and Gelucire 44/14 decrease Pglycoprotein mediated efflux of Rhodamine 123 partially due to modifying P-glycoprotein protein expression within Caco-2 Cells. J. Pharm. Pharm. Sci. 10: 319–331.
São Pedro, A., Santo, I.E., Silva, C.V., Detoni, C. and Albuquerque, E. 2013. The use of nanotechnology as an approach for essential oil-based formulations with antimicrobial activity. pp. 1364–1374. *In*: Méndez-Vilas, A. (ed.). Microbial Pathogens and Strategies for Combating Them: Science, Technology and Education. Microbiology Book Series. Formatex Research Center, Spain.
Sato, K. and Garti, N. 1988. Crystallization and polymorphism of fats and fatty acids. p. 464. *In*: Garti, N. (ed.). Surfactant Science No. 31. Taylor & Francis, New York.
Sekhon, B.S. 2013. Surfactants: Pharmaceutical and medicinal aspects. Journal of Pharmaceutical Technology, Research and Management 1: 11–36.
Shegokar, R., Singh, K.K. and Müller, R.H. 2011. Production & stability of stavudine solid lipid nanoparticles—From lab to industrial scale. Int. J. Pharm. 416: 9.
Souto, E.B. and Müller, R.H. 2006. Investigation of the factors influencing the incorporation of clotrimazole In Sln and Nlc prepared by hot high pressure homogenization. J. Microencapsul. 23: 377–388.
Wissing, S.A., Kayserb, O. and Muller, R.H. 2004. Solid lipid nanoparticles for parenteral drug delivery. Adv. Drug Deliv. Rev. 56: 5.
Wissing, S.A. and Muller, R.H. 2002. The influence of the crystallinity of lipid nanoparticles on their occlusive properties. Int. J. Pharm. 242: 377–379.
Zuzarte, M. and Salgueiro, L. 2015. Essential Oils Chemistry. Springer Int. Publishing, Switzerland.

13

Essential Oil-Based Nanomedicines against Trypanosomatides

Maria Jose Morilla and *Eder Lilia Romero**

Introduction

Neglected Tropical Diseases (NTDs) are defined as diseases suffering from a lack of attention by the public health community. The NTDs are associated with poverty and neglected by media and policy makers, affecting more than one billion people—one-sixth of the world's population-mostly in rural areas of low-income countries (WHO. Neglected diseases). In the Latin American and Caribbean (LAC) region, the NTDs are among the most common conditions afflicting the estimated 99 million people, who live on less than US$2 per day (Hotez et al. 2012), the burden of disease closely approximating or even exceeding that resulting from HIV/AIDS (Hotez et al. 2008). The lost ability to attend school or work, retardation of growth in children, impairment of cognitive skills and development in young children as well as on the health of girls and women are typical consequences of NTDs affecting endemic populations (Hotez et al. 2006, 2009). Trypanosomiasis, parasitic infections caused by protozoans of the Trypanosomatid family (a diverse group of flagellated parasites that show similar cellular structures and undergo morphological alterations during their life cycles), are considered NTDs (Doctors without borders and Drugs for Neglected Disease initiative).

Trypanosomatids affect 20 million people and cause 100,000 deaths per year, primarily in the tropical and subtropical areas of the world (Stuart et al. 2008). Trypanosomatids cause the human diseases American trypanosomiasis (or Chagas disease), leishmaniasis, African trypanosomiasis (or sleeping sickness) and Surra, a disease affecting animals.

The drugs available to treat Chagas disease, leishmaniasis and surra, are toxic and ineffective. The main drawback of current pharmacotherapy is the lack of treatments capable of selectively and safely eliminating intracellular parasites. Long term treatments are required, that in case of leishmaniasis may generate resistance and can be expensive. Developing new specific parasiticidal drugs on the other hand, is a long and highly costly process that, because of the lack of economical profit, no pharmaceutical company consider worth to face (Wilkinson and Kelly 2009).

One approach to circumvent the development of specific new drugs would be to screen the parasiticidal activity of different Essential Oils (EOs). Since ancient times, EOs have been widely used to relieve a wide variety of human maladies including bronchitis, pneumonia, pharyngitis,

Programa de Nanomedicinas, Departamento de Ciencia y Tecnología, Universidad Nacional de Quilmes, Roque Saenz Peña 352, Bernal, B1876BXD, Buenos Aires, Argentina.
* Corresponding author: elromero@unq.edu.ar

diarrhea, periodontal disease, wounds, because of their analgesic, sedative and anti-inflammatory activity (Gurib-Fakim 2006, Handbook of essential oils 2012). EOs also hold other relevant medicinal properties such as bactericidal, virucidal, fungicidal, antiparasitical and insecticidal activity (Burt 2004). EOs, however, are volatile, insoluble in water and easily decompose, owing to direct exposure to heat, humidity, light or oxygen (Turek and Stintzing 2013). Another major issue is their low oral bioavailability (Sharma et al. 2004, Anand et al. 2007).

As defined by the International Organization for Standardization (ISO), the term 'essential oil' is reserved for a "product obtained from vegetable raw material, either by distillation with water or steam, or from the epicarp of citrus fruits by a mechanical process, or by dry distillation" (ISO 9235, 1997), that is, by physical means only. Essential oils (EOs) are volatile, natural, aromatic oily liquids that can be obtained from several parts of the plants especially the aerial ones as leaves and flowers. They are derived from complex metabolic pathways to protect the plant organism from diverse pathogenic microorganisms, to repel insects that act as plague vectors, to reduce the appetite of some herbivorous by conferring unpleasant taste to the plant (Bakkali et al. 2008). Constituents are lipophilic and highly volatile secondary plant metabolites, reaching a mass below a molecular weight of 300 Da, that can be physically separated from other plant components or membranous tissue (Sell 2010). The EOs are generally complex mixtures that include hydrocarbons (terpenes and sesquiterpenes) and oxygenated compounds (alcohols, esters, ethers, aldehydes, ketones, lactones, phenols and phenol ethers) (Guenther 1972).

A significant amount of toxicity data is available for not only the oils but also the individual components such that many are generally regarded as safe (GRAS) by the FDA. GRAS status has permitted the use of EOs as flavouring agents in food and as additives to cosmetics, perfumes and cleaning products (Boire et al. 2013).

The constituents of EOs act synergistically (phyto-synergic interactions) by different mechanisms which significantly reduce the possibility of occurrence of resistance. The synergism may involve the protection of an active substance from degradation by enzymes from the pathogen or the circumvention of numerous multi-drug resistance mechanisms based on ionic bomb-mediated expulsion of antibiotic molecules (Bassolé and Juliani 2012).

Nanomedicines, on the other hand, are a powerful tool to selectively target a drug, to reduce its toxicity, and to increase its efficacy. Nanomedicines may also solve the classical problems of physical stability, low solubility, low permeability and bioavailability of EOs. The aim of this chapter is to provide an overview on the preclinical applications of EO (or some of their isolated components) based-nanomedicines used as trypanocidal agents, discussing the pros and cons of each experimental approach as well the feasibility of its future translation.

EOs-based Trypanocidal Nanomedicines

Intracellular amastigotes are a target common to all trypanosomatides. Cytoplasmic amastigotes are difficult to eliminate because of their hidden location within the host cell, and in case of Chagas, are responsible for the perpetuation of the disease. However, eradicating intracellular targets be it with highly hydrophobic or hydrophilic drugs would not be feasible by increasing drug's bioavailability or modifying its pharmacokinetics. Instead, radical changes to the drug's ADMET (Absorption, Distribution, Metabolization, Excretion, Toxicity) far beyond those provided by conventional drugs depots or micronization are needed to selectively deliver therapeutic amounts of drugs into the cell cytoplasm (Mitragotri et al. 2014). A potential tool to meet these requirements could be provided by nanomedicines. Nanomedicines are defined as the combination of nanoparticles of any chemical nature (lipidic, such as liposomes, solid lipid or nanostructured nanoparticles; micellar; polymeric; metallic, metallic oxides, etc.) and Active Pharmaceutical Ingredients (API) (low molecular weight molecules, or macromolecules such as nucleic acids, polysaccharides or proteins) (ETP Nanomedicine).

Nanomedicines are classified as 'nonbiological complex drugs' (NBCDs), that is, drugs showing inherent complexity that determines their pharmacologic activity and ADMET profile, but being of nonbiological, that is, synthetic origin. The pharmacologic activity of NBCDs is governed by the complexity of their structures (comprising not only nanoparticles but also complex mixtures of macromolecules and even small molecules that cannot be fully characterized) (Borchard 2014). In other words, changes in pharmacokinetics, biodistribution and intracellular traffic of API (as being part of NBCDs) is achieved without modifying API's chemical structure but rationally designing the nano-object's structure (Couvreur 2013).

In theory, nanomedicines may protect carried APIs against oxidation and/or hydrolysis and enhance their intracellular delivery. The most succeeding uses of nanomedicines result from their convective extravasation and passive accumulation, in a site selective fashion, at sites with increased vascular permeability, as occurs in certain solid tumours fed by a dense and leaky neovasculature. The combination of highly permeable vessels that allows nanoparticles to site specifically extravasate by convection, with blocked lymphatic drainage usually associated to solid tumours, is known as 'enhanced permeation and retention effect' (EPR effect). The EPR effect is responsible for the passive targeting—of circulating nanomedicines, that can remain site-specifically accumulated for months (Maeda et al. 2000). The permeability of vessels irrigating inflamed infected sites is increased as well, and although the lymph drainage is not impaired, a certain degree of site specific or passive targeting and potential accumulation of circulating nanoparticles is also expected (Maeda 2012). Once in the neighbourhood of target cells, nanomedicines may slowly release the carried API, that would enter the cells in its free form, as for doxorubicinin Doxil (Barenholz 2012). Alternatively, the entire nanomedicine (nanoparticle + API) may be endocytically taken up by the target cell, in this case enabling a massive intracellular API delivery. Numerous preclinical strategies against trypanosomatides rely on this approach (Morilla et al. 2005, Montanari et al. 2010, Hernández et al. 2012, Perez et al. 2014, 2016). Most importantly, it is to be noted that the access of nanomedicines to deep organs such as liver or spleen, different to skin or mucosa, is only achieved by nanoparticulate material distribution upon endovenous route administration. A topical route enables the access of nanomedicines to the skin, eventually, to the dermis as the deeper target; the oral route enables accessing the gastrointestinal mucosa; the inhalatory route to the epithelial cells of respiratory tract. None of these routes grants the access of nanomedicines to systemic circulation. If administered by topical, oral and mucosa route, nanomedicines may only act as depots at the site of administration from where free APIs are released to enter blood circulation. Clearly in that case the selective tissue targeting and potential intracellular delivery provided by nanomedicines, is lost. It is the endocytic uptake of nanomedicines the only actual mechanism ensuring a massive intracellular delivery of APIs. The energy dependent endocytic uptake replaces the diffusion-driven by a concentration gradient-cell entry of API's. In this way, higher intracellular concentrations of API's may be achieved without increasing its dose. In the nanomedical approach, the performance of APIs can be radically altered with no need of chemical/enzymatic synthesis. Nanomedicines on the other hand, may be used to solve the classical problems of low solubility/permeability/bioavailability of APIs. The selective targeting (only available after intravenous administration) is responsible for lower toxicity and may account for a higher efficacy. Because of these reasons, nanomedicines could speed the search for therapeutic alternatives to eradicate the intracellular form of trypanosomatides.

Chagas Disease: Main Features, Current Treatment, and Unmet Needs

Etiological agent: the etiological agent of the American trypanosomiasis or Chagas disease is the *Trypanosoma cruzi*, a protozoan of the Sarcomastigophora phylum, Mastigophora subphylum, Kinetoplastida order, and Trypanosomatidae family. *T. cruzi* has a flagellum and its single mitochondrion contains the kinetoplast, an extranuclear DNA network corresponding to the parasite's mitochondrial genome, which is localized near the flagellate's basal body.

Life cycle: the life cycle of *T. cruzi* is complex, with several developmental forms in insect vectors and mammalian hosts (Rassi et al. 2010). The parasite *T. cruzi* is commonly transmitted to humans and other mammals by an insect vector. The vectors (triatomine bugs, also called reduviid bugs, 'kissing' bugs, cone-nosed bugs, and blood suckers) can live indoors, in cracks and holes of substandard housing, or in a variety of outdoor settings and usually become infected by sucking blood from animals or human beings who have circulating parasites (trypomastigote forms). In the digestive tract of triatomines, the trypomastigotes differentiate into epimastigotes (multiplicative form) and then to metacyclic trypomastigotes in the final portion of the intestine. Infection of mammals occurs when they are exposed to the infective metacyclic forms of the parasite that are eliminated with the faeces of triatomines after feeding. This contact occurs through the mucosa or through injury. Once in the vertebrate host, the metacyclic trypomastigotes invade the local reticuloendothelial and connective cells, and differentiate into amastigotes that begin replicating by binary fission. When the cell is swollen with amastigotes, they transform back into trypomastigotes by growing flagellate. The trypomastigotes lyse the cells, invade adjacent tissues, and spread via the lymphatics and bloodstream to distant sites, mainly muscle cells (cardiac, smooth and skeletal) and ganglion cells, where they undergo further cycles of intracellular multiplication. The cycle of transmission is completed when circulating trypomastigotes are taken up in blood meals by vectors (Rassi et al. 2010). This disease can also be transmitted by blood transfusion, organ transplantation, from a mother to her foetus and by ingestion of food contaminated with the parasites (Benchimol Barbosa 2006, Bern et al. 2007).

Pathogenesis: Chagas disease occurs in two phases: acute and chronic. Acute phase usually is clinically manifested in children younger than 12 years. Initial infection at the site of parasite entry is characterized by the presence of infective trypomastigotes in leukocytes and cells of subcutaneous tissues, and by the development of interstitial edema, lymphocytic infiltration, and reactive hyperplasia of adjacent lymph nodes. After dissemination through the lymphatic system and the bloodstream, parasites concentrate mainly in the muscles (including the myocardium) and ganglion cells. The characteristic pseudocysts that are present in some tissues are intracellular aggregates of multiplying forms (amastigotes). The clinical manifestations of the acute phase occur in < 1% of patients, and include acute myocarditis, pericardial effusion and/or meningoencephalitis (Prata 2001, Bern et al. 2011). After the acute phase, patients enter an indeterminate phase defined by the presence of infection and confirmed by either serological or parasitological tests; or reported to have chronic Chagas disease with no detectable pathology (Mitelman 2011). One-third of these patients will develop chronic symptomatic Chagas disease 2 decades after the initial infection (Coura and Borges-Pereira 2010). Chagas disease is the most severe parasitic infection of the heart (Hidron et al. 2010) and the heart is the organ most often affected in individuals with chronic *T. cruzi* infection (Rassi et al. 2000). Rather than autoimmune mechanisms, myocardial damage in chronic *T. cruzi* infection is due to the persistence of parasites and the accompanying chronic inflammation. Cardiac denervation (mainly parasympathetic), and abnormalities in the coronary microvasculature might also contribute to the pathogenesis of chronic lesions (Marin-Neto et al. 2007, Rassi et al. 2012). Approximately one third of patients can develop dilation of the gastrointestinal tract (megacolon, megaesophagus, megastomach, megaduodenum, megajejunum, megagallbladder, megacholedochus) and gastrointestinal motor disorders, such as achalasia of the cardia, disturbances of gastric emptying, altered intestinal transit and colon and gall bladder motor disorders (Köberle 1968). Chagas disease is known to cause both central nervous system and enteric nervous system injury. The chronic gastrointestinal manifestations of Chagas disease are mainly a result of enteric nervous system injury caused by *T. cruzi* infection (Matsuda et al. 2009).

Epidemiology and geographic distribution: *T. cruzi* is restricted to South America, Central America, and parts of North America (Mexico and Southern United States). Historically transmission and morbidity were concentrated in this region, but migration has brought chronic infected individuals to cities both in and outside of Latin America, making Chagas disease a public health problem of global concern. According to the Pan American Health Organization (PAHO) 10,000 to 14,000 deaths per year are caused by Chagas disease (Salvatella 2006). Two remaining countries with prevalence of about 1% (Brazil and Mexico), together with Argentina, are home to almost 60% of all people infected with *T. cruzi* in Latin America (Rassi et al. 2010). Last WHO data based on 2010 indicators, show that in the 21 Latin America countries, almost 6 million people are infected with *T. cruzi*, two thirds of them living in the South Cone (Chagas disease in Latin America 2015). Overall, 13% of the entire population in Latin America is at risk of acquiring Chagas's disease (Molina et al. 2016). The United States ranks seventh worldwide for the total number of people infected with *T. cruzi*: in 2009 an estimated 300,167 infected people lived in the United States (Bern and Montgomery 2009). Moreover, many individuals with Chagas disease have emigrated from Latin America to countries other than the United States. But by far the largest population of these infected immigrants live in Spain (47,000–67,000), with most originating from Ecuador, Argentina, Bolivia and Peru.

In Latin American countries, Chagas disease is one of the costliest NTDs (Rassi et al. 2010). The early mortality and substantial disability caused by this disease, which often occurs in the most productive population, young adults, results in a significant economic loss (Franco-Paredes et al. 2007). A recent study estimates global costs of US$7.19 billion/year, exceeding many prominent diseases globally, such as cervical cancer and rotaviruses (Lee et al. 2013). More than 10% of these costs emanate from the United States and Canada, where Chagas disease is not endemic and is not recognized as a significant health problem (Pereira Nunes et al. 2013).

Current treatments and unmet needs: Only two drugs discovered in the late 1960s and early 1970s, the 5-nitrofuran nifurtimox (NFX; Lampit™, Bayer HealthCare AG, Leverkusen, Germany) and the 2-nitroimidazole benznidazole (BNZ; now produced by LAFEPE as Benznidazol LAFEPE® in Brazil and by Maprimed/ELEA in Argentina as Abarax®) (Grunberg et al. 1967) are available for specific antiparasitic treatment. Neither BNZ nor NFX are approved by the US FDA and both are contraindicated in pregnancy. Both have significant activity in the acute phase (up to 80% of parasitological cures, defined as negativization of parasitological and serological tests), with efficacy varying according to the geographical area (Rodriques Coura and de Castro 2002, Urbina 2010). BNZ reduces the severity of the associated inflammatory processes of chronic patients, probably by reducing the parasite loads in infected tissues. Instead, NFX displays potent and specific toxicity on heart and pancreas of rats while BNZ is devoid of such activities, suggesting that the latter may pose a lesser risk to Chagas disease patients with cardiac compromise.

Despite its toxicity, currently all patients are recommended to take BNZ in the chronic phase. However, recent double blind randomized clinical trials, where the effect of BNZ was compared against placebo on patients affected with chagasic cardiomyopathy (BENEFIT), or with mild cardiac compromise (TRAENA) have been discouraging (Riarte 2013, Morillo et al. 2015).

Recent guidelines posted by the Drugs for Neglected Diseases initiative (DNDi) recommend to speed the finding of new antichagasic drugs for the chronic phase (DNDi 2016). The new drugs should be intended for oral route, in a dosing regimen comparable to systemic antifungals, ideally once daily for 30 days, with a safety superior to BNZ and efficacy not inferior or ideally superior to BNZ; ideally against chronic and acute reactivation caused by all parasite strains. However, one of the main difficulties in determining the effect of any antichagasic drug is evaluating the cure of treated patients. The only accepted criteria are negativization of serologies. In patients affected by chronic Chagas disease the lapsed time to the negativization may be up to 20 years, complicating the patients monitoring and making difficult the evaluation of new drugs in clinical trials (WHO 2002a).

Despite such difficulties, the search for new antichagasic drugs is predicted not to be difficult: different to leishmaniasis, there is no evidence of induced drug resistance (although natural resistance to these drugs exists), and the ecology of the infection makes the likelihood of drug resistance to be seemingly low (humans with chronic infections—who would be the primary target of drug treatment —are mostly poor transmitters to insect vectors). Consequently, a single, highly effective drug is likely to last a long time—if not forever (Tarleton 2016). Nonetheless, drugs in development such as cysteine protease (Cruzipain) inhibitors, inhibitors of trypanothione synthesis and metabolism and inhibitors of purine salvage extensively discussed in (Barrett and Croft 2012), are still in preclinical tests. This means the medium time remaining to reach the market is 10 years at best (Paul et al. 2010). Moreover, new triazols registered as antifungal against invasive fungal infections that inhibit the ergosterol biosynthesis by potent and selective inhibition of fungal and protozoan cytochrome P-450-dependent C14α sterol demethylase (CYP51), posaconazole and ravuconazole, have shown excellent preclinical antichagasic activity. However, a set of recent clinical trials (ClinicalTrials.goviii identifiers NCT01489228, NCT01377480, and NCT01162967), failed in finding any advantage in their combined use with BNZ (Bustamante and Tarleton 2014, Molina et al. 2014, Khare et al. 2015, Morilla and Romero 2015).

Up to now unfortunately, the antichagasic activity of BNZ has not been surpassed by any other known drug (Morilla and Romero 2015); searching for novel sources of APIs displaying anti-*T. cruzi* activity is an urgent need. Not surprisingly, the wide spectrum of antimicrobial activity of EO described in folk medicine, includes an important anti-*T. cruzi* activity. The anti-*T. cruzi* activity of EOs or isolated components within different nano-drug delivery systems will be now described.

Isolated EOs components as anti-*T. cruzi* nanomedicines

Up to the moment, the only component isolated from EOs, loaded in nano-drug delivery systems and tested as anti-chagasic agent is the lychnopholide, a sesquiterpene lactone isolated from *Lychnophora trichocarpha* (Table 13.1). The *in vitro* anti-*T. cruzi* activity of lychnopholide was first reported by Oliveira et al. 1996. The therapeutic application of lychnopholide however, is limited due to its poor aqueous solubility, high lipophilicity and chemical instability in alkaline media that hamper its oral administration (Branquinho et al. 2012). The *in vivo* anti-chagasic activity of lychnopholide, loaded into nanocapsules was tested on the acute and chronic phase of Chagas disease infection, upon endovenous (Branquinho et al. 2014) and oral administration (de Mello et al. 2016).

In the first work, the effect of lychnopholide within polymeric nanocapsules was determined in mice acutely infected with two *T. cruzi* strains having different susceptibility patterns to benznidazole (Branquinho et al. 2014). Polymeric nanocapsules consisted of an oil droplet that contains benznidazole surrounded by a polymeric shell stabilized by surfactants. In this work, two types of polymeric nanocapsules were used: plain poly-caprolactone nanocapsules, that are rapidly cleared from blood circulation by macrophages of the mononuclear phagocyte system, and pegylated nanocapsules, that because of the steric stabilization which reduces uptake by phagocytes, keeps in blood circulation for longer time. Animals infected with the CL strain (sensitive to benznidazole) and Y strain (partially resistant to benznidazole) were treated by daily endovenous injections starting 24 hours after infection for 10 days or starting on the 7th day (the prepatent period) for 20 days. Parasitological cure was defined as negativization of parasitological and serological tests (blood examination of number of parasites, hemoculture, PCR, and enzyme-linked immunosorbent assay).

Animals infected with the CL strain treated for 10 days exhibited 100% and 50% parasitological cure when treated with oral benznidazole and plain nanocapsules, respectively. On the other hand, all treatments (oral benznidazole, plain and pegylated nanocapsules) in the prepatent period for 20 days produced 100% cure.

In animals infected with the Y strain, only those treated with plain nanocapsules for 10 days were cured, but animals treated in the prepatent period for 20 days showed 100, 75, and 62.5% cure when treated with pegylated nanocapsules, oral benznidazole, and plain nanocapsules, respectively. Free lychnopholide reduced the parasitemia and improved mice survival, but no mice were cured. Lychnopholide-loaded nanocapsules showed higher cure rates, reduced parasitemia, and increased survival compared with free lychnopholide; and could be useful in the treatment of infection with

Table 13.1. Essential oils and components as anti-Trypanosomatides nanomedicines.

Trypano-somatide	Essential oil and their components	Assays performed	Nanoparticle type and properties	Reference
L. amazonen sis	*Curcuma longa* cortex hexanic extract	*In vitro* promastigotes	Multilamellar liposomes soybean phosphatidyl-choline:cholesterol:Tween 20 (10:1.3:0.15 mg/ml)	Amaral et al. 2014
L. tropica	*Nigella sativa* oil	*In vitro* promastigotes and intracellular amastigotes	TiO_2Ag nanoparticles 90 nm	Abamor et al. 2016
L. amazonen sis	*Pterodon pubescens* fruit extracts	*In vitro* promastigotes and intracellular amastigotes	Nanoemulsion 5% w/w of extract, 5% w/w soybean lecithin, 7.5%w/w poly-ethylene glycol (PEG)-40 hydrogenated castor oil/ sorbitanoleate 200 nm	Da Silva et al. 2016
L. donovani	Andrographolide	*In vivo* visceral leishmaniasis hamster model sc injection	Mannose-grafted multillamelar liposomes phosphatidylethanolamine: colesterol:dicetyl phosphate 7:1:1 molar ratio	Sinha et al. 2000
L. donovani	Andrographolide	*In vitro* intracellular amastigotes	PLGA (50:50) nanoparticles stabilized with 4% of polyvinyl alcohol 173 nm, –34.8 mV Z potential	Roy et al. 2010
L. donovani	Andrographolide	*In vitro* intracellular amastigotes	PLGA (50:50) nanoparticles stabilized by vitamin E TPGS 180 nm, –37.6 mV Z potential	Mondal et al. 2013
L. donovani	Artemisinin	*In vitro* intracellular amastigotes	PLGA nanoparticles 220 nm, 9 mV Z potential	Want et al. 2014
L. donovani	Artemisinin	*In vivo* visceral leishmaniasis murine model Intraperitoneal injection	PLGA nanoparticles	Want et al. 2015
L. donovani	Piperolactam A	*In vitro*	2-hydroxypropyl-cyclodextrin	Bhattacharya et al. 2016
T. cruzi	Lychnopholide	*In vivo* murine model of acute phase Chagas intravenous	Nanocapsules poly-caprolactone ~ 180 nm poly(lactic acid)-co-polyethylene glycol ~ 100 nm	Branquinho et al. 2014
T. cruzi	Lychnopholide	*In vivo* murine model of acute and chronic phase Chagas intravenous and oral	Nanocapsules	de Mello et al. 2016
T. evansi	Andiroba oil (*Carapa guaianensis*) and aroeira oil (*Schinusmolle*)	*In vitro*	Nanoemulsions span 80 as lipophilic surfactant and Tween 80 or Tween 20 as hydrophilic surfactant Andiroba oil: 240 nm, –55 mV Z potential Aroeira oil: 120 nm, –29 mV Z potential	Baldissera et al. 2013

Table 13.1 contd....

...Table 13.1 contd.

Trypano-somatide	Essential oil and their components	Assays performed	Nanoparticle type and properties	Reference
T. evansi	Tea tree oil - *Melaleuca alternifolia*	*In vitro* and *in vivo* tests oral	Solid lipid nanoparticles cetylpalmitate as solid lipid and polysorbate 80 as surfactant 290 nm, –14 mV Z potential	Baldissera et al. 2014a
T. evansi	*Achyrocline Satureioide* soil	*In vivo* *rats* *oral*	Nanocapsules poly-ε-caprolactone, sorbitan monostearate and polysorbate 80 260 nm, –30 mV Z potential	Baldisiera et al. 2014b
T. evansi	*Achyrocline Satureioide* soil	*In vivo* *rats* *oral*	Nanocapsules	Do Carmo et al. 2015
T. evansi	Curcumin	*In vitro* and *in vivo* *rats* *intraperitoneal*	Lipid-core nanocapsules poly (ε-caprolactone) and sorbitan monostearate, polysorbate 80 200 nm, –12 mV Z potential	Gressler et al. 2015
T. evansi	Nerolidol	*In vitro* and *in vivo* tests oral	Nanospheres Eudragit RS 100, sorbitan monooleate and polysorbate 80 150 nm, –12.8 mV Z potential	Baldissera et al. 2016a
T. evansi	α-Bisabolol	*In vitro* and *in vivo* tests oral	Solid lipid nanoparticles shea butter and sorbitan monooleate as solid core and polysorbate 80 as surfactant 190 nm, –7.7 mV Z potential	Baldissera et al. 2016b

benznidazole resistant strains. The authors suggest that the long-circulating property of pegylated nanocapsules improves lychnopholide efficacy as it was more effective in reaching the parasite *in vivo*.

More recently, the same group evaluated the oral treatment with lychnopholide-loaded nanocapsules in the acute phase of the disease and compared it with nanocapsules administered by the oral and intravenous routes during the chronic phase in animals infected with the Y strain of *T. cruzi* (de Mello et al. 2016).

The cure rates in the acute phase (treatment started at 4th day after infection up to 20 consecutive days) and chronic phase (treatment started at 90th day after infection up to 20 consecutive days) were 62.5 and 55.6%, respectively, upon oral administration of pegylated nanocapsules and 57.0 and 30.0%, respectively, with plain nanocapsules. These cure rates were significantly higher than that of free lychnopholide, which did not cure any animals. If well plain nanocapsules administered orally during the acute phase showed same cure rates than benznidazole, only plain nanocapsules cured mice in the chronic phase. Similar results were achieved with intravenous treatment during the chronic phase. The authors suggest that the higher cure rates obtained with pegylated nanocapsules may be due to the smaller particle size of these nanocapsules compared with plain ones (105 nm vs. 180 nm) and the presence of polyethyleneglycol (Peg), which influence tissue diffusion and the controlled release of lychnopholide. The authors suggest that pegylated nanocapsules may improve the stability of the drug in the gastrointestinal tract. In this case, the nanoparticles may only enhance the lychnopholide bioavailability, enabling its access to systemic circulation in free form. Besides, the oxidation products of terpenes, especially oxidated sesquiterpenes with lactone rings, terpenoids and other plant metabolites have been demonstrated by high allergenic activity, mainly potent contact reactions (Sköld et al. 2002, Hammer et al. 2006, Vigan 2010).

Leishmaniasis Main Features, Current Treatment, and Unmet Needs

Etiological agent: Leishmaniasis is a vector-borne disease that is caused by several species of obligating intra-macrophage protozoan parasite (Oryan et al. 2008, Shirian et al. 2013); it is caused by about 20 Leishmania species and is transmitted to humans by approximately 70 species of female sandflies belonging to the genus Phlebotomus (Old World) or Lutzomyia (New World) (Oryan et al. 2013, Shirian et al. 2014). It is estimated that 350 million people are at risk in 88 countries, with a global incidence of 1–1.5 million cases of cutaneous and 500,000 cases of visceral leishmaniasis. The key control measures mainly rely on early case detection and chemotherapy, which has been hampered by the toxicity of drugs, side-effects and by the emergence of drug resistance in parasites. Control of reservoir host and vector is difficult due to operational difficulties and frequent relapses in the host (Pinto et al. 2001).

Life cycle: The infected female sandflies inject the infective stage (i.e., pathogenic metacyclic promastigotes) from their proboscis during blood meals, into the subdermal layers of the skin. Promastigotes that reach the puncture wound are phagocytized by macrophages and other types of mononuclear phagocytic cells (neutrophils, dendritic cells and fibroblasts), but disease is propagated primarily within macrophages. Internalized promastigotes are delivered to the mature phagolysosome compartment of host macrophages where they differentiate to the non-motile amastigote stage, which multiply by simple division and proceed to infect other mononuclear phagocytic cells. Sandflies become infected by ingesting infected cells during blood meals. In sandflies, amastigotes transform into promastigotes, develop in the gut (in the hindgut for leishmanial organisms in the *Viannia* subgenus; in the midgut for organisms in the *Leishmania* subgenus), and migrate to the proboscis. *Leishmania* are stealthy pathogens that either avoid or actively suppress macrophage microbicidal processes (Naderer and McConville 2011, CDC leishmania).

Pathogenesis: Parasite species, the immune host response raised against the infection, and other factors affect whether the infection becomes symptomatic and whether cutaneous or visceral leishmaniasis results. For instance, parasites can persist long term in the host, even after apparent resolution of the disease; this may lead to long-term immunity to subsequent infection, or be the cause of re-occurring infections in immunocompromised individuals (Naderer and McConville 2011).

The clinical manifestation of the disease varies from self-limiting ulcerative skin lesions developing at the site of sand fly bite (localized cutaneous leishmaniasis), multiple non-ulcerative nodules (diffuse cutaneous leishmaniasis), destructive mucosal inflammation (mucosal leishmaniasis, known as espundia), and disseminated visceral infection (visceral leishmaniasis, commonly known as kala-azar) (WHO 1990). VL is responsible for significant morbidity and mortality in the developing world (Davies et al. 2003) and if untreated VL carries a mortality of 75–79% (Piscopo and Mallia 2006).

Epidemiology and geographic distribution: Leishmaniasis is a public health problem in more than 88 countries (Davies et al. 2003, Alidadi and Oryan 2014). The estimated world prevalence of all forms of the disease is 12 million, with 1.5–2 million added new cases annually of cutaneous, and 500000 cases of visceral leishmaniasis and about 50,000 deaths from the disease each year (Desjeux 2004, Oryan et al. 2007), a death toll that is surpassed between the parasitic diseases only by malaria (WHO 2002b). About 200 million people are at risk of VL in 70 countries with estimated annual incidence of 500,000 and 50,000 deaths each year (Desjeux 2004). More than 90% of VL cases occur in just six countries: Bangladesh, India, Nepal, Sudan, Ethiopia and Brazil. Cutaneous Leishmaniasis (CL) is endemic in the tropics and neotropics region with more than 70 countries worldwide (Reithinger et al. 2007); 90% cases of CL occur in Afghanistan, Algeria, Brazil, Pakistan, Peru, Saudi Arabia and Syria. The global number of cases has increased during the past decade because of inadequate vector or reservoir control and increased opportunistic infections with HIV/AIDS (Cruz et al. 2006). Mucosal leishmaniasis is usually limited to South America with few exceptions (Morsy et al. 1995).

Ninety percent of the cases occur in Brazil, Bolivia and Peru. Twenty percent of leishmaniasis patients in Brazil develop MCL (Reithinger et al. 2007).

The increase in worldwide incidence and prevalence of leishmaniasis is mostly attributed to several man-made risk factors. Generally, environmental conditions such as deforestation, socio-economic status, demographic and human behaviour such as urbanization and great migration, together with immunosuppression pose major risks for human leishmaniasis (Votýpka et al. 2012, Reveiz et al. 2013). The combination of these factors may increase the human exposure to infected sandflies (Assimina et al. 2008).

Current treatment and unmet needs: Early detection, diagnosis and treatment are crucial for individual patients and for the community. Untreated leishmaniasis patients are a reservoir for parasites and therefore provide disease transmission in anthroponotic leishmaniasis regions (Guerin et al. 2002). Pentavalent antimonials (Sb^v) such as intravenous- or intramuscular-injected sodium stibogluconate (Pentostam, GlaxoSmithKline, UK, or the generic product from Albert David, India) meglumine antimoniate (Glucantime, Aventis, France) and pentamidine have been the first-line drugs for human leishmaniasis in many countries of the world for more than 70 years (Ameen 2007, Croft 2008). These are toxic drugs with serious-life-threatening in some cases-side effects such as cardiac arrhythmia and acute pancreatitis. Patients under the age of two and age over 45 years with symptoms of advanced disease and with severe malnutrition are at higher risk of death during treatment with antimonial compounds owing to drug toxicity, slowness of drug action or a combination of these factors (Collin et al. 2004, Haldar et al. 2011). Later, miltefosine, paromomycin and liposomal amphotericin B, alone or in combination became the drug of choice in recent decades to prevent the emergence of resistance to antimonials (Croft et al. 2006). Antimonials are still applied in many poor countries.

Amphotericin B is currently first-line treatment for disease in some countries where antimonials failure rates are high (Sundar et al. 2000). This is, however, another highly toxic drug that is caused from fever, chills and rigour, to life-threatening side effects including hypokalemia and nephrotoxicity. In the first-dose of this drug, anaphylaxis is not uncommon. Moreover, amphotericin B is costly and requires a complicated regimen (15 slow infusions on alternate days). Liposomal amphotericin B on the other hand, is considered as the best existing drug against visceral leishmaniasis, and is used as the first-line choice in the United States and Europe. Until recently, its use in developing countries was prevented by its high market price (Bern et al. 2006).

Miltefosine is a teratogenic drug, of use strictly forbidden in pregnant women or in women who could become pregnant within two months of treatment. Miltefosine has a long half-life and parasite resistance is easily induced. Non-adherence to the recommended regimen could lead to prevalent parasite resistance. The increasing use of miltefosine in veterinary for canine leishmaniasis might also increase the development of miltefosine resistance (Dorlo et al. 2012).

An approach based combination therapy is proposed to increase efficacy of treatments, prevent the development of drug resistance against leishmaniasis, making treatments shorter and potentially cheaper (Oryan and Akbari 2016). As a complementary approach to this view, since several EOs and isolated components have previously shown leishmanicidal activity, the detailed trypanocidal activity of EOs within different nano-drug delivery systems will be described and discussed next.

EOs and EOs Components as Anti-leishmania Nanomedicines

Cutaneous Leishmaniasis. The effect of EOs loaded-nanoparticles against parasites causing cutaneous leishmaniasis has been tested *in vitro* only (Table 13.1).

In the first work, liposomes were loaded with *Curcuma longa* EOs (Amaral et al. 2014). The EOs of turmeric, whose main constituents are sesquiterpenoids like turmerones (Murakami et al. 2013) has shown anti-inflammatory, anticancer, antimicrobial and anti-leishmania activity (Araujo and Leon 2001). Amaral et al. (2014), showed that the main components of the hexane fraction of *Curcuma longa* obtained from the crude methanol extract of the turmeric cortex were β-turmerone

(17.5%), ar-tumerone (15.8%), α-zingiberene (13.6%), β-sesquiphellandrene (12.5%), curlone (7.2%), germacrone (6.8%) and dehydrocurdione (6.8%). The extract loaded in multilamellar liposomes showed higher anti-*L. amazonensis* promastigote activity than the free extract and ar-turmerone, as seen by the minimum inhibitor concentration (MIC) (5.5 μg/mL, 125 μg/mL and 50 μg/mL, respectively). After 48 hours of incubation, liposomes caused the complete inhibition of parasites growth at 2.75 μg/mL with half maximal inhibitory concentration (IC50)/48 hours = 0.4 μg/mL, while the IC50 of the free extract was nine-fold higher (35.4 μg/mL). In this preliminary work, however, liposomes were tested against the extracellular form of the parasite only.

In the second study, non-toxic concentrations of silver doped titanium dioxide nanoparticles were combined with commercially obtained *Nigella sativa* oil (Abamor et al. 2016). *N. sativa* is a therapeutic plant used in alternative medicine, and its EOs and major constituent thymoquinone has shown activity against *L. infantum* and *L. tropica* parasites (Mahmoudvand et al. 2015). On the other hand, TiO_2 and Ag nanoparticles show great antimicrobial activity against several bacteria, viruses, and fungi (Jesline et al. 2015). The complex between TiO_2 and Ag nanoparticles (TiO_2Ag nanoparticles, TiAg Nps) have shown higher antimicrobial activity when compared with Ag and TiO_2 nanoparticles alone (Gupta et al. 2013). TiAg Nps has shown antiproliferative activity against *L. infantum* and *L. tropica* promastigotes in a dose-dependent manner (Allahverdiyev et al. 2013). However, concentrations of TiAg Nps that are effective against *Leishmania* promastigotes are also toxic for host macrophages. On the other hand, nontoxic concentrations of nanoparticles are ineffective against promastigotes, while they partially killed *Leishmania* amastigotes. Several studies show that combination of metallic nanoparticles with antibiotics or other antimicrobial agents prevent their toxicity, while enhancing their bactericidal effects on various types of bacteria (Allahverdiyev et al. 2011). Abamor et al. 2016, showed that non-toxic concentrations of TiAg Nps and *N. sativa* oil alone did not show any significant effects on proliferation rates of promastigotes. However, the combination of TiAg Nps and *N. sativa* oil reduced the proliferation rates in a dose-dependent manner. Combination of 20 µg/ml TiAg Np with 50 µg/ml *N. sativa* oil produced the major reduction of proliferation of *L. tropica* promastigotes and produced nearly four-folds decreased in metabolic activity rates. This combination caused a nearly 20-folds decrease in the infection index compared to control, which indicates the clearance of intracellular amastigotes without causing any toxic effects on macrophages. Metabolic activity rates of amastigotes treated with TiAg Nps-*N. sativa* combination diminished nearly five-folds in contrast to groups treated with each agent alone. The combination increased nine-folds nitric oxide generation by infected macrophages. The authors suggest that TiAg Nps induced an oxidative stress by producing large amounts of reactive oxygen species which impairs functions of vital enzymes, while *N. sativa* oil disrupt cellular membranes leading to mitochondrial mediated apoptosis.

More recently, nanoemulsions of *Pterodon pubescence* fruit extract were prepared (da Silva et al. 2016). Pterodon fruits are rich in geranylgeraniol derivatives (about 60%), which have antileishmanial activity (Lopes et al. 2012). EOs and fractions of *P. emarginatus* fruit have shown antileishmanial effect (Dutra et al. 2009). However, the activity of *P. pubescens* fruit extracts has not been investigated. Nanoemulsions of the hexane and the supercritical extract were prepared to improve the solubility and the activity against *L. amazonensis* intracellular amastigotes and to reduce its cytotoxicity. Supercritical extract contained higher levels of geranylgeraniol derivatives (linear diterpenes) compared to the hexane extract that was richer in vouacapans (cyclic diterpenes). Although both extracts showed cytotoxicity, the supercritical extracts were more effective against *L. amazonensis* promastigotes and amastigotes than hexane extracts. This was attributed to the high content of the geranylgeraniol derivative in the supercritical extracts. The nanoemulsion was more selective against leishmania than to macrophages. The nanoemulsions showed better activity against parasites (IC50: 2.7 µg/mL for nanoemulsion of hexane extract; IC50: 1.9 µg/mL for nanoemulsion of supercritical extract) than miltefosine (0.7 µg/mL).

Visceral Leishmaniasis. Six works tested the *in vitro* and *in vivo* activity of andrographolide, artemisinin and piperolactam A against the parasites that causes visceral leishmaniasis (Table 13.1).

Andrographolide (AG), a diterpenoid lactone extracted from the leaves of the Indian medicinal plant *Andrographis paniculata*, has shown a strong *in vitro* antileishmanial activity (Sinha et al. 2000). However, AG is sparingly soluble in water, unstable in alkaline and acidic conditions and has a very short biological half-life ($t½$ = 2 hours).

The activity of AG within mannose-grafted multilamellar liposomes was initially tested in hamsters infected with *L. donovani* parasites (Sinha et al. 2000). Upon subcutaneous injection, every 3 days for a total of six doses, AG loaded liposomes caused 94% reduction of splenic parasitic burden. In comparison, plain liposomes and free AG caused 67 and 37% reduction of splenic parasitic burden, respectively. The authors explained the succeeding results to the increased uptake of the mannosylated liposomes by infected macrophages of the mononuclear phagocytic system through the mannosyl-fucosyl receptors (Banerjee et al. 1996). However, a subcutaneous administration of multilamellar liposomes does not favour another role than as a subcutaneous drug depot. No convincing explanations are presented about how the liposomes would be taken up by infected macrophages in the spleen. This work was discontinued and 10 years later, when the AG received renewed attention, it was loaded in 50:50 poly DL-lactide-co-glycolic acid (PLGA) nanoparticles and its *in vitro* activity against intracellular amastigotes was tested (Roy et al. 2010). PLGA nanoparticles showed a five-fold increase in anti-intracellular amastigote activity compared with free AG (IC50 28 μM vs. 140 μM) and eight-fold compared with Sb^V (IC50 246 μM). Later, the same group tested the effect of AG within PLGA nanoparticles stabilized with vitamin E TPGS (D-α-tocopherylpolyethyleneglycol 1000 succinate, a potent efflux pump inhibitor), on sensitive and drug resistant *L. donovani* amastigotes (Mondal et al. 2013). AG-loaded nanoparticles stabilized with TPGS showed higher antileishmanial activity that free AG and nanoparticles without TPGS. Selectivity index in wild-type and drug resistant strains, was found to be 10-fold and three-fold higher for nanoparticles with TPGS than free AG and nanoparticles without TPGS, respectively. Cytotoxicity of AG loaded nanoparticles on macrophages with and without TPGS was significantly less than amphotericin B, paromomycin and sodium stibogluconate. The higher activity of AG-loaded nanoparticles is owed to its higher uptake by infected macrophages as compared to the free AG.

Artemisinin, a sesquiterpenoid lactone, is a secondary metabolite of the Chinese herb *Artemisia annua*, which has shown potential in the treatment of experimental cutaneous and visceral leishmaniasis (Sen et al. 2010); however, its efficacy is compromised due to its low bioavailability and short half-life (Chen et al. 2009, Sen et al. 2010).

The activity of artemisinin loaded into PLGA nanoparticles was tested *in vitro* and *in vivo*. Artemisinin-loaded PLGA nanoparticles at non-cytotoxic concentration for macrophages, significantly reduced the number of amastigotes per macrophage and the percent of infected macrophages compared to free artemisinin (Want et al. 2014). Artemisinin-loaded PLGA nanoparticles, intraperitoneally administered in mice model of visceral leishmaniasis showed higher antileishmanial efficacy compared with free artemisinin (Want et al. 2015). Upon artemisinin loaded nanoparticles treatment at 20 mg/kg body weight, animals showed a significant reduction in hepatosplenomegaly as well as in parasite load in the liver (85%) and spleen (82%) compared to free artemisinin (70% in liver and 62.7% in spleen). The protection was associated with a Th1-biased immune response to the treatment, as evident from a positive delayed-type hypersensitivity reaction, the increase IgG2a levels, lymphoproliferation and enhancement of proinflammatory cytokines production (IFN-γ and IL-2) with a significant suppression of Th2 cytokines (IL-10 and IL-4) after *in vitro* re-stimulation, compared to infected control and free artemisinin treatment.

Roots of *Piper betle*, extensively used in traditional herbal preparations contain different bioactive secondary metabolites such as ursolic acid, diosgenin, sitosterol, stigmasterol, resorcinol and aristolactams (Lin et al. 2013). Piperolactam A, an aristolactam alkaloid isolated from the most potent ethyl acetate fraction, exhibited leishmanicidal activity against wild type strains of *L. donovani in vitro* (Silva et al. 2012). Piperolactam A, however, is poorly water soluble. Bhattacharya et al. (2016) prepared inclusion complex of piperolactam A with 2-hydroxypropyl-cyclodextrin to enhance its solubility and bioavailability. Free piperolactam A showed high anti-axenic amastigote activities in wild-type (IC50 36 μM), sodium stibogluconate resistant (IC50 103 μM), paromomycin resistant (IC50 91 μM) and field isolated resistant (IC50 72 μM) strains, together with low cytotoxicity (CC50 900 μM) in macrophages. Inclusion of piperolactam A in cyclodextrin resulted in 10-fold and 4–10-fold increase in selectivity indexes (CC50/IC50) for wild-type and drug resistant strains, respectively.

Trypanosomiasis affecting animals

Trypanosoma evansi is an hemoflagellated protozoan belongs to the genus Trypanosoma, subgenus Trypanozoon (salivarian section) of veterinary importance of worldwide distribution. It can affect a very large range of domestic and wild hosts including camelids, equines, cattle, buffaloes, sheep, goats, pigs, dogs and other carnivores, deer, gazelles and elephants (Desquesnes et al. 2013). In Latin America *T. evansi* found a new large range of wild and domestic hosts, including reservoirs (capybaras) and biological vectors (vampire bats simultaneously a host, reservoir, and vector of *T. evansi*). *T. evansi* causes surra, a major disease in camels, equines and dogs, in which it can be fatal in the absence of treatment, and exhibits nonspecific clinical signs (anaemia, loss of weight, abortion, and death), which are variable from one host and one place to another. The word 'surra' comes from the Hindi and means 'rotten', which qualifies the state of the animals after chronic evolution of the disease (Vittoz 1955); this especially fits to the evolution of the disease in camels.

Therapy is based on several drugs such as suramine, diminazeme aceturate, quinapyramine, melarsoprol, homidium chloride and isometamidium chloride, however, some cases of resistance of parasites to these drugs have been reported (Brun et al. 1998, Maudlin et al. 2004). Diminazene aceturate is the most common drug used in the treatment of domestic animals in Brazil, since it has a higher therapeutic index and low costs compared with other drugs. Diminazene aceturate at a single oral dose (3.5 mg/kg) is capable to eliminate the parasites in the bloodstream a few hours after administration. However, it has no curative efficacy since trypanosomes may pass through the blood-brain barrier, reaching the central nervous system, where it refuges. Diminazene aceturate does not cross the blood-brain barrier in amounts sufficient to eliminate all the parasites (Masocha et al. 2007). Besides, diminazene causes hepatotoxicity and nephrotoxicity (Silva et al. 2002).

There are six studies in which the authors evaluated the anti-*T. evansi* activity of different nano-drug delivery systems loaded with andiroba and aroeira EOs, tea tree oil EO; achyrocline satureioides oil, curcumin, nerolidol and α-Bisabolol (Table 13.1).

Andiroba oil is extracted from nuts of *Carapa guaianensis* plant. It has popular use in South America as an analgesic, anti-inflammatory, antibacterial, antiparasitic, anticancer and anti allergy (Penido et al. 2006, Santos et al. 2012). The aroeira oil (*Schinus molle* L.), on the other hand, is obtained from leaves of a plant belonging to the family Anacardiaceae. Its EOs is used in folk medicine due to its antimicrobial, antifungal, antipyretic, anti-inflammatory, as well as its healing activities (Marongiu et al. 2004).

The *in vitro* trypanocidal action of andiroba and aroeira oils in pure and as nanoemulsions was tested (Baldissera et al. 2013). Andiroba oil contains essential fatty acids, triterpenes, tannins and alkaloids; while aroeira oil contains mono and sesquiterpenes non-oxygenated. Nanoemulsion of andiroba and aroeira oils showed faster trypanocidal effect than the pure form of oils and aceturate diminazene.

Tea tree (TTO) EOs is derived from an Australian native plant known as *Melaleuca alternifolia*. Medicinal TTO is a small Myrtaceous tree with sub-dermal foliar oil glands containing EOs composed mainly by monoterpernes (Keszei et al. 2010). TTO has documented insecticidal, acaricidal, repellent (Iori et al. 2005, Callander and James 2012), antibacterial, antifungical, anti-inflammatory and analgesic properties (Hart et al. 2000, Carson et al. 2006).

In vivo anti-*T. evansi* activity of TTO loaded into Solid Lipid Nanoparticles (SLN) was tested (Baldissera et al. 2014a). Treatment efficacy was determined by the number of mice that did not show clinical signs of *T. evansi* infection: survival at the end of experiment, negative PCR and blood smears. The main components of TTO were terpinen-4-ol (42%), c-terpinene (20%), a-terpinene (10%), 1.8-cineole (6%) and terpinolene (4%). Oral treatment (three doses starting the day after infection) of *T. evansi* infected mice with TTO-SLN at 1 mL/kg extend animal survival, but had no curative efficacy. However, when free TTO was combined with diminazene aceturate 100% cure was observed, a much better result than the diminazene aceturate treatment alone (33.3%). Unfortunately, the combination of TTO-SLN with diminazene aceturate, was skipped.

Achyrocline satureioides is one of the most relevant species belonging to the family Asteraceae, commonly known as 'macela or marcela', and is used as a medicinal plant in Brazil and other nearby countries. Various medicinal properties have been attributed to macela, such as hepatoprotective, antioxidant, antimicrobial, antitumoural, antiviral and antimicrobial (Kadarian et al. 2002, Calvo et al. 2006, Sabini et al. 2012). *A. satureioides* have shown *in vitro* antiparasitic effect against *Giardia lamblia* (Brandelli et al. 2009) and *T. cruzi* (Rojas de Arias et al. 1995). *In vivo* activity of *A. satureioides* EOs loaded into poly-ε-caprolactone nanocapsules and its antioxidative effect in liver and kidney of rats infected with *T. evansi* was tested (Baldissera et al. 2014b). The most representative components in *A. satureioides* EOs were: α-pinene (36.19%), β-caryophyllene (25.65%), β-ocineme (7.36%), 1.8-cineole (5.93%) and γ-eudesmol (4.73%). Oral treatment started 2 hours after inoculation of trypomastigotes and kept once a day during the next 5 days. If however, the *A. satureioides* EOs nanocapsules were not able to completely eliminate the parasites from the bloodstream, they reduced the number of trypanosomes. Infected rats showed a significant increase in thiobarbituric acid reactive substances (TBARS, index of lipid peroxidation) levels in liver and kidney. Animals treated with both free and nanoencapsulated *A. satureioides* EOs, showed lower TBARS levels than the infected non-treated animals. Besides, both treatments reduced the histological damage in the liver samples. These results suggested that *A. satureioides* EOs protected the kidney and liver against free radicals induced by infection. Of the components of *A. satureioides* EOs, β-caryophyllene has demonstrated inhibition of lipid peroxidation and hepatoprotective activity, mainly due to restoring the values of plasma markers of liver damage, such as alanine aminotransferase (ALT), aspartate aminotransferase (AST), alkaline phosphatase and lactate dehydrogenase (LDH) in a carbontetrachloride-induced fibrosis murine model (Calleja et al. 2012).

Then, the *in vivo* activity of free and nanoencapsulated *A. Satureioides* Eos was compared with that of diminazene aceturate and hematological and biochemical parameters were studied in rats infected with *T. evansi* (do Carmo et al. 2015). Treatment with *A. satureioides* Eos nanocapsules avoided the sharp reduction in number of red blood cells, minimizing the anaemia caused by *T. evansi* infection. Treatment with free and nanoencapsulated *A. satureioides* Eos also produced a normalization in the number of total leucocytes, lymphocytes and monocytes of infected animals. Opposing, untreated infected rats displayed reduced number of leucocytes. On the other hand, infection with *T. evansi* causes high liver damage with an intense local inflammatory response, measured as increased of ALT levels and decreased serum albumin levels. Treatment with free *A. satureioides* EOs increased ALT and AST levels, demonstrating liver damage; however, nanoencapsulated EOs did not cause elevation of these enzymes. Finally, both treatments inhibited the increase in creatinine levels caused by infection for *T. evansi*. In summary, the nanoencapsulated *A. satureioides* EOs showed better activity on the trypanosome; it did not cause liver toxicity and prevented renal damage and minimized the effects of anaemia.

Curcumin, the active principle of turmeric, isolated from the rhizomes of *Curcuma longa* is a yellow-orange powder, with high antioxidant activity but low aqueous solubility, chemical instability and low bioavailability (Subramanian et al. 1994, Ireson et al. 2002). *In vitro* and *in vivo* activity of free and curcumin-loaded lipid-core nanocapsules was tested against *T. evansi* (Gressler et al. 2015). Intraperitoneal treatment started 12 hours post-infection and kept twice a day during 6 days. Both, free and nanoencapsulated curcumin, significantly reduced the parasitaemia. Infected untreated rats displayed increased protein peroxidation and nitric oxideserum levels, whereas these variables were reduced on curcumin treated animals. Besides, infected treated rats showed a reduction in ALT and creatinine levels.

Nerolidol (3,7,11-trimethyl-1,6,10-dodecatrien-3-ol), also known as peruviol, is an aliphatic sesquiterpene alcohol present in EOs of many plants. Nerolidol has shown antileishmanial (Arruda et al. 2005), antimalarial (Lopes et al. 1999), antischistosomal (Silva et al. 2014) and antitrypanosomal (Mohd-Shukri and Zainal-Abidin 2011) activities. *In vitro* and *in vivo* activity of nerolidol loaded into nanospheres was recently tested (Baldissera et al. 2016a). *In vitro*, free and nanoencapsulated nerolidol showed a faster trypanocidal effect compared to diminazene aceturate. *In vivo, T. evansi* infected mice were treated by oral administration of 1.0 mL/kg/day nerolidol-loaded nanospheres for 5 days before infection and 5 days after infection. Nerolidol-loaded nanospheres showed an increased in mice survival (66.66%) compared to mice treated with free nerolidol (0%) and diminazene aceturate (33.33%). Free and nanoencapsulated nerolidol showed no cytotoxicity *in vitro* against peripheral blood mononuclear cells, an important marker used in toxicological surveys.

α-Bisabolol ((–)-6-methyl-2-(4-methyl-3-cyclohexen-1-yl)5-heptein-2-ol), also known as levomenol, is an sesquiterpene alcohol present in EOs of several plants. α-Bisabolol has shown strong antileishmanial activity (Rottini et al. 2015, Corpas-López et al. 2015). α-Bisabolol was loaded into SLN and its *in vitro* and *in vivo* activity was tested (Baldisera et al. 2016b). *In vitro,* free and α-bisabolol-loaded SLN showed faster trypanocidal effect when compared to diminazene aceturate. *In vivo,* treatment with 1.0 mL/kg α-Bisabolol-SLN once a day by 7 days via oral gavage or diminazene aceturate were able to increase the pre-patent period and longevity when compared to untreated animals, but showed no curative efficacy. Combination of α-Bisabolol-SLN and diminazene aceturate, however showed a curative efficacy of 50%.

Conclusions

Undoubtedly, EOs are a source of natural resources having singular and powerful medicinal properties still underexploited. But, despite their extensive use of EO in folk medicine along thousands of years, the idea of counting on EOs based nanomedicines may be far from experiencing a succeeding bench to bedside pathway. There are several uncertainties that raise doubts on the short-term access of such approach to pharmaceutical market. The first is related to the regulatory status of EOs: since as per its intended use, the FDA has only approved EOs as cosmetics, food additives and in aromatherapy (FDA Cosmetics, products, ingredients). This means that from a regulatory point of view, no official records guarantee the use of EOs as therapeutic agents. It also means that reports accurate enough to endorse the therapeutic efficacy of EOs are still lacking. The second is related to their analytical heterogeneity: nanoparticles loaded with EOs (consisting of several dozens of different low molecular weight molecules and especially if volatile molecules at low concentration are present), instead of single API, will result in a new, very difficult to be structurally characterized NBCDs. On the other hand, EOs isolated components will lose the synergic property displayed by the original EOs; the loss must be compensated by the improved activity proportioned by the structure of the nanomedicine. This means that only those EOs based nanomedicines with significantly improved performance compared to conventional treatments, would be useful. The third is a problem related to the administration route: both in folk medicine and in FDA approved uses, EOs are administered by oral, topical or inhalatory routes. Up to now parenteral (intravenous, intramuscular, subcutaneous, peritoneal) administration of EOs has only been practiced in preclinical models. As previously discussed, intravenous administration is mandatory to take advantage of the strong changes in pharmacokinetics, biodistribution and even intracellular targeting offered by nanomedicines. The outcome of the first in human effect of intravenous EOs based nanomedicines therefore, is largely unpredictable. Finally, most of the research complied in this chapter has surveyed the trypanocidal activity of nanomedicines with *in vitro* assays. Such assays are not predictive of *in vivo* performance, because the anatomical barriers that could impair the physical access of nanoparticulate material to the target sites, are absent in cell free assays or bidimensional cell cultures. *In vivo*, most of the administration routes of nanomedicines compiled in this chapter are not intravenous. If nanomedicines cannot be distributed by blood circulation, their physical access to target sites is largely impaired. There are several results where no clear differences are found between free or within nanoparticles EOs, may be due to the use of a sub optimal administration route. In these situations, the nanoparticulate material can remain trapped, acting as a local depot, a small fraction accessing to blood circulation at best. In practice, almost no studies have determined the structural stability of nanomedicines upon administration in terms of EOs or components fraction associated to nanoparticles, neither their shelf life (especially for those carrying volatile components) was systematically determined. All these aspects, if not urgently addressed, may slow down or even impair their regulatory approval as therapeutic agents. To meet such goals, a substantial number of predictive assays, much more specific than current preclinical data, must be gathered. Finally, most of the research groups focused in developing therapeutic nanomedicines against trypanosomatides surveyed in this chapter, display an extensive expertise in isolation, purification and study or chemical and therapeutic properties of EOs, but not in designing nanomedical therapeutic strategies. Overall, a deeper insight in nanomedical knowhow is required to face the challenge of designing efficient EOs-based trypanocidal nanomedicines.

References

Abamor, E.S. and Allahverdiyev, A.M. 2016. A nanotechnology based new approach for chemotherapy of cutaneous leishmaniasis: TiO_2@Ag nanoparticles - *Nigella sativa* oil combinations. Exp. Parasitol. 66: 150e163.

Alidadi, S. and Oryan, A. 2014. Cutaneous leishmaniasis and the strategies for its prevention and control. Trop. Med. Surg. 2: e114.

Allahverdiyev, A.M., Kon, K.V., Abamor, E.S., Bagirova, M. and Rafailovich, M. 2011. Coping with antibiotic resistance: combining nanoparticles with antibiotics and other antimicrobial agents. Expert Rev. Anti Infect. Ther. 9(11): 1035e1052.

Allahverdiyev, A.M., Abamor, E.S., Bagirova, M., Baydar, S.Y., Ates, S.C., Kaya, F., Kaya, C. and Rafailovich, M. 2013. Investigation of antileishmanial activities of TiO_2@Ag nanoparticles on biological properties of *L. tropica* and *L. infantum* parasites, *in vitro*. Exp. Parasitol. 135(1): 55e63.

Amaral, A.C.F., Gomes, L.A., Rocha de, A., Silva, J., Ferreira, J.L.P., de, S., Ramos, A., do Socorro, M., Rosa, S., Vermelho, A.B. and Rodrigues, I.A. 2014. Liposomal formulation of turmerone-rich hexane fractions from *Curcuma longa* enhances their antileishmanial activity. BioMed. Research International 2014. http://dx.doi.org/10.1155/2014/694934.

Ameen, M. 2007. Cutaneous leishmaniasis: therapeutic strategies and future directions. Expert. Opin. Pharmacother. 8: 2689–2699.

Anand, P., Kunnumakkara, A.B., Newman, R.A. and Aggarwal, B.B. 2007. Bioavailability of curcumin: problems and promises. Mol. Pharm. 4(6): 807–818.

Araujo, C.A.C. and Leon, L.L. 2001. Biological activities of *Curcuma longa* L. Memorias do Instituto Oswaldo Cruz. 96: 723–728.

Arruda, D.C., D'Alexandri, F.L., Katzin, A.M. and Uliana, S.R.B. 2005. Antileishmanial activity of the terpene nerolidol. Ant. Agents Chem. 49: 1679e1787.

Assimina, Z., Charilaos, K. and Fotoula, B. 2008. Leishmaniasis: an overlooked public health concern. Health Sci. J. 2: 196–205.

Bakkali, F., Averbeck, S., Averbeck, D. and Idaomar, M. 2008. Biological effects of essential oils—a review. Food Chem. Toxicol. 46: 446–475.

Baldissera, M.D., Da Silva, A.S., Oliveira, C.B., Zimmermann, C.E.P., Vaucher, R.A., Santos, R.C.V., Rech, V.C., Tonin, A.A., Mattos, J.L., Koester, L., Santurio, J.M. and Monteiro, S.G. 2013. Trypanocidal activity of the essential oils in their conventional and nanoemulsion forms: *In vitro* tests. Exp. Parasitol. 134: 356–361.

Baldissera, M.D., Da Silva, A.S., Oliveira, C.B., Santos, R.C.V., Vaucher, R.A., Raffin, R.P., Gomes, P., Dambros, M.G.C., Miletti, L.C., Boligon, A.A., Athayde, M.L. and Monteiro, S.G. 2014a. Trypanocidal action of tea tree oil (Melaleuca alternifolia) against *Trypanosoma evansi in vitro* and *in vivo* used mice as experimental model. Exp. Parasitol. 141: 21–27.

Baldissera, M.D., Oliveira, C.B., Rech, V.C., Peres Rezer, J.F., Sagrillo, M.R., Alves, M.P., da Silva, A.P.T., Leal, D.B.R., Boligon, A.A., Athayde, M.L., da Silva, A.S., Mendes, R.E. and Monteiro, S.G. 2014b. Treatment with essential oil of *Achyrocline satureioides* in rats infected with *Trypanosoma evansi*: Relationship between protective effect and tissue damage. Pathology - Research and Practice 210: 1068–1074.

Baldissera, M.D., Grando, T.H., Souza, C.F., Cossetin, L.F., Sagrillo, M.R., Nascimento, K., da Silva, A.P.T., Dalla Lana, D.F., Da Silva, A.S., Stefani, L.M. and Monteiro, S.G. 2016a. Nerolidol nanospheres increases its trypanocidal efficacy against *Trypanosoma evansi*: New approach against diminazene aceturate resistance and toxicity. Exp. Parasitol. 166: 144e149a.

Baldissera, M.D., Grando, T.H., de Souza, C.F., Cossetin, L.F., da Silva, A.P.T., Giongo, J.L. and Monteiro, S.G. 2016b. A nanotechnology based new approach for *Trypanosoma evansi* chemotherapy: *in vitro* and *in vivo* trypanocidal effect of (–)-α-Bisabolol. Exp. Parasitol. doi: 10.1016/j.exppara.2016.09.018.

Banerjee, G., Nandi, G., Mahato, S.B., Pakrashi, A. and Basu, M.K. 1996. Drug delivery system: targeting of pentamidines to specic sites using sugar grafted liposomes. J. Antimicrob. Chemother. 38: 145–150.
Barenholz, Y. 2012. Doxil®-the first FDA-approved nano-drug: lessons learned. J. Control Release 160(2): 117–34.
Barrett, M.P. and Croft, S.L. 2012. Management of trypanosomiasis and leishmaniasis. Br. Med. Bull. 104: 175–196.
Bassolé, I.H.N. and Juliani, H.R. 2012. Essential oils in combination and their antimicrobial properties. Molecules 17: 3989–4006.
Benchimol Barbosa, P.R. 2006. The oral transmission of Chagas disease: an acute form of infection responsible for regional outbreaks. Int. J. Cardiol. 112: 132–3.
Bern, C., Adler-Moore, J., Berenguer, J., Boelaert, M., den Boer, M., Davidson, R.N., Figueras, C., Gradoni, L., Kafetzis, D.A., Ritmeijer, K., Rosenthal, E., Royce, C., Russo, R., Sundar, S. and Alvar, J. 2006. Liposomal amphotericin B for the treatment of visceral leishmaniasis. Clin. Infect. Dis. 43: 917–924.
Bern, C., Montgomery, S.P., Herwaldt, B.L., Rassi, A., Jr., Marin-Neto, J.A., Dantas, R.O., Maguire, J.H., Acquatella, H., Morillo, C., Kirchhoff, L.V., Gilman, R.H., Reyes, P.A., Salvatella, R. and Moore, A.C. 2007. Evaluation and treatment of Chagas disease in the United States: A systematic review. JAMA 298: 2171–81.
Bern, C. and Montgomery, S.P. 2009. An estimate of the burden of Chagas disease in the United States. Clin. Infect. Dis. 49: e52–4.
Bern, C., Martin, D.L. and Gilman, R.H. 2011. Acute and congenital Chagas disease. Adv. Parasitol. 75: 19–47.
Bhattacharya, P., Mondal, S., Basak, S., Das, P., Saha, A. and Bera, T. 2016. *In vitro* susceptibilities of wild and drug resistant *Leishmania donovani* amastigotes to piperolactam A loaded hydroxypropyl-β-cyclodextrin nanoparticles. Acta Trop. 158: 97–106.
Boire, N.A., Riedel, S. and Parrish, N.M. 2013. Essential oils and future antibiotics: new weapons against emerging 'Superbugs'? J. Anc. Dis. Prev. Rem. 1: 105.
Borchard, G. 2014. Complex molecules – current developments. GaBi J. 3(2): 54–55.
Brandelli, C.L.C., Giordani, R.B., De Carli, G.A. and Tasca, T. 2009. Indigenous traditional medicine: *in vitro* anti-giardial activity of plants used in the treatment of diarrhoea. Parasitol. Res. 104: 1345–1349.
Branquinho, R.T., Mosqueira, V.C., Kano, E.K., de Souza, J., Dorim, D.D., Saúde-Guimarães, D.A. and Lana, M. 2012. HPLC-DAD and UV spectrophotometry for the determination of lychnopholide in nanocapsule dosage form: validation and application to release kinetic study. J. Chromatogr. Sci. 52(1): 19–26.
Branquinho, R.T., Mosqueira, V.C., de Oliveira-Silva, J.C., Simões-Silva, M.R., Saúde-Guimarães, D.A. and de Lana, M. 2014. Sesquiterpene lactone in nanostructured parenteral dosage form is efficacious in experimental Chagas disease. Antimicrob. Agents Chemother 58(4): 2067–75.
Brun, R., Hecker, H. and Lun, Z. 1998. *Trypanosoma evansi* and *T. equiperdum*: distribuition, biology, treatment and phylogenetic relationship (a review). Vet. Parasitol. 79: 95–107.
Burt, S. 2004. Essential oils: their antibacterial properties and potential applications in foods—a review. Int. J. Food Microbiol. 94: 223–253.
Bustamante, J.M. and Tarleton, R.L. 2014. Potential new clinical therapies for Chagas disease. Expert Rev. Clin. Pharmacol. 7: 317–325.
Callander, J.T. and James, P.J. 2012. Insecticidal and repellent effects of tea tree (*Melaleuca alternifolia*) oil against *Lucilia cuprina*. Vet. Parasitol. 184: 271–278.
Calleja, M.A., Vieites, J.M., Meterdez, T.M., Torres, M.I., Faus, J., Gil, A. and Suárez, A. 2012. The antioxidant effect of β-caryophyllene protects rat liver from carbontetrachloride-induced fibrosis by inhibiting hepatic stellate cell activation. Br. J. Nutr. 109: 394–401.
Calvo, D., Cariddi, L.N., Grosso, M., Demo, M.S. and Maldonado, A.M. 2006. *Achyrocline satureioides* (LAM.) DC (Marcela): antimicrobial activity on *Staphylococcus* spp. and immunomodulating effects on human lymphocytes. Rev. Latinoam. Microbiol. 48: 247–255.
Carson, C.F., Hammer, K.A. and Riley, T.V. 2006. Melaleuca alternifolia (tea tree) oil: a review of antimicrobial and other medicinal properties. Clin. Microbiol. Rev. 19: 50–62.
CDC leishmania. https://www.cdc.gov/parasites/leishmaniasis/biology.html.
Chagas disease in Latin America: An epidemiological update based on 2010 estimates. 2015. Wkly Epidemiol. Rec. 2015; 90: 33–43.
Chen, Y., Lin, X., Park, H. and Greever, R. 2009. Study of artemisinin nanocapsules as anticancer drug delivery systems. Nanomedicine 5(3): 316–322.
Collin, S., Davidson, R., Ritmeijer, K., Keus, K., Melaku, Y., Kipngetich, S. and Davies, C. 2004. Conflict and kala-azar: determinants of adverse outcomes of kala-azar among patients in southern Sudan. Clin. Infect. Dis. 38: 612–619.
Corpas-López, V., Morillas-Márquez, F., Navarro-Moll, M.C., Merino-Espinosa, G., Díaz-Sáez, V. and Martín-Sánchez, J. 2015. (–) α-bisabolol, a promising oral compound for the Treatment of visceral leishmaniasis. J. Nat. Prod. 78: 1202–1207.
Coura, J.R. and Borges-Pereira, J. 2010. Chagas disease: 100 years after its discovery. A systemic review. Acta Trop. 115: 5–13.
Couvreur, P. 2013. Nanoparticles in drug delivery: past, present and future. Adv. Drug. Deliv. Rev. 65(1): 21–23.
Croft, S.L., Sundar, S. and Fairlamb, A.H. 2006. Drug resistance in leishmaniasis. Clin. Microbiol. Rev. 19: 111–126.
Croft, S.L. 2008. Kinetoplastida: new therapeutic strategies. Parasite 15: 522–7.
Cruz, I., Neito, J., Moreno, J., Canavate, C., Desjeux, P. and Alvar, J. 2006. Leishmania/HIV coinfections in the second decade. Indian J. Med. Res. 123: 357–88.

da Silva Santos, E., Garcia, F.P., Outuki, P.M., Hoscheid, J., Nunes de Goes, P.R., Cardozo-Filho, L., Nakamura, C.V. and Carvalho Cardoso, M.L. 2016. Optimization of extraction method and evaluation of antileishmanial activity of oil and nanoemulsions of *Pterodon pubescens benth*. Fruit extracts, Experimental Parasitol. doi:10.1016/j.exppara.2016.10.004.
Davies, C.R., Kaye, P., Croft, S.L. and Sundar, S. 2003. Leishmaniasis: new approaches to disease control. BMJ 326: 377–382.
de Mello, C.G., Branquinho, R.T., Oliveira, M.T., Milagre, M.M., Saúde-Guimarães, D.A., Mosqueira, V.C. and Lana, M.D. 2016. Efficacy of lychnopholide polymeric nanocapsules after oral and intravenous administration in murine experimental Chagas disease. Antimicrob. Agents Chemother 60(9): 5215–22.
Desjeux, P. 2004. Leishmaniasis: current situation and new perspectives. Comp. Immunol. Microbiol. Infect. Dis. 27: 305–318.
Desquesnes, M., Holzmuller, P., Lai, D.-H., Dargantes, A., Lun, Z.-R. and Jittaplapong, S. 2013. Trypanosoma evansi and Surra: a review and perspectives on origin, history, distribution, taxonomy, morphology, hosts, and pathogenic effects. BioMed. Research International 2013: 22.
do Carmo, G.M., Baldissera, M.D., Vaucher, R.A., Rech, V.C., Oliveira, C.B., Sagrillo, M.R., Boligon, A.A., Athayded, M.L., Alves, M.P., França, R.T., Lopes, S.T.A., Schwertz, C.I., Mendes, R.E., Monteiro, S.G. and Da Silva, A.S. 2015. Effect of the treatment with *Achyrocline satureioides* (free and nanocapsules essential oil) and diminazene aceturate on hematological and biochemical parameters in rats infected by *Trypanosoma evansi.* Exp. Parasitol. 149: 39–46.
Doctors without borders and Drugs for Neglected Disease initiative. http://www.doctorswithoutborders.org/sites/usa/files/MSF-DNDi-NTDs-Policy-Paper.pdf.
Dorlo, T.P.C., Balasegaram, M., Beijnen, J.H. and de Vrie, P.J. 2012. Miltefosine: a review of its pharmacology and therapeutic efficacy in the treatment of leishmaniasis. J. Antimicrob. Chemother. 67(11): 2576–97.
DNDi. 2016. Chagas http://www.dndi.org/wp-content/uploads/2016/10/Factsheet_2015_Chagas_disease.pdf.
Dutra, R.C., Braga, F.G., Coimbra, E.S., Silva, A.D. and Barbosa, N.R. 2009. Atividades antimicrobiana e leishmanicida das sementes de *Pterodon emarginatus Vogel*. Rev. Bras. Farmacogn 19: 429–435.
ETP Nanomedicine http://www.etp-nanomedicine.eu/public/about-nanomedicine/what-is-nanomedicine.
FDA Cosmetics, products, ingredients. http://www.fda.gov/Cosmetics/ProductsIngredients/Products/ucm127054.htm.
Franco-Paredes, C., Von, A., Hidron, A., Rodríguez-Morales, A.J., Tellez, I., Barragán, M., Jones, D., Náquira, C.G. and Mendez, J. 2007. Chagas disease: an impediment in achieving the Millennium Development Goals in Latin America. BMC Int. Health Hum. Rights 7: 7.
Gressler, L.T., Oliveira, C.B., Coradini, K., Dalla Rosa, L., Grando, T.H., Baldissera, M.D., Zimmermann, C.E., Da Silva, A.S., Almeida, T.C., Hermes, C.L., Wolkmer, P., Silva, C.B., Moreira, K.L., Beck, R.C., Moresco, R.N., Da Veiga, M.L., Stefani, L.M. and Monteiro, S.G. 2015. Trypanocidal activity of free and nanoencapsulated curcumin on *Trypanosoma evansi*. Parasitology 142(3): 439–48.
Grunberg, E., Beskid, G., Cleeland, R., De Lorenzo, W.F., Titsworth, E., Scholer, H.J., Richle, R. and Brener, Z. 1967. Antiprotozoan and antibacterial activity of 2-nitroimidazole derivatives. Antimicrob. Agents Chemother. 7: 513–9.
Guenther, E. 1972. The Essential Oils, Krieger Publishing Company, Malabar, Fla, USA.
Guerin, P.J., Olliaro, P., Sundar, S., Boelaert, M., Croft, S.L., Desjeux, P., Wasunna, M.K. and Bryceson, A.D. 2002. Visceral leishmaniasis: current status of control, diagnosis, and treatment, and a proposed research and development agenda. Lancet Infect. Dis. 2: 494–501.
Gupta, K., Singh, R.P., Pandey, A. and Pandey, A. 2013. Photocatalytic antibacterial performance of TiO_2 and Ag-doped TiO_2 against *S. aureus*, *P. aeruginosa* and *E. coli*. Beilstein J. Nanotechnol. 4: 345e351.
Gurib-Fakim, A. 2006. Medicinal plants: Traditions of yesterday and drugs of tomorrow. Mol. Aspects Med. 27: 1–93.
Haldar, A.K., Sen, P. and Roy, S. 2011. Use of antimony in the treatment of leishmaniasis: current status and future directions. Molecular Biology International 2011: 23.
Hammer, K.A., Carson, C.F., Riley, T.V. and Nielsen, J.B. 2006. A review of the toxicity of *Melaleuca alternifolia* (tea tree) oil. Food Chem. Toxicol. 44: 616–625.
Handbook of essential oil 2012. http://ttngmai.files.wordpress.com/2012/09/handbookofessentionaloil.pdf.
Hart, P.H., Brand, C., Carson, C.F., Riley, T.V., Prager, R.H. and Finlay-Jones, J.J. 2000. Terpin-4-ol, the main component of the essential oil of Melaleuca alternifolia (tea tree oil), suppresses inflammatory mediator production by activated human monocytes. Inflamm. Res. 49: 619–626.
Hernández, I.P., Montanari, J., Valdivieso, W., Morilla, M.J., Romero, E.L. and Escobar, P. 2012. *In vitro* phototoxicity of ultradeformable liposomes containing chloroaluminum phthalocyanine against New World Leishmania species. J. Photochem. Photobiol. B. 117C: 157–163.
Hidron, A., Vogenthaler, N., Santos-Preciado, J.I., Rodriguez-Morales, A.J., Franco-Paredes, C. and Rassi, A. Jr. 2010. Cardiac involvement with parasitic infections. Clin. Microbiol. Rev. 23: 324–49.
Hotez, P.J., Molyneux, D.H., Fenwick, A., Ottesen, E., Ehrlich Sachs, S. and Sachs, J.D. 2006. Incorporating a rapid-impact package for neglected tropical diseases with programs for HIV/AIDS, tuberculosis, and malaria. PLoS Med. 3: e102.
Hotez, P.J., Bottazzi, M.E., Franco-Paredes, C., Ault, S. and Periago, M.R. 2008. The neglected tropical diseases of Latin America and the Caribbean: a review of disease burden and distribution and a roadmap for control and elimination. PLoS Negl. Trop. Dis. 2: e300.
Hotez, P.J., Fenwick, A., Savioli, L. and Molyneux, D.H. 2009. Rescuing the bottom billion through control of neglected tropical diseases. Lancet 373: 1570–1575.
Hotez, P.J., Dumonteil, E., Woc-Colburn, L., Serpa, J.A., Bezek, S., Edwards, M.S., Hallmark, C.J., Musselwhite, L.W., Flink, B.J. and Bottazzi, M.E. 2012. Chagas disease: the new HIV/AIDS of the Americas. PLoS Negl. Trop. Dis. 6(5): e1498.

Iori, A., Grazioli, D., Gentile, E., Marano, G. and Salvatore, G. 2005. Acaricidal properties of the essential oil of Melaleuca alternifolia Cheel (tea tree oil) against nymphs of Ixodes ricinus. Vet. Parasitol. 129: 173–176.

Ireson, C.R., Jones, D.J., Orr, S., Coughtrie, M.W., Boocock, D.J., Williams, M.L., Farmer, P.B., Steward, W.P. and Gescher, A.J. 2002. Metabolism of the cancer chemopreventive agent curcumin in human and rat intestine. Cancer Epidemiol. Biomarkers Prev. 11: 105–111.

ISO 9235: 1997. Aromatic natural raw materials—Vocabulary.

Jesline, A., John, N.P., Narayanan, P.M., Vani, C. and Murugan, S. 2015. Antimicrobial activity of zinc and titanium dioxide nanoparticles against biofilm-producing methicillin-resistant *Staphylococcus aureus*. Appl. Nanosci. 5(2): 157e162.

Kadarian, C., Broussalis, A.M., Miño, J., Lopez, P., Gorzalczany, S., Ferraro, G. and Acevedo, C. 2002. Hepatoprotective activity of *Achyrocline satureioides* (Lam.) DC. Pharmacol. Res. 45: 57–61.

Keszei, A., Brubaker, C.L. and Foley, W.J. 2010. A molecular perpective on a terpene variation in Australian Myrtaceae. Aust. J. Bot. 56: 197–213.

Khare, S., Liu, X., Stinson, M., Rivera, I., Groessl, T., Tuntland, T., Yeh, V., Wen, B., Molteni, V., Glynne, R. and Supek, F. 2015. Antitrypanosomal treatment with benznidazole is superior to posaconazole regimens in mouse models of Chagas disease. Antimicrob. Agents Chemother. 59: 6385–6394.

Köberle, F. 1968. Chagas' disease and Chagas' syndromes: the pathology of American trypanosomiasis. Adv. Parasitol. 6: 63–116.

Lee, B.Y., Bacon, K.M., Bottazzi, M.E. and Hotez, P.J. 2013. Global economic burden of Chagas disease: a computational simulation model. Lancet Infect. Dis. 13: 342–8.

Lin, C.F., Hwang, T.L., Chien, C.C., Tu, H.Y. and Lay, H.L. 2013. A new hydroxychavicol dimer from the roots of Piper betle. Molecules 18(3): 2563–70.

Lopes, M.V., Desoti, V.C., Caleare Ade, O., Ueda-Nakamura, T., Silva, S.O. and Nakamura, C.V. 2012. Mitochondria superoxide anion production contributes to geranylgeraniol-induced death in Leishmania amazonensis. Evid. Based Complement. Alternat. Med. 2012 http://dx.doi.org/10.1155/2012/298320.

Lopes, N.P., Kato, M.J., Andrade, E.H., Maia, J.G., Yoshida, M., Planchart, A.R. and Katzin, A.M. 1999. Antimalarial use of volatile oil from leaves of Vitola surinamensis (Rol.) Warb. By Waiapi Amazon Indians. J. Ethnoph. 67: 313e319.

Maeda, H., Wu, J., Sawa, T., Matsumura, Y. and Hori, K. 2000. Tumor vascular permeability and the EPR effect in macromolecular therapeutics: a review. J. Control Release 65(1-2): 271–84.

Maeda, H. 2012. Vascular permeability in cancer and infection as related to macromolecular drug delivery, with emphasis on the EPR effect for tumor-selective drug targeting. Proc. Jpn. Acad. Ser. B Phys. Biol. Sci. 88(3): 53–71.

Mahmoudvand, H., Tavakoli, R., Sharififar, F., Minaie, K., Ezatpour, B., Jahanbakhsh, S. and Sharifi, I. 2015. Leishmanicidal and cytotoxic activities of Nigella sativa and its active principle, thymoquinone. Pharm. Biol. 53(7): 1052e1057.

Marin-Neto, J.A., Cunha-Neto, E., Maciel, B.C. and Simões, M.V. 2007. Pathogenesis of chronic Chagas heart disease. Circulation 115: 1109–23.

Marongiu, B., Alessandra, P.S.P., Casu, R. and Pierucci, P. 2004. Chemical composition of the oil and supercritical CO_2 extracts of *Schinnus molle* L. Flavour Frag. J. 19: 554–558.

Masocha, W., Rottenberg, M.E. and Kristensson, K. 2007. Migration of African trypanosomes across the blood-brain barrier. Physiol. Behav. 92: 110e114.

Matsuda, N.M., Miller, S.M. and Barbosa, R.S. 2009. Evora Clinics (Sao Paulo). Dec. 64(12): 1219–1224.

Maudlin, I., Holmes, P. and Miles, M.A. 2004. The Trypanosomiases. CABI Publishing, 387 Wallingford, p. 1150.

Mitelman. 2011. Consenso de Enfermedad de Chagas-Mazza. Sociedad Argentina de Cardiología. Rev. Argent. Cardiol. 79: 544–64.

Mitragotri, S., Burke, P.A. and Langer, R. 2014. Overcoming the challenges in administering biopharmaceuticals: formulation and delivery strategies. Nat. Rev. Drug Discov. 13: 655–672.

Mohd-Shukri, H.B. and Zainal-Abidin, B.A.H. 2011. The effects of nerolidol, allicin and berenil on the morphology of Trypanosoma evansi in mice: a comparative study using light and electron microscopic approaches. Malays. Appl. Biol. J. 40: 25e32.

Molina, I., Gómez, J., Salvador, F., Treviño, B., Sulleiro, E., Serre, N., Pou, D., Roure, S., Cabezos, J., Valerio, L., Blanco-Grau, A., Sánchez-Montalvá, A., Vidal, X. and Pahissa, A. 2014. Randomized trial of posaconazole and benznidazole for chronic Chagas' disease. N. Engl. J. Med. 370: 1899–1908.

Molina, I., Salvador, F. and Sánchez-Montalvá, A. 2016. Actualización en enfermedad de Chagas. Enferm. Infecc. Microbiol. Clin. 34(2): 132–138.

Mondal, S., Roy, P., Das, S., Halder, A., Mukherjee, A. and Bera, T. 2013. *In vitro* susceptibilities of wild and drug resistant *leishmania donovani* amastigote stages to andrographolide nanoparticle: role of vitamin E derivative TPGS for nanoparticle efficacy. PLoS One. 8(12): e81492.

Montanari, J., Salomón, C., Esteva, M., Maidana, C., Morilla, M.J. and Romero, E.L. 2010. Sunlight triggered photodynamic ultradeformable liposomes against *Leishmania braziliensis* are also leishmanicidal in the dark. J. Control. Release 147(3): 368–76.

Morilla, M.J., Montanari, J., Frank, F., Malchiodi, E., Petray, P. and Romero, E.L. 2005. Etanidazole in pH-sensitive liposomes: design and *in vitro/in vivo* anti-*Trypanosome cruzi* activity. J. Control. Release 103: 599–607.

Morilla, M.J. and Romero, E.L. 2015. Nanomedicines against Chagas disease: an update on therapeutics, prophylaxis and diagnosis. Nanomedicine (Lond.) 10(3): 465–481.

Morillo, C.A., Marin-Neto, J.A., Avezum, A., Sosa-Estani, S., Rassi, A., Rosas, F. et al. 2015. Randomized trial of benznidazole for chronic Chagas' cardiomyopathy. N. Engl. J. Med. 373: 1295–306.

Morsy, T.A., Khalil, N.M., Salama, M.M., Hamdi, K.N., Shamrany, Y.A. and Abdalla, K.F. 1995. Mucosal leishmaniasis caused by Leishmania tropica in Saudi Arabia. J. Egypt. Soc. Parasitol. 25: 73–9.

Murakami, A., Furukawa, I., Miyamoto, S., Tanaka, T. and Ohigashi, H. 2013. Curcumin combined with turmerones, essential oil components of turmeric, abolishes inflammation-associated mouse colon carcinogenesis. BioFactors 39: 221–232.

Naderer, T. and McConville, M.J. 2011. Intracellular growth and pathogenesis of *Leishmania* parasites. Essays in Biochemistry 51: 81–95.

Oliveira, A.B., Saúde, D.A., Perry, K.S.P., Duarte, D.S., Raslan, D.S., Boaventura, M.A.D. and Chiari, E. 1996. Trypanocidal sesquiterpenes from *Lychnophora* species. Phytother. Res. 10: 292–295.

Oryan, A., Mehrabani, D., Owji, S.M., Motazedian, M.H. and Asgari, Q. 2007. Histopathologic and electron microscopic characterization of cutaneous leishmaniasis in Tatera indica and *Gerbillus* spp. infected with *Leishmania major*. Comp. Clin. Pathol. 16: 275–279.

Oryan, A., Mehrabani, D., Owji, S.M., Motazedian, M.H., Hatam, G.H. and Asgari, Q. 2008. Morphologic changes due to cutaneous leishmaniosis in BALB/c mice experimentally infected with *Leishmania major*. J. Appl. Anim. Res. 34: 87–92.

Oryan, A., Shirian, S., Tabandeh, M.R., Hatam, G.R., Randau, G. and Daneshbod, Y. 2013. Genetic diversity of Leishmania major strains isolated from different clinical forms of cutaneous leishmaniasis in southern Iran based on minicircle kDNA. Infect. Genet. Evol. 19: 226–231.

Oryan, A. and Akbari, M. 2016. Worldwide risk factors in leishmaniasis. Asian Pac. J. Trop. Med. 9(10): 925–932.

Paul, S.M., Mytelka, D.S., Dunwiddie, C.T., Persinger, C.C., Munos, B.H., Lindborg, S.R. and Schacht, A.L. 2010. How to improve R&D productivity: the pharmaceutical industry's grand challenge. Nat. Rev. Drug Discov. 9(3): 203–214.

Penido, C., Costa, K.A., Costa, M.F., Pereira, J.F., Siani, A.C. and Henriques, M.G. 2006. Inhibition of allergen-induced eosinophil recruitment by natural tetranortriterpenoids is mediated by the suppression of IL-5, CCL11/eotaxin and NF kappa B activation. Int. Immunopharmacol. 6: 109–121.

Pereira Nunes, M.C., Dones, W., Morillo, C.A., Encina, J.J. and Ribeiro, A.L. 2013. Chagas disease an overview of clinical and epidemiological aspects. J. Am. Coll. Cardiol. 62: 767–76.

Perez, A.P., Casasco, A., Defain Tesoriero, M.V., Pappalardo, J.S., Altube, M.J., Duempelmann, L., Higa, L., Morilla, M.J., Petray, P. and Romero, E.L. 2014. Enhanced photodynamic leishmanicidal activity of hydrophobic zinc phthalocyanine within archaeolipids containing liposomes. Int. J. Nanomed. 9: 3335–3345.

Perez, A.P., Altube, M.J., Schilrreff, P., Apezteguia, G., Santana Celes, F., Zacchino, S., Indiani de Oliveira, C., Romero, E.L. and Morilla, M.J. 2016. Topical amphotericin B in ultradeformable liposomes: formulation, skin penetration study, antifungal and antileishmanial activity *in vitro*. Colloids Surf. B: Biointerfaces 139: 190–198.

Pinto, M.C., Campbell-Lendrum, D.H., Lozovei, A.L., Teodoro, U. and Davies, C.R. 2001. Phlebotomine sandfly responses to carbon dioxide and human odour in the field. Med. Vet. Entomol. 15: 132–9.

Piscopo, T.V. and Mallia, A.C. 2006. Leishmaniasis. Postgrad. Med. J. 82: 649–57.

Prata, A. 2001. Clinical and epidemiological aspects of Chagas disease. Lancet Infect. Dis. 1: 92–100.

Rassi, A. Jr., Rassi, A. and Little, W.C. 2000. Chagas' heart disease. Clin. Cardiol. 23: 883–9.

Rassi, A. Jr., Rassi, A. and Marin-Neto, J.A. 2010. Chagas disease. Lancet 375: 1388–402.

Rassi, A. Jr., Rassi, A. and de Rezende, J.M. 2012. American Trypanosomiasis (Chagas Disease). Infect. Dis. Clin. N. Am. 26: 275–291.

Reithinger, R., Dujardin, J.L., Louzir, H., Pirmez, C., Alexander, B. and Brooker, S. 2007. Cutaneous leishmaniasis. Lancet Infect. Dis. 7: 581–96.

Reveiz, L., Maia-Elkhoury, A.N., Nicholls, R.S., Romero, G.A. and Yadon, Z.E. 2013. Interventions for American cutaneous and mucocutaneous leishmaniasis: a systematic review update. PLoS One. 8: e61843.

Riarte, A. 2013. Placebo-controlled evaluation of impact of benznidazole treatment on long-term disease progression in adults with chronic Chagas disease. *In*: 62nd Annual Meeting of the American Society of Tropical Medicine and Hygiene, Washington, DC, USA.

Rodriques Coura, J. and de Castro, S.L. 2002. A critical review on Chagas disease chemotherapy. Mem. Inst. Oswaldo. Cruz. 97: 3–24.

Rojas de Arias, A., Ferro, E., Inchausti, A., Ascurra, M., Acosta, N., Rodriguez, E. and Fournet, A. 1995. Mutagenicity, insecticidal and trypanocidal activity of some Paraguayan Asteraceae. J. Ethnopharmacol. 45: 35–41.

Rottini, M.M., Amaral, A.C.F., Ferreira, J.L.P., Silva, J.R.A., Taniwaki, N.N., de Souza, C.S.F., d'Escoffier, L.N., Almeida-Souza, F., Hardoim, D.J., da Costa, S.C.G. and Calabrese, K.S. 2015. *In vitro* evaluation of (–) α-bisabolol as a promising agent against *Leishmania amazonensis*. Exp. Parasitol. 148: 66–72.

Roy, P., Das, S., Bera, T., Mondol, S. and Mukherjee, A. 2010. Andrographolide nanoparticles in leishmaniasis: characterization and *in vitro* evaluations. Int. J. Nanomed. 5: 1113–1121.

Sabini, M.C., Escobar, F.M., Tonn, C.E., Zanon, S.M., Contiagini, M.S. and Sabini, L.I. 2012. Evaluation of antiviral activity of aqueous extracts from *Achyrocline satureioides* against Western equine encephalitis virus. Nat. Prod. Res. 26: 405–415.

Salvatella, R. 2006. Estimación cuantitativa de la enfermedad de Chagas en las Américas. Report no. OPS/HDM/CD/425-6. Montevideo (Uruguay): Organización Panamericana de la Salud.

Santos, R.C., dos Santos Alves, C.F., Schneider, T., Lopes, L.Q., Aurich, C., Giongo, J.L., Brandelli, A. and Vaucher, R.A. 2012. Antimicrobial activity of Amazonian oils against *Paenibacillus* species. J. Invertebrate Pathol. 109: 265–268.

Sell. 2010. Chemistry of essential oils. pp. 121–150. *In*: Baser, K.H. and Buchbauer, G.C. (eds.). Handbook of Essential Oils. Science. Technology, and Applications. CRC Press, Boca Raton, Fla, USA.
Sen, R., Ganguly, S., Saha, P. and Chatterjee, M. 2010. Efficacy of artemisinin in experimental visceral leishmaniasis. Int. J. Antimicrob. Agents 36: 43–49.
Sharma, R.A., Euden, S.A., Platton, S.L., Cooke, D.N., Shafayat, A., Hewitt, H.R., Marczylo, T.H., Morgan, B., Hemingway, D., Plummer, S.M., Pirmohamed, M., Gescher, A.J. and Steward, W.P. 2004. Phase I clinical trial of oral curcumin: biomarkers of systemic activity and compliance. Clin. Cancer Res. 10(20): 6847–6854.
Shirian, S., Oryan, A., Hatam, G.R. and Daneshbod, Y. 2013. Three *Leishmania*/L. species–*L. infantum*, *L. major*, *L. tropica*—as causative agents of mucosal leishmaniasis in Iran. Pathog. Glob. Health. 107: 267–272.
Shirian, S., Oryan, A., Hatam, G.R., Panahi, S. and Daneshbod, Y. 2014. Comparison of conventional, molecular, and immunohistochemical methods in diagnosis of typical and atypical cutaneous leishmaniasis. Arch. Pathol. Lab. Med. 138: 235–240.
Silva, F.M.A.D., Koolen, H.H.F., Lima, J.P.S.D., Santos, D.M.F., Jardim, I.S., Souza, A.D.L.D. and Pinheiro, M.L.B. 2012. Leishmanicidal activities of fractions rich in aporphine alkaloids from Amazonian Unonopsis species. Rev. Bras. Farmacogn. 22(6): 1368–1371.
Silva, M.P.N., Oliveira, G.L.S., de Carvalho, R.B.F., de Souza, D.P., Freitas, R.M., Pinto, P.L.S. and de Moraes, J. 2014. Antischistosomal activity of the terpene nerolidol. Molecules 19: 3793e3803.
Silva, R.A.M., Seidl, A., Ramirez, L. and D'avila, A.M.R. 2002. *Trypanosoma evansi* e *Trypanosoma vivax*: Biologia, Diagnostico e controle. Embrapa Pantanal, Corumb a, p. 137.
Sinha, J., Mukhopadhyay, S., Das, N. and Basu, M.K. 2000. Targeting of liposomal andrographolide to *L. donovani*-infected macrophages *in vivo*. Drug Del. 7: 209–213.
Sköld, M., Börje, A., Matur, M. and Karlberg, A.-T. 2002. Studies on the autoxidation and sensitizing capacity of the fragance chemical linalool, identifying a linalool hydroperoxide. Contact Dematitis 46: 267–272.
Stuart, K., Brun, R., Croft, S., Fair-lamb, A., Gürtler, R.E., McKerrow, J., Reed, S. and Tarleton, R. 2008. Kinetoplastids: related protozoan pathogens, different diseases. J. Clin. Invest. 118: 1301–1310.
Subramanian, M., Sreejayan Rao, M.N., Devasagayam, T.P. and Singh, B.B. 1994. Diminution of singlet oxygen-induced damage by curcumin and related antioxidants. Mutation Research 311: 249–255.
Sundar, S., More, D.K., Singh, M.K., Singh, V.P., Sharma, S., Makharia, A., Kumar, P.C. and Murray, H.W. 2000. Failure of pentavalent antimony in visceral leishmaniasis in India: report from the center of the Indian. Dis. 31: 1104–1107.
Tarleton, R.L. 2016. Chagas Disease: a solvable problem, ignored. Trends Mol. Med. 22: 10 835.
Turek, C. and Stintzing, F.C. 2013. Stability of essential oils: a review. Compr. Rev. Food. Sci. Food. Saf. 12: 40–53.
Urbina, J.A. 2010. Specific chemotherapy of Chagas disease: Relevance, current limitations, and new approaches. Acta Trop. 115: 55–68.
Vigan, M. 2010. Essential oils: renewal of interest and toxicity. Eur. J. Dermatol. 20: 685–692.
Vittoz, R. 1955. Prophylaxie du surra en Asie. Bulletin de l'Office International des Epizooties 44: 83–106.
Votýpka, J., Kasap, O.E., Volf, P., Kodym, P. and Alten, B. 2012. Risk factors for cutaneous leishmaniasis in Cukurova region, Turkey. Trans. R. Soc. Trop. Med. Hyg. 106: 186–190.
Want, M.Y., Islamuddin, M., Chouhan, G., Dasgupta, A.K., Chattopadhyay, A.P. and Afrin, F. 2014. A new approach for the delivery of artemisinin: formulation, characterization, and *ex vivo* antileishmanial studies. J. Colloid Interface Sci. 432: 258–69.
Want, M.Y., Islamuddin, M., Chouhan, G., Ozbak, H.A., Hemeg, H.A., Dasgupta, A.K., Chattopadhyay, A.P. and Afrin, F. 2015. Therapeutic efficacy of artemisinin-loaded nanoparticles in experimental visceral leishmaniasis. Colloids Surf. B Biointerfaces. 130: 215–21.
Wilkinson, S.R. and Kelly, J.M. 2009. Trypanocidal drugs: mechanisms, resistance and new targets. Expert Rev. Mol. Med. 11: e31.
WHO. 1990. World Health Organization technical report series; no. 793; 1990.
WHO. 2002a. World Health Organization Expert Committee. World Health Organ Tech. Rep. Ser. 2002; 905: 1–109.
WHO. 2002b. Urbanization: an increasing risk factor for leishmaniasis. Wkly Epidemiol. Rec. 2002; 77(44): 365.
WHO. Neglected diseases. http://www.who.int/neglected_diseases/en.

14

Combining Inorganic Antibacterial[#] Nanophases and Essential Oils

Recent Findings and Prospects

Mauro Pollini,[1] *Alessandro Sannino,*[1] *Federica Paladini,*[1,*]
Maria Chiara Sportelli,[2] *Rosaria Anna Picca,*[2] *Nicola Cioffi,*[2]
Giuseppe Fracchiolla[3] and *Antonio Valentini*[4]

Introduction

Nanotechnology, an important field of modern research deals with materials ranging from 1 to 100 nm in size, is gaining attention in a large number of fields such as health care, cosmetics, biomedical, food, environment, health, etc. (Ahmed et al. 2016a, Hussain et al. 2016). Nanotechnology is an immensely developing field due to its extensive range of applications in different areas of technology and science, and has been considered as a useful tool for solving biomedical problems (Ahmed et al. 2016b, Scandorieiro et al. 2016).

The high incidence of bacterial infections and the growing resistance of bacteria to conventional antibiotics, have posed serious concerns related to the need for novel antimicrobial agents. Bacterial antimicrobial resistance to most conventional antibiotics has become a clinical and public health problem, and the control of infections associated to multidrug-resistant microorganisms can lead to high treatment costs, therapeutic failure and death (Seil and Webster 2012, Scandorieiro et al. 2016). Some natural materials, such as zinc, silver and copper, possess remarkable antibacterial properties at nanometre regime, due to the increased surface to volume ratio (Seil and Webster 2012, Chatterjee et al. 2014). In recent years, noble metal nanoparticles have emerged in the field of biology, medicine and electronics and, among them, silver nanoparticles (AgNPs) have received particular attention for their great potential in antimicrobial activity, therapeutics, bio-molecular

[1] University of Salento, Department of Engineering for Innovation, Via per Monteroni, 73100 Lecce, Italy.
[2] University of Bari "Aldo Moro", Department of Chemistry, Via Orabona 4, 70126 Bari, Italy.
[3] Universityof Bari "Aldo Moro", Dipartimento di Farmacia - Scienze del Farmaco, Via Orabona 4, 70126 Bari, Italy.
[4] University of Bari "Aldo Moro", Department of Physics "M. Merlin", Via Amendola 173, 70126 Bari, Italy.
* Corresponding author: federica.paladini@unisalento.it

[#] *This chapter was prepared by three research groups that equally contributed to the work. The first author of each sub-list representing the first author for his/her own research group.*

detection, silver nano-coated medical devices and optical receptors (Solgi and Taghizadeh 2012, Ahmed et al. 2016a, Velusamy et al. 2016). Several methods have been used for the synthesis of nanoparticles. Generally, nanoparticles are prepared by a variety of chemical and physical methods, which can be quite expensive and potentially hazardous to the environment (Ahmed et al. 2016a). During the last decade, research efforts have been focused on the development of simple, clean, non-toxic, cost effective and eco-friendly protocols for the synthesis of nanoparticles (Ahmed et al. 2016a, 2016b, Velusamy et al. 2016). Extracts from natural substances are emerging as sustainable and eco-friendly natural resources for production of NPs for biological and medical applications through environment-friendly approaches of green chemistry (Ahmed et al. 2016a, Hussain et al. 2016, Velusamy et al. 2016). A number of biomolecules in extracts have been shown to successfully act as reducing agents in the green synthesis of NPs. For example, Essential Oils (EOs) extracted from the fresh leaves of *Anacardium occidentale* have been used for the reduction of auric acid to Au nanoparticles, while black tea, lemon, Geranium (*Pelargonium graveolens*) leaves, basil plant, *O. tenuiflorum, S. tricobatum, S. cumini, C. asiatica* and *C. sinensis* extracts, olive leaf and oak fruit bark extracts have been used for the synthesis of silver nanoparticles (Shankar et al. 2003, Ahmad et al. 2010, Sheny et al. 2012, or Vankar and Shukla 2012 as given in reference section; Abdel-Aziz et al. 2014, Khalil et al. 2014, Logeswari et al. 2015, Ali et al. 2016, Veisi et al. 2016). Along with their more recent application as reducing agent in the green synthesis of metal nanoparticles, essential oils have been largely employed for their properties observed in nature, and in particular for their insecticidal, antifungal and antibacterial activities against a wide spectrum of pathogenic bacterial strains (Inouye et al. 2001, Bilia et al. 2014). The European Pharmacopoeia has defined the EO as "odorant product, generally of a complex composition, obtained from a botanically defined plant raw material, either by driving by steam of water, either by dry distillation or by a suitable mechanical method without heating" (Asbahani et al. 2015). At present, approximately 3000 EOs are known, 300 of which are commercially important especially for the pharmaceutical, agronomic, food, sanitary, cosmetic and perfume industries (Inouye et al. 2001, Bilia et al. 2014). Moreover, as potential sources of anti-microbial compounds, plant essential oils have been also used in traditional medicine for many years (Utchariyakiat et al. 2016). Synergistic and additive drug interactions may be potential strategies for controlling bacterial resistance evolution, because the administration of multiple drugs may disrupt several bacterial functions, thus minimizing selection of resistant strains (Scandorieiro et al. 2016).

Although antibacterial inorganic nanophases and EOs are well known and the number of their application fields is enormously increasing, their combination in hybrid antibacterial systems is still an unexplored topic.

This chapter aims to provide an overview of the most recent research efforts in the definition of antibacterial strategies against bacterial resistance through nanotechnology and biotechnology approaches, with a special focus on the huge potential of metal nanoparticles and their possible combination with antibacterial EOs.

Antibacterial Essential Oils

In the last decades, research has been devoted to the development of new products for enhancing the quality of human life. The antimicrobial agents, generally used to avoid the increase of potential health risks due to bacterial infections, or the production of unpleasant odour, and to preserve food, are among these products. Therefore, various antibacterial finishing and disinfection techniques are being developed to be applied from hospital environment (medical clothes, protective garments, etc.) to everyday household and clothing (Windler et al. 2013, Morais et al. 2016). Almost every class of chemical products, ranged from the very simple substances such as halogen ions and metals to very complex compounds, has been tested to confer the antibacterial activity to one material. The use of many of these antimicrobial agents without a tight control over their bioactivity has been stopped because of their possible undesired harmful or toxic effects (Giannossa et al. 2013). In particular, antimicrobials are powerful drugs that are generally safe and very helpful in fighting bacterial infections. Moreover, these are among the most commonly prescribed drugs used in conventional medicine both for human and animals to prevent, control and treat disease, and to promote the growth of food-producing animals (European Commission 2016). Moreover, one of the most widespread problems related to the indiscriminate use of antibiotics consist in the onset of antimicrobial resistance (AMR) that makes these drugs inactive (Blair et al. 2015). Many forms of resistance spread with remarkable speed and infections from resistant bacteria are now too common, and some pathogens have even become resistant to multiple classes of antibiotics (TATFAR - CDC Europe 2016). Bacteria will inevitably develop antibiotic resistance using several biological mechanisms (Blair et al. 2015). Among all the possible mechanisms, some bacterial strains are able to reorganize themselves through the formation of bacterial biofilms. The microbes in biofilms are attached together by a self-produced biopolymer matrix and the bacterial consortium can consist of one or more species living in a socio-microbiological way. The matrix is important since it provides structural stability and protection to the biofilm. At that stage, the biofilm shows maximum resistance to antibiotics ranging from 100- to 1000-fold respect to the isolated cells. In the clinical context, it has been estimated that about 60% of all microbial infections such as dental caries, gastroenteritis, endocarditis or bronchitis involve bacterial biofilms. The formation of bacterial biofilms lead to an increase of the resistance against the antibiotic therapy and a rapid evolution and chronicity of these infections, too. In addition, it was observed that different bacterial strains, which are not able to determine serious infections under normal conditions, could attack inert surfaces as medical devices through the growth of bacterial biofilms. This type of aggregation promotes the infection of the host and the spread of multi-resistant bacterial strains (Arber et al. 1974, Pace et al. 2005). With the increase in antibiotic-resistant bacteria and the lack of new antibiotics being brought into the market, alternative strategies need to be found to cope with infections resulting from drug-resistant bacteria (Spellberg et al. 2008, Spellberg et al. 2011). Recently, the market for natural products from plant extracts, dry extracts and essential oils, and their use as 'green advanced materials', is undergoing a rapid expansion following the now recognized antimicrobial action spectra extremely large of these substances. Natural products with bioactive molecules in their composition are capable of performing various pharmacological activities including antimicrobial, anticancer and antioxidant (Burt 2004, Bayala et al. 2014). Thus, during recent years, the interest towards the plethora of EOs pharmacological properties is fully renewed coming into the phytomedicine research focus (Buckle 1999, Mimica-Dukic et al. 2004, Pichersky et al. 2006, Sylvestre et al. 2006, Bakkali et al. 2008). EOs are a complex mixture of small molecules, whose synthesis occurs during the secondary metabolism pathway by secretory cells of aromatic plants. These secondary metabolites are located in different plant glands such as leaves, seeds, barks and roots. From the organoleptic point of view, the EOs have characteristic fragrances and are usually colourless or pale-yellow oils. The physicochemical properties concern a generally low boiling point, very low solubility in water and high solubility in alcohols, oils and organic solvents, and excluding few examples, such as the *Cinnamomum* EO, their density is lower than water (Pichersky et al. 2006, Can Baser and Buchbauer 2015). The EOs could be obtained by different extraction methods ranging from distillation techniques to supercritical fluids or ultrasound and microwave-assisted extraction (Santoyo et al. 2005, Kimbaris et al. 2006). Among all the extraction methods, hydrodistillation and steam distillation techniques remain the most used extraction methods for EOs in commercial and medicinal use. According to environmental and plant living conditions, climate, soil, cultivation techniques and to the level of expertise and care given by farmers and distillers during the extraction steps, the same plant species may show significant chemical differences in their qualitative and quantitative EO composition, which is defined as EO chemotype (Lahlou and Berrada 2003, Lahlou 2004, Waseem and Low 2015).

Three main structure-related compounds are identified as EO constituents classified as terpenes and terpenoids, aromatic compounds (especially phenols and phenylpropanoids) and aliphatic derivatives. All these compounds differ for the functional groups (terpenes, alcohols, aldehydes, ketones, phenols, esters, ethers and other moieties), which confer a specific pattern of physicochemical properties and biological activity profile. In the mixture, such compounds are present at different concentrations. In particular, terpenes, terpenoids and phenols have a higher concentration compared with aliphatic compounds, which are present in low or trace quantities (Burt 2004, Pichersky et al. 2006, Bakkali et al. 2008). Some of these compounds are present in their geometric or optically active isomeric forms. The determination of EO chemotype is a mandatory step for the quality assurance and standardization of pure EOs in the production chain. Among all the chromatographic methods, the conventional analytical methods based on Gas Chromatography (GC), also coupled to Mass Spectrometry (GC/MS), are frequently applied to the qualitative and quantitative classification of EOs based on the selectivity of their constituents. Such methods generally require an analysis time ranging from 5 to 60 minutes depending on the column length, heating rate and mixture complexity to perform an overall analytical cycle. These methods are generally characterized by high performance in repeatability and accuracy as usually requested in this field (Bicchi et al. 2004, Waseem and Low 2015). To assess the EO antibiotic activity, a plethora of recognized parameters and methods are reported in the literature, such as the evaluation of the Minimum Inhibitory Concentration (MIC) and the Minimum Bactericidal Concentration (MBC) on planktonic cells and biofilms. The evaluation of EOs and antibiotic drug synergies are generally carried out by checkerboard method, to evaluate the Fractional Inhibitory Concentration (FIC) index. It is interesting to note that the antimicrobial spectrum is broader than that of antibiotics on the market, MIC values ranging between 0.05 and 5 mg/ml. In addition, essential oils are very active toward AMR-bacterial strains that fail in adopting appropriate mechanisms of resistance, probably due to the synergistic mechanisms exerted by EO phytocomplex (set of phytochemical compounds) (Pichersky et al. 2006, Bakkali et al. 2008, Can Baser and Buchbauer 2015).

Although the chemical structure of the single EO components affect their precise mode of action and antibacterial activity, it is most likely that EOs antibacterial activity strictly depends on the phytocomplex, and is the result of separate and different mechanisms determined by synergic interactions with different cell targets (Carson et al. 2002). The lipophilicity of EO components promotes a permeability increase of the bacterial cell membrane with an extensive leakage of ions and other important molecules, which leads to bacteriostatic effect or to cell death. The presence of the hydroxyl group in phenolic compounds (carvacrol and thymol) play a pivotal role in increasing the cell membrane permeability to ATP after the disintegration of the outer membrane of gram-negative bacteria and lipopolysaccharides (LPS) release (Lambert et al. 2001). In particular, *Thymus* and *Origanum* EOs were found to present a high percentage of phenolic terpens, ranging from 60 to 70%, with carvacrol (oregano aroma) and thymol as major constituents. About 20 to 30% was determined as aliphatic and alicyclic hydrocarbons such as α-pinene, *p*-cymene and cariofillene, which play a key role for the determination of antimicrobial activity. The *Origanum* EO has been used to prolong the shelf life of food products by reducing the rate of microbial growth and reducing the spoilage of fatty acids (De Falco et al. 2013, Teixeira et al. 2013). In fact, cyclic hydrocarbons appear to act on cell proteins embedded in the cytoplasmic membrane such as ATPases. These lipophilic molecules could accumulate in the lipid bilayer disrupting the protein-protein and protein-lipid interactions; alternatively, direct interaction of the lipophilic compounds with hydrophobic parts of the protein were possible (Juven et al. 1994). The *Cinnamomum* EO chemotype was characterized by cinnammaldehyde, in the 70–80% range, and other compounds which were classified as hydrocarbon terpenes as carifillene and terpenoidalcohl. The aldehyde compounds have been shown to interfere with oxidative process-related enzymes of Gram–negative bacterial strain inhibiting the amino acid decarboxylase enzyme. This antimicrobial activity mechanism was promoted by a low concentration of hydrocarbon terpenes (1–3%) and terpenoidalchols (0.5–1%) in the *Cinnamomum* EO that interfere with membrane bilayer by means of a synergic mechanism with the same aldehyde compounds (Helander et al. 1998, Ranasinghe et al. 2013). Among all EOs Tea Tree Oil (TTO), distilled from leaves of the native Australian plant *Melaleuca alternifolia*, has been reported to have a variety of therapeutic properties (e.g., anti-inflammatory and antiseptic) and is a popular ingredient in a number of natural cosmetic products. TTO, capable of eliminating a large number of microorganisms, showed a potential use for the treatment of bacterial and fungal infections. Although TTO, when topically used, is generally considered safe, it can be toxic when ingested, producing a variety of negative effects (e.g., vomiting, diarrhea and hallucinations).

TTO antibacterial activity is determined by its chemotype, characterized by Terpinen-4-ol in 32–38%, gamma terpinene in 15–18% range and 4% of cymene isomers, all these compounds interfere with cell viability mechanism and permeability of the cell membranes (Carson et al. 2006, Hammer et al. 2012). However, it is reported that carvacrol, cinnamaldehyde, cinnamic acid, eugenol and thymolchemotype EOs can have a synergistic effect in combination with antibiotics (Rosato et al. 2010, Aleksic et al. 2014, Liu et al. 2015, Utchariyakiat et al. 2016). Several modes of action have been put forward by which antibiotics and the essential oil components may act synergistically, such as by affecting multiple targets; by physicochemical interactions and inhibiting antibacterial-resistance mechanisms (Rosato et al. 2007, Honório et al. 2015).

Combining Inorganic Nanophases and Essential Oils

Numerous studies have demonstrated that plant extracts and EOs contain diverse bioactive components that can control or inhibit bacterial growth (Hammer et al. 1999, Raut and Karuppayil 2014). The metabolites produced by plants are a promising and cost-effective alternative to common disinfectant or antibacterial compounds. Emergence of drug resistant strains of pathogens, increase in the immunocompromised population and limitations of the available antibiotics/drugs have motivated people to use the complementary and alternative therapies, including the use of EOs (Lopez-Romero et al. 2015, WHO—World Health Organization 2015). Combination of EOs with antimicrobial nanoparticles (NPs), in principle, could allow for the production of composites with synergic biocidal activity. Anyhow, in most cases, EOs are used as reducing agents for NP biosynthesis rather than as active biocidal components. In fact, as a mixture of many natural, volatile, and aromatic compounds (i.e., phenols, alkaloids, lectins/polypeptides and polyacetylenes), they can exert a certain reducing power, thus allowing for the reduction of metal ions in solution through hydrothermal protocols (Mittal et al. 2013). Encapsulation of metal NPs in bioactive essential oils is still a poorly explored issue. To the best of our knowledge, few examples are reported in literature dealing with the engineering and the development of organic-inorganic NP-EO antimicrobial composites with combined biocidal activity (da Rosa et al. 2015). There are also some works that study interactions of EO components with polymeric NPs, used for delivering these oil components into the site of infection (Allahverdiyev et al. 2011). As a cheap and natural polymer, chitosan (CS) is the most widespread compound used in this case. Recently, synthesis of CSNPs loaded with cinnamon (Hu et al. 2015), summer savory (Feyzioglu and Tornuk 2016) and oregano (Hosseini et al. 2016) antimicrobial essential oils has been reported. Loaded CSNPs are generally prepared using the nano-encapsulation ionic gelation one-pot approach. It is well known that plant derived EOs are very labile compounds that can be easily decomposed or evaporated at processing conditions when incorporated into food or packaging material with the effect of high temperature, pressure, etc. (Hosseini et al. 2013). One of the most effective methods for protection of EOs from degradation caused by extreme processing conditions is their encapsulation. Encapsulation of bioactive substances not only protects them from adverse environmental conditions, thereby extending the shelf-life of the product, but also enables a controlled release of active compound (Hosseini et al. 2015). A similar approach towards nano-encapsulation of essential oils was applied to cellulose acetate nanocapsules loaded with both copper-ferrite NPs and lemongrass EOs. In this case, the synergic antimicrobial effect is exerted simultaneously by organic and inorganic components, on *S. aureus* microbial cells (Liakos et al. 2016). In some studies, EOs are reported to play a dual role: on the one hand, they are responsible for the biosynthesis of metal/metal oxide NPs; on the other hand, those molecules that are not involved in the redox process, can improve NP antimicrobial action. One of the first examples regards magnetite (Fe_3O_4) NP biosynthesis using either cinnamon or *Melissa officinalis* EOs (Anghel et al. 2014, Grumezescu et al. 2014). Recently, orange oil was successfully used to synthesize ZnONPs for the production of wound dressings with lowered risk towards skin reactions and allergies (Rădulescu et al. 2015). During the last year, multi-metal nanostructures were also produced, using antimicrobial EOs. Azizi et al. (Azizi et al. 2016) report on the biosynthesis of ZnO-Ag core shell nanocomposites using wild ginger EO. Antimicrobial assays performed on both Gram-positive and Gram-negative bacteria highlighted the importance of a synergic biocidal action performed by both organic and inorganic composite components. Au/Ag alloy NPs were prepared using *Coleus aromaticus* essential oil (Vilas et al. 2016). To the best of our knowledge, it is the first example in which an EO is used for the synthesis of alloy NPs. Agar wheel diffusion method showed a pronounced antimicrobial efficacy for EO@Ag/AuNPs in respect to bare NPs and EO. Preparation of NP-EO composites is a little explored issue. A couple of examples show how the well-known antimicrobial properties of AgNPs and corresponding ions can be joined with those of EOs. In 2014, Ahmad et al. showed the potentialities of peppermint/Ag^+ composites against *E. coli*, *S. aureus* and *C. albicans*. Analogously, synergistic and additive effect of oregano EO and AgNPs was demonstrated against multidrug-resistant bacterial strains (Scandorieiro et al. 2016). Anti-biofilm capabilities of peppermint EO have been recently confirmed by Duncan et al. (2015), who used it to modify silica nanocapsules. This composite was effective against clinically isolated pathogenic bacteria strains. Moreover, EO@SiO_2NPs, in contrast to their antimicrobial action, selectively promoted fibroblast proliferation in a mixed bacteria/mammalian cell system, making them promising for wound healing applications. Vanilla, patchouli and ylang-ylang EOs have been successfully tested against *S. aureus* and *K. pneumoniae* clinical strains, in association with myristic acid-stabilized Fe_3O_4NPs (Bilcu et al. 2014). $Zn(OH)_2$ (Ghaedi et al. 2015) and ZnONPs (Arfat et al. 2014), as nontoxic and biocompatible materials, are widely used for the production of antimicrobial food packaging (Sportelli et al. 2014). Their efficacy can be improved by their encapsulation in natural polymers, along with basil (Arfat et al. 2014) and sage (Ghaedi et al. 2015) EOs.

Nanoantimicrobials vs. AMR: A Critical Overview and Possible Roadmaps

In the 20th century, the advent of antibiotics was regarded as one of the most successful stories in medicine saving an impressive number of lives around the world. On the other hand, due to their overuse or underuse, it was observed that some pathogens (especially among bacteria strains) developed resistance to specific drugs intended for treatment of infections caused by them (WHO—World Health Organization 2015). As a result, the concept of AMR (European Commission 2016) was introduced as referred to Multi-Drug Resistant (MDR) microorganisms (Davies and Davies 2010). This phenomenon is nowadays considered as a global threat not only in terms of public health, but also for its economic burden due to health care costs and productivity losses (European Commission 2015). In the European Union (EU), 25000 people die each year due to infections caused by resistant bacteria (European Commission 2016). This results in economic losses of about €1.5 billion per year (European Commission 2016). Similarly, the Center for Disease Control and prevention (CDC) estimates that each year in the USA at least 2 million people are infected with 'superbugs' (Huh and Kwon 2011), and at least 23000 people die as a direct result of these infections (Centers for Disease Control and prevention (CDC) 2015). Moreover, AMR greatly affects food safety as antibiotics are overused in food producing animals and aquaculture (European Commission 2016). To this aim, in 2015 the World Health Organization (WHO) has launched a global action plan on AMR to tackle it (World Health Organization 2015) highlighting the need for a 'One Health' approach as already proposed and started by EU in 1999 (Geoghegan-Quinn 2014). Under the Horizon 2020 European program, AMR represents one of the key research areas of Health theme fostering the development of new effective antimicrobials or alternatives for its treatment. Analogously, in 2014 the White House has launched the National strategy for combating antibiotic-resistant bacteria outlining "basic and applied research and development of new antibiotics/therapeutics" as one of the goals (White House

2014). CDC started implementing this initiative thanks to US$160 million funded by the Congress for year 2016. Furthermore, transnational strategies were coordinated to respond AMR global threat, such as the Transatlantic Taskforce on Antimicrobial Resistance (TATFAR) created in 2009. TATFAR aims at improving cooperation between the U.S. and the EU in three key areas: (1) appropriate therapeutic use of antimicrobial drugs in medical and veterinary communities, (2) prevention of healthcare and community-associated drug-resistant infections, and (3) strategies for improving the pipeline of new antimicrobial drugs (TATFAR - CDC Europe 2016). However, production of new drugs may be an expensive (either in terms of costs and time) process (Piddock 2012), which may lead to low-efficiency products as horizontal gene transfer in bacteria play a predominant role in the evolution and transmission of resistance (Davies and Davies 2010). Considering this scenario, both academia and industrial stakeholders are seeking nanotechnology as a new paradigm for developing antimicrobial agents (Pelgrift and Friedman 2013, Singh et al. 2015, Rai et al. 2016, Shimanovich and Gedanken 2016). In particular, nanoantimicrobials (NAMs) based on metals or metal oxides (Cioffi and Rai 2012, Sportelli et al. 2014, Beyth et al. 2015) represent a class of nanomaterials which exert enhanced bioactivity, mainly due to their high surface-to-volume ratio and reduced size (Sportelli et al. 2014, Rai et al. 2016). Moreover, other factors including their surface chemistry (Le Ouay and Stellacci 2015, Shimanovich and Gedanken 2016) are extremely important in determining the antimicrobial action (Daima and Bansal 2015). Since their advent in the early 2000s, the potentialities of antimicrobial engineered nanomaterials have been envisaged as demonstrated by the number of publications and patents related to this topic (Scopus Database 2016). NAMs show multiple mechanisms of antimicrobial action thus making the development of AMR unlikely (Huh and Kwon 2011, Pelgrift and Friedman 2013); however, recently it has been considered that a synergistic approach based on the combination of NAMs, conventional (e.g., antibiotics (Allahverdiyev et al. 2011, Panáček et al. 2016)) and/or alternative antimicrobial compounds (e.g., essential oils (Milovanovic et al. 2013, Scandorieiro et al. 2016)) might provide a solution to combat AMR (Allahverdiyev et al. 2011, Aruguete et al. 2013, Pelgrift and Friedman 2013, Shimanovich and Gedanken 2016). In fact, several strategies can be offered by matter manipulation at the nanoscale to develop 'nanoantibiotics' (Huh and Kwon 2011), roughly classified as 'nanocarriers' for targeted drug delivery (Moazeni et al. 2016, Shimanovich and Gedanken 2016) and multicomponent functional nanostructures (Li et al. 2005, Karthick Raja Namasivayam and Ganesh 2013, Li et al. 2014). Both approaches can be applied to overcome AMR and restore bioactivity offering several advantages including decrease in therapy dose and frequency, sustained and controlled release, enhancement of the concentration at the site of infection with reduced systemic side effects (Khan et al. 2016, Shimanovich and Gedanken 2016), easy implementation in commercial products (Sportelli et al. 2014, Singh et al. 2015). On the other hand, the use of nanomaterial implies a deep study of possible critical issues that can emerge about toxicity for humans and environment. Although nanomaterials (e.g., ZnO, Ag nanoparticles) are already present in some commercial products (Khan et al. 2016), studies about their safety (European Scientific Committee on Consumer Safety (SCCS) 2012, European Scientific Committee on Emerging and Newly Identified Health Risks (SCENIHR) 2014) and fate are still in their infancy. Clinical studies are required for long-term exposure to nanomaterials for unraveling their mechanisms of action *in vivo* (Padovani et al. 2015). The path towards the development of cost-effective, safe, green and efficient nanoantibiotics will involve the knowledge about their toxicokinetics, life cycle and risk assessment, thus conveying expertise in several fields (e.g., chemistry, materials science, biology, toxicology, medicine) for a successful story.

Industrial Applications and Perspectives

Scientists and industry stakeholders have identified potential uses of nanotechnology in segments of the food industry, from agriculture to food processing, food packaging and nutrient supplements (Siegrist et al. 2007, 2008, Mu and Sprando 2010, Duncan et al. 2015). In particular, in the food industry, nanotechnology may provide stronger, high-barrier packaging materials, sensors for detection of trace contaminants and more effective antimicrobial agents (Siegrist et al. 2007, 2008, Duncan et al. 2015). Food contamination and growth of pathogenic microorganisms can occur when exposed to the environment, and nowadays, people are getting more alert on safety of synthetic food additives (Sung et al. 2013, Prakash et al. 2015). Indeed, primarily synthetic antimicrobial agents are commonly incorporated directly into the food to reduce food spoilage by microorganisms, and the related concerns about possible side-effects are increasing (Kuorwel et al. 2011). Therefore, the interest in inorganic disinfectants such as metal nanoparticles (NPs) and in antimicrobial agents derived from essential oils is growing for active packaging systems aiming to protect food products from microbial contamination (Ahmad et al. 2010, Kuorwel et al. 2011, Hajipour et al. 2012, Klein et al. 2013, Salarbashi et al. 2013).

Although many works are devoted to the development of antimicrobial packaging through the introduction of metal nanoparticles or essential oils, however, their synergistic effect has not yet been fully explored yet. Interesting works describe the use of metal NAMs for food packaging application, demonstrating considerable potential of nanocomposite materials based on copper, silver and zinc oxide against food pathogens.

A new type of nanomaterial developed as antibacterial additive for food packaging applications was composed of copper nanoparticles embedded in polylactic acid, thus combining the antibacterial properties of copper nanoparticles with the biodegradability of the polymer matrix (Cioffi et al. 2005). The potential of silver nanoparticles against foodborne pathogens able to cause health hazards to human beings has also been explored, with encouraging results as effective bactericidal covering material (Rajeshkumar and Malarkodi 2014). A research work by Esmailzadeh et al. (2016) has demonstrated the antibacterial effect of low density polyethylene (LDPE) containing ZnO nanoparticles on *Bacillus subtilis* and *Enterobacter aerogenes* for application as nanocomposite packaging. *In situ* preparation of ZnO nanoparticles onto starch-coated polyethylene film was also investigated against *Escherichia coli* and a new active packaging film incorporating ZnO nanoparticles was developed and tested to extend the shelf life for food applications (Tankhiwale and Bajpai 2012, Salarbashi et al. 2016). The results obtained by Morsy et al. (2014) have demonstrated that essential oils of rosemary and oregano, as well as silver and ZnO nanoparticles incorporated into pullulan films were effective against pathogenic microorganism. Plant-derived EOs have shown remarkable antimicrobial potency against spoilage and pathogenic microorganisms in meat and meat products (Jayasena and Jo 2013). Basil, thyme, oregano and cinnamon essential oils are well suited to be utilized as preservatives in foods and as potential alternatives for synthetic food additives (Kuorwel et al. 2011, Manso et al. 2013). Although their huge potential in food industry, some issues related to the application of the essential oils in food packaging still limit their use. Essential oils are recognized as safe substances by the Food and Drug Administration but the required concentration for an effective biocide action can be higher than that of a standard antibiotic (Varona et al. 2013). Considering the great variability of the EOs and their main compounds, a case-by-case evaluation is needed to assure their safe use in food packaging (Llana-Ruiz-Cabello et al. 2015). Moreover, essential oils must be adequately formulated to protect them from degradation, evaporation, to provide a controlled release and to limit their intense aroma (Jayasena and Jo 2013, Varona et al. 2013, Tongnuanchan and Benjakul 2014). Due to their flavour and fragrance, essential oils have been also used since early times in perfumery, pharmaceutical, cosmetic and as antiseptic, healing and therapeutic ingredients in traditional medicine and aromatherapy (Manou et al. 1998, Xiao et al. 2014). Aromatherapy textiles are manufactured with the use of microcapsule encapsulation of perfumes, fragrance compounds and essential oils. Obtaining coatings of these oils by complex coacervation has become an alternative method for achieving antibacterial textiles, which are developed to transfer antibacterial agents to the human body via close contact with the skin (Xiao et al. 2014). Nowadays, consumers around the world are increasingly focused on health and beauty (Carvalho et al. 2016). The use of essential oils in the production of cosmetics may have several advantages, such as enhancing the dermato-cosmetic properties and preservation, as well as the marketing image of the final product (Muyima et al. 2002). The renewed consumer interest in natural cosmetic products creates the demand for new products with botanical and functional ingredients (Muyima et al. 2002). Cosmetic products are suitable media for growth of checked microorganisms because of the materials in their compositions. Therefore, against the possibility of contamination, the addition of some preservatives in to the cosmetic products can prevent microbial growth (Yorgancioglu and Bayramoglu 2013). In pharmaceutics, cosmetic and textile applications, the use of silver nanoparticles has also become usual because of their antimicrobial effects (Solgi et al. 2009). Silver nanoparticles can be considered a safe and stable preservative for pharmaceutical or cosmetic preparations, and effective against a broad spectrum of microorganisms (Kokura et al. 2010). Moreover, the interest in the biomedical application of the silver nanoparticles is enormously growing due to the bacterial resistance to many antibiotics. Silver nanoparticles combined with essential oils has potential to be applied in cosmetics, food and pharmaceutical industry and in clinical and hospital settings in the treatment of wounds and burn infections or for disinfecting hospital (Scandorieiro et al. 2016). Essential oils, propolis and silver nanoparticles have been presented for their high potential for controlling and prevention candidiasis, which is the fourth leading cause of nosocomial infections (Szweda et al. 2015). Both essential oils and metal ions interact with many different intracellular components, thereby resulting in the disruption of vital cell functions and cell death. The application of essential oils and heavy metal ions, particularly tea tree oil and silver ions, has been proposed as alternative antimicrobial agents for the treatment of chronic and infected wounds (Low et al. 2016). The tea tree and the geranium oils exhibit strong antibacterial activity against Gram-negative pathogens responsible for difficult-to-treat wound infections, and several studies have suggested the uses of tea tree oil for the treatment of acne vulgaris, seborrheic dermatitis and chronic gingivitis (Pazyar et al. 2013, Sienkiewicz et al. 2014). The effect of silver nanoparticles (AgNPs) and EOs as novel antimicrobial agents has been also explored in agricultural applications. The studies carried by Solgi et al. (2009) suggested the potential application of EOs or AgNPs as alternative agents to common chemicals used in preservative solutions in extending the vase-life of gerbera flowers. For several years, essential oils have been widely used for insect control in pest management to reduce pesticide treatments based on toxic chemicals because of their repellent, ovicidal and antifeedant efficacy (Chung et al. 2013, Kim et al. 2013, Werdin González et al. 2014, Xiao et al. 2014). Nanomaterials in different forms can be used for efficient management of insect pests and formulations of potential insecticides and pesticides for enhanced agricultural productivity (Rai and Ingle 2012).

Conclusions

Nanotechnology and green chemistry approaches have demonstrated a great potential in the definition of novel strategies to develop environment-friendly antibacterial systems which can be applied to many industrial application fields.

Food, textile, biomedical fields and agriculture represent just some examples of the application of essential oils and inorganic nanophases. Although extremely important and valuable, most of the reported investigations have been carried at the laboratory scale and should be transferred to large scale to fully satisfy the future demand for antimicrobial products. More efforts are necessary to further explore the enormous range of biological activities of essential oils and their potential industrial applications. Future research should overcome the gap between conditions at research level and demands for large-scale applications, thus improving the existing manufacturing technologies (Xiao et al. 2014, Calo et al. 2015).

References

Abdel-Aziz, M.S., Shaheen, M.S., El-Nekeety, A.A. and Abdel-Wahhab, M.A. 2014. Antioxidant and antibacterial activity of silver nanoparticles biosynthesized using Chenopodium murale leaf extract. J. Saudi Chem. Soc., SI: Nanomaterials for Energy and Environmental Applications 18: 356–363.

Ahmad, A., Khan, A., Samber, N. and Manzoor, N. 2014. Antimicrobial activity of *Mentha piperita* essential oil in combination with silver ions. Synergy 1: 92–98.

Ahmad, N., Sharma, S., Alam, M.K., Singh, V.N., Shamsi, S.F., Mehta, B.R. and Fatma, A. 2010. Rapid synthesis of silver nanoparticles using dried medicinal plant of basil. Colloids Surf. B Biointerfaces 81: 81–86.

Ahmed, S., Ahmad, M., Swami, B.L. and Ikram, S. 2016a. A review on plants extract mediated synthesis of silver nanoparticles for antimicrobial applications: A green expertise. J. Adv. Res. 7: 17–28.

Ahmed, S., Annu, Ikram, S. and Yudha S. 2016b. Biosynthesis of gold nanoparticles: A green approach. J. Photochem. Photobiol. B 161: 141–153.

Aleksic, V., Mimica-Dukic, N., Simin, N., Nedeljkovic, N.S. and Knezevic, P. 2014. Synergistic effect of Myrtus communis L. essential oils and conventional antibiotics against multi-drug resistant *Acinetobacter baumannii* wound isolates. Phytomed. 21: 1666–1674.

Ali, Z.A., Yahya, R., Sekaran, S.D. and Puteh, R. 2016. Green synthesis of silver nanoparticles using apple extract and its antibacterial properties. Adv. Mat. Sci. Eng. 2016: e4102196.

Allahverdiyev, A.M., Kon, K.V., Abamor, E.S., Bagirova, M. and Rafailovich, M. 2011. Coping with antibiotic resistance: combining nanoparticles with antibiotics and other antimicrobial agents. Expert Review of Anti-infective Therapy 9: 1035–1052.

Anghel, A.G., Grumezescu, A.M., Chirea, M., Grumezescu, V., Socol, G., Iordache, F., Oprea, A.E., Anghel, I. and Holban, A.M. 2014. MAPLE fabricated Fe_3O_4@*Cinnamomum verum* antimicrobial surfaces for improved gastrostomy tubes. Molecules 19: 8981–8994.

Arber, W., Haas, R., Henle, W., Hofschneider, P.H., Humphery, J.H., Jerne, N.K., Koldovsky, P., Koprowski, H., Maaløe, O., Rott, R., Schweiger, H.G., Sela, M., Syruček, L., Vogt, P.K. and Wecker, E. 1974. Current Topics in Microbiology and Immunology. Springer-Verlag.

Arfat, Y.A., Benjakul, S., Prodpran, T., Sumpavapol, P. and Songtipya, P. 2014. Properties and antimicrobial activity of fish protein isolate/fish skin gelatin film containing basil leaf essential oil and zinc oxide nanoparticles. Food Hydrocolloids 41: 265–273.

Aruguete, D.M., Kim, B., Hochella, M.F., Ma, Y., Cheng, Y., Hoegh, A., Liu, J. and Pruden, A. 2013. Antimicrobial nanotechnology: its potential for the effective management of microbial drug resistance and implications for research needs in microbial nanotoxicology. Environ. Sci.: Processes Impacts 15: 93–102.

Asbahani, A.E., Miladi, K., Badri, W., Sala, M., Addi, E.H.A., Casabianca, H., Mousadik, A.E., Hartmann, D., Jilale, A., Renaud, F.N.R. and Elaissari, A. 2015. Essential oils: From extraction to encapsulation. Int. J. Pharmaceutics 483: 220–243.

Azizi, S., Mohamad, R., Rahim, R.A., Moghaddam, A.B., Moniri, M., Ariff, A., Saad, W.Z. and Namvab, F. 2016. ZnO-Ag core shell nanocomposite formed by green method using essential oil of wild ginger and their bactericidal and cytotoxic effects. Appl. Surf. Sci. 384: 517–524.

Bakkali, F., Averbeck, S., Averbeck, D. and Idaomar, M. 2008. Biological effects of essential oils—A review. Food and Chemical Toxicology 46: 446–475.

Bayala, B., Bassole, I.H., Scifo, R., Gnoula, C., Morel, L., Lobaccaro, J.-M.A. and Simpore, J. 2014. Anticancer activity of essential oils and their chemical components—a review. Am. J. Cancer Res. 4: 591–607.

Beyth, N., Houri-Haddad, Y., Domb, A., Khan, W. and Hazan, R. 2015. Alternative antimicrobial approach: Nano-antimicrobial materials. Evidence-Based Complementary and Alternative Medicine 2015: 16.

Bicchi, C., Brunelli, C., Cordero, C., Rubiolo, P., Galli, M. and Sironi, A. 2004. Direct resistively heated column gas chromatography (Ultrafast module-GC) for high-speed analysis of essential oils of differing complexities. J. Chromat. A 1024: 195–207.

Bilcu, M., Grumezescu, A.M., Oprea, A.E., Popescu, R.C., Mogoşanu, G.D., Hristu, R., Stanciu, G.A., Mihailescu, D.F., Lazar, V., Bezirtzoglou, E. and Chifiriuc, M.C. 2014. Efficiency of Vanilla, Patchouli and Ylang Ylang essential oils stabilized by iron oxide@C14 nanostructures against bacterial adherence and biofilms formed by *Staphylococcus aureus* and *Klebsiella pneumoniae* Clinical strains. Molecules 19: 17943–17956.

Bilia, A.R., Guccione, C., Isacchi, B., Righeschi, C., Firenzuoli, F. and Bergonzi, M.C. 2014. Essential oils loaded in nanosystems: a developing strategy for a successful therapeutic approach. Evid. Based Complement. Alternat. Med. 2014: 651593.

Blair, J.M.A., Webber, M.A., Baylay, A.J., Ogbolu, D.O. and Piddock, L.J.V. 2015. Molecular mechanisms of antibiotic resistance. Nat. Rev. Micro. 13: 42–51.

Buckle, J. 1999. Use of aromatherapy as a complementary treatment for chronic pain. Altern. Ther. Health Med. 5: 42–51.

Burt, S. 2004. Essential oils: their antibacterial properties and potential applications in foods—a review. Int. J. Food Microbiol. 94: 223–253.

Calo, J.R., Crandall, P.G., O'Bryan, C.A. and Ricke, S.C. 2015. Essential oils as antimicrobials in food systems—a review. Food Control. 54: 111–119.

Can Baser, K.H. and Buchbauer, G. 2015. Handbook of essential oils: science, technology, and applications, 2nd ed, Natural Product Chemistry. CRC Press.

Carson, C.F., Mee, B.J. and Riley, T.V. 2002. Mechanism of action of *Melaleuca alternifolia* (Tea Tree) oil on *Staphylococcus aureus* determined by time-kill, lysis, leakage, and salt tolerance assays and electron microscopy. Antimicrob. Agents Chemother. 46: 1914–1920.

Carson, C.F., Hammer, K.A. and Riley, T.V. 2006. *Melaleuca alternifolia* (Tea Tree) oil: a review of antimicrobial and other medicinal properties. Clin. Microbiol. Rev. 19: 50–62.

Carvalho, I.T., Estevinho, B.N. and Santos, L. 2016. Application of microencapsulated essential oils in cosmetic and personal healthcare products—a review. Int. J. Cosmet. Sci. 38: 109–119.

Centers for Disease Control and prevention (CDC). 2015. About antimicrobial resistance [WWW Document]. URL http://www.cdc.gov/drugresistance/about.html.

Chatterjee, A.K., Chakraborty, R. and Basu, T. 2014. Mechanism of antibacterial activity of copper nanoparticles. Nanotech. 25: 135101.

Chung, S.K., Seo, J.Y., Lim, J.H., Park, H.H., Yea, M.J. and Park, H.J. 2013. Microencapsulation of essential oil for insect repellent in food packaging system. J. Food Sci. 78: E709–E714.

Cioffi, N., Ditaranto, N., Torsi, L., Picca, R.A., Sabbatini, L., Valentini, A., Novello, L., Tantillo, G., Bleve-Zacheo, T. and Zambonin, P.G. 2005. Analytical characterization of bioactive fluoropolymer ultra-thin coatings modified by copper nanoparticles. Anal. Bioanal. Chem. 381: 607–616.

Cioffi, N. and Rai, M. 2012. Nano-Antimicrobials: Progress and Prospects. Springer-Verlag, Berlin Heidelberg.

da Rosa, C.G., de Oliveira Brisola Maciel, M.V., de Carvalho, S.M., de Melo, A.P., Jummes, B., da Silva, T., Martelli, S.M., Villetti, M.A., Bertoldi, F.C. and Barreto, P.L.M. 2015. Characterization and evaluation of physicochemical and antimicrobial properties of zein nanoparticles loaded with phenolics monoterpenes. Colloids Surf. A Physicochem. Eng. Asp. 481: 337–344.

Daima, H.K. and Bansal, V. 2015. Influence of physico-chemical properties of nanomaterials on their antibacterial applications. pp. 151–166. *In*: Rai, M. and Kon, K. (eds.). Nanotechnology in Diagnosis, Treatment and Prophylaxis of Infectious Diseases. Academic Press/Elsevier Inc., Boston.

Davies, J. and Davies, D. 2010. Origins and evolution of antibiotic resistance. Microbiology and Molecular Biology Reviews 74: 417–433.

De Falco, E., Mancini, E., Roscigno, G., Mignola, E., Taglialatela-Scafati, O. and Senatore, F. 2013. Chemical composition and biological activity of essential oils of *Origanum vulgare* L. subsp. *vulgare* L. under different growth conditions. Molecules 18: 14948–14960.

Duncan, B., Li, X., Landis, R.F., Kim, S.T., Gupta, A., Wang, L.-S., Ramanathan, R., Tang, R., Boerth, J.A. and Rotello, V.M. 2015. Nanoparticle-stabilized capsules for the treatment of bacterial biofilms. ACS Nano 9: 7775–7782.

Esmailzadeh, H., Sangpour, P., Shahraz, F., Hejazi, J. and Khaksar, R. 2016. Effect of nanocomposite packaging containing ZnO on growth of *Bacillus subtilis* and *Enterobacter aerogenes*. Mat. Sci. Eng. C 58: 1058–1063.

European Commission. 2015. Progress Report EU Action Plan against AMR.

European Commission. 2016. Antimicrobial resistance [WWW Document]. URL http://ec.europa.eu/dgs/health_food-safety/amr/index_en.htm.

European Scientific Committee on Consumer Safety (SCCS). 2012. Opinion on ZnO (nano form).

European Scientific Committee on Emerging and Newly Identified Health Risks (SCENIHR). 2014. Final opinion on Nanosilver: safety, health and environmental effects and role in antimicrobial resistance.

Feyzioglu, G.C. and Tornuk, F. 2016. Development of chitosan nanoparticles loaded with summer savory (*Satureja hortensis* L.) essential oil for antimicrobial and antioxidant delivery applications. LWT - Food Sci. Tech. 70: 104–110.

Geoghegan-Quinn, M. 2014. Funding for antimicrobial resistance research in Europe. The Lancet 384: 1186.

Ghaedi, M., Naghiha, R., Jannesar, R., dehghanian, N., Mirtamizdoust, B. and Pezeshkpour, V. 2015. Antibacterial and antifungal activity of flower extracts of *Urtica dioica, Chamaemelum nobile* and *Salvia officinalis*: Effects of Zn[OH]2 nanoparticles and Hp-2-minh on their property. J. Ind. Eng. Chem. 32: 353–359.

Giannossa, L.C., Longano, D., Ditaranto, N., Nitti, M.A., Paladini, F., Pollini, M., Rai, M., Sannino, A., Valentini, A. and Cioffi, N. 2013. Metal nanoantimicrobials for textile applications. Nanotech: Rev. 2: 307–331.

Grumezescu, A.M., Andronescu, E., Oprea, A.E., Holban, A.M., Socol, G., Grumezescu, V., Chifiriuc, M.C., Iordache, F. and Maniu, H. 2014. MAPLE fabricated magnetite@*Melissa officinalis* and poly lactic acid: chitosan coated surfaces with anti-staphylococcal properties. J. Sol-Gel Sci. Technol. 73: 612–619.

Hajipour, M.J., Fromm, K.M., Akbar Ashkarran, A., Jimenez de Aberasturi, D., Larramendi, I.R. de, Rojo, T., Serpooshan, V., Parak, W.J. and Mahmoudi, M. 2012. Antibacterial properties of nanoparticles. Trends Biotechnol. 30: 499–511.

Hammer, K.A., Carson, C.F. and Riley, T.V. 1999. Antimicrobial activity of essential oils and other plant extracts. J. Appl. Microbiol. 86: 985–990.

Hammer, K.A., Carson, C.F. and Riley, T.V. 2012. Effects of *Melaleuca alternifolia* (Tea Tree) essential oil and the major monoterpene component terpinen-4-ol on the development of single- and multistep antibiotic resistance and antimicrobial susceptibility. Antimicrob. Agents Chemother 56: 909–915.

Helander, I.M., Alakomi, H.-L., Latva-Kala, K., Mattila-Sandholm, T., Pol, I., Smid, E.J., Gorris, L.G.M. and von Wright, A. 1998. Characterization of the action of selected essential oil components on gram-negative bacteria. J. Agric. Food Chem. 46: 3590–3595.

Honório, V.G., Bezerra, J., Souza, G.T., Carvalho, R.J., Gomes-Neto, N.J., Figueiredo, R.C.B.Q., Melo, J.V., Souza, E.L. and Magnani, M. 2015. Inhibition of *Staphylococcus aureus* cocktail using the synergies of oregano and rosemary essential oils or carvacrol and 1,8-cineole. Front. Microbiol. 6.

Hosseini, S.F., Rezaei, M., Zandi, M. and Ghavi, F.F. 2013. Preparation and functional properties of fish gelatin–chitosan blend edible films. Food Chem., ASSET 2011 136: 1490–1495.
Hosseini, S.F., Rezaei, M., Zandi, M. and Farahmandghavi, F. 2015. Fabrication of bio-nanocomposite films based on fish gelatin reinforced with chitosan nanoparticles. Food Hydrocoll. 44: 172–182.
Hosseini, S.F., Rezaei, M., Zandi, M. and Farahmandghavi, F. 2016. Development of bioactive fish gelatin/chitosan nanoparticles composite films with antimicrobial properties. Food Chem. 194: 1266–1274.
Hu, J., Wang, X., Xiao, Z. and Bi, W. 2015. Effect of chitosan nanoparticles loaded with cinnamon essential oil on the quality of chilled pork. LWT - Food Sci. Tech. 63: 519–526.
Huh, A.J. and Kwon, Y.J. 2011. "Nanoantibiotics": A new paradigm for treating infectious diseases using nanomaterials in the antibiotics resistant era. J. Controlled Release 156: 128–145.
Hussain, I., Singh, N.B., Singh, A., Singh, H. and Singh, S.C. 2016. Green synthesis of nanoparticles and its potential application. Biotechnol. Lett. 38: 545–560.
Inouye, S., Takizawa, T. and Yamaguchi, H. 2001. Antibacterial activity of essential oils and their major constituents against respiratory tract pathogens by gaseous contact. J. Antimicrob. Chemother. 47: 565–573.
Jayasena, D.D. and Jo, C. 2013. Essential oils as potential antimicrobial agents in meat and meat products: A review. Trends Food Sci. Tech. 34: 96–108.
Juven, B.j., Kanner, J., Schved, F. and Weisslowicz, H. 1994. Factors that interact with the antibacterial action of thyme essential oil and its active constituents. J. Appl. Bacteriol. 76: 626–631.
Karthick Raja Namasivayam, S. and Ganesh, S. 2013. Evaluation of improved antibacterial activity of chitosan coated silver nanoparticles—azithromycin nanoconjugate against clinical isolate of human pathogenic bacteria. J. Pure Appl. Microbiol. 7: 1131–1140.
Khalil, M.M.H., Ismail, E.H., El-Baghdady, K.Z. and Mohamed, D. 2014. Green synthesis of silver nanoparticles using olive leaf extract and its antibacterial activity. Arab. J. Chem. 7: 1131–1139.
Khan, S.T., Musarrat, J. and Al-Khedhairy, A.A. 2016. Countering drug resistance, infectious diseases, and sepsis using metal and metal oxides nanoparticles: Current status. Colloids Surf B: Biointerfaces 146: 70–83.
Kim, I.-H., Han, J., Na, J.H., Chang, P.-S., Chung, M.S., Park, K.H. and Min, S.C. 2013. Insect-resistant food packaging film development using cinnamon oil and microencapsulation technologies. J. Food Sci. 78: E229–E237.
Kimbaris, A.C., Siatis, N.G., Daferera, D.J., Tarantilis, P.A., Pappas, C.S. and Polissiou, M.G. 2006. Comparison of distillation and ultrasound-assisted extraction methods for the isolation of sensitive aroma compounds from garlic (*Allium sativum*). Ultrasonics Sonochem. 13: 54–60.
Klein, G., Rüben, C. and Upmann, M. 2013. Antimicrobial activity of essential oil components against potential food spoilage microorganisms. Curr. Microbiol. 67: 200–208.
Kokura, S., Handa, O., Takagi, T., Ishikawa, T., Naito, Y. and Yoshikawa, T. 2010. Silver nanoparticles as a safe preservative for use in cosmetics. Nanomedicine 6: 570–574.
Kuorwel, K.K., Cran, M.J., Sonneveld, K., Miltz, J. and Bigger, S.W. 2011. Essential oils and their principal constituents as antimicrobial agents for synthetic packaging films. J. Food Sci. 76: R164–R177.
Lahlou, M. and Berrada, R. 2003. Composition and niticidal activity of essential oils of three chemotypes of *Rosmarinus officinalis* L. acclimatized in Morocco. Flavour Fragr. J. 18: 124–127.
Lahlou, M. 2004. Essential oils and fragrance compounds: bioactivity and mechanisms of action. Flavour Fragr. J. 19, 159–165.
Lambert, R.J.W., Skandamis, P.N., Coote, P.J. and Nychas, G.-J. e. 2001. A study of the minimum inhibitory concentration and mode of action of oregano essential oil, thymol and carvacrol. J. Appl: Microbiol. 91: 453–462.
Le Ouay, B. and Stellacci, F. 2015. Antibacterial activity of silver nanoparticles: A surface science insight. Nano Today 10: 339–354.
Li, P., Li, J., Wu, C., Wu, Q. and Li, J. 2005. Synergistic antibacterial effects of β-lactam antibiotic combined with silver nanoparticles. Nanotechnology 16: 1912–1917.
Li, X., Robinson, S.M., Gupta, A., Saha, K., Jiang, Z., Moyano, D.F., Sahar, A., Riley, M.A. and Rotello, V.M. 2014. Functional gold nanoparticles as potent antimicrobial agents against multi-drug-resistant bacteria. ACS Nano 8: 10682–10686.
Liakos, I.L., Abdellatif, M.H., Innocenti, C., Scarpellini, A., Carzino, R., Brunetti, V., Marras, S., Brescia, R., Drago, F. and Pompa, P.P. 2016. Antimicrobial lemongrass essential oil—copper ferrite cellulose acetate nanocapsules. Molecules 21: 520.
Liu, Q., Niu, H., Zhang, W., Mu, H., Sun, C. and Duan, J. 2015. Synergy among thymol, eugenol, berberine, cinnamaldehyde and streptomycin against planktonic and biofilm-associated food-borne pathogens. Lett. Appl. Microbiol. 60: 421–430.
Llana-Ruiz-Cabello, M., Pichardo, S., Maisanaba, S., Puerto, M., Prieto, A.I., Gutiérrez-Praena, D., Jos, A. and Cameán, A.M. 2015. *In vitro* toxicological evaluation of essential oils and their main compounds used in active food packaging: a review. Food Chem. Toxicol. 81: 9–27.
Logeswari, P., Silambarasan, S. and Abraham, J. 2015. Synthesis of silver nanoparticles using plants extract and analysis of their antimicrobial property. J. Saudi Chem. Soc. 19: 311–317.
Lopez-Romero, J.C., Gonzales-Rìos, H., Borges, A., Sìmones, M. and Borges, A. 2015. Antibacterial effects and mode of action of selected essential oils components against *Escherichia coli* and *Staphylococcus aureus*. Evid. Based Complement. Alternat. Med. 2015: 795435–9 pages.
Low, W.-L., Kenward, K., Britland, S.T., Amin, M.C. and Martin, C. 2016. Essential oils and metal ions as alternative antimicrobial agents: a focus on tea tree oil and silver. Int. Wound J. 1–16.

Manou, I., Bouillard, L., Devleeschouwer, M.J. and Barel, A.O. 1998. Evaluation of the preservative properties of *Thymus vulgaris* essential oil in topically applied formulations under a challenge test. J. Appl. Microbiol. 84: 368–376.
Manso, S., Cacho-Nerin, F., Becerril, R. and Nerín, C. 2013. Combined analytical and microbiological tools to study the effect on *Aspergillus flavus* of cinnamon essential oil contained in food packaging. Food Control 30: 370–378.
Milovanovic, S., Stamenic, M., Markovic, D., Radetic, M. and Zizovic, I. 2013. Solubility of thymol in supercritical carbon dioxide and its impregnation on cotton gauze. J. Supercrit Fluids 84: 173–181.
Mimica-Dukic, N., Bozin, B., Sokovic, M. and Simin, N. 2004. Antimicrobial and antioxidant activities of *Melissa officinalis* L. (Lamiaceae) essential oil. J. Agric. Food Chem. 52: 2485–2489.
Mittal, A.K., Chisti, Y. and Banerjee, U.C. 2013. Synthesis of metallic nanoparticles using plant extracts. Biotechnol. Adv. 31: 346–356.
Moazeni, M., Kelidari, H.R., Saeedi, M., Morteza-Semnani, K., Nabili, M., Gohar, A.A., Akbari, J., Lotfali, E. and Nokhodchi, A. 2016. Time to overcome fluconazole resistant *Candida* isolates: Solid lipid nanoparticles as a novel antifungal drug delivery system. Colloids Surf B: Biointerfaces 142: 400–407.
Morais, D.S., Guedes, R.M. and Lopes, M.A. 2016. Antimicrobial approaches for textiles: from research to market. Materials 9: 498.
Morsy, M.K., Khalaf, H.H., Sharoba, A.M., El-Tanahi, H.H. and Cutter, C.N. 2014. Incorporation of essential oils and nanoparticles in pullulan films to control foodborne pathogens on meat and poultry products. J. Food. Sci. 79: M675.
Mu, L. and Sprando, R.L. 2010. Application of nanotechnology in cosmetics. Pharm. Res. 27: 1746–1749.
Muyima, N.Y.O., Zulu, G., Bhengu, T. and Popplewell, D. 2002. The potential application of some novel essential oils as natural cosmetic preservatives in an aqueous cream formulation. Flavour Fragr. J. 17: 258–266.
Pace, J.L., Rupp, M.E. and Finch, G.F. 2005. Biofilms, Infection, and Antimicrobial Therapy. CRC Press.
Padovani, G.C., Feitosa, V.P., Sauro, S., Tay, F.R., Durán, G., Paula, A.J. and Durán, N. 2015. Advances in dental materials through nanotechnology: facts, perspectives and toxicological aspects. Trends Biotechnol. 33: 621–636.
Panáček, A., Smékalová, M., Večeřová, R., Bogdanová, K., Röderová, M., Kolář, M., Kilianová, M., Hradilová, Š., Froning, J.P., Havrdová, M., Prucek, R., Zbořil, R. and Kvítek, L. 2016. Silver nanoparticles strongly enhance and restore bactericidal activity of inactive antibiotics against multiresistant Enterobacteriaceae. Colloids Surf B: Biointerfaces 142: 392–399.
Pazyar, N., Yaghoobi, R., Bagherani, N. and Kazerouni, A. 2013. A review of applications of tea tree oil in dermatology. Int. J. Dermatol. 52: 784–790.
Pelgrift, R.Y. and Friedman, A.J. 2013. Nanotechnology as a therapeutic tool to combat microbial resistance. Adv. Drug Deliv. Rev. 65: 1803–1815.
Pichersky, E., Noel, J.P. and Dudareva, N. 2006. Biosynthesis of plant volatiles: nature's diversity and ingenuity. Science 311: 808–811.
Piddock, L.J. 2012. The crisis of no new antibiotics—what is the way forward? Lancet Infect. Dis. 12: 249–253.
Prakash, B., Kedia, A., Mishra, P.K. and Dubey, N.K. 2015. Plant essential oils as food preservatives to control moulds, mycotoxin contamination and oxidative deterioration of agri-food commodities – Potentials and challenges. Food Control. 47: 381–391.
Rădulescu, M., Andronescu, E., Cirja, A., Holban, A., Mogoantă, V., Bălşeanu, T., Cătălin, B., Neagu, T., Lascăr, I., Florea, D.A., Grumezescu, A.M., Ciubuca, B. and Lazăr, V. 2015. Antimicrobial coatings based on zinc oxide and orange oil for improved bioactive wound dressings and other applications. Rom. J. Morphol. Embryol. 57: 107–114.
Rai, M. and Ingle, A. 2012. Role of nanotechnology in agriculture with special reference to management of insect pests. Appl. Microbiol. Biotechnol. 94: 287–293.
Rai, M., Ingle, A.P., Gaikwad, S., Gupta, I., Gade, A. and da Silva, S.S. 2016. Nanotechnology based anti-infectives to fight microbial intrusions. J. Appl. Microbiol. 120: 527–542.
Rajeshkumar, S. and Malarkodi, C. 2014. *In vitro* antibacterial activity and mechanism of silver nanoparticles against foodborne pathogens. Bioinorg. Chem. Appl. 2014: e581890.
Ranasinghe, P., Pigera, S., Premakumara, G.S., Galappaththy, P., Constantine, G.R. and Katulanda, P. 2013. Medicinal properties of "true" cinnamon (*Cinnamomum zeylanicum*): a systematic review. BMC Complementary Alternative Med. 13: 275.
Raut, J.S. and Karuppayil, S.M. 2014. A status review on the medicinal properties of essential oils. Industrial Crops and Products 62: 250–264.
Rosato, A., Vitali, C., De Laurentis, N., Armenise, D. and Antonietta Milillo, M. 2007. Antibacterial effect of some essential oils administered alone or in combination with Norfloxacin. Phytomed. 14: 727–732.
Rosato, A., Piarulli, M., Corbo, F., Muraglia, M., Carone, A., Vitali, M.E. and Vitali, C. 2010. *In vitro* synergistic antibacterial action of certain combinations of gentamicin and essential oils. Curr. Med. Chem. 17: 3289–3295.
Salarbashi, D., Tajik, S., Ghasemlou, M., Shojaee-Aliabadi, S., Shahidi Noghabi, M. and Khaksar, R. 2013. Characterization of soluble soybean polysaccharide film incorporated essential oil intended for food packaging. Carbohydrate Pol. 98: 1127–1136.
Salarbashi, D., Mortazavi, S.A., Noghabi, M.S., Fazly Bazzaz, B.S., Sedaghat, N., Ramezani, M. and Shahabi-Ghahfarrokhi, I. 2016. Development of new active packaging film made from a soluble soybean polysaccharide incorporating ZnO nanoparticles. Carbohydrate Pol. 140: 220–227.
Santoyo, S., Cavero, S., Jaime, L., Ibañez, E., Señoráns, F.J. and Reglero, G. 2005. Chemical composition and antimicrobial activity of *Rosmarinus officinalis* L. essential oil obtained via supercritical fluid extraction. J. Food Protection 68: 790–795.

Scandorieiro, S., de Camargo, L.C., Lancheros, C.A.C., Yamada-Ogatta, S.F., Nakamura, C.V., de Oliveira, A.G., Andrade, C.G.T.J., Duran, N., Nakazato, G. and Kobayashi, R.K.T. 2016. Synergistic and additive effect of oregano essential oil and biological silver nanoparticles against multidrug-resistant bacterial strains. Front. Microbiol. 760.

Scopus Database. 2016. Scopus - Document search & Database [WWW Document]. Scopus 2016. Elsevier B.V. URL https://www.scopus.com/(accessed 6.17.16).

Seil, J.T. and Webster, T.J. 2012. Antimicrobial applications of nanotechnology: methods and literature. Int. J. Nanomedicine 7: 2767–2781.

Shankar, S.S., Ahmad, A. and Sastry, M. 2003. Geranium leaf assisted biosynthesis of silver nanoparticles. Biotechnology Progress 19: 1627–1631.

Sheny, D.S., Mathew, J. and Philip, D. 2012. Synthesis characterization and catalytic action of hexagonal gold nanoparticles using essential oils extracted from *Anacardium occidentale*. Spectrochim. Acta A Mol. Biomol. Spectrosc. 97: 306–310.

Shimanovich, U. and Gedanken, A. 2016. Nanotechnology solutions to restore antibiotic activity. J. Mater. Chem. B 4: 824–833.

Siegrist, M., Cousin, M.-E., Kastenholz, H. and Wiek, A. 2007. Public acceptance of nanotechnology foods and food packaging: The influence of affect and trust. Appetite 49: 459–466.

Siegrist, M., Stampfli, N., Kastenholz, H. and Keller, C. 2008. Perceived risks and perceived benefits of different nanotechnology foods and nanotechnology food packaging. Appetite 51: 283–290.

Sienkiewicz, M., Poznańska-Kurowska, K., Kaszuba, A. and Kowalczyk, E. 2014. The antibacterial activity of geranium oil against Gram-negative bacteria isolated from difficult-to-heal wounds. Burns 40: 1046–1051.

Singh, B.N., Prateeksha, C.V.R., Rawat, A.K.S., Upreti, D.K. and Singh, B.R. 2015. Antimicrobial nanotechnologies: what are the current possibilities? Curr. Sci. 108: 1210.

Solgi, M., Kafi, M., Taghavi, T.S. and Naderi, R. 2009. Essential oils and silver nanoparticles (SNP) as novel agents to extend vase-life of gerbera (*Gerbera jamesonii* cv. "Dune") flowers. Postharvest Biol. Technol. 53: 155–158.

Solgi, M. and Taghizadeh, M. 2012. Silver nanoparticles ecofriendly synthesis by two medicinal plants. Int. J. Nanomat. Biostructures 2: 60–64.

Spellberg, B., Guidos, R., Gilbert, D., Bradley, J., Boucher, H.W., Scheld, W.M., Bartlett, J.G., Edwards, J., Infectious Diseases Society of America. 2008. The epidemic of antibiotic-resistant infections: a call to action for the medical community from the Infectious Diseases Society of America. Clin. Infect. Dis. 46: 155–164.

Spellberg, B., Blaser, B., Guidos, R.J., Boucher, H.W., Eisenstein, B.I. and Reler, L.B. 2011. Combating antimicrobial resistance: Policy recommendations to save lives. Clin. Infect. Dis. 52: S397–S428.

Sportelli, M.C., Picca, R.A. and Cioffi, N. 2014. Nano-antimicrobials based on metals. pp. 181–218. *In*: Phoenix, D.A., Harris, F. and Dennison, S.R. (eds.). Novel Antimicrobial Agents and Strategies. Wiley-VCH Verlag GmbH & Co. KGaA.

Sung, S.-Y., Sin, L.T., Tee, T.-T., Bee, S.-T., Rahmat, A.R., Rahman, W.A.W.A., Tan, A.-C. and Vikhraman, M. 2013. Antimicrobial agents for food packaging applications. Trends Food Sci. Tech. 33: 110–123.

Sylvestre, M., Pichette, A., Longtin, A., Nagau, F. and Legault, J. 2006. Essential oil analysis and anticancer activity of leaf essential oil of *Croton flavens* L. from Guadeloupe. J. Ethnopharmacology 103: 99–102.

Szweda, P., Gucwa, K., Kurzyk, E., Romanowska, E., Dzierżanowska-Fangrat, K., Zielińska Jurek, A., Kuś, P.M. and Milewski, S. 2015. Essential oils, silver nanoparticles and propolis as alternative agents against Fluconazole Resistant *Candida albicans, Candida glabrata* and *Candida krusei* clinical isolates. Indian J. Microbiol. 55: 175–183.

Tankhiwale, R. and Bajpai, S.K. 2012. Preparation, characterization and antibacterial applications of ZnO-nanoparticles coated polyethylene films for food packaging. Colloids Surf. B Biointerfaces 90: 16–20.

TATFAR - CDC Europe. 2016. TATFAR | Antimicrobial Resistance | CDC [WWW Document]. Centers for desease control and prevention. URL http://www.cdc.gov/drugresistance/tatfar/(accessed 11.18.16).

Teixeira, B., Marques, A., Ramos, C., Serrano, C., Matos, O., Neng, N.R., Nogueira, J.M.F., Saraiva, J.A. and Nunes, M.L. 2013. Chemical composition and bioactivity of different oregano (*Origanum vulgare*) extracts and essential oil. J. Sci. Food Agric. 93: 2707–2714.

Tongnuanchan, P. and Benjakul, S. 2014. Essential oils: extraction, bioactivities, and their uses for food preservation. J. Food Sci. 79: R1231–1249.

Utchariyakiat, I., Surassmo, S., Jaturanpinyo, M., Khuntayaporn, P. and Chomnawang, M.T. 2016. Efficacy of cinnamon bark oil and cinnamaldehyde on anti-multidrug resistant *Pseudomonas aeruginosa* and the synergistic effects in combination with other antimicrobial agents. BMC Complement. Altern-Med. 16.

Vankar, P.S. and Shukla, D. 2012. Biosynthesis of silver nanoparticles using lemon leaves extract and its application for antimicrobial finish on fabric. Appl. Nanosci. 2: 163–168.

Varona, S., Rodríguez Rojo, S., Martín, Á., Cocero, M.J., Serra, A.T., Crespo, T. and Duarte, C.M.M. 2013. Antimicrobial activity of lavandin essential oil formulations against three pathogenic food-borne bacteria. Ind. Crops Products 42: 243–250.

Veisi, H., Hemmati, S., Shirvani, H. and Veisi, H. 2016. Green synthesis and characterization of monodispersed silver nanoparticles obtained using oak fruit bark extract and their antibacterial activity. Appl. Organometal. Chem. 30: 387–391.

Velusamy, P., Kumar, G.V., Jeyanthi, V., Das, J. and Pachaiappan, R. 2016. Bio-inspired green nanoparticles: Synthesis, mechanism, and antibacterial application. Toxicol. Res. 32: 95–102.

Vilas, V., Philip, D. and Mathew, J. 2016. Biosynthesis of Au and Au/Ag alloy nanoparticles using *Coleus aromaticus* essential oil and evaluation of their catalytic, antibacterial and antiradical activities. J. Mol. Liquids 221: 179–189.

Waseem, R. and Low, K.H. 2015. Advanced analytical techniques for the extraction and characterization of plant-derived essential oils by gas chromatography with mass spectrometry. J. Sep. Science 38: 483–501.

Werdin González, J.O., Gutiérrez, M.M., Ferrero, A.A. and Fernández Band, B. 2014. Essential oils nanoformulations for stored-product pest control - characterization and biological properties. Chemosphere 100: 130–138.
White House. 2014. National strategy for combating antibiotic-resistant bacteria (CARB).
WHO—World Health Organization. 2015. WHO- Antimicrobial resistance fact sheets n°194 [WWW Document]. World Health Organization. URL http://www.who.int/en/index.html.
Windler, L., Height, M. and Nowack, B. 2013. Comparative evaluation of antimicrobials for textile applications. Environment Intern. 53: 62–73.
World Health Organization. 2015. Global Action Plan on Antimicrobial Resistance, WHO Press. ed. Switzerland.
Xiao, Z., Liu, W., Zhu, G., Zhou, R. and Niu, Y. 2014. A review of the preparation and application of flavour and essential oils microcapsules based on complex coacervation technology. J. Sci. Food Agric. 94: 1482–1494.
Yorgancioglu, A. and Bayramoglu, E.E. 2013. Production of cosmetic purpose collagen containing antimicrobial emulsion with certain essential oils. Ind. Crops Products 44: 378–382.

Section III

Antimicrobial Activity Testing

15

Antimicrobial Activity Testing Techniques

Estefanía Butassi, Marcela Raimondi, Agustina Postigo, Estefanía Cordisco and *Maximiliano Sortino**

Introduction

The introduction of antibiotics in the 1940s, beginning with penicillin, opened a new era in the treatment of infectious diseases, described as the 'golden age' of antibiotic research (1940–1962). Then, with the discovery of widely used antibiotics, such as tetracyclines, cephalosporins, aminoglycosides and macrolides, many common bacterial diseases could be cured (Penesyan 2015). However, in recent years the existing compounds are losing their efficacy due to the increase in microbial resistance, the number of infections caused by resistant bacteria is increasing globally and has become a global concern to public health (Blair 2015). On the other hand, since the early 1980s, fungal infections have become the leading cause of diseases in humans, especially among immunocompromised and hospitalized with serious illnesses (Eggimann et al. 2003, Diekema and Pfaller 2004, Nicolato 2016). Because morbidity and mortality are associated with these infections (Gudlaugsson et al. 2003), mycoses have emerged as major public health problems (Pfaller 2012a). Although there are several antifungal drugs, their use is limited by a number of factors, such as low potency, poor solubility, emergence of resistant strains and drug toxicity.

For the reasons exposed above, there is a continuing need to investigate and develop safer and more effective antibiotic and antifungal agents, ideally with a broad-spectrum of action and minimal side effects (Carrillo-Muñoz 2006). Nanotechnology is an alternative formulation considering that nanoparticles containing antibiotic (Shahverdi 2007, Fayaz 2010) and antifungal drugs (Sinha 2013, Qiu 2015, Moazeni 2016, Tutaj 2016) maintained or increased the activity of the drugs and reduced undesirable effects. This innovation was also used for new natural antifungal candidates, such as curcumin, allicin, copaiba oil and eugenol (Scorzoni 2016). Besides, the combination of antimicrobial nanoparticles (i.e., silver and zinc nanoparticles) with antibiotics has been considered a potential method to overcome bacterial drug resistance (Iram 2015, Deng 2016).

In pharmacological investigation of new antimicrobial agents, susceptibility tests are used to determine their activity and efficacy. Ideally, assays should be rapid, simple and easy to implement and should produce rapid results at low cost.

Farmacognosia, Facultad de Ciencias Bioquímicas y Farmacéuticas, Universidad Nacional de Rosario, Rosario, Argentina, Suipacha 531.

* Corresponding author: msortino@fbioyf.unr.edu.ar

Taking into account that antimicrobial susceptibility test methods may lack of clearly defined testing conditions that can lead to low reproducibility, their standardization has advanced in recent years. The Clinical Laboratory Standards Institute (CLSI) and the European Committee on Antimicrobial Susceptibility Testing (EUCAST) have set the benchmark methodology by providing laboratory tested, reproducible, and peer-reviewed standards to commercial antifungal (CLSI 2008a, 2008b, 2009, 2010, EUCAST 2015a, 2015b) and antibacterial drugs (CLSI 1998, 2012, EUCAST 2015c).

In this chapter, we focused on the methodologies for the evaluation of nanoparticles with antimicrobial potential, describing the procedures of the most known and commonly used *in vitro* and *in vivo* and including reference and standardized methodologies recommendations.

Source of the Strains Employed in Antimicrobial Assays

A universal panel of test organisms does not exist and is largely determined by specific drug detection objectives. Microorganisms employed in the studies should be selected as pathogenic representative of different localizations or illness, according to the objectives of the study. The following bacteria and fungi are usually included in the trials:

- Clinically important Gram-positive cocci from the genus *Staphylococcus* (e.g., *S. aureus*), *Enterococcus* (i.e., *E. faecalis*), etc.; and Gram-negative bacilli such as *Escherichia coli*, *Pseudomonas aeruginosa* and others.
- Assays with *S. aureus* must include both sensitive and resistant strains such as methicillin-resistant *S. aureus* (MRSA).
- Pathogens from the group known as ESKAPE (*Enterococcus faecium*, *Staphylococcus aureus*, *Klebsiella pneumoniae*, *Acinetobacter baumannii*, *Pseudomonas aeruginosa* and *Enterobacter* spp.) capable of escaping the biocidal action of antibiotics that represent new paradigms in pathogenesis, transmission and resistance (Pendleton 2013).
- Food borne most common pathogens such as diarrhoeagenic serotypes of *E. coli*, *Salmonella* and *Listeria monocytogenes* (Liébana 2016).
- *Candida albicans*, the most common fungal pathogen, and non-albicans *Candida* species, such as *C. tropicalis*, *C. parapsilosis*, *C. krusei* and *C. glabrata*, that are becoming increasingly more common (Kullberg and Arendrup 2015). The assays may also include strains previously detected as susceptible and resistant to known antifungal drugs.
- Dermatophytes (*Trichophyton* spp., *Microsporum* spp. and *Epidermophyton floccosum*) that are the most common cause of fungal infections of the skin and nails around the world (Ohst 2016).
- *Aspergillus* species (*A. fumigatus*, *A. niger*, *A. flavus*, *A. terreus*) considering that invasive aspergillosis is a frequent complication in people with hematological malignancies and hematopoietic cell transplantation (Marr 2015).
- *Cryptococcus neoformans*, a human opportunistic fungal pathogen causing severe disseminated meningoencephalitis, mostly in HIV-patients (Perfect and Bicanic 2015).

They can be obtained from culture collections, such as the American Type Culture Collection (ATCC) or other reference centers.

Inoculum Preparation

An inoculum is defined as the number of bacteria, yeasts or fungal propagules (conidia or spores) suspended in a certain volume and expressed as Colony Forming Units per milliliter (CFU/mL). The reference documents describe standardized inoculum preparation methodologies for bacteria, yeast or mold fungus. The procedures described in the documents of both institutions, CLSI and EUCAST, are essentially the same with minor modifications that are detailed below. It is important to consider that the handling of microorganisms must be performed in the biological safety cabinet when the risk of substantial splash or aerosolization is present.

Bacteria

Microorganisms should be cultured overnight in an appropriated non-selective medium to ensure purity and viability. The inoculum is prepared by making a suspension of the organism in saline and by adjusting its density to McFarland 0.5 turbidity standard (approximately corresponding to 1–2 $\times 10^8$ CFU/mL for *E. coli*) using a photometric device or visually with a standard of McFarland turbidity. Then, the inoculum is adjusted to the desired concentration required for each methodology. The suspension should optimally be used within 15 minutes and always within 60 minutes of preparation. Inoculum quantitation can be performed by plating 0.01 mL of a 1:100 dilution of the adjusted inoculum on the appropriate culture medium and incubated at optimal conditions (time, temperature, atmosphere, etc.) to determine the viable number of CFU/mL.

Yeasts

Microorganisms are sub-cultured onto blood agar or Sabouraud Dextrose Agar (SDA) at 35°C (± 2°C) to ensure purity and viability. Twenty four-hour-old culture of *Candida* species or 48-hour-old culture of *Cryptococcus* species are employed to obtain fresh viable yeast cells. The inoculum is prepared by picking five distinct colonies (1 mm in diameter) and suspending them in 5 mL of sterile 0.85% saline (0.145 mol/L; 8.5 g/L NaCl). The resulting suspension is vortexed for 15 seconds and its turbidity is adjusted to that produced by a 0.5 McFarland standard yielding a yeast stock suspension of approximately 1×10^6 to 5×10^6 cells/mL. The adjustment can be made visually or with a spectrophotometer at 530 nm according to the CLSI guidelines (CLSI 2008a, CLSI 2009). However, EUCAST, considered that the actual cell count obtained after dilution is confirmed by using a Neubauer Chamber (EUCAST 2015a) because of lack of correlation between turbidity and concentration is observed, especially due to the size of yeast. Then, the inoculum is adjusted to the desired concentration required for each methodology. Inoculum quantitation can be performed by plating 0.01 mL of a 1:100 dilution of the adjusted inoculum on SDA and incubated at 28 to 30°C for 24–48 hours to determine the viable number of CFU/mL.

Molds

Fungi should be cultured on growth media that induced prolific conidia and spore production such as V8 juice, oatmeal or potato dextrose agar (Espinel-Ingroff 1991). Colonies are covered with approximately 1 mL of sterile 0.85% saline, and a suspension is prepared by gently probing the colonies with the tip of a transfer pipette. One drop (approximately 0.01 mL) of Tween 20 can be added to facilitate the preparation of some inocula (e.g., *Aspergillus* spp.). The resulting mixture of conidia or spores and hyphal fragments is withdrawn and transferred to a sterile tube. After heavy particles are allowed to settle for 3 to 5 minutes, the upper homogeneous suspension is transferred to a sterile tube, the cap tightened and mixed with a vortex mixer for 15 seconds. If a significant number of hyphae is detected (> 5% of fungal structures), transfer 5 mL of the suspension to a sterile syringe attached to a sterile filter with a pore diameter of 11 μm, filter and collected in a sterile tube. This step removes hyphae and yields a suspension composed of conidia. If clumps are detected, the inoculum is shaken again in a vortex mixer for a further 15 seconds. This step is repeated as many times as necessary, until clumps are no longer encountered (EUCAST 2015b). The cells are enumerated by using a Neubauer Chamber, assuring reproducible and suitable preparation independent of the color and size of conidia, and the inoculum is adjusted to the desired concentration required for each methodology. Inoculum quantitation can be performed by plating 0.01 mL of a 1:100 dilution of the adjusted inoculum on SDA and incubated at 28–30ºC and observed daily for the presence of fungal colonies to determine the viable number of CFU/mL. A purity control of inoculum may be performed by streaking it on a suitable agar plate and incubated until there is sufficient visible growth to detect mixed cultures.

Agar Diffusion Methods

Agar diffusion technique is a semi-quantitative test consisting of applying a quantity of sample with a known concentration (the plant extract, extract fraction, pure substance or nanoparticle) to the surface of the agar previously seeded with a standardized sample inoculum. The sample diffuses into the agar medium, creating a gradient of circular concentration. If it inhibits microbial growth or kills it, there will be, after incubation, the zone of growth inhibition around the disk, which is measured in millimeters (Scorzoni 2016).

The disadvantages of agar diffusion methodology are that it does not allow determination of Minimum Inhibitory Concentration (MIC) and cannot distinguish bactericidal and bacteriostatic or fungicidal and fungistatic effects. However, considering that the concentration of the compound is the highest next to the disk, and decreases when the distance from the disk increases, the larger zones generally correlate with smaller MIC values. According to the size of the inhibition zone, the activity can be classified qualitatively into three categories: total inhibition, partial inhibition or non-inhibition (Scorzoni 2016).

The inhibition zone may be affected by different factors related to the compound (lipophilicity, volatilization and concentration), agar medium (composition, pH and volume), microbial agent (preparation and concentration of the inoculum, time and temperature of the incubations) and others (Scorzoni 2016). The CLSI documents for disk diffusion susceptibility testing M02-A11, M44-A2 and M51-A (for bacteria, yeast and non-dermatophyte filamentous fungi, respectively) (CLSI 2012, 2009, 2010) considered the Mueller-Hinton Agar (MHA) medium suitable for diffusion assays because it is available, shows acceptable reproducibility and adequate fungal growth.

There are two variants of agar diffusion methodologies that differ in the way the sample is applied: 'disk diffusion assay' in which the sterile paper disks are impregnated with the sample; and 'agar well diffusion assay' in which the samples are placed in wells made in agar medium.

Agar Disk Diffusion Assay

Procedure

The culture medium is prepared by dissolving the appropriate amount of each component in sterile water and being poured into sterile Petri dishes. After solidification, the agar plates are inoculated with a standardized inoculum of the test microorganism by streaking for 2–3 times by rotating the plate at an angle of 60° for each streak to ensure uniform distribution of the inoculum. The agar plates should be dried until excess moisture (usually 10 to 30 minutes) has evaporated. Then, the filter paper disks (6 mm diameter), containing the test compound at a desired quantity, are placed on the agar surface of the inoculated plates using sterile forceps. Plates containing microbes and compounds are incubated in a thermostatic chamber at the appropriate temperature and time required for the development of microorganism (Fig. 15.1). Some microorganisms require special conditions, for example *Streptococcus* spp., *Haemophilus* spp., *Moraxella catarrhalis*, *Listeria monocytogenes* and *Pasteurella multocida* must be incubated in 4–6% CO_2 atmosphere (EUCAST 2015c). After the plates are incubated, they are examined for zones of inhibition, which appear as a clear zone around wells and disks (Cheesbrough 2000). The diameter of such zones is measured using a meter ruler or by digital caliper (Ghaedi 2015), and the mean value for each organism is recorded and expressed in millimeters (Hammond and Lambert 1978). To minimize experimental error, the experiment must be repeated several times (Sharma and Ghose 2015, Hameed 2015).

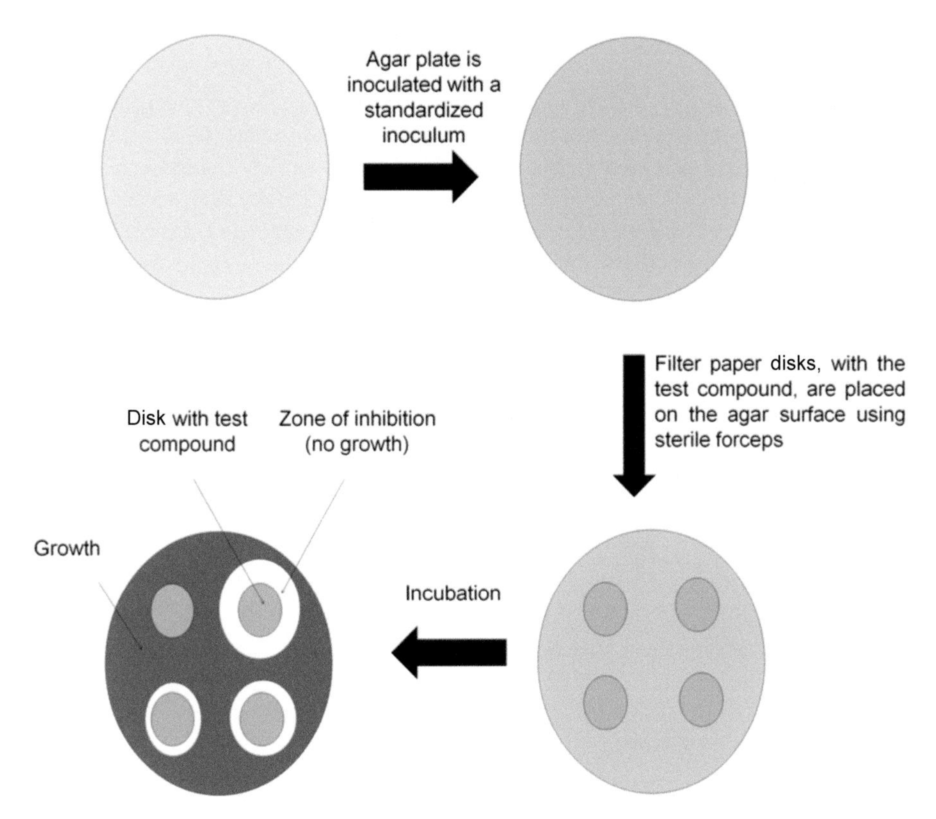

Figure 15.1. Agar Disk Diffusion method.

In each experiment, the following controls should be included:

- Positive control: Regardless of the susceptibility test adopted, it is necessary to use a reference drug to validate the methodology (Cos 2006). The antimicrobials can be chosen according to the purpose of the study, for example; β-lactamics, aminoglycosides or tetracyclines as representative of antibacterial drugs, amphotericin B is a fungicidal agent widely used in the treatment of serious systemic infections and fluconazole is used in the treatment of infections caused by dermatophytes and *Candida* species (Kim 2008). CLSI documents specify the concentration of antifungal drugs, which will be used in the disks.
- Negative control: Disk containing sterilized physiological saline or the maximum amount of DMSO employed for dissolving the compounds.
- Quality control: Evaluation of quality control strains (i.e., *Candida parapsilosis* ATCC 22019, *Candida krusei* ATCC 6258, *Aspergillus flavus* ATCC 204304, *Aspergillus fumigatus* ATCC 204305) against antifungals (i.e., amphotericin B, fluconazole, itraconazole, etc.) should be included to monitor the precision (repeatability) and accuracy (trueness) of the results obtained. Acceptable zone diameter limits are listed in each document. Antibacterial methodologies included susceptible and resistant strains of most representative species. For example, two strains of *E. coli* are included in quality control in the document by EUCAST:ATCC 25922, susceptible, wild-type; and ATCC 35218 TEM-1 ß-lactamase, ampicillin resistant (EUCAST 2015c).
- For nanoparticle evaluations, it is necessary to include disks containing both the antimicrobial alone, the components of the nanoparticle formulation and stabilizers.

Agar Well Diffusion Method

The procedure is similar to the disk diffusion method, but, instead of using paper filter disk, the test compounds are added to agar wells. After solidification of the culture medium on agar plates, wells of the desired diameter are prepared with a sterilized stainless steel cork borer or glass tube. Next, using sterile micropipette, the volume of each compound is added to the agar well and the plates are incubated under suitable conditions depending on the test microorganism (Fig. 15.2). The zones of inhibition are measured as described above.

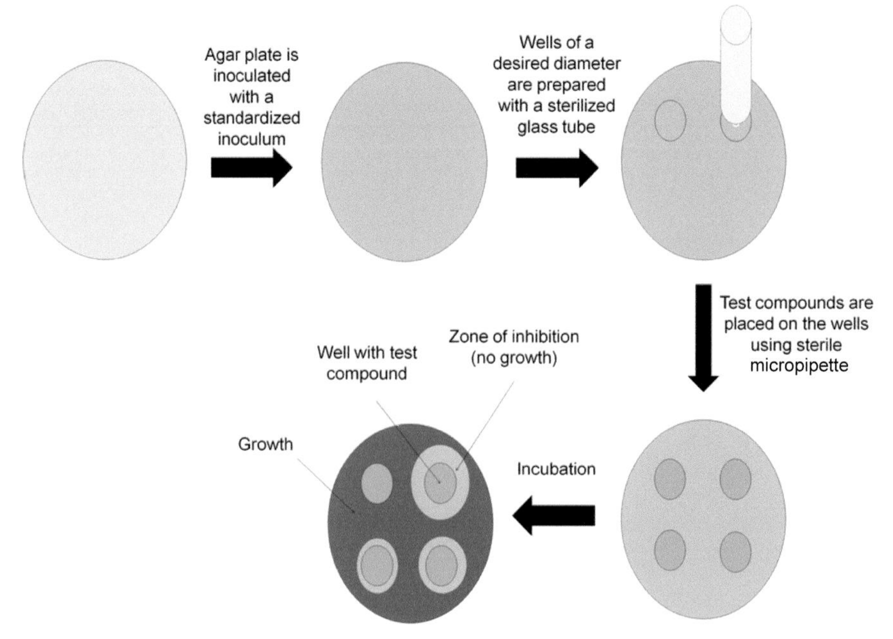

Figure 15.2. Agar Well Diffusion method.

Dilution Methods

In the dilution methods, the microorganisms are evaluated for their ability to produce sufficient growth in culture media containing serial dilutions of the antimicrobial agents. They are the most appropriate methods since they allow the determination of MIC that is defined as the lowest concentration of a test compound that inhibits growth of a microorganism, usually expressed in μg/mL or mg/L. The methodologies can be classified accordingly whether it is carried out in the agar (agar dilution method) or broth medium (broth macrodilution or microdilution methods). Broth and agar dilution method may be used to quantitatively measure the antimicrobial activity *in vitro* against microorganisms.

The **agar dilution method** involves the incorporation of the antimicrobial agent (at different concentrations) in a molten agar medium (MHA for bacteria and SDA for fungi), habitually using serial two-fold dilutions. The medium is allowed to solidify and is seeded with microbial inoculum onto the agar surface. The MIC endpoint is recorded as the lowest concentration of antimicrobial agent that completely inhibits growth under suitable incubation conditions. Agar dilution standardized method M26-A is recommended for fastidious organisms such as anaerobes and *Helicobacter* species (CLSI 1998).

The **broth dilution methods** are the most widely used assays to determine the antifungal activity and standardized methodologies are described in the documents from CLSI and the EUCAST. For bacteria that grow aerobically, CLSI documents M07-A10 recommend Mueller Hinton broth as culture medium (CLSI 2015).

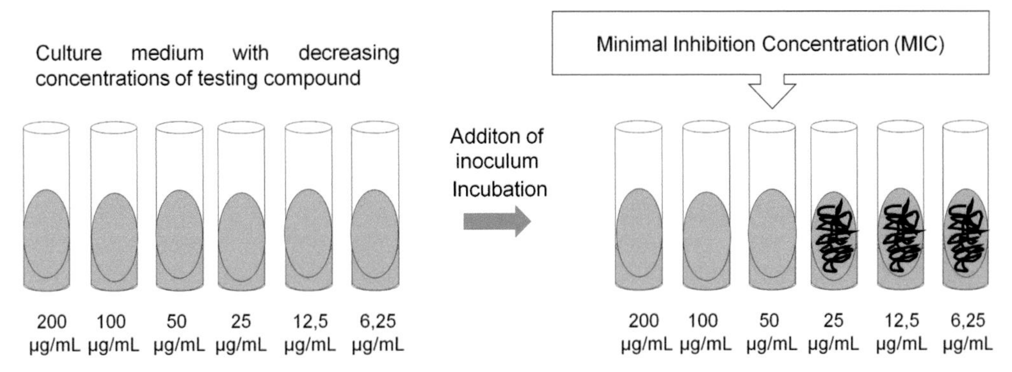

Figure 15.3. Agar dilution method.

For fungi, the documents are M27-A3 and M38-A2 (CLSI 2008a, 2008b), for yeasts and filamentous fungi, respectively; and EUCAST ones, E.DEF 7.3 and E.DEF 9.3 (EUCAST 2015a, 2015b), for yeasts and molds, respectively. Both organizations recommend as culture medium synthetic RPMI 1640 with L-glutamine, without bicarbonate, and with phenol red as a pH indicator, buffered to pH 7.0 using a 0.165 M solution of MOPS (3-N-Morpholinepropanesulfonic acid) which is an excellent buffer for many biological systems at near-neutral pH and with 2% of glucose that has demonstrated better growth and facilitate the determination of endpoints (Denning 1997).

Broth dilution methods can be performed in tubes containing a minimum volume of 2 mL (macrodilution) or with smaller volumes up to 200 μL using 96-well microtitration plate (microdilution). The last one is recommended for the screening of compounds because it has the high throughput potential, is less expensive (uses less media, inoculum, and less of the compound), and provides highly reproducible results. Unlike the microdilution method, the main disadvantages of the macrodilution method are the tedious, manual undertaking, the risk of errors in the preparation of antimicrobial solutions for each test, and the comparatively large quantity of reagents and space required (Balouiri 2016). Thus, the reproducibility and the economy of reagents and space that occurs due to the miniaturization of the test are the main advantages of the microdilution method.

Procedure

The procedure involves the preparation of two-fold dilutions of the antimicrobial agent at a desired concentration range in a liquid growth medium (Mueller Hinton broth for bacteria and RPMI-1640 for fungi). In the microdilution technique, 100 μl of each specific concentration is added to the respective well of the microtiter plates, and then 100 μl of inoculum is added. For this reason, the medium is prepared to double strength to allow a 50% dilution [1:1] once the inoculum is added. After mixing well, inoculated tubes or 96-well microtiter plates are incubated (mostly without agitation) under suitable conditions depending on the test microorganism (Fig. 15.4).

For antifungals, CLSI and EUCAST dilution methods differ in the inoculum size. CLSI guidelines recommend an inoculum adjusted to 5.0×10^2–2.5×10^3 for yeast (CLSI 2008a) and 0.4×10^4–5×10^4 CFU/mL for filamentous fungi (CLSI 2008b); meanwhile, in EUCAST assay, the inoculum is adjusted to 0.5–2.5×10^5 and 2–5×10^5 CFU/mL for yeast and filamentous fungi, respectively (Arikan 2007).

The following controls must be included:

- **Growth control:** Mueller Hinton broth or RPMI 1640 without antimicrobial to assess the viability of the test organisms. With the broth tests, the growth control also serves as a turbidity control for reading endpoints.

- **Positive control:** See above (disk diffusion assay). Antimicrobial drugs must be evaluated at the concentration range specified in standard documents (i.e., amphotericin B: 0.0313 to 16 µg/mL; fluconazole: 0.125 to 64 µg/mL).
- **Negative control:** In cases that may be considered necessary, the maximum amount of DMSO used to dissolve the compounds.
- **Blanks:** culture medium with the compounds at the different concentrations are evaluated and the physiological solution instead of inoculum serves as control for the self-turbidity of the antimicrobials and are considered for reading end points. Blank with culture medium without any compound and physiological solution instead of inoculum serves as blank of culture medium.
- **Quality control:** See above (disk diffusion assay). MIC values of quality control strains against commercial antimicrobials must fall into the expected ranges listed in the respective documents.
- For evaluation of nanoparticles, it is necessary to evaluate the active ingredient alone along with the remaining components and stabilizers of nanoparticles formulation at the same concentration that are present in the nanoparticles.

The MIC is recorded as the lowest concentration of the agent inhibiting the visible growth of microorganisms (Panáček 2009). For bacteria and yeast, fungal biomass can be assessed spectrophotometrically by monitoring absorption at 405 nm. % *inhibition* = $100–[(100 \times (T–B_t))/((G–B_m))]$ where T is the A_{405} of the test well and B_t the blank at this concentration; G, the growth control well and B_m, the blank of culture medium. For molds, the growth in each MIC well is compared to growth control with the aid of a reading mirror. M38-A2 document ranked with a numerical score each well growth: 4: no reduction in growth; 3: slight reduction in growth or approximately 75% of the growth control (drug-free medium); 2: prominent reduction in growth or approximately 50% of the growth control; 1: slight growth or approximately 25% of the growth control; and 0-optically clear or absence of growth (CLSI 2008b).

However, considering that some species do not generate enough biomass, they can be estimated using colorimetric substances. Tetrazolium salts, 3-(4,5-dimethylthiazol-2-yl)-2,5-diphenyltetrazolium bromide (MTT),3-bis {2-methoxy-4-nitro-5-[(sulfenylamino) carbonyl]-2H-tetrazolium-hydroxide}

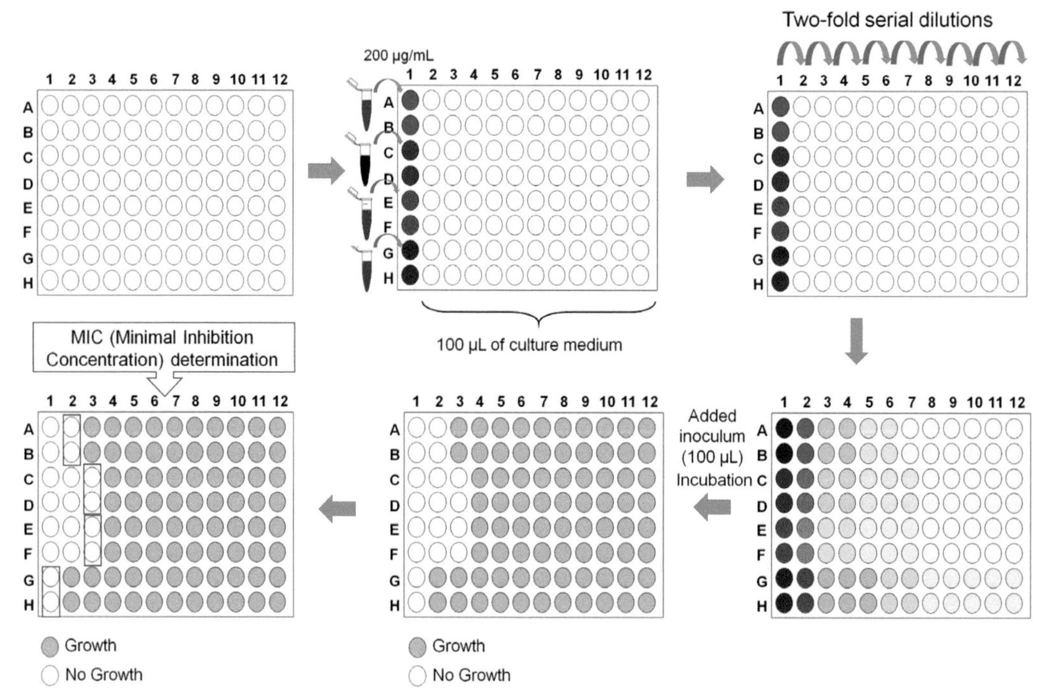

Figure 15.4. Broth microdilution method.

(XTT), and Alamar blue dye (resazurin) are often used in the MIC endpoint determination. MTT, XTT and resazurin are used as oxidation-reduction indicators in cell viability assays. Tetrazolium salts can rapidly penetrate cells and the dehydrogenase activity is reduced to formazan violet crystals by receiving electrons enzymatically from substances in the hydrogen transport system or non-enzymatically from artificial electron carriers (phenazine methosulfate and menadione). Crystals are quantified spectrophotometrically by estimating the number of living cells in the sample. MTT, a yellow tetrazole, is reduced to purple formazanin living cells, by NAD(P) H-dependent cellular oxidoreductase enzymes A solubilization solution (usually either dimethyl sulfoxide, an acidified ethanol solution, or a solution of the detergent sodium dodecyl sulfate in diluted hydrochloric acid) is added to dissolve the insoluble purple formazan product into a colored solution. The absorbance of this colored solution can be quantified by measuring at a certain wavelength (usually between 500 and 600 nm) by a spectrophotometer. The degree of light absorption depends on the solvent. XTT has been proposed to replace MTT, producing greater sensitivity and a higher dynamic range. The formed formazan dye is water-soluble, avoiding a final solubilization step. Resazurin is a blue dye, itself weakly fluorescent until it is irreversibly reduced to the pink colored and highly red fluorescent resorufin in the presence of NAD(P) H-dependent cellular oxidoreductase enzymes (Balouiri 2016).

Minimum Bactericidal Concentration and Minimum Fungicidal Concentration Determination

The determination of Minimum Bactericidal Concentration (MBC) and Minimum Fungicidal Concentration (MFC), also known as the Minimum Lethal Concentration (MLC), are the most common estimate of fungicidal activity. They are defined as the lowest concentration lethal to the microorganism and is performed to determine whether the substance kills or simply inhibit it, -cide or -static activity respectively. If its value is the same as the MIC, the compound is considered bactericide or fungicide, but if the MFC is higher than the MIC, then it is bacteriostatic or fungistatic (CLSI 2008a).

Procedure

Small aliquots (10 μL) of each of the broth dilution tests (where no growth were observed) are sub-cultured on a rich solid medium (SDA) and incubated at appropriate time and temperature, depending on the species being tested. MBC or MFC are defined as the lowest concentration of the substance in which no visible growth of subculture occurs (99.9% of the final inoculum is killed) (CLSI 1998, Arikan 2007).

The following controls must be included:

- **Positive control:** standard drugs with bactericidal (e.g., fluoroquinolones) or fungicidal (e.g., amphotericin B) activity.
- **Negative control:** standard drugs bacteriostatic (e.g., chloramphenicol) or fungistatic (e.g., fluconazole) activity.

Time-kill Study

Time-kill test is convenient to determine the bactericidal and fungicidal effect. It is a strong tool for obtaining information about the dynamic interaction between the antimicrobial agent and the microbial strain. The time-kill test reveals a time-dependent or a concentration-dependent antimicrobial effect (Pfaller 2012b).

Procedure

The microorganism is inoculated into the culture broth medium supplemented with the compound and a concentration corresponding to MFC at the appropriate temperature. At different times, every minute or hour depending on the kinetic of the compound, a 30 μL aliquot is withdrawn for the determination of the colony count. The aliquot must be serially diluted 10-fold with sterile water to provide counts ranging 50–500 CFU/plate. The agar plates are incubated at the appropriate time and temperature, depending on the species being tested. The total number of CFUs/mL is enumerated. The percentage of dead cells is calculated in relation to the growth control, and death curves are plotted against time. The assays were performed on three separate occasions and results were averaged (Klepser 1998, Esteban-Tejeda 2009, Jain 2009, Monteiro 2011). For bacteria, the standardized method is described in M26-A document of CLSI (1998).

Colony Size Method

The assay is also known as 'Poisoned food method'. It is frequently used to evaluate the antifungal effect against filamentous fungi (Ali-Shtayeh and Abu Ghdeib 1999).

Procedure

Assays are performed on Petri dishes and the culture medium is usually Potato Dextrose Agar (PDA) with a uniform agar depth of 4 mm. The nanoparticles to be evaluated are incorporated into the molten agar culture medium at a desired final concentration, well mixed and then are poured into Petri dishes. Plates without nanoparticles must be incorporated to the assay (growth control). Then, fungal mycelium agar plugs of uniform size with a diameter of 6 mm, obtained from the actively growing edge of the fungus with a sterilized stainless steel cork borer or glass tube, are placed upside down on the center of each Petri dish. The plates are incubated under appropriate conditions according to the fungal strain tested. The diameter of the mycelial colony that develops on the plates containing nanoparticles is measured with a caliper and expressed as average diameter (mm). The antifungal activity is measured employing the Antifungal Index that compares the colony diameter on the nanoparticles containing plates with one obtained in control plates (Fig. 15.5). The Antifungal Index is calculated as follows: *Antifungal index* (%) = $(1 - D_t/(D_c) \times 100)$ where D_C is the diameter of the mycelium growth zone in the control plate and D_t is the diameter of the mycelium growth zone in the test plate. The antifungal index can be obtained when the mycelia reaches the edges of the control plate (in fast growing fungi), after 192 hours if the mycelia does not reach the edges of the control plate (slow growing fungi) or can measured every 12 hours for growth kinetics determinations.

Morphological changes after treatment can be observed by cutting pieces of mycelial material from the edge of the fungal cultures and observing them by microscopy (He 2011).

In vivo Experiments

In vivo experiments are crucial for confirmation of *in vitro* data and are necessary to understand if the antimicrobials have the same activity when they are subjected to complex systems, where degradation and/or modification of the compounds can occur and change its potential (Scorzoni 2016). Animal mammalian models require the use of immunosuppressive agents to fully replicate infections and their utilization is highly regulated by ethical committees and good laboratory practices (Marine 2010). Considering that, alternative animal models have been emerged, such as the moth *Galleria mellonella* and the nematode worm *Caenorhabditis elegans* (Scorzoni 2016).

G. mellonella larvae provide a number of advantages as a pre-screen for mammalian infection models to study host interaction and antimicrobial efficacy. They have innate and humoral immune

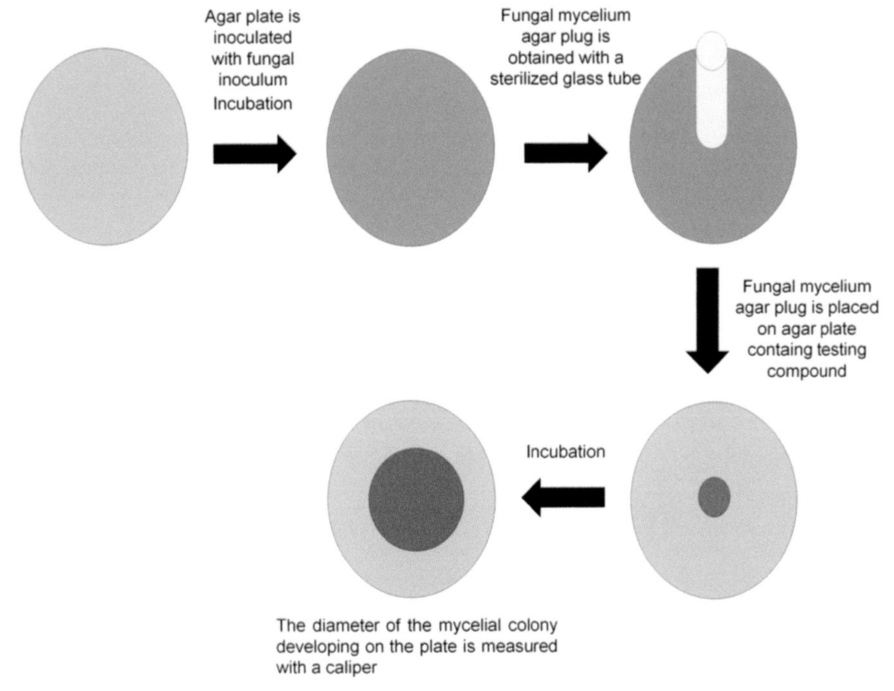

Figure 15.5. Colony size method.

responses remarkably similar to vertebrates, they can be given a precise dosage, are cheap and easy to maintain, are extremely manageable in high-throughput studies and have no ethical issues attached to their use (Li 2013, Benthall 2015).

Procedure

G. mellonella caterpillars in the final larval stage are selected to have a similar in size (approximately 0.33 g) and absence of gray marks. The inoculum of microorganism (5 μL) is injected directly into the last left pro-leg using a Hamilton syringe (the area must be cleaned using an alcohol swab). A control group received 5 μL phosphate buffered saline instead of the inoculum.

Antimicrobial drugs (different doses diluted in sterile water) are injected using the same technique. A mock inoculation with PBS is performed to monitor death due to physical injury or infection by pathogenic contaminants. The larvae are then incubated at 37°C in plastic containers, the number of dead larvae is scored daily and death curves are plotted.

Conclusions

In recent years, the development of new antimicrobial drugs has become an important factor in pharmacological research due to the increase in the number of microbial infections, therapeutic failure, the development of resistant strains, and the low availability of existing drugs. Nanotechnology has emerged as a novel alternative for antimicrobial treatment with the aim of reducing toxicity and increasing the bioavailability; and consequently, the efficacy of traditional and novel drugs.

The methods for the detection of antimicrobial susceptibility agents have been widely used for the discovery of new ones and can also be utilized for the evaluation of the activity of nanoparticles. CLSI and EUCAST have developed standardized protocols for some techniques such as diffusion and dilution assays, which require special modifications when evaluating nanoparticles. Several reports have demonstrated the efficacy of nanoparticles as antimicrobial drugs by using the methodologies described in this chapter.

References

Ali-Shtayeh, M.S. and Abu Ghdeib, S.I. 1999. Antifungal activity of plant extracts against Dermatophytes. Mycoses 42: 665–672.
Arikan, S. 2007. Current status of antifungal susceptibility testing methods. Med. Mycol. 45: 569–587.
Balouiri, M., Sadiki, M. and Ibnsouda, S.K. 2016. Methods for *in vitro* evaluating antimicrobial activity: A review. J. Pharm. Anal. 6: 71–79.
Benthall, G., Touzel, R.E., Hind, C.K., Titball, R.W., Sutton, J.M., Thomas, R.J. and Wand, M.E. 2015. Evaluation of antibiotic efficacy against infections caused by planktonic or biofilm cultures of *Pseudomonas aeruginosa* and *Klebsiella pneumoniae* in *Galleria mellonella*. Int. J. Antimicrob. Agents 46: 538–545.
Blair, J.M., Webber, M.A., Baylay, A.J., Ogbolu, D.O. and Piddock, L.J. 2015. Molecular mechanisms of antibiotic resistance. Nat. Rev. Microbiol. 13: 42–51.
Carrillo-Muñoz, A.J., Giusiano, G., Ezkurra, P.A. and Quindós, G. 2006. Antifungal agents: mode of action in yeast cells. Rev. Esp. Quimioter. 19: 130–139.
Cheesbrough, M. 2000. District Laboratory Practice in Tropical Countries. The press syndicate of the University of Cambridge, Trumpington St Cambridge, pp. 157–206.
CLSI. 1998. Methods for Determining Bactericidal Activity of Antimicrobial Agents. Approved Guideline, CLSI document M26-A. Clinical and Laboratory Standards Institute, 950 West Valley Road Suite 2500, Wayne, Pennsylvania 19087, USA.
CLSI. 2008a. Reference Method for Broth Dilution Antifungal Susceptibility Testing of Yeasts; Approved. Standard—Third Edition. CLSI document M27-A3. Clinical and Laboratory Standards Institute, 950 West Valley Road, Suite 2500, Wayne, Pennsylvania 19087, USA.
CLSI. 2008b. Reference Method for Broth Dilution Antifungal Susceptibility Testing of Filamentous Fungi; Approved. Standard-Second Edition. CLSI document M38-A2. Clinical and Laboratory Standards Institute, 950 West Valley Road, Suite 2500, Wayne, Pennsylvania 19087, USA.
CLSI. 2009. Method for antifungal disk diffusion susceptibility testing of yeast: Approved Guideline-Second Edition. CLSI document M44-A2. Clinical and Laboratory Standards Institute, 950 West Valley Road, Suite 2500, Wayne, Pennsylvania 19087, USA.
CLSI. 2010. Method for antifungal disk diffusion susceptibility testing of non-dermatophyte filamentous fungi: Approved Guideline-Second Edition. CLSI document M51-A. Clinical and Laboratory Standards Institute, 950 West Valley Road, Suite 2500, Wayne, Pennsylvania 19087, USA.
CLSI. 2012. Performance standards for antimicrobial disk susceptibility tests, approved standard, 7th ed., CLSI document M02-A11. Clinical and Laboratory Standards Institute, 950 West Valley Road, Suite 2500, Wayne, Pennsylvania 19087, USA.
CLSI. 2015. Susceptibility tests for Bacteria that grow aerobically; approved standar-10th edition. CLSI document M07-A10. Clinical and Laboratory Standards Institute, 950 West Valley Road, Suite 2500, Wayne, Pennsylvania 19087, USA.
Cos, P., Vlietinck, A.J., Berghe, D.V. and Maes, L. 2006. Anti-infective potential of natural products: how to develop a stronger *in vitro* 'proof-of-concept'. J. Ethnopharm. 106: 290–302.
Deng, H., McShan, D., Zhang, Y., Sinha, S.S., Arslan, Z., Ray, P.C. and Yu, H. 2016. Mechanistic study of the synergistic antibacterial activity of combined silver nanoparticles and common antibiotics. Env. Sci. Technol. 50: 8840–8848.
Denning, D.W., Radford, S.A., Oakley, K.L., Hall, L., Johnson, E.M. and Warnock, DW. 1997. Correlation between *in vitro* susceptibility testing to itraconazole and *in vivo* outcome of *Aspergillus fumigatus* infection. J. Antimicrob. Chemother. 40: 401–414.
Diekema, D.J. and Pfaller, M.A. 2004. Nosocomial candidemia: an ounce of prevention is better than a pound of cure. Infect Control Hosp. Epidemiol. 25: 624–626.
Eggimann, P., Garbino, J. and Pittet, D. 2003. Epidemiology of *Candida* species infections in critically ill non-immunosuppressed patients. Lancet Infect. Dis. 3: 685–702.
Espinel-Ingroff, A. and Kerkering, T.M. 1991. Spectrophotometric method of inoculum preparation for *the in vitro* susceptibility testing of filamentous fungi. J. Clin. Microbiol. 29: 393–394.
Esteban-Tejeda, L., Malpartida, F., Esteban-Cubillo, A., Pecharromán, C. and Moya, J.S. 2009. Antibacterial and antifungal activity of a soda-lime glass containing copper nanoparticles. Nanotech. 2: 505701.
EUCAST. 2015a. EUCAST definitive document E.DEF 7.3 Method for the determination of broth dilution minimum Inhibitory concentrations of antifungal agents for yeasts. The European Committee on Antimicrobial Susceptibility Testing.
EUCAST. 2015b. EUCAST definitive document E.DEF 9.3 Method for the determination of broth dilution minimum inhibitory concentrations of antifungal agents for conidia forming moulds. The European Committee on Antimicrobial Susceptibility Testing.
EUCAST. 2015c. EUCAST definitive document E.DEF 9.3 Disk Diffusion Method for Antimicrobial Susceptibility Testing. The European Committee on Antimicrobial Susceptibility Testing.
Fayaz, A.M., Balaji, K., Girilal, M., Yadav, R., Kalaichelvan, P.T. and Venketesan, R. 2010. Biogenic synthesis of silver nanoparticles and their synergistic effect with antibiotics: a study against gram-positive and gram-negative bacteria. Nanomedicine: Nanomedicine: NBM 6: 103–109.
Ghaedi, M., Naghiha, R., Jannesar, R. and Mirtamizdoust, B. 2015. Antibacterial and antifungal activity of flower extracts of *Urtica dioica*, *Chamaemelum nobile* and *Salvia officinalis*: Effects of Zn [OH] 2 nanoparticles and Hp-2-minh on their property. J. Ind. Eng. Chem. 32: 353–359.

Gudlaugsson, O., Gillespie, S., Lee, K., Vande Berg, J., Hu, J., Messer, S., Herwaldt, L., Pfaller, M. and Diekema, D. 2003. Attributable mortality of nosocomial candidemia, revisited. Clin. Infect. Dis. 37: 1172–1177.

Hameed, A.S.H., Karthikeyan, C., Kumar, V.S., Kumaresan, S. and Sasikumar, S. 2015. Effect of Mg^{2+}, Ca^{2+}, Sr^{2+} and Ba^{2+} metal ions on the antifungal activity of ZnO nanoparticles tested against *Candida albicans*. Mater. Sci. Eng. C 52: 171–177.

Hammond, S.M. and Lambert, P.A. 1978. Antimicrobial Actions. Edward Arnold Ltd, London, pp. 8–9.

He, L., Liu, Y., Mustapha, A. and Lin, M. 2011. Antifungal activity of zinc oxide nanoparticles against *Botrytis cinerea* and *Penicillium expansum*. Microbiol. Res. 166: 207–215.

Iram, S., Nadhman, A., Akhtar, N., Hameed, A., Zulfiqar, Z. and Yameen, M.A. 2015. Potentiating efficacy of antibiotic conjugates with zinc oxide nanoparticles against clinical isolates of *Staphylococcus aureus*. Dig. J. Nanomater. Biostructure 10: 901–14.

Jain, J., Arora, S., Rajwade, J.M., Omray, P., Khandelwal, S. and Paknikar, K.M. 2009. Silver nanoparticles in therapeutics: development of an antimicrobial gel formulation for topical use. Mol. Pharm. 6: 1388–1401.

Kim, K.J., Sung, W.S., Moon, S.K., Choi, J.S., Kim, J.G. and Lee, D.G. 2008. Antifungal effect of silver nanoparticles on dermatophytes. J. Microbiol. Biotechnol. 18: 1482–1484.

Klepser, M.E., Ernst, E.J., Lewis, R.E., Ernst, M.E. and Pfaller, M.A. 1998. Influence of test conditions on antifungal time-kill curve results: proposal for standardized methods. Antimicrob. Agents Chemother. 42: 1207–1212.

Kullberg, B.J. and Arendrup, M.C. 2015. Invasive candidiasis. N. Engl. J. Med. 373: 1445–1456.

Li, D.D., Deng, L., Hu, G.H., Zhao, L.X., Hu, D.D., Jiang, Y.Y. and Wang, Y. 2013. Using *Galleria mellonella–Candida albicans* infection model to evaluate antifungal agents. Biol. Pharm. Bull. 36: 1482–1487.

Liébana, S., Brandão, D., Cortés, P., Campoy, S., Alegret, S. and Pividori, M.I. 2016. Electrochemical genosensing of *Salmonella*, *Listeria* and *Escherichia coli* on silica magnetic particles. Anal. Chim. Acta 904: 1–9.

Marine, M., Pastor, F.J. and Guarro, J. 2010. Efficacy of posaconazole in a murine disseminated infection by *Candida tropicalis*. Antimicrob. Agents Chemother. 54: 530–532.

Marr, K.A., Schlamm, H.T., Herbrecht, R., Rottinghaus, S.T., Bow, E.J., Cornely, O.A. and Lee, D.G. 2015. Combination antifungal therapy for invasive aspergillosis: a randomized trial. Ann. Intern. Med. 162: 81–89.

Moazeni, M., Kelidari, H.R., Saeedi, M., Morteza-Semnani, K., Nabili, M., Gohar, A. and Nokhodchi, A. 2016. Time to overcome fluconazole resistant *Candida* isolates: Solid lipid nanoparticles as a novel antifungal drug delivery system. Colloids Surf. B Biointerfaces 142: 400–407.

Monteiro, D.R., Gorup, L.F., Silva, S., Negri, M., de Camargo, E.R., Oliveira, R. and Henriques, M. 2011. Silver colloidal nanoparticles: antifungal effect against adhered cells and biofilms of *Candida albicans* and *Candida glabrata*. Biofouling 27: 711–719.

Nicolato, A., Nouér, S.A., Garnica, M., Portugal, R., Maiolino, A. and Nucci, M. 2016. Invasive fungal diseases in patients with acute lymphoid leukemia. Leuk. Lymphoma 57: 2084–2089.

Ohst, T., Kupsch, C. and Gräser, Y. 2016. Detection of common dermatophytes in clinical specimens using a simple quantitative real-time TaqMan polymerase chain reaction assay. Brit. J. Derm. 174: 602–609.

Panáček, A., Kolář, M., Večeřová, R., Prucek, R., Soukupová, J., Kryštof, V. and Kvítek, L. 2009. Antifungal activity of silver nanoparticles against *Candida* spp. Biomaterials 30: 6333–6340.

Pendleton, J.N., Gorman, S.P. and Gilmore, B.F. 2013. Clinical relevance of the ESKAPE pathogens. Expert Rev. Anti. Infect. Ther. 11: 297–308.

Penesyan, A., Gillings, M. and Paulsen, I.T. 2015. Antibiotic discovery: combatting bacterial resistance in cells and in biofilm communities. Molecules 20: 5286–5298.

Perfect, J.R. and Bicanic, T. 2015. Cryptococcosis diagnosis and treatment: what do we know now? Fungal Genet Biol. 78: 49–54.

Pfaller, M.A. 2012a. Antifungal drug resistance: mechanisms, epidemiology, and consequences for treatment. Am. J. Med. 125: S3–S13.

Pfaller, M., Neofytos, D., Diekema, D., Azie, N., Meier-Kriesche, H.U., Quan, S.P. and Horn, D. 2012b. Epidemiology and outcomes of candidemia in 3648 patients: data from the Prospective Antifungal Therapy (PATH Alliance®) registry, 2004–2008. Diagn. Microbiol. Infect. Dis. 74: 323–331.

Qiu, L., Hu, B., Chen, H., Li, S., Hu, Y., Zheng, Y. and Wu, X. 2015. Antifungal efficacy of itraconazole-loaded TPGS-b-(PCL-ran-PGA) nanoparticles. Int. J. Nanomed. 10: 1415–1423.

Scorzoni, L., Sangalli-Leite, F., de Lacorte Singulani, J., Costa-Orlandi, C.B., Fusco-Almeida, A.M. and Mendes-Giannini, M.J.S. 2016. Searching new antifungals: The use of *in vitro* and *in vivo* methods for evaluation of natural compounds. J. Microbiol. Meth. 123: 68–78.

Sharma, R.K. and Ghose, R. 2015. Synthesis of zinc oxide nanoparticles by homogeneous precipitation method and its application in antifungal activity against *Candida albicans*. Ceram Int. 41: 967–975.

Shahverdi, A.R., Fakhimi, A., Shahverdi, H.R. and Minaian, S. 2007. Synthesis and effect of silver nanoparticles on the antibacterial activity of different antibiotics against *Staphylococcus aureus* and *Escherichia coli*. Nanomedicine: NBM 3: 168–171.

Sinha, B., Mukherjee, B. and Pattnaik, G. 2013. Poly-lactide-co-glycolide nanoparticles containing voriconazole for pulmonary delivery: *in vitro* and *in vivo* study. Nanomedicine: NBM 9: 94–104.

Tutaj, K., Szlazak, R., Szalapata, K., Starzyk, J., Luchowski, R., Grudzinski, W. and Gruszecki, W.I. 2016. Amphotericin B-silver hybrid nanoparticles: synthesis, properties and antifungal activity. Nanomedicine: NBM 12: 1095–1103.

Index

A

Acinetobacter 40, 54, 69, 71, 79, 80, 298
African trypanosomiasis 258
Aldehydes 25, 29
American trypanosomiasis (or Chagas disease) 258
anethol or estragole 40
anti-adhesion 209–216
anti-adhesives 213
Antibacterial 8, 9, 192, 193, 196–202, 298, 301
Antibacterial activity 100, 106, 107, 109, 112, 115–117
Antibacterial Essential Oils 280
antibiotic adjuvants 144, 146
antibiotic-resistant 211
antibiotics 143–146, 150, 153, 154, 163, 165, 168–170
Antifungal 297, 298, 301–303, 306
Antifungal Drugs 127–129, 132
anti-infectious agents 146
anti-infectious coatings 210
anti-infective medicines 145
anti-infectives 211
antimicrobial 144–151, 153, 154, 209–216
antimicrobial activity 40, 41, 54, 71, 214, 215, 279, 282
antimicrobial agent 210, 212–216
antimicrobial coating surfaces 211
antimicrobial coatings 209, 214
antimicrobial drug 223, 228, 233, 235, 236
Antimicrobial drug discovery 145, 146, 153
antimicrobial drug resistance 213, 215, 216
Antimicrobial mechanism 144, 175–186
antimicrobial surfaces 210–214, 216
antimicrobial target 145, 146, 150
antimicrobials 144, 146, 154
antiviral drug resistance 144
Apoptosis 175, 176, 179–184, 186
Apoptosis-like response 182–184, 186
aromatic plant 160, 161
Aspergillus 298, 299, 301

B

Bacillus cereus 14, 21
Bacteria 14–19, 21, 23–34, 99–117, 297–300, 302–304, 306
bacterial biofilms 281
Bioactives 3, 4, 9
biocides 144, 145, 148, 150, 154
biocompatibility 225, 228–230, 234, 236
Biofilm 7, 8, 106, 146, 148, 159, 160, 164–170, 191, 192, 194, 195, 196, 198–200, 202, 209–211, 213–216
Biofunctionalization 197, 198
biological activity 149

C

C. perfringens 14
Caenorhabditis elegans 306
Campylobacter jejuni 14, 18, 22, 30
Candida 298, 299, 301
carcinogenicity 43, 51, 52
carvacrol 16, 18, 21–33
Chagas disease 258, 260–263
Chirality 16
Cinnamaldehyde 16, 18, 22–28, 30–33
Clostridium botulinum 14
CLSI 298–307
coating 209–216
Colony Size 306, 307
Crohn's disease 18
Cryptococcus 298, 299

D

demethylenation 43, 51
dillapiole 39, 40, 46, 55, 56, 58, 59, 63, 65, 68–71, 73, 77, 83
Dilution 299, 300, 302–305, 307
Disk Diffusion 300–302, 304
drug delivery 221–236
drug resistant organisms 211

E

edible films 33
Enantiomers 109
Encapsulation 283, 284, 286
Enterobacter 298
Enterococcus 298
Epidermophyton 298
Escherichia coli 14, 18, 30
essential oils 1–13, 39–98, 127–174, 272–257
ethnomedical practices 53
EUCAST 298–303, 307
Eugenol 162, 163, 166
extensively drug-resistant (XDR) 144
extracts 280, 281, 283

F

FDA 16
Fixed Oils 249–251
Foodborne 14–34

Fruit 18, 32, 34
fungal infections 127–138
fungal pathogens 127, 132, 135, 138

G

Galleria mellonella 306
Gram-negative 18, 26, 33
Gram-positive 18, 26, 33
GRAS 16, 30, 246, 248, 250–253, 255

H

H. pylori 54, 69, 71
herbal 163
heterocyclic amines 31, 34
High Pressure Homogenization 245
Hospital-acquired Infections (HAIs) 209–211, 216

I

In vivo 298, 306
Infections 209, 211, 213, 214, 216, 279–281, 283–285, 287
Inoculum 298–300, 302–305, 307
inorganic nanophases 280, 283, 287
Intracellular 258–261, 264, 268, 269, 272
Intracellular amastigotes 259, 264, 268, 269
intracellular parasites 258

J

Johne's disease 18
Juice 26–28, 34

K

K. pneumonia 54, 69, 71, 76
Klebsiella 298

L

Leafy Greens 24, 32–34
leishmaniasis 258, 262, 264, 266–269, 273
lipid-based drug delivery systems 242
Liposomes 6–8, 10, 210, 221–226, 230–232, 234–236, 242, 255, 259, 264, 267–269
Liquid Lipid 243, 244, 246, 249, 251, 253, 254
Listeria 298, 300
Listeria monocytogenes 14

M

MBC 305
Meats 29–33
Mechanism of Actions 137
Membrane 14, 25, 26, 34
Membrane disruption 175–177, 186
Membranes 107, 108, 115
metabolic pathway 43, 51
metal nanoparticles 279, 280, 286
methicillin-resistant *S. aureus* (MRSA) 68
methyleugenol 39–41, 45, 52, 54–65, 67–71, 74, 76, 77
MFC 305, 306
MIC 16, 17, 26
Micelles 222, 223, 225
microbial adhesion 212, 213, 215
microbial anti-adhesion 210, 212, 213
microbial contact-killing 210
Microbial infections 191, 192, 196
microbial targets 211
microbial-repelling 211
microbicidal antibiotics 150
Microemulsion 245, 246, 251
Microsporum 298
Miglyol 244, 247, 249, 251
Minimum Inhibitory Concentration (MIC) 300
MLC 305
Molds 299, 303, 304
mono-terpenes 16
Multi drug Resistance 146, 191, 192, 198
multidrug-resistant microorganisms (MDR) 143–146
multi-target drugs 145
Mycobacterium aviumparatuberculosis 14
myristicin 40, 43, 46, 52, 54–61, 63, 65, 68–71, 74, 77

N

Nano and Microemulsions 227, 228
Nanoantimicrobials 284, 285
Nanoemulsion 6, 8–10, 159–170, 210
Nanomaterials 191, 192, 195, 196
Nanoparticle 161, 162, 169, 170, 221–236, 259, 260, 264, 265, 267–270, 272, 273
Nanostructured Lipid Carrier (NLCs) 243–247, 250, 251, 253, 255
nanotechnology 160–162, 164, 166, 168–170, 209–220, 279, 280, 282, 284, 285, 287
Natural Products 132, 138, 209–216
Newport 26, 32, 33
NLC Matrices 243
Non-lamellar Liquid Crystalline Lipid Phases 223, 232, 233
nosocomial 164, 167, 169

O

Oysters 32, 34

P

pandrug-resistant (PDR) 144
Pathogenic microorganisms 175–186
Pathogens 14, 16, 18, 21, 26, 28, 29, 31, 33, 34
pharmacokinetics 221–223, 236
Phase Inversion Temperature 245
Phenolic 21, 25, 29, 30
Phospholipids 248, 250, 254
physicochemical 43, 44, 46, 48, 50
Phytochemical 163, 165
polymeric nanoparticles 242, 245, 255
post-antimicrobial 154
Prop-1(2)-enylbenzene Derivatives 39–84
Properties 16, 28, 33
Protein Nanoparticles 222, 223, 225, 228, 234
Pseudomonas 298

R

resistance-modifying agent (RMA) 144
ROS 194, 195, 201

S

Salmonella 14, 18, 24, 26, 29–34, 298
Salmonella enterica 14, 18, 24, 26, 29–31
Scale-up 253
second generation of lipid nanoparticles 255
secondary metabolites 160, 164
Sensory 16, 28
sessile cells 159, 164, 166–168, 170
Silver nanoparticles 170
SLNs 242, 243, 255
Solid Lipid 242–246, 248, 249, 251–253
Solid lipid nanoparticles (SLN) 6, 7, 10, 230, 235
Solvent Displacement 245
Spoilage 28
Spores 21, 30
Staphylococcus aureus 14, 21, 298
Surfactants 243, 245, 248, 252, 254, 255
Surra 258, 270
synergists 51

T

targeting 221–226, 228–236
Terpenes 15, 16, 25, 26
Terpenoids 15
thymol 16, 18, 22, 23, 25, 26, 30
Time-kill 305
traditional and therapeutic uses 40, 52
Traditional and Therapeutic uses of plants 52
Translational Medicine 145
Trichophyton 298
Type of NLCs 243

V

Vapors 33
Vegetable 18, 32, 34

W

water-insoluble drugs 221–235
Well 300, 302–304, 306
Wine marinades 29
World Health Organization (WHO) 144, 148
wound healing 214, 216

Y

Yeasts 298, 299, 303